D0889791

E. V. Dehmlow/S. S. Dehmlow

Phase Transfer Catalysis

Monographs in Modern Chemistry

Series Editor: Hans F. Ebel

The series is to be continued

E. V. DEHMLOW/
S. S. DEHMLOW

PHASE TRANSFER CATALYSIS

Verlag Chemie
Weinheim · Deerfield Beach, Florida · Basel
1980

Eckehard V. Dehmlow
Sigrid S. Dehmlow
Fakultät für Chemie
Universität Bielefeld
Universitätsstr. 25
D-4810 Bielefeld, Germany

CIP-Kurztitelaufnahme der Deutschen Bibliothek

Dehmlow, Eckehard, V.:
Phase transfer catalysis/E. V. Dehmlow; S. S.
Dehmlow.—Weinheim, Deerfield Beach (Florida)
Basel: Verlag Chemie, 1980
(Monographs in modern chemistry; Vol. 11)
ISBN 3-527-25820-5 (Weinheim, Basel)
ISBN 0-89573-024-3 (Deerfield Beach)
NE: Dehmlow, Sigrid S.

Composition: William Clowes (Great Yarmouth) Limited, England
Printer: D. Betz GmbH, D-6100 Darmstadt 12
Bookbinder: Georg Kränke, D-6148 Heppenheim
Printed in West Germany

Preface

One hundred and fifty years of organic synthesis since Wöhler's first successful experiments have provided the chemist with a confusingly large arsenal of sophisticated methods. However, considering the ease with which a living organism produces complex structures in essentially aqueous surroundings at temperatures only a little above room temperature, the chemist realizes that he has not progressed very far and endeavors to advance further. Thus, organic chemists are constantly on the alert for faster, simpler, and cheaper preparative methods. This book, therefore, is an attempt to collect the somewhat scattered examples of a new technique that has emerged only in the last decade or so. In many cases, the new procedure eliminates the customary requirement of running an organic reaction in a homogeneous, often "absolute" (i.e., perfectly dry) medium.

In phase transfer catalysis (PTC) a substrate in an organic phase is reacted chemically with a reagent present in another phase which is usually aqueous or solid. Reaction is achieved by means of a transfer agent; this agent or catalyst is capable of solubilizing or extracting inorganic and organic ions, in the form of ion pairs, into organic media.

This book attempts a comprehensive survey of work done so far in the area. It developed out of two review articles published in 1974 and 1977 [E. V. Dehmlow, *Angew. Chem.*, **86**, 187 (1974); *Angew. Chem. Int. Ed. Engl.*, **13**, 170 (1974), reprinted in "New Synthetic Methods," Vol. 1, p. 1, Verlag Chemie, Weinheim, 1975, and in slightly shortened form in *Chemical Technology*, **1975**, 210; and E. V. Dehmlow, *Angew. Chem.*, **89**, 521 (1977); *Angew. Chem. Int. Ed. Engl.*, **16**, 493 (1977)]. A review booklet containing much theoretical derivation on extraction equilibria, along with some applications, by A. Brändström, is available in mimeographed form. [A. Brändström, "Preparative Ion Pair Extraction—An Introduction to Theory and Practice," Apotekersocieteten/Hässle Läkemedel, Stockholm, 1974.]

Of the many possible phase transfer catalysts, quaternary ammonium and phosphonium salts are most widely used. In the past such salts have been utilized for purposes other than those of interest presented in this work:

(a) as **catalysts** for a multitude of reactions in homogeneous media;
(b) as aqueous/organic **extractants** mostly for quantitative-analytical, but to some extent for preparative purposes also;
(c) —less frequently—as **reagents** in homogeneous media; and
(d) in micellar catalysis.

We can exclude (a) as not pertaining to the subject covered here. The material under (b) and (c) has a strong bearing on our subject, and will be considered here to some extent, since PTC has to do with both extraction into the organic phase and any chemical transformations occurring therein. Micellar catalysis is mechanistically and preparatively different from standard PTC. The demarcation between PTC and micellar catalysis will be elucidated.

The present book is written with the practicing organic chemist in mind. Typical experimental procedures for the more frequently used PTC reactions are given, and care is taken to mention pertinent data of reaction conditions even in the less important cases.

Since the subject is actively pursued from many sides and since many of its potential fruits have not been harvested yet, the following will be a preliminary rather than a definite mapping. PTC is defined in the introductory Chapter 1. This is followed by a presentation of the fundamental theoretical aspects of ion pairs in general and the factors influencing the aqueous/organic extraction equilibria. Chapter 2 concerns the mechanism of PTC in different fields, the evaluation of catalysts, and its differences from micellar catalysis. The main part of the book, Chapter 3, deals with applications, grouped according to various reaction types with tabular surveys and typical practical procedures in detail.

Originally, a comprehensive literature coverage had been aimed at through the end of 1977, but—due to production delays—as many references as came to the attention of the authors in 1978/79 (a few hundreds altogether!) could be included. All additions, however, had to be concise and were put at the end of paragraphs. The journal literature has been searched on the basis of Chemical Abstracts Indexes, Science Citation Index, and by browsing through the 1973–79 issues of about 50 of the more important primary journals. Literature evaluation in this field is not straight forward since relevant keywords have only recently been introduced and since in the early years isolated groups in the field did not cite each other. Thus, undoubtedly interesting observations are buried in old volumes of journals and patents, and may not have been recognized. Patent literature was searched with the aid of a computer program through the facilities of IDC, Internationale Dokumentationsgesellschaft für Chemie m.b.H., Frankfurt, W. Germany.

Because of the multitude of possible catalysts and applications the authors are quite certain that they have not located all relevant references especially those pertaining to the patent literature. Thus, they apologize beforehand to any researcher who does not find his contribution listed in the appropriate place. The authors would be grateful for any additional information sent to them which might be used in a possible second edition.

The contributions of some individuals and institutions helped to make this book become reality. Thanks are due to a number of persons, for private communications and information given prior to publication, notably to M. Mąkosza and A. Brändström. We are grateful to IDC, Frankfurt (H. Ziegenhirt), for generous help with the machine research and to Schering A. G., Berlin (Dokumentationszentrale, K. Specht, and Forschungsabteilung, H. Vorbrüggen) for providing many photocopies of Derwent patent abstracts, journal articles, and patent specifications. Help in typing the manuscript from our not always easy-to-read handwriting was given by

I. Bodammer, M. Slopianka, and K. Wrentschur. The experimental work from our laboratory mentioned in this book was executed together with co-workers whose names appear in the citation. Thanks are due to all of them, especially Manfred Lissel and Marion Slopianka. The first reading and collecting of material for this book was done in the summer of 1975 during a sabbatical term as guests at the University of East Anglia, Norwich, England. Our sincerest thanks are due to A. R. Katritzky, not only for his kind hospitality at the School of Chemical Sciences, but also because of the way he and his family made the stay of the five Dehmlows in Norwich so very enjoyable. Last but not least, we wish to express our appreciation for the improved spelling and style of this English text to Edeline Wentrup-Byrne.

Eckehard V. Dehmlow
Sigrid S. Dehmlow

Frequently Used Abbreviations

Adogen 464	methyltrioctylammonium chloride (technical grade)
Aliquat 336	same product as Adogen 464
Bu	*n*-butyl
Hex	*n*-hexyl
Hep	*n*-heptyl
Oct	*n*-octyl
Pent	*n*-pentyl
Prop	*n*-propyl
PTC, PT	phase transfer catalysis, phase transfer
TEBA	benzyltriethylammonium chloride
TEBA-bromide	benzyltriethylammonium bromide

Contents

1. Ion Pairs and Ion Pair Extraction

1.1. Introduction: The Nature of Phase Transfer Catalysis

Phase transfer catalysis (PTC) concerns reactions between, on the one hand, salts dissolved in water or present in the solid state, and, on the other, substances dissolved in organic media. Without a catalyst such reactions are usually slow and inefficient or do not occur at all. The traditional procedure would involve dissolving the reactants in a homogeneous medium. If a hydroxylic solvent is used, the reaction may be slow because of extensive solvation of the anion. Solvolytic side reactions sometimes reduce the efficiency even further. Polar aprotic solvents are often superior, but they are usually expensive and difficult to remove after the reaction and may present environmental problems in large scale operations. However, in some cases such as O *vs.* C alkylation with ambident anions, polar aprotic solvents may hinder rather than promote the reaction by predominant formation of the undesired product.

Phase transfer catalysis permits or accelerates reactions between ionic compounds and organic, water-insoluble substrates in solvents of low polarity. The catalysts most commonly used are onium salts or complexing agents which can mask and thereby solubilize alkali metal ions. The basic function of the catalyst is to transfer the anions of the reacting salt into the organic medium in the form of ion pairs. In aprotic solvents these are virtually unsolvated and unshielded (except, perhaps, by their counterions) and consequently are very reactive.

It is clear, therefore, that PTC has considerable advantages over conventional procedures:

—expensive anhydrous or aprotic solvents no longer required
—improved reaction rates and/or
—lower reaction temperatures
—in many cases easier work-up
—aqueous alkali metal hydroxides can be employed instead of alkoxides, sodamide, sodium hydride, or metallic sodium.

Further special advantages are also found, *e.g.*:

—occurrence of reactions that do not otherwise proceed
—modification of the selectivity
—modification of the product ratio (*e.g.*, O *vs.* C alkylation)
—increased yields through the suppression of side reactions.

PTC is a relatively new field of chemistry that originated in the research work of three independent groups. The foundations were laid in the mid to late 1960s by M. Mąkosza, C. M. Starks, and A. Brändström.

Reactions involving phase transfer phenomena were, of course, performed even earlier, and undoubtedly a considerable number of such reactions are buried in the older literature [1, 2, 3] and especially in patents [4], the oldest one presently known being from 1913. Some of the original authors entered the field more or less incidentally and did not apparently reflect on the mechanisms involved in such catalytic reactions. None of them, however, realized the potential and scope of the new technique.

PTC techniques as we know them today originated in the work of Mąkosza and coworkers in 1965 [5]. They began a systematic exploration of alkylations and subsequently of other reactions in two-phase systems containing mainly concentrated aqueous alkali metal hydroxides. The descriptive terms used by them were "catalytic two-phase reactions," "catalytic alkylation of anions," "catalytic generation of carbenes," *etc.* Mąkosza prefers these terms even now for mechanistic reasons in many cases. This work became more widely known with the publication, in *Tetrahedron Letters* 1969 [6], of his dichlorocarbene generation discovery.

Brändström started from a more physicochemical and analytical point of view. His first papers relevant to our subject appeared in 1969 [7], followed by an early review in 1970 [8], entitled "Ion Pair Extraction in Preparative Organic Chemistry." Brändström uses the term "extractive alkylation" for alkylation reactions in a two-phase mixture in the presence of molar amounts of catalyst.

The term "phase transfer catalysis" was coined by Starks and first used in patents in 1968 [9]. The recognition of the new technique and the term "phase transfer catalysis" probably originates from Starks' 1971 paper published in the *Journal of the American Chemical Society* [10]. For the first time the scope of the method was clearly outlined and extended beyond the original applications, *i.e.*, alkylation and carbene generation. Furthermore, a unifying mechanistic concept for all of these reactions was proposed. This provided an enormous impact to the development of the field. Since then a flood of papers has appeared expanding PTC to new types of reactions, *i.e.*, liquid-solid PTC and polymer-bound catalysis, and developing new catalytic reagents such as crown ethers. These will be discussed in subsequent sections.

The term "phase transfer catalysis" is now the most widely accepted name. It is used in general textbooks and works of reference, in abstract services and thesauri, as well as in Organic Syntheses. Certain modifications are necessary in order to apply the above-mentioned unifying mechanism to many PTC reactions.*) Mąkosza prefers to exclude some reactions from "true" PTC reactions and calls them "so-called PTC reactions" or "catalytic two-phase reactions." For the purpose of this review, however, the present authors prefer to include all two-phase reactions between solid or aqueous salts, acids, or bases and substrates in organic solvents catalyzed by onium salts, crown ether/alkali metal salts, or similar chelated salts under the heading "phase transfer catalysis." This permits classification of reactions involving similar catalytic effects or similar catalysts, irrespective of whether the mechanism is known or is always the same. This is a practical, phenomenological definition. It does not

*) *E.g.*, deprotonation often occurs at the phase boundary of two-phase systems.

imply a specific action of the catalyst and depends only on the presence of two phases and the catalyst. Reactions occurring exclusively or primarily at the interphase are included, provided they are accelerated by a catalyst. Many such reactions are known.

The question may arise as to whether micellar catalysis should be included in a monograph on PTC. Where must one draw the borderline? Micellar reactions have been reviewed extensively [29–31]. They are normally performed in a homogeneous or pseudohomogeneous phase. These reactions, therefore, do not come under our definition of PTC reactions and consequently have been largely omitted here. However, a PTC reaction as defined in this book could proceed mechanistically in a micellar or inverted micellar environment. Although this would not be a typical PTC reaction, and no authentic examples have been documented, there are some indications that such processes can occur.

Previous reviews of PTC are found in Brändström's booklet "Preparative Ion Pair Extraction" [11], two articles by one of the present authors [12, 13], two conference papers and an article by Mąkosza [14, 15, 114], and a survey by Dockx including other uses of quaternary ammonium compounds [16]. Shorter introductions are also found in the literature [17–20, 115]. Brändström recently published a chapter on the principles of PTC in "Advances in Physical Organic Chemistry" [112]. Finally, many applications of PTC were compiled in the recent books "Phase Transfer Catalysis in Organic Synthesis" by Weber and Gokel [113] and "Phase Transfer Catalysis; Principles and Techniques" by Starks and Liotta [116].

1.2. Ion Pairs in Organic Media

In many cases, PTC consists of the extraction of ionic molecules into an organic solvent or their solubilization therein. It is of interest, therefore, to know something about the state and properties of such solutions. A complete review of the subject is beyond the scope of this book. However, in this section a concise and qualitative summary will be presented. For a more thorough, in-depth treatment of physicochemical concepts, methods, and results the reader is referred to textbooks of physical chemistry, physical organic chemistry [*e.g.*, 21], or recent monographs [22, 23, 39]. Structure and reactivity of carbanions in ion pairs and of carbanionoid metal organic compounds have been treated in a review [40] and special monographs [41–43].

Typical inorganic sodium and potassium salts do not dissolve in nonpolar organic solvents. The same is true for salts of inorganic anions with small organic cations, *e.g.*, tetramethylammonium. Such ammonium salts can, however, often be dissolved in dichloromethane and chloroform. Moreover, the use of rather large organic anions may make alkali metal salts soluble in solvents such as benzene. For example, sodio diethyl-*n*-butylmalonate gives a 0.14 M solution in benzene which does not show a measurable freezing point depression thus indicating a high degree of association. Similarly, large onium cations (*e.g.*, tetra-*n*-hexylammonium) solubilize salts with even small, organophobic anions (*e.g.*, hydroxide ions) in hydrocarbons.

Ionophores, *i.e.*, molecules made up of ions in the crystal lattice, dissociate (or partially dissociate) into solvated cations and anions in solvents of high dielectric constant. Such solutions in water are good conductors. In other, less polar, solvents even strong electrolytes may dissolve to give solutions of low electrical conductance, indicating that only a fraction of the dissolved salt is dissociated into free ions. To account for this behavior Bjerrum advanced the ion pair hypothesis in 1926. Subsequently his hypothesis was refined by Fuoss [38] and a number of other researchers. Ion pairs are associations of oppositely charged ions to give neutral entities. Basically, ion pairs are held together by Coulombic forces, but occasionally strong interactions with the medium also contribute. Ion pairs are thermodynamically distinct species coexisting in equilibrium with the free ions:

$$[Q^{\oplus}X^{\ominus}] \rightleftharpoons Q^{\oplus} + X^{\ominus}$$

An important distinction between free ions and ion pairs is that solutions containing only the latter do not conduct electricity. Thus, conductivity measurements allow one to estimate the proportion of free ions. As regards cryoscopy and vapor pressure measurements ion pairs behave as single entities. The dissociation constants of ion pairs in many solvents are known. As a general rule of thumb, at low concentrations, solvents of dielectric constant greater than 40 contain mainly dissociated ions. In solvents of dielectric constant lower than 10–15, even at high dilution, almost no free ions exist.

In any given solvent, the larger the ions the greater the degree of dissociation. For example, the equilibrium constant of ion pair formation K in nitrobenzene is $K = 80$ for tetraethylammonium chloride; $K = 62$ for the bromide; too small to estimate for tetra-*n*-butylammonium pikrate and $K = 7$ for tetraethylammonium pikrate [26].

The behavior and structure of ion pairs and higher complexes have been studied extensively by physical techniques like conductivity, UV, visible, IR, and Raman spectroscopy, electron and nuclear magnetic resonance. A review of methods and results is available [22].

Solvents are not continuous structureless substances. Ion-dipole interactions between solute and solvent molecules occur on contact. The solvation number is an indication of the extent of such interactions. The closer the contact between solute and solvent, the greater the interaction. Dipole, dispersion, induction, and hydrogen bonding forces interplay with the Coulombic forces and together determine the stability and properties of the ion pair. Hence, the nature of both the dissolved compound and the solvent are important. Solvation shells decrease the mobilities and diffusion coefficients of both ions and ion pairs. The solvating power of an aprotic solvent does not depend on the dielectric constant but rather on its ability to donate or accept electron density. Solvent effects in organic chemistry have been reviewed extensively [24].

As mentioned above, the formation of ion pairs and their physical and chemical properties are strongly influenced by interaction with the solvent. In this respect, solvents can be classed into three groups:

(1) Polar protic solvents readily solvate both anions and cations. Inorganic cations interact with the free electron pairs while anions are solvated by means of hydrogen bonds. Large quaternary ammonium ions are unsolvated [37] or at least not specifically solvated, *i.e.*, there is no strong, direct interaction. In these solvents there is a high degree of dissociation into free solvated ions. However, many anions show a relatively low reactivity (nucleophilicity) because of strong shielding by the solvation shell.
(2) Polar aprotic solvents (*e.g.*, DMSO, DMF) readily solvate cations. Anions, however, are only poorly solvated since the positive end of the solvent dipole cannot be approached easily, and reaction rates are therefore high. Salts are highly dissociated in such solvents.
(3) PTC reactions are usually carried out in aprotic solvents of low polarity. Their dielectric constants range from 8.9 (methylene chloride), 4.7 (chloroform), and 4.2 (diethyl ether) to 2.3 (benzene) and 1.9 (hexane). Although the solubility of typical inorganic salts in these solvents is negligible, organic quaternary ammonium, phosphonium, and other onium salts, as well as organically "masked" alkali metal salts, are often quite soluble, especially in methylene chloride and chloroform. In these solvents the concentration of free ions is negligible, ion pairs being the dominant species. Since interactions between the ion pairs and solvent molecules are weak, reaction with electrophiles in the organic medium is fast, and some weak nucleophiles (*e.g.*, acetate) appear strong. These reactivity differences are sometimes very striking, and such ion pairs have occasionally been described as "bare" or "naked" ions. This phrase is, however, somewhat misleading since in less polar solvents ion pairs are present. Also, the effect of the solvent on relative and absolute reaction rates is still very marked even in "nonsolvating" media.

Self-association between ion-pairs leads to the formation of aggregates, *e.g.*, dimers, trimers, or quadruplets. Such associations are energetically favorable and occur extensively in nonpolar media unless the solutions are infinitely dilute. Association takes place to a measurable extent at concentrations as low as 0.001 mol/l. For example, the cryoscopic association numbers (ratio of the experimental molecular weight to formula weight) for tetra-*n*-butylammonium thiocyanate in benzene is 2.5 at 0.0013 mols per 1000 g solvent, rises to 31.9 at 0.281 mols per 1000 g solvent, and decreases slightly again at even higher concentrations (22.7 at 0.753 mols per 1000 g solvent) [25]. This ion pair association has a very profound effect on the extraction of salts from aqueous to organic phases (*cf.*, Section 1.3.1). The degree of association depends on the cation, anion, solvent, and concentration. Trimers of monovalent ions are charged species and conduct electricity in the same way as ion pairs which contain polyvalent ions.

From both physical and kinetic evidence it is thought that two types of ion pairs exist, for which various names have been given.

(1) loose, external, or solvent separated ion pairs;
(2) tight, internal, intimate, or contact ion pairs.

A solvated free ion can approach its counter ion without difficulty up to the point where the two solvation shells touch, thus forming a loose ion pair. If the two ions come even closer together and squeeze out the solvent molecules separating them, a contact ion pair is formed. Depending on the nature and concentration of the cation, anion, and solvent, and within certain limits of temperature, both types of ion pairs can exist as thermodyamically distinct species in a dynamic equilibrium. In ethereal solvents, the formation of solvent separated ion pairs is usually exothermic. If these two types of ion pairs coexist, their dissociation into free ions is described by three interrelated equilibria. The position of these equilibria is greatly influenced by steric effects of the cation. Contact ion pairs are more likely to show stereospecificity, for example, in H/D exchange reactions [28]. Crown ethers are known to convert many (but not all, *cf.*, for instance [27]) contact ion pairs of alkali metal cations to solvent separated ion pairs. The latter then react less specifically [28]. The influence of various ethereal solvents (*e.g.*, polyethyleneglycol ethers or added crown ethers) on the structure of ion pairs has been reviewed [32].

The role of ion pairs in nucleophilic aliphatic substitution does not directly concern us here; however, the following references may prove useful [33–36].

1.3. Extraction of Ion Pairs from Aqueous Solution

1.3.1. Principles

Solvent extraction of ionic substances from aqueous into organic media is a field well known to analytical and industrial chemists. The extraction of metals and acids has been thoroughly investigated [71–74].

Even the most simple two-phase (water/organic) substitution reaction between the anion of a salt and an organic substrate involves a number of equilibria:

(a) Overall reaction:

$$Na_{aq}^{\oplus}Y_{aq}^{\ominus} + RX_{org} \xrightarrow[\text{PT catalyst}]{[Q^{\oplus}X^{\ominus}]} Na_{aq}^{\oplus}X_{aq}^{\ominus} + RY_{org}$$

where aq and org denote the aqueous and organic phases, resp.; $Q^{\oplus}$ is the catalyst cation, and $[Q^{\oplus}X^{\ominus}]$, etc., ion pairs. The overall equation can be broken down into two contributions:

(b) Chemical reaction in the organic phase:

$$RX_{org} + [Q^{\oplus}Y^{\ominus}]_{org} \rightleftharpoons RY_{org} + [Q^{\oplus}X^{\ominus}]_{org}$$

(c) Extraction equilibrium:

$$[Q^{\oplus}X^{\ominus}]_{org} + Na_{aq}^{\oplus} + Y_{aq}^{\ominus} \rightleftharpoons [Q^{\oplus}Y^{\ominus}]_{org} + Na_{aq}^{\oplus} + X_{aq}^{\ominus}$$

Information concerning the factors determining equation (c) may be obtained by considering the following simpler extractions

(d) $Q_{aq}^{\oplus} + X_{aq}^{\ominus} \rightleftharpoons [Q^{\oplus}X^{\ominus}]_{org}$

(e) $Q_{aq}^{\oplus} + Y_{aq}^{\ominus} \rightleftharpoons [Q^{\oplus}Y^{\ominus}]_{org}$

The latter equation is essential in order to understand all PTC. Fortunately, experimental methods for the investigation of these extractions have been developed, and many numerical values for the development of analytical processes have been determined [refs., see Tables].

Schill and Modin [44, 45] define the stoichiometric extraction constant E_{QX} of equation (d) as

$$E_{QX} = \frac{[Q^{\oplus}X^{\ominus}]_{org}}{[Q^{\oplus}]_{aq} \cdot [X^{\ominus}]_{aq}}$$

It should be noted that E_{QX} is not the more familiar distribution constant. Since the denominator consists of a product, the extraction equilibrium is influenced by both the anion and cation concentrations in the aqueous phase. For precise determinations in not too dilute solution, activities instead of concentrations would have to be used.*) Because the extraction system is usually influenced by a number of factors other than the equilibrium depicted in equation (d), the situation is even more complicated. Among these are:

(i) Association or dissociation of the ion pairs in the organic phase, which lowers the concentration of $[Q^{\oplus}X^{\ominus}]_{org}$ and thus aids in the extraction;
(ii) association effects in the aqueous phase, resulting in a decrease in extraction;
(iii) pH-dependent equilibria in the aqueous phase, which influence the effective anion concentration

$$X^{\ominus} + H_3O^{\oplus} \rightleftharpoons HX + H_2O$$

or the effective cation concentration if $Q^{\oplus}$ is a primary, secondary, or tertiary ammonium ion $HB^{\oplus}$

$$HB^{\oplus} + H_2O \rightleftharpoons B + H_3O^{\oplus}$$

(iv) extraction of uncharged species such as HX and B into the organic layer and in addition the possible formation of ion associates $[Q^{\oplus}HX_2^{\ominus}]$ and $[Q^{\oplus}X^{\ominus}, 2HX]$ in the organic phase.

Schill and Modin [44, 45] consider equation (d) as the main process and all others (i–iv) as "side processes." They define a conditional extraction constant E^*_{QX} which includes all competing "side reactions" and which is related to the stoichiometric extraction constant E_{QX} thus

$$E^*_{QX} = E_{QX} \cdot \frac{\alpha_{QX}}{\alpha_{Q(X)} \cdot \alpha_{Q(X)}}$$

The α-coefficients serve as correction factors. They deviate from unity with increasing influence of the "side process" and equal unity when their influence is negligible.

Although dissociation constants of quaternary ammonium compounds are of the order 10^{-4} to 10^{-5} in methylene chloride or chloroform, their influence cannot be

*) The use of activities leads to the thermodynamic extraction constant, which is rarely measured.

neglected in the dilute solutions often employed. Ion pair association in the organic phase is desirable because it favors extraction. Thus, more concentrated solutions are an advantage. If the anion is brought into the system partly as inorganic salt, NaX, a high concentration and excess of NaX in the aqueous phase enhances the extraction of $[Q^{\oplus}X^{\ominus}]$ into the organic phase. At the same time, a possible ion association of the inorganic salt in the aqueous phase does not adversely affect the overall process in the majority of cases.

Processes (iii) and (iv) are common and where necessary must be taken into account. In fact, under certain conditions one or more of the "side processes" become increasingly important. In principle, all processes of this type can be measured, and their influence calculated. Brändström reviews the necessary equations [11, 112] for such calculations using logarithmic extraction diagrams, including many cases of very complicated equilibria. The results of these calculations help answer both analytical and preparative questions of the type:

How to perform a selective extraction?
How to select optimal conditions for quantitative extraction, *e.g.*, pH, concentration?
What are the ideal conditions for the quantitative separation of the mixture?

In most cases of PTC work, however, a qualitative idea of the feasibility of an extraction is sufficient. In practice, a competitive extraction of two or more anions occurs. In the case of the displacement reaction (a), (see p. 6), the extraction equilibrium (c) must be considered. Using the fundamental extraction equation for $[Q^{\oplus}X^{\ominus}]$ and $[Q^{\oplus}Y^{\ominus}]$ one arrives at the following equation:

$$K = \frac{E_{QY}}{E_{QX}} = \frac{[Q^{\oplus}Y^{\ominus}]_{org}}{[Q^{\oplus}X^{\ominus}]_{org}} \cdot \frac{[X^{\ominus}]_{aq}}{[Y^{\ominus}]_{aq}}$$

If E_{QX} and E_{QY} are known, it is possible to calculate whether or not the extraction of the anion $Y^{\ominus}$ into the organic medium can compete with the extraction of the leaving group $X^{\ominus}$ for any given concentration of $X^{\ominus}$, $Y^{\ominus}$, or the catalyst cation, at any stage of the reaction. For this purpose, it is assumed that concentrations rather than activities can be used and that the "side processes" described above do not disturb or can at least be evaluated. Furthermore, it is necessary to assume that the cation $Q^{\oplus}$ is lipophilic to the extent that it is present virtually only in the organic phase. Under these conditions it is possible to calculate which fraction of the catalyst cation $Q^{\oplus}$ is paired with $X^{\ominus}$ and which fraction is paired with $Y^{\ominus}$ [13]. Hence, the outcome of the overall reaction may be predicted. In most cases this will be a rough estimate but may suffice for preparative work. For optimizing extraction studies or for analytical work, a more costly and time-consuming study of the "side processes" is necessary [11, 112].

It should be emphasized strongly that "side processes" must not be neglected, and a simple estimation of conditional extraction constants may be misleading. Brändström gives details of the distribution of NBu_4Cl between water and methylene chloride at 25 °C over a wide concentration range [112]. In this case, activity coefficients in the aqueous layer are known, activity coefficients in the organic layer can be calculated, and the dissociation and dimerization constants in the organic phase are obtainable. Thus, the calculated and experimental total concentrations in both phases can be compared, while assuming simultaneous extraction, dissociation, and dimerization. In the concentration range studied, the conditional extraction constant first decreases from 3.54 to 1.51 and then increases to 7.54 with increasing dilution [112]!

The magnitude of stoichiometric extraction constants depends not only on the organic solvent but also on the size and structure of the anion and cation. These factors are considered in the following sections. It should be noted that a very large variety of extraction constants is possible through judicious choice of ions: even the most hydrophilic anions (*e.g.*, $OH^{\ominus}$) can be extracted using very lipophilic cations, and the most hydrophilic cations such as $NMe_4^{\oplus}$, are transported to the organic medium by very lipophilic anions. Of the many cations capable of extracting anions—substituted ammonium, phosphonium, sulfonium, arsonium ions, *etc.*, as well as crown ether complexed alkali metal cations—only the ammonium ions have been investigated in a systematic manner. Our discussion, therefore, is confined to these, but structure effects for the extraction of other cations is expected to be similar.

1.3.2. Influence of the Solvent

Brändström [46, 112] has determined a large number of apparent extraction constants between water and various solvents for a standard quaternary ammonium salt, tetra-*n*-butylammonium bromide (Table 1–1). A solvent useful for PTC work should be immiscible with water because otherwise highly hydrated "shielded" ion pairs of low reactivity are present. In order to avoid hydrogen bonding to the ion-pair anion, the solvent should also be aprotic. Inspection of the data in Table 1–1 shows the very large range of extraction constants possible. Solvents in the right hand column of the table are generally unsuitable for PTC work. Some of them are partially miscible with water, others are too reactive and would interfere in many processes. It can be seen, however, that for the standard salt in question, which is of medium lipophilicity, all of these solvents are in the extraction range good to excellent. Structurally related, somewhat polar compounds (*e.g.*, homologs) should have a similar high extracting capability for ion pairs. This leads to an important generalization: If compounds of the general type shown in the third column of Table 1–1 are used as reagents in PTC reactions, *e.g.*, alkylations, no additional organic solvent is necessary since the extraction of ion pairs into the pure organic phase is very satisfactory.

Turning now to the entries in the left hand and central columns of Table 1–1 we notice a large variation of extraction capability with seemingly small structural changes (*cf.*, *cis*- and *trans*-1,2-dichlorethylene, 1,1,2-trichloroethane, 1,1,2,2-tetrachloroethane, and pentachloroethane). Specific interactions between solvent and solute must play a role even in these supposedly "nonsolvating" solvents. What is more important from a practical point of view is that the low boiling, chlorinated hydrocarbons (chloroform, methylene chloride, and to a lesser extent 1,2-dichloroethane) appear to be the best solvents. Not only do they exhibit a high extraction

Table 1–1. Apparent Extraction Constants $E_{NBu_4Br} = [NBu_4Br]_{org}/\{[NBu_4^{\oplus}]_{aq} \cdot [Br^{\ominus}]_{aq}\}$ (calculated from the distribution of 0.1 M tetra-*n*-butylammonium bromide between water and the solvents listed) [46].

Solvent	E_{NBu_4Br}	Solvent	E_{NBu_4Br}	Solvent	E_{NBu_4Br}
CH_2Cl_2	35	$CH_3—CHCl—CH_2Cl$	0.5	$C_2H_5COCH_3$	14
$CHCl_3$	47	$ClCH_2—CHCl—CH_2Cl$	6.1	CH_3NO_2	168
$CDCl_3$	41	C_6H_5Cl	<0.1	*n*-$C_3H_7NO_2$	9.0
CCl_4	<0.1	*o*-$Cl_2C_6H_4$	<0.1	$ClCH_2COOC_2H_5$	1.4
CH_3CHCl_2	0.5	$CH_2=CCl_2$	<0.1	$NCCH_2—COOCH_3$	54
$ClCH_2—CH_2Cl$	6.1	*trans*-$ClCH=CHCl$	<0.1	*n*-C_4H_9OH	69
$ClCH_2—CHCl_2$	8.6	*cis*-$ClCH=CHCl$	33	$CH_3CHOH—C_2H_5$	23
$Cl_2CH—CHCl_2$	145	$ClCH=CCl_2$	0.2	$ClCH_2CN$	17,000
$Cl_2CH—CCl_3$	<0.1	$C_2H_5OC_2H_5$	<0.1	$Cl_3C—CN$	2.3
n-C_3H_7Cl	<0.1	$ClCH_2CH_2—O—CH_2CH_2Cl$	2.8	$CH_2=CHCN$	130
$Cl—(CH_2)_3—Cl$	2.9	$CH_3COOC_2H_5$	0.2	$ClCH_2CH_2CN$	940
$Cl—(CH_2)_4—Cl$	0.3	$C_2H_5COC_2H_5$	1.1	*n*-C_3H_7CN	13.7
		n-$C_3H_7COCH_3$	1.7	$CH_2=CH—CH_2CN$	67
				$CH_3O—CH_2CN$	84
				$CH_3O—CH_2—CH_2CN$	91
				$C_2H_5O—CH_2—CH_2CN$	38

capability for our standard salt (for others see Sections 1.3.3 and 1.3.4), but they are also cheap and easily removable. One drawback of such solvents is that they could give rise to side reactions, but most PTC reactions are so fast that this is not a big danger.

It should be noted that as expected diethyl ether and ethyl acetate have a low extraction capability. Less understandable is the poor performance of chlorobenzene and *o*-dichlorobenzene. These solvents are often used when there is danger of side reactions with chloroform or methylene chloride. A low extraction constant does not necessarily exclude a solvent as unsuitable, but since this means that at any given time during the reaction, only a small percentage of the theoretically possible amount of ion pairs is in the organic phase, the reaction will be slowed down. If, however, a larger, more lipophilic cation is used (see next Section), this effect is to some extent counteracted. *Salting-Out Effects.* Extraction constants are not only influenced by the solvent system, but also by foreign salts. Most extraction constants found in the literature were determined at constant ionic strength. There is, however, a very strong salting-out effect. Brändström examined the conditional extraction constants of NBu_4Cl and NBu_4Br between water and methylene chloride in the presence of potassium carbonate [112]. Linear parallel dependences on the molality of K_2CO_3 were found. There is no competitive extraction of carbonate or hydrogencarbonate observed, so a genuine salting-out must occur. Two mols of K_2CO_3 per liter increased the extraction constants about a thousandfold. This salting-out effect is of great importance for PTC, especially with concentrated (50%) aqueous sodium hydroxide. In this medium almost all quaternary ammonium salts are sparingly soluble and easily extracted. In addition, it acts as a desiccant for the organic phase. In other cases solvents miscible with water, *e.g.*, acetonitrile or THF, can be used if foreign salts make two-phase systems possible. Thus the action of the salt is twofold: it separates the organic solvent and salts out the ion pairs.

1.3.3. Influence of the Onium Cation

It is well known that high molecular weight amines can be extracted, as ammonium salt ion pairs using various counter ions, from aqueous solutions into media such as chloroform. Recently, a number of analytical methods [44, 47–51, 54, 58] and separation procedures [7, 52, 53] have been based on selective extractions of this type. Quaternary ammonium salts have been similarly applied [55–57]. As already mentioned in Section 1.3.1 and excellently reviewed by Brändström [11, 112], extremely complicated multi-constant equilibria, which depend on the structure of the anion, cation, and solvent, as well as on the pH, ionic strength, and concentrations, can exist. Such equilibria have been determined in cases of analytical interest. From these physicochemical and analytical studies it would appear that there is a relation between cation size and extraction constant. This is the important aspect for PTC work.

It is obvious that an increasing number of C atoms surrounding the central N atom of the ammonium cations will increase their lipophilicity thus raising the extraction

Table 1–2. Extraction Constants $E_{NR_4pic} = [NR_{4\,pic}]_{org}/\{[NR_4^{\oplus}]_{aq} \cdot [pic^{\ominus}]_{aq}\}$ of Quaternary Ammonium Picrates, Extraction from Water into an Organic Solvent [53,49].

Solvent	Cation				
	$N(CH_3)_4$	$N(C_2H_5)_4$	$N(n\text{-}C_3H_7)_4$	$N(n\text{-}C_4H_9)_4$	$N(n\text{-}C_5H_{11})_4$
CH_2Cl_2	1.5	220	2.9×10^4	4.8×10^6	2.45×10^8
$CHCl_3$	0.22	21	4.4×10^3	8.1×10^5	
C_6H_6		0.22	35	3.9×10^3	7.9×10^5
CCl_4				87	2.9×10^4

constant E_{QX}. Gustavii [53] observed a linear relationship between log E_{QX} and n, the number of C atoms in ammonium ions. He extracted picrates into methylene chloride using primary amines as well as symmetrical secondary and tertiary amines and symmetrical quaternary ammonium salts. The following relations were found:

Ammonium ions from
primary amines: $\log E_{Qpicrate} = -2.40 + 0.63 \cdot n$
secondary amines: $\log E_{Qpicrate} = -1.35 + 0.61 \cdot n$
tertiary amines: $\log E_{Qpicrate} = 0.10 + 0.54n$
quaternary ammonium: $\log E_{Qpicrate} = -2.00 + 0.54n$

For PTC purposes, quaternary ammonium ions are of special interest because they are less likely to interfere in reactions. It is important, therefore, to test whether relations similar to the one above are also valid for other solvents and counter ions or whether they are restricted to symmetrical cations. Gustavii [49, 53] gives extraction constants for picrates in a variety of solvents (Table 1–2). Again an average increase of log E_{QX} of about 0.54 units per C atom is evident. Methylene chloride and chloroform are again the best solvents for extraction although their relative order is reversed from that in Table 1–1.

The effect of cation size on the extraction of other anions has also been determined. To a first approximation ("side processes" and other experimental errors cannot be easily excluded), the same rise of about 0.54 units per added C atom was found again. The following are examples of extractions of anions by symmetrical quaternary ions into chloroform which obey this rule*): Acetate, phenylacetate, benzoate, salicylate 3-hydroxybenzoate [55], p-toluenesulfonate, β-naphthalenesulfonate, trinitrobenzenesulfonate, 2,4-dinitro-α-naphthol, dipicrylamine, sulfonate group carrying azodyes [35], nitrite, perchlorate, chloride, bromide, iodide [58]. The results of Czapkiewicz-Tutaj and Czapkiewicz [59] concerning the extraction of quaternary ammonium bromides into 1,2-dichloroethane are found in graphical form. Calculating log E_{QX} values from these results, we obtained the data found in Table 1–3. Although the error is undoubtedly large because of the limited results available, the data presented below should prove useful.

It can be seen that even for unsymmetrical cations in yet another solvent, log E_{QX}

*) Going for example from $NR_4^{\oplus}$, R = methyl to R = ethyl, increases the number of C atoms by 4.

Table 1–3. Logarithms of Extraction Constants (log E_{QX}) for Ammonium Bromides, Extraction from Water into 1,2-Dichloroethane (calculated from diagrams in ref. [59]).

Salt	Number of Cations	log *E*
$N(n\text{-}C_{16}H_{33})(CH_3)_3Br$	19	3.88
$N(n\text{-}C_{15}H_{31})(CH_3)_3Br$	18	3.28
$N(n\text{-}C_{14}H_{29})(CH_3)_3Br$	17	2.66
$N(n\text{-}C_{12}H_{25})(CH_3)_3Br$	13	0.34
$N(n\text{-}C_{14}H_{29})(C_2H_5)_3Br$	20	3.72
$N(n\text{-}C_{12}H_{25})(C_2H_5)_3Br$	18	2.54
$N(n\text{-}C_{10}H_{21})(C_2H_5)_3Br$	16	1.36
$N(n\text{-}C_{10}H_{21})(n\text{-}C_3H_7)_3Br$	19	3.91
$N(n\text{-}C_9H_{19})(n\text{-}C_3H_7)_3Br$	18	3.30
$N(n\text{-}C_{10}H_{21})Bu_3Br$	22	4.15
NBu_4Br	16	0.79
$N(C_2H_5)Bu_3Br$	14	−0.17
$N(n\text{-}C_5H_{11})_4Br$	20	2.98
$N(n\text{-}C_6H_{13})_4Br$	24	5.16

rises by 0.54 to 0.61 units per C atom if one of the substrate chains is elongated. As expected, however, the number of C atoms is not the only factor controlling the extraction constant, thus calculations of unknown extraction constants would appear safe for related series only: either with homologous symmetrical ions, or when the length of only one of the four carbon chains is changed. Benzyl groups, for example, are much less lipophilic than *n*-heptyl groups and their contribution to the extraction capability lies somewhere between butyl and propyl. Thus, benzyltriethylammonium chloride (TEBA) is partitioned mainly into the aqueous phase unless a "salting-out effect," as in the case of concentrated sodium hydroxide is operative.

In summing up, therefore, the logarithms of extraction constants of homologous series of quaternary ammonium salts rise by a more or less constant factor of about 0.54 per added C atom, irrespective to the nonpolar aprotic solvent and anion.

1.3.4. Influence of the Anion

Partial or complete anion exchange by equilibrating an organic quaternary ammonium salt solution [$Q^{\oplus}X^{\ominus}$] with an aqueous NaY solution have been performed repeatedly. From such studies, scales of lipophilicities may be constructed. Clifford and Irving [63] arrived at the following order of extractabilities, going from the lipophilic ion $ClO_4^{\ominus}$ to the most hydrophilic ion $PO_4^{3\ominus}$, for chloroform/water:

$$ClO_4^{\ominus} \gg ClO_3^{\ominus} > NO_3^{\ominus} > Cl^{\ominus} \gg HSO_4^{\ominus} > OH^{\ominus} > SO_4^{2\ominus} > CO_3^{2\ominus} > PO_4^{3\ominus}$$

Other results are:

for water/nitrobenzene [76], $I^{\ominus} > Br^{\ominus} > Cl^{\ominus}$;
for aqueous acids/CCl_4 [64], $Br^{\ominus} > NO_3^{\ominus} > Cl^{\ominus} > HSO_4^{\ominus} > CH_3CO_2^{\ominus} > F^{\ominus}$;
and for water/toluene or water/CH_2Cl_2 [65], $ClO_4^{\ominus} > I^{\ominus} > NO_3^{\ominus} > C_6H_5CO_2^{\ominus} > Br^{\ominus} > Cl^{\ominus} > HSO_4^{\ominus} > HCO_3^{\ominus} > CH_3CO_2^{\ominus} > F^{\ominus} > OH^{\ominus}$.

Finally, Mąkosza and co-workers include cyanide [66]:

$$I^{\ominus} > Br^{\ominus} > CN^{\ominus} > Cl^{\ominus} > OH^{\ominus} > F^{\ominus} > SO_4^{2\ominus}.$$

In their experiments, the organic phase was benzene, chlorobenzene, or *o*-dichlorobenzene and the aqueous phase concentrated sodium hydroxide. Other anions that could be exchanged for chloride include $BH_4^{\ominus}$, $SCN^{\ominus}$, $[Fe(CN)_6]^{3\ominus}$, $MnO_4^{\ominus}$, $NO_2^{\ominus}$, and $N_3^{\ominus}$ [66].

It is clear that the same order is found in the various polar, nonprotic solvents. This is plausible for dilute solutions, where the differences in solvation energies for the anions $X^{\ominus}$ and $Y^{\ominus}$ in the organic solvents and water are the major factors governing extractability. It is possible, however, that for some of the salts used—since different cations were employed—solubility limits in either phrase, as well as hydration, dissociation, or association behavior will change the order of extractability. One of these factors appears to be responsible for the unexpected position of benzoate in the third series above (*cf.*, Table 1–4), and the order of hydroxide and fluoride seems uncertain. Combining these values with further, quantitative data (Table 1–4) we arrive at the following order:

$$\sim ClO_3^{\ominus} \sim$$
$$\text{Picrate} \gg ClO_4^{\ominus} > I^{\ominus} > \text{toluene sulfonate} > NO_3^{\ominus} > Br^{\ominus} >$$
$$\sim CN^{\ominus} \sim$$
$$> \text{benzoate} > Cl^{\ominus} > HSO_4^{\ominus} > HCO_3^{\ominus} > \text{acetate} > F^{\ominus}, OH^{\ominus} >$$
$$> SO_4^{2\ominus} > CO_3^{2\ominus} > PO_4^{3\ominus}$$

Table 1–4. Extraction Constants E_{QX} of Tetra-*n*-butylammonium Ion Pairs distributed between Water and Chloroform and—in parentheses—between Water and Methylene Chloride.

Anion	E_{QX}	Reference	Anion	E_{QX}	Reference
$Cl^{\ominus}$	0.78 (0.35)	[53, 58]	$CH_3CO_2^{\ominus}$	7.6×10^{-3}	[55]
$Br^{\ominus}$	19.5 (17)	[53, 58]	$C_6H_5—CH_2—CO_2^{\ominus}$	1.86	[55]
$I^{\ominus}$	1023 (2188)	[53, 58]	$C_6H_5CO_2^{\ominus}$	2.45	[55]
$ClO_4^{\ominus}$	3020 (4.37×10^4)	[53, 58]	salicylate	263	[55]
$NO_3^{\ominus}$	24.5 (79)	[53, 58]	3-hydroxybenzoate	0.03	[55]
			phenoxide	0.93	[55]
			picrate	8.1×10^5 (4.8×10^6)	[49, 53]
			p-toluenesulfonate	214	[45]
			naphthalene-2-sulfonate	2818	[45]
			anthracene-2-sulfonate	1.3×10^5	[62]
			trinitrobenzenesulfonate	2.9×10^4	[45]
			2,4-dinitro-1-naphthoxide	2.8×10^6	[45]

The same order of lipophilicites has usually been observed with tetraphenylphosphonium, -arsonium, and triphenylsulfonium cations [75]. One very important fact obvious from this series should be noted for future reference: If anions of di- or tribasic acids are present they are much more difficult to extract than the related hydrogen-monobasic, -dibasic, or dihydrogen-monobasic anions as long as small cations like $NBu_4^{\oplus}$ are used, *i.e.*,

$$HSO_4^{\ominus} > SO_4^{2\ominus}; \quad H_2PO_4^{\ominus} > HPO_4^{2\ominus} > PO_4^{3\ominus};$$
$$HO_2C{-}(CH_2)_4{-}CO_2^{\ominus} > {}^{\ominus}O_2C{-}(CH_2)_4CO_2^{\ominus}$$

When $Q^{\oplus}$ is larger, however, the extractability of Q_2SO_4, for example, rises faster than that of $QHSO_4$. With $NHex_4$ it is necessary to have the solutions strongly acidic to prevent the extraction of $(NHex_4)_2SO_4$ along with $NHex_4HSO_4$ [11].

A practical conclusion from these facts is that for medium-sized cations hydrogensulfates are not only very good starting materials for the preparation of many onium salts but are also very useful PT catalysts. The addition of one molar equivalent of sodium hydroxide will transform the hydrogen sulfate anion into neutral sulfate which cannot interfere, because it is less easily extractable than almost any other inorganic or organic anion. Chloride is the second choice of anion. The more hydrophilic quaternary ammonium salts such as acetate, fluoride, or hydroxide are difficult to prepare, highly hygroscopic, and/or instable.

Turning now to the more quantitative aspects of PTC, many extraction constants for water/chloroform or dichloromethane have been determined [45, 53, 55, 60–62]. Relevant PTC data are collected in Table 1–4. For comparison purposes, the cation tetrabutylammonium is kept constant, although in many cases constants for related symmetrical ions are also known. The determination of further anionic constants would be highly desirable.

The wide range of extraction constant values in the halogenide series is remarkable. Obviously iodide is preferentially extracted, as is perchlorate. An immediate conclusion must be that if only small amounts of cation are present, these are extracted largely as ion pairs with the very lipophilic anions. This effect is possibly detrimental to a planned PTC reaction, especially if iodide ions are liberated during the course of the reaction. Iodides (and to a lesser extent bromides) should not, therefore, as a rule be used as PTC catalysts.

Of the organic anions, acetate and formate seem to be the most hydrophilic ones. The strong structural effect of salicylate (internal hydrogen bond!) and 3-hydroxybenzoate should be noted when comparing extraction constants. In principle, similar effects are found for homologous series of anions. Every CH_2 group added makes the anion more lipophilic. Besides alkyl groups, other lipophilic substituents like nitro, chloro, bromo, *etc.* increase the anion extraction constants strongly.

It should be mentioned in passing that it is possible to extract acids HX as ion pairs $Q^{\oplus}X^{\ominus}\cdots HX$. Preparative applications of this "extraction by hydrogen bonding" for $HF_2^{\ominus}$, $HCl_2^{\ominus}$, and $X^{\ominus}\cdots HOOH$ are known (*cf.*, Chapter 3).

Since many PTC reactions are carried out in the presence of strong alkali metal hydroxides the relative order of magnitude of quaternary ammonium hydroxide extraction constants is of special interest. Although there are several references to $OH^{\ominus}$ extraction constants in the literature, most of them are not reliable. Some have been determined in solvents partially miscible with water and hence are of no interest to us here. Others are reported for chloroform or dichloromethane as the organic phase [55, 56], but it can be shown that these react rapidly with hydroxide [67]. On the other hand, quaternary ammonium hydroxides are extremely insoluble in nonpolar solvents unless a residual concentration of protic solvents is present. "Tetrabutylammoniumhydroxide in toluene" for instance is frequently used as a reagent. It is prepared by diluting the commercial methanol solution with toluene and distilling off some (but not all) of the methanol. Hydroxide ion is very strongly hydrated and always coextracts some water or other hydroxylic solvents (*cf.*, [96]). Thus, for solubility reasons the determination of extraction constants by measuring the partitioning of quaternary ammonium hydroxides between water and nonpolar solvents is limited. Furthermore, $NR_4^{\oplus}OH^{\ominus}$ solutions are rather unstable (*cf.*, Section 3.1.2). Degradation leads to NR_3 which can be cotitrated acidimetrically thus leading to wrong conclusions concerning the concentration of hydroxide in either phase.

In the case of methylene chloride, it can be shown that freshly prepared aqueous $NBu_4^{\oplus}OH^{\ominus}$ can be extracted into it to some extent. The titer of the solution decreases quickly, however, due to reaction with the solvent [67]. By this solvent, if a solution of tetrabutylammonium hydrogensulfate is equilibrated with concentrated sodium hydroxide, an appreciable, but again fast decreasing amount, of $NBu_4^{\oplus}OH^{\ominus}$ is extracted. This is in agreement with the order of extraction series where $OH^{\ominus} > SO_4^{2\ominus}$. Already after 10 minutes 67% of the chloride equivalent to the ammonium hydroxide is formed from attack by the solvent:

$$2OH^{\ominus} + CH_2Cl_2 \rightarrow CH_2O + H_2O + 2Cl^{\ominus} \quad [67].$$

A rough estimate of the extraction constant for tetra-*n*-heptylammonium hydroxide between benzene and water is 1; that is of the order of 10^{-4} times the value of the chloride [67]. In typical PTC experiments with sodium hydroxide as reagent, the difference in extraction constants is partially offset by the use of a big molar excess of NaOH in the presence of small, catalytic amounts of ammonium chlorides. When equilibrating benzene solutions of very lipophilic cations (tetra-*n*-hexylammonium or tetra-*n*-heptylammonium) originally brought in as chloride with a large excess of sodium hydroxide, one-quarter to one-third of the ammonium ions in the organic phase will be paired with hydroxide, the rest with chloride. With less lipophilic cations (*e.g.*, tetra-*n*-butylammonium or benzyltriethylammonium) both the solubility of hydroxides in benzene and the extractability of the ion pairs are low. Thus, only a small percentage of the ammonium ions, mainly as chloride (96–99.5%), will be in the organic phase [67].

In conclusion, it should be mentioned that many anions are extracted into organic media along with a certain amount of water of solvation. The extent of ion-pair hydration will depend on anion, cation, solvent, as well as reaction conditions. Various studies of these effects have been made [*e.g.*, 68–70].

Mąkosza and co-workers give a preparative method of obtaining anhydrous benzyltributylammonium salts of many anions *via* anion exchange from chlorides in the presence of concentrated sodium hydroxide [66].

1.4. Crown Ethers, Cryptates, and Other Chelating Agents as Extractants

"Crowns" are defined as macroheterocycles usually containing the basic unit $(\text{—Y—CH}_2\text{—CH}_2)_n$, where Y is O, S, or N. Crown ethers in particular have met with immense interest in the last decade, and various aspects of their preparation and chemistry have been reviewed [77–82]. Because the systematic nomenclature of these compounds is very clumsy, common names are generally used. These are exemplified by 18-crown-6 (**1**) in which 18 indicates the number of atoms in the ring, crown the

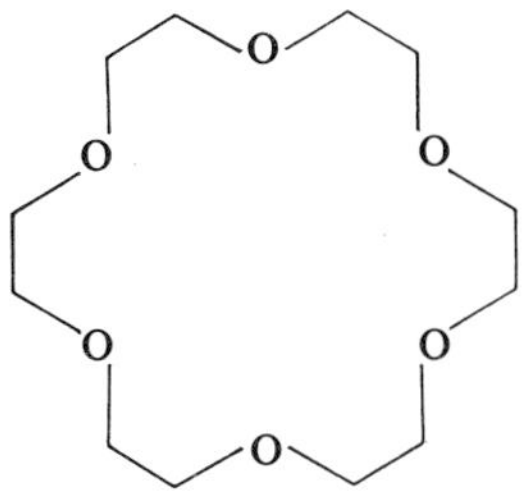

1 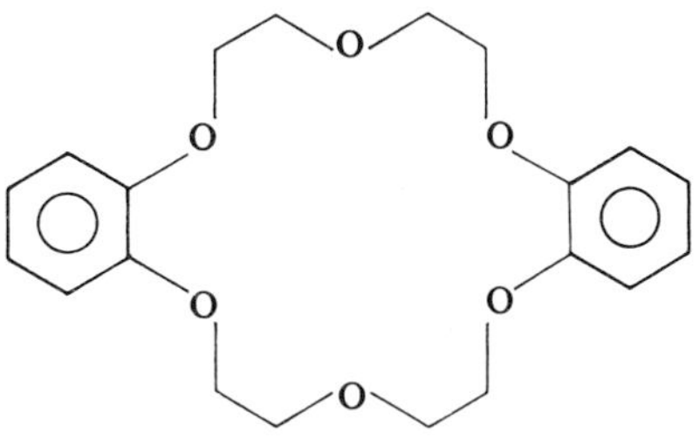 **2**

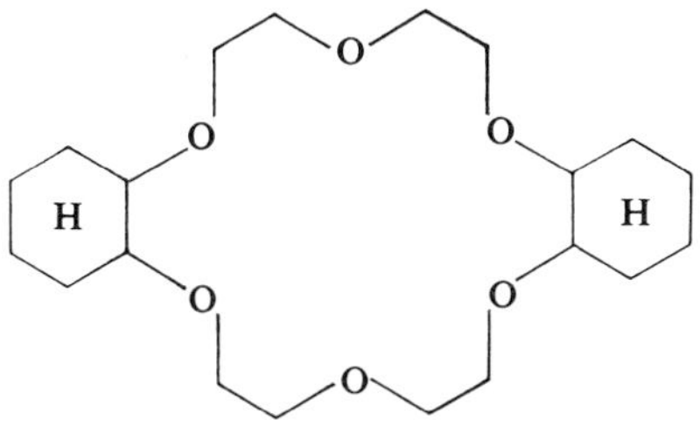

3 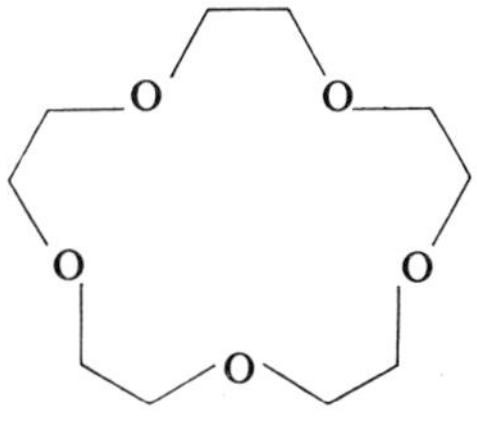 **4**

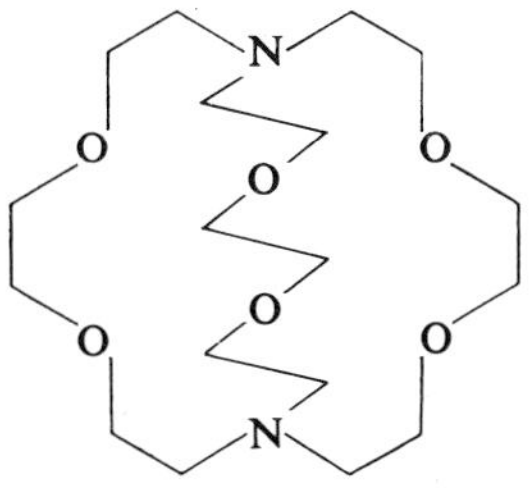

5

Scheme 1–1

class, in this case crown ether, and 6 the number of oxygen atoms in the relation *1,4,7, etc.* Other common, commercially available crown ethers are dibenzo-18-crown-6 (**2**), dicyclohexano-18-crown-6 (**3**) (often named incorrectly "dicyclohexyl-18-crown-6"), and 15-crown-5 (**4**).

New crown ethers, aza-analogues, polyoxa-polyaza-macrocycles, analogues containing annelated heterocycles, and bi- and polycyclic analogues are being reported. One of these in particular has found special attention. Compound **5**, designated as cryptate [2.2.2]*) by Lehn and co-workers [83] is sold under the trade name Kryptofix® 222. A common feature of all crowns and related substances is a central hole or cavity. By chelation within this hole†) complexes of varying stability can be formed with other species depending on the appropriate radii and electronic configurations. Cations, anions, neutral zerovalent metals, and neutral molecules such as nitriles are capable of such behavior [108].

The cation complexes of general interest for PTC are those formed with potassium and sodium metal cations. The most stable potassium complexes contain the 18-membered rings of compounds **1**, **2**, **3** or **5**, while sodium is chelated preferentially by **4** and other similar 15-membered crowns. Among the other complexing cations are the hydronium ion, $H_3O^{\oplus}$ [106], ammonium ions [84], and diazonium ions [91, 111]. Cram and co-workers have shown that when this "host-guest-complexation" is extended to chiral crown ethers and substituted racemic primary ammonium salts, it can be utilized for optical resolutions [84]. They used optically active binaphthyl units, but many other diastereometric crowns with potential or actual optical activity are known [85]. Physical methods including X-ray analysis have in many cases established the accurate complex structures.

Alkali metal ion/crown chelates are of particular importance when discussing the present subject. In addition, complexes with molecules such as sodium amide [86] and potassium hydride [87] may find application in future PTC work. Other related phenomena, *e.g.*, the introduction of anions into cryptates [88], the dissolution of alkali metals in ether solvents by means of crowns [89], the isolation of a stable crystalline $Na^{\ominus}$/cryptate $Na^{\oplus}$ salt [90], and the formation of radical anions from aromatic hydrocarbons, crowns, and alkali metals [95], are not of immediate interest in connection with PTC.

Returning to complex formation between alkali metal ions and crowns or cryptates, these have multifarious effects:

(a) "Organic masking" of the alkali metal provides an "onium ion"-like entity that can be extracted or solubilized with the accompanying anion, just like the onium salts themselves, into nonpolar organic solvents. Although their stabilities are lower in most cases, open chain polyethers and various polyamines form similar complexes and can be used to extract salts too [97]. Crown ether complexes can also be used as models for the transport and differentiation of ions through liquid membranes [98].

---

*) The numbers indicate the three two-oxygen-atom bridges.
†) There are also complexes where the metal ion is situated above the plain of the crown.

(b) The anion part of an ion pair present in an organic solvent is "activated," *i.e.*, more reactive than without a cation-complexing agent, even in cases where the solubilizing action of the crown ether is not necessary because the inorganic salt itself is soluble. In dipolar aprotic solvents, where a high degree of dissociation occurs, this effect is most notable. Such systems have been described as involving the "reaction of naked anions." This picturesque phrase is misleading, however, because solvent–solute interactions are strong even in "weakly solvating" or "nonsolvating" media [99].

The 18-crown-6-mediated "anion activation" in acetonitrile (dielectric constant 39) has been investigated [99], and a leveling of nucleophilicites was observed. The rate constants for displacements with benzyl tosylate for $N_3^{\ominus}$, $Ac^{\ominus}$, $CN^{\ominus}$, $F^{\ominus}$, $Cl^{\ominus}$, $Br^{\ominus}$, and $I^{\ominus}$ exhibit a total variation of less than a factor of 10. Acetate and fluoride displayed a remarkably increased reactivity compared to "normal" reactions in hydroxylic solvents. Although this "anion activating" effect has been used frequently for reactions in homogeneous media, we shall cite only one striking example. Merrifield and co-workers selectively cleaved protected amino acids and peptides from oxyacyl resins using potassium cyanide in DMF, N-methylpyrrolidone, or acetonitrile at room temperature [100]. In the presence of dicyclohexano-18-crown-6 (**3**) (see page 17) 90–97% reaction was observed after 8–16 hours, whereas in the absence of the crown ether, only 5–6% reaction occurred after 72 hours.

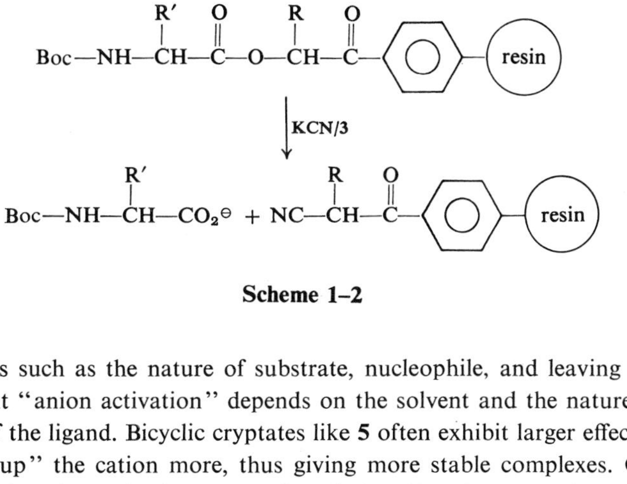

$$\text{Boc—NH—CH(R')—C(=O)—O—CH(R)—C(=O)—C}_6\text{H}_4\text{—resin}$$

$$\downarrow \text{KCN/3}$$

$$\text{Boc—NH—CH(R')—CO}_2^{\ominus} + \text{NC—CH(R)—C(=O)—C}_6\text{H}_4\text{—resin}$$

Scheme 1–2

When factors such as the nature of substrate, nucleophile, and leaving group are kept constant "anion activation" depends on the solvent and the nature and concentration of the ligand. Bicyclic cryptates like **5** often exhibit larger effects because they "wrap up" the cation more, thus giving more stable complexes. Crowns in dipolar aprotic solvents lead to more dissociation. In other cases (*e.g.*, potassium *tert*-butoxide in DMSO) ionic aggregates are broken up by crown complexation leading to a stronger basicity of the alkoxide as measured by the rate of proton abstraction [101]. In less polar media such as THF or dioxane, ion pairs are the dominant species. Here, crown ethers may favor the formation of solvent-separated, looser ion pairs [32, 81] of much higher reactivity [102]. Even in hydroxylic solvents remarkable crown effects are observed because the structure and composition of the solvation shell around the ion pairs is changed and ionic aggregates are partially broken down. For example, the *syn/anti* ratios of base catalyzed eliminations may be influenced strongly [103].

Knöchel and co-workers have studied anion activation extensively in polar aprotic media under both homogeneous and PTC conditions [93, 104, 105, 107]. For the reaction in homogeneous absolute acetonitrile between benzyl choride and acetate ion, generated from several alkali acetates and various series of ligands, cyclic aminopolyethers were superior to crowns as rate accelerators. Using ^{1}H-NMR spectroscopy, this behavior was correlated to the amount of ion pair separation [104]. If substoichiometric catalytic amounts of the complexants are used, the same reaction is converted to a solid-liquid PTC system. Under these circumstances, anion activation itself is interplaying with other factors: dissolution and complexation of the salt, anion exchange, and transport phenomena. The performance of a considerable number of ligands in this standard two-phase reaction was compared [93, 107]. As expected, there is no simple relation between anion activation, solubilization, and the overall reaction rate.

In conclusion, it should be emphasized that anion activation is rapidly suppressed by traces of water in the medium because of ion pair or anion solvation thus reducing the nucleophilicity [107, 109, 110]. In addition, crowns tend to carry water into solvents even as nonpolar as chloroform [110].

1.5. Solid-Liquid Anion Exchange

Originally PTC was performed exclusively using an aqueous and an organic phase. This technique is called liquid-liquid PTC by some authors. As mentioned in previous sections, coextraction of some water of hydration often occurs, which may interfere with the desired reaction through suppression and/or diversion. It was therefore reasonable to suggest that the use of water should be abandoned in such cases, and that PTC should be carried out with solid salts. A case in point is the generation of dichlorocarbene from sodium trichloroacetate, a reaction normally carried out in absolute dimethoxyethane (which is expensive):

$$Na^{\oplus\ominus}O_2CCCl_3 \longrightarrow NaCl + CO_2 + [CCl_2] \longrightarrow \text{further reactions}$$

Sodium trichloroacetate is decomposed with water according to the following reaction:

$$Na^{\oplus\ominus}O_2CCCl_3 + H_2O \longrightarrow NaHCO_3 + HCCl_3$$

and although trichloroacetate ions can be extracted by quaternary ammonium salts from aqueous solution into chloroform containing an olefin as carbene acceptor, the coextracted water of hydration suppresses the carbene reaction. If, however, solid sodium trichloroacetate, catalytic amounts of a quaternary onium salt, chloroform, and an olefin are stirred at 25–80 °C, very satisfactory yields of dichlorocyclopropanes are obtained [92]. This version of the reaction is generally called solid-liquid PTC, and it was assumed, even in a recent review [81], that crown ethers are the only suitable catalysts for this purpose. It is argued that a crown behaving as a two-dimensional system with multiple polar sites can approach a crystal lattice so closely that the required movement of the cation from the lattice to the ligand is small. In contrast, onium cations have a sterically shielded positive key atom, so that they cannot approach the lattice closely. Furthermore, it is speculated that the efficiency of crowns as catalysts may be attributable in part to the intermittent liberation of free, uncomplexed ligands from the complexed product which may first precipitate and then liberate the crown to complex new reactant salt.

In many cases in fact, onium salts are very effective solid-liquid PT catalysts, and there are even cases where crowns are less efficient. This stems from the fact that initial ligand/cation complexation may be fast only in homogeneous solution in hydroxylic solvents. The undesirable solvent must then be displaced with the aid of a nonpolar solvent. This undoubtedly leaves a more or less substantial solvation envelope on the ion pairs. Since alkali cation crown complexes are rather stable, it is doubtful whether or not the crown is ever freed again. However, catalytic processes are very similar for both onium ions and crown-complexes: the dissolved ion pair simply exchanges anions on the surface of the crystal lattice. It is possible, therefore, that some factor as yet unevaluated is important in these systems. Possible factors are the speed of stirring, the size of the reactant salt, and the danger of product salt deposition on the surface of the salt to be dissolved. High shear stirring or ball milling during the PTC process may be helpful.

Besides crown ethers, cryptates, and onium salts, open chain polyethers, diamines, and polyamines can also be used to catalyze the reaction between solid alkali metal salts and alkylating agents [93, 94].

In conclusion, the major advantages of solid-liquid PTC techniques will be stressed again. "Anion activation" is possible in the more polar solvents which cannot be used with water, *e.g.*, acetonitrile. Since the solubilized salt is dissociated to a large extent in such a solvent and the anions are not solvated tightly, reactions are fast. This is the case even for anions considered as weak nucleophiles in hydroxylic solvents (*e.g.*, acetate, fluoride).

2. Mechanism of Phase Transfer Catalysis

2.1. Mechanistic Investigations *)

PTC was defined in Section 1.1 as a two-phase reaction between salts in solid or aqueous form, acids, or bases and substrates in organic solvents in the presence of so-called phase transfer catalysts. Typically these agents are onium salts or complexing agents for alkali metal cations such as crown ethers, cryptates, or their open chain analogues. As already stated in Section 1.1, our definition of PTC is not based on any one mechanism but rather on the effects observed. Extensive investigations have led to the mechanistic elucidation of many PTC reactions though.

Reactions are divided into those occurring under neutral or acidic conditions, and those which occur in the presence of highly concentrated inorganic bases. The former will be discussed first. Numerous displacement reactions, oxidations, and reductions occur under neutral or acidic conditions. (For preparative applications *cf.*, Sections 3.2, 3.4 and 3.5.)

2.1.1. Mechanism under Neutral Conditions

The original PTC mechanism for displacement reactions proposed by Starks [2] is shown below. An ion pair, formed by the extraction of anion $Y^{\ominus}$ into the organic phase by the onium salt cation $Q^{\oplus}$, undergoes a fast displacement with RX. The new salt $[Q^{\oplus}X^{\ominus}]$ then returns to the aqueous phase, where $Q^{\oplus}$ picks up a new $Y^{\ominus}$ ion for the next cycle.

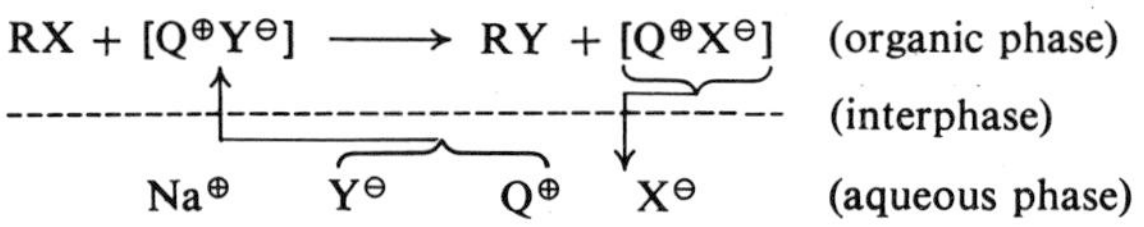

$$RX + [Q^{\oplus}Y^{\ominus}] \longrightarrow RY + [Q^{\oplus}X^{\ominus}] \quad \text{(organic phase)}$$

--- (interphase)

$$Na^{\oplus} \quad Y^{\ominus} \quad Q^{\oplus} \quad X^{\ominus} \quad \text{(aqueous phase)}$$

Scheme 2–1

From what is known about the ease of extraction of ion pairs into nonpolar organic media and the high reactivity of onium salts in homogeneous solutions of this kind [*cf.*, Chapter 1) the proposed mechanism is very reasonable, but a rigorous proof is necessary.

Since PT catalysts of the onium type can be tensides or resemble surfactants it should first be established that the reaction actually occurs in the organic medium and not at the interphase or in the aqueous phase.

The influence of stirring rates on reaction kinetics has been investigated by several

---

*) A recent review was published by Brändström [1].

authors. It is known that the reaction rate of a typical interfacial reaction is proportional to the rate of stirring when the latter lies between 600 and 1700 rpm [6]. In contrast, PTC reactions are not influenced by the rate of stirring apart from a certain minimum value necessary for breaking up concentration gradients on both sides of the interphase. Using standard laboratory equipment (magnetic stirrer rods or flat-bladed mechanical stirrers in round-bottomed flasks), it was found that the rates of reactions involving neutral reagents were constant above 250 [2], 200 [4], or 350 [5] rpm. In the presence of concentrated sodium hydroxide a larger range of stirring rates was found necessary for the results to be reproducible. In a typical ether synthesis from alcohol, alkyl chloride, and NaOH/catalyst only 80 rpm was required [7], whereas the reaction between alcohols, dimethyl sulfate, and NaOH/catalyst needed much more effective intermixing [8]. Dihalocarbene reactions with haloform, 50% aqueous sodium hydroxide, and TEBA require 750–800 rpm for reproducible results [9]. These differences may be mechanistically relevant (*vide infra*) but obviously under neutral conditions reaction does not occur at the interphase. The fact that both the organic substrate and the lipophilic catalysts are more or less insoluble in the aqueous phase suggests that reaction occurs either in the organic phase or possibly in a micellar phase. Further findings corroborate these conclusions. Studies using ion-selective electrodes support the suggestion made in Scheme 2–1, that the establishment of extraction equilibria is fast. The rate limiting step must therefore occur in the organic phase. The reaction of 1-bromooctane with benzenethiolate ion in benzene/water has been investigated using various catalysts [3, 4]. A plot of $\ln\{[RBr]_{org}/[C_6H_5S^{\ominus}]_{aq}\}$ against time gave straight lines (characteristic of normal second order kinetics) as expected for the absence of any phase boundary. Furthermore, the rate constant was linearly dependent on the catalyst concentration over a 20-fold change in concentration. Thus, the rate constants for various catalysts were normalized.

Similarly, rate constants for the reaction between a large excess of aqueous chloride, bromide, or iodide and *n*-octyl methanesulfonate in chlorobenzene have been determined. Tributylhexadecylphosphonium halides were catalysts here. The reaction was found to be pseudo-first order

$$\text{rate} = k_{app} \cdot [ROSO_2CH_3]$$

with k_{app} linearly dependent on the catalyst concentration [5].
The overall reaction

$$RX_{org} + Y_{aq}^{\ominus} \xrightarrow{PTC} RY_{org} + X_{aq}^{\ominus} \qquad (1)$$

can be broken down according to the PTC postulate into an extraction step [Eq. (2)] and the chemical reaction proper [Eq. (3)]:

$$[Q^{\oplus}X^{\ominus}]_{org} + Y_{aq}^{\ominus} \overset{K}{\rightleftharpoons} [Q^{\oplus}Y^{\ominus}]_{org} + X_{aq}^{\ominus} \qquad (2)$$

$$[Q^{\oplus}Y^{\ominus}]_{org} + RX \xrightarrow{k_2} [Q^{\oplus}X^{\ominus}]_{org} + RY \qquad (3)$$

It was shown in Section 1.3.1 that

$$K = \frac{[Q^{\oplus}Y^{\ominus}]_{org} \cdot [X^{\ominus}]_{aq}}{[Q^{\oplus}X^{\ominus}]_{org} \cdot [Y^{\ominus}]_{aq}} = \frac{E_{QY}}{E_{QX}} \tag{4}$$

If q_0 is the original concentration of catalyst it follows*) that

$$[Q^{\oplus}Y^{\ominus}]_{org} \cdot [X^{\ominus}]_{aq} = K \cdot (q_0 - [Q^{\oplus}Y^{\ominus}]_{org}) \cdot [Y^{\ominus}]_{aq} \tag{5}$$

$$[Q^{\oplus}Y^{\ominus}]_{org} = \frac{K \cdot q_0 \cdot [Y^{\ominus}]_{aq}}{[X^{\ominus}]_{aq} + K \cdot [Y^{\ominus}]_{aq}} \tag{6}$$

The rate expression for the organic phase is

$$d[RX]/dt = k_2 \cdot [Q^{\oplus}Y^{\ominus}]_{org} \cdot [RX] \tag{7}$$

or in its integrated form

$$k_2 \cdot t = \frac{1}{c_{0(QY)} - c_{0(RX)}} \cdot \ln \frac{c_{0(RX)}}{c_{0(QY)}} \cdot \frac{[Q^{\oplus}Y^{\ominus}]_{org}}{[RX]} \tag{8}$$

where $c_{0(QY)}$ and $c_{0(RX)}$ are the initial concentrations of $[Q^{\oplus}Y^{\ominus}]$ and RX in the organic phase.

Substituting Eq. (6) into Eq. (8) gives

$$k_2 \cdot t = \frac{1}{c_{0(QY)} - c_{0(RX)}} \cdot \ln \frac{c_{0(RX)}}{c_{0(QY)}} \cdot \frac{K \cdot q_0 \cdot [Y^{\ominus}]_{aq}}{\{[X^{\ominus}]_{aq} + K \cdot [Y^{\ominus}]_{aq}\} \cdot [RX]} \tag{9}$$

or, in a simplified form

$$\text{constant} \cdot t = \text{constant} \cdot \ln \frac{[Y^{\ominus}]_{aq}}{\{[X^{\ominus}]_{aq} + K \cdot [Y^{\ominus}]_{aq}\} \cdot [RX]} \tag{10}$$

Thus, plotting $\ln([Y^{\ominus}]_{aq}/[RX])$ (or its inverse, *cf.*, [4]) against time gives a straight line, if

$$[X^{\ominus}]_{aq} + K \cdot [Y^{\ominus}]_{aq} = \text{constant} \tag{11}$$

Eq. (11) is fulfilled strictly only if $K = 1$, as for the most trivial case of isotope exchange. If K does not vary much from unity the deviation may not be very large. Second-order rate profiles were calculated under fixed conditions for a variation in K between 1000 and 0.001 [10]. $K \gg 1$ causes unnaturally large concentrations of $Y^{\ominus}$ in the organic phase, and the resultant curves give the appearance of autocatalysis. $K < 0.1$ stalls the reaction. This is the undesired "catalyst poisoning effect" already mentioned in Chapter 1: the onium cation associates more and more with the anion of the leaving group as its concentration increases during the course of the reaction. This poisoning can be avoided, if molar amounts of catalyst are present. For $K = 0.01$, a raise in the concentration of the catalyst from 1 to 20 mol-% is sufficient to keep the conversion running [10].

Pseudo-first-order kinetics of PTC processes have been repeatedly investigated (*cf.*, [5]). However, first-order behavior can be expected only if the magnitude of K and an

---

*) Square brackets in the law of mass action type equations signify concentrations, otherwise ion pairs are meant.

Table 2–1. Calculated Minimum Molar Ratios $Y^{\ominus}/RX$ Sufficient for Observation of Pseudo-First-Order Kinetics [10].

K	Minimum $Y^{\ominus}/RX$ Ratio
1	10
1.33	8
2	4
10	2
100	0.8
1000	0.4

excess of $[Y^{\ominus}]_{aq}$ combine to make $[Q^{\oplus}Y^{\ominus}]_{org}$ effectively constant [Eq. (6)]. The excess of $[Y^{\ominus}]_{aq}/[RX]$ required has been calculated (*cf.*, Table 2–1). When K is very large a lower than equivalent concentration of $Y^{\ominus}$ is required for pseudo-first-order behavior, at least until $Y^{\ominus}$ is exhausted, and the reaction is quenched [10].

It must be stressed that the extraction constants E_{QX} and E_{QY}, and hence K, are influenced not only by the activity of the ions but also by the many factors considered in Section 1.3 under the heading "side processes." Thus, K [2] and k_2 [2, 4] rise sharply whenever the aqueous phase is close to saturation in either reactant or product salt.

Pseudo-first-order kinetics with respect to alkyl halide are observed in octane with aqueous sodium cyanide, a reaction catalyzed by tributylhexadecylphosphonium bromide [2].

A linear relationship between the rate constant and a 20-fold change in catalyst concentration is observed [2]. If, however, tetrabutylphosphonium bromide is employed as the catalyst, no simple rate law is applicable. The reaction starts slowly and proceeds more rapidly with time. The authors noted that, whereas in an aqueous/bromooctane two-phase system tributylhexadecylphosphonium bromide is partitioned mainly into the organic phase, the tetrabutylphosphonium salts stay mainly in the aqueous phase. During the course of the reaction increasing amounts of the product, octyl cyanide (nonanenitrile), are formed into which tetrabutylphosphonium salts are increasingly extracted. Consequently, the reaction rate increases gradually. In order to verify that the concentration of the tetrabutylphosphonium salt in the organic phase increases with increasing conversion of 1-bromooctane, ^{14}C-labeled $PBu_4^{\oplus}Br^{\ominus}$ was used. Samples of the organic phase were analyzed during the course of the reaction for starting material, product, and ^{14}C. From these data, first-order rate constants and the concentration of $PBu_4^{\oplus}Br^{\ominus}$ were calculated and plotted. A linear double logarithmic relationship with the slope equal to unity was obtained. The sum of these experiments shows unequivocally that

(a) the reaction is occurring in the organic phase, and
(b) the rate is proportional to the amount of catalyst cation in the organic phase.

These results exclude the involvement of micelles. Having established the organic phase as the reaction site, Starks and Owens [2] next considered the possibility of inverted micelles being involved. To this end the state of aggregation and hydration of the onium salts in organic solvents was investigated. It was found that tributylhexadecylphosphonium and -ammonium bromides and tridecylmethylammonium chloride in benzene and 1-bromopropane had aggregation numbers of between 1.2 and 5.4 in a concentration range of 0.04 to 0.25 M. Similar values are found for the long-chain ammonium salts of fatty acids in nonpolar solvents, and these are considered as being "inverted micelles" [11]. Within the concentration range mentioned above, a considerable change in aggregation numbers occurs, especially in benzene, even though the reaction rate remains linearly dependent on the concentration of the catalyst. Furthermore, an extraction of sodium cyanide which is imperative for an inverted micellar mechanism cannot be demonstrated: the concentration of sodium cyanide in the organic layer of a mixture of 1-chlorooctane/aqueous NaCN at 0.01–0.1 mol/l quaternary salt was less than 10^{-5} mol/l.

Solutions of $(n\text{-}C_{16}H_{33})PBu_3^{\oplus}X^{\ominus}$ in toluene or 1-cyanooctane were shaken with tritiated water and the extent of hydration was measured [2]. Average hydration numbers for $X^{\ominus} = NO_3^{\ominus}$, $Cl^{\ominus}$, and $CN^{\ominus}$ were 0.4, 4, and 5. Using the same cation and $X^{\ominus} = Cl^{\ominus}$, $Br^{\ominus}$, and $I^{\ominus}$ the amount of water extracted into chlorobenzene was found to be 3.4, 2.1, or 1.1 equivalents, respectively. Subsequently, rates of nucleophilic substitution reactions between *n*-octyl methanesulfonate and the halides under homogeneous conditions were compared with those obtained for reactions under PTC conditions*). Under anhydrous homogeneous conditions absolute rates are higher. The relative rates found for $Cl^{\ominus}:Br^{\ominus}:I^{\ominus}$ are 6.5:2.5:1.0. Under PTC conditions, both the absolute and relative rates are modified by hydration, and the rate ratio $Cl^{\ominus}:Br^{\ominus}:I^{\ominus}$ was found to be 0.6:1.1:1.0. When the reaction was carried out in homogeneous chlorobenzene with the addition of as much water as was shown to be coextracted in the two-phase experiments by the phosphonium halides and the substrate, both the absolute and relative rates of the PTC process could be simulated [5]. Thus, the rates of the overall reaction are influenced by ion pair hydration involving a limited number of water molecules.

In their paper on the theory of PTC kinetics, Gordon and Kutina [10] discuss in detail the implications of the interplay between extraction and chemical reaction [Eqs. (1) and (2)]. Some of their calculations and plots refer to rather special situations. These might give guidelines in the choice of systems for rate measurements that are likely to obey simple rate laws. Thus, catalyst comparisons using different onium ion and /or alkali metal ion concentrations in similar experiments are seen to merit special attention. Of more general importance, however, is the situation where the overall reaction is reversible

$$RX_{org} + Y_{aq}^{\ominus} \overset{K^{\bullet}}{\rightleftharpoons} RY_{org} + X_{aq}^{\ominus} \tag{1a}$$

*) 0.15 mol H_2O per mol octyl methanesulfonate is also extracted.

The component equilibria may then be expressed by Eqs. (2) (unchanged) and (3a):

$$[Q^{\oplus}X^{\ominus}]_{org} + Y_{aq}^{\ominus} \underset{}{\overset{K}{\rightleftharpoons}} [Q^{\oplus}Y^{\ominus}]_{org} + X_{aq}^{\ominus} \quad (2)$$

$$[Q^{\oplus}Y^{\ominus}]_{org} + RX \underset{}{\overset{K_{org}}{\rightleftharpoons}} [Q^{\oplus}X^{\ominus}]_{org} + RY \quad (3a)$$

It follows that

$$K^* = K \cdot K_{org} \quad (12)$$

However, on closer inspection it is seen that the equilibrium expressed in Eq. (12) can generally be reached only if K approaches unity or if a molar amount of catalyst is present. If K is very different from 1 (say 1000 or 0.001) the actual equilibrium position will be determined by the amount of catalyst used, at least from a practical standpoint in a limited period of time. Let us consider a reaction employing 1% of catalyst (0.01 mol), $K_{org} = 10{,}000$, $K = 0.001$, and 1 M solutions of one mol each of RX and $Y_{aq}^{\ominus}$. As more and more $X^{\ominus}$ is liberated, thus "poisoning" the catalyst, the forward reaction (1a) becomes slower and slower. At the same time, the reverse reaction depicted in Eq. (1a) gains momentum. Using Eq. (4), the number of moles of catalyst z, paired with $Y^{\ominus}$, at 10% conversion can be calculated.*)

$$\frac{z \cdot 0.1}{[0.01 - z] \cdot 0.9} = 0.001$$

$$z = 8.9 \cdot 10^{-5}$$

Thus, after 10% conversion, 99.1% of the catalyst is paired with $X^{\ominus}$ rather than $Y^{\ominus}$, and the forward reaction comes to a virtual standstill. The actual equilibrium must lie close to 10% conversion, although K^* can have a maximum value of 10! Further examples of such calculations are reported [10].

As depicted in the general scheme for PTC (*cf.*, p. 22), the catalyst cation migrates back and forth with various counterions between two phases. This possibility has been neglected in the transformation of Eqs. (1) or (1a) into Eqs. (2) and (3) or (3a), respectively, since this effect cancels out in the process. Nevertheless, it has been questioned whether or not this physical movement of cations between the layers is necessary at all. Depending on the solvent and on the lipophilicity of the onium salt used, the catalyst is partitioned between the organic and aqueous phase to a greater or lesser extent. Independently, Brändström and Montanari showed that the concomitant transfer of the catalyst cation is not a requisite for PTC. A drop of a dichloromethane solution of a quaternary ammonium bromide was introduced below the aqueous surface of a tall column filled with an aqueous solution of sodium 4-(2,4-dinitrophenylazo)-phenoxide [12]. In water this indicator is red in the anionic form, and blue in dichloromethane as an ion pair with $Q^{\oplus}$. During its passage through the column, the drop turned blue, thus indicating the anion exchange. If $Q^{\oplus}$ was being coextracted into the aqueous layer it could not return because the rapidly falling drop would leave it behind. In another experiment a U tube containing two independent

*) The usual assumptions of Section 1.3 (no "side processes," concentrations instead of activities) are made.

organic phases separated by an aqueous solution was used [13]. Various catalysts were employed, varying from very lipophilic ones that partitioned 100% into the organic phase to others that were present partly in the aqueous layer. *n*-Octyl methanesulfonate was introduced into one of the organic phases, and the catalyst into the other organic phase. With the less lipophilic catalysts, the PTC reaction started after a certain time for diffusion through the middle phase had elapsed and the rate increased from zero to a constant value. In contrast, the 100% lipophilic catalysts ($NOct_4Br$ and $PBu_3C_{16}H_{33}Br$) were incapable of initiating a PTC reaction in the other organic layer. But a fast reaction occurred if these were added to the organic phase containing the substrate.

From these experiments it is clear that Scheme 2–1 has to be modified somewhat. Scheme 2–2, therefore, depicts the actual PTC process as it occurs with lipophilic catalysts.

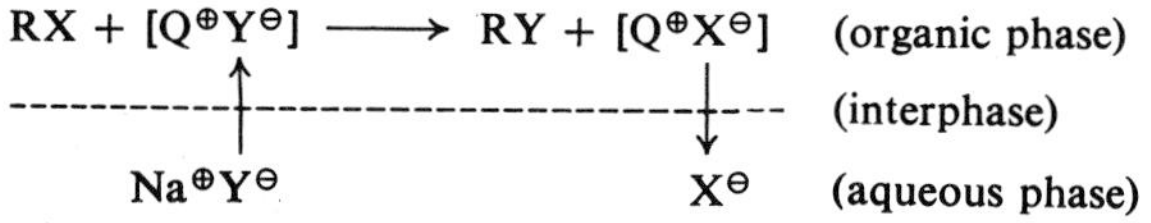

$$
\begin{array}{ccccl}
RX + [Q^{\oplus}Y^{\ominus}] & \longrightarrow & RY + [Q^{\oplus}X^{\ominus}] & & \text{(organic phase)} \\
\uparrow & & \downarrow & & \text{(interphase)} \\
Na^{\oplus}Y^{\ominus} & & X^{\ominus} & & \text{(aqueous phase)}
\end{array}
$$

Scheme 2–2

Another mechanistically important question is whether the reaction in the organic medium is occurring *via* an ion pair or the free anion. It was stated in Chapter 1 that dissociation constants of ion pairs in the nonpolar media of PTC are of the order 10^{-5}. Since the concentration of free anions is very small, reactions from the latter is highly unlikely. Nevertheless, Brändström investigated the kinetics of PTC halide exchange [14]. If the free anion were responsible for the reaction, the rate would be proportional to the square root of the concentration of $[Q^{\oplus}X^{\ominus}]$ in the organic phase. Since the velocity was found to be directly proportional to $[Q^{\oplus}X^{\ominus}]$, ion pairs must be the dominating nucleophiles in the concentration range examined. In a second extensive study on the alkylation of β-diketones using the extractive alkylation (PTC) method [15], it was found that, in principle, both free anions and ion pairs, with tetrabutylammonium counterion, can function as a nucleophile. Again, however, under preparative conditions in dichloromethane the ion pair is virtually the only active nucleophile.

2.1.2. Mechanisms in the Presence of Bases

We have just seen that the mechanism of PTC under neutral conditions has been fully explored and clarified. All the observed effects can be satisfactorily explained by Scheme 2–1 or 2–2 and the pertinent equilibria and kinetic expressions. Turning our attention next to PTC reactions performed with additional base we shall see that the mechanistic picture is not quite as clearcut.

Among the reactions carried out in the presence of strong base are various conversions such as C-, O-, and N-alkylations, isomerizations, H/D-exchanges, additions, β-eliminations, α-eliminations, hydrolysis reactions and many others. The practical applications of these PTC reactions, which are probably among the most numerous and interesting preparatively, are considered in Section 3.3.

Let us first consider alkylations, which must involve a gradual change depending on the pK_a of the C—H, N—H, or O—H acid. Relatively strong acids, acetylacetone (pK_a ≈ 9) for example, dissolve in sodium hydroxide. The action of the phase transfer catalyst is therefore to reextract the anion, in the form of an ion pair, into the organic phase where either C- or O-alkylation occurs (*cf.*, Section 3.3.5 on the direction of alkylation of ambident anions). In other words, the mechanism of Scheme 2–2 still holds.

With weaker acids, for example, aliphatic alcohols (pK_a ≈ 18), it is easy to show that both the unchanged substrate and its anion are extracted into the organic medium. The following equilibria are involved primarily:

$$Na^{\oplus}OH^{\ominus} + ROH \rightleftharpoons Na^{\oplus}OR^{\ominus} + H_2O \qquad \text{(aqueous phase)}$$

$$ROH_{aq} \rightleftharpoons ROH_{org} \qquad \text{(both phases)}$$

$$[Q^{\oplus}X^{\ominus}]_{org} + Na^{\oplus}OR_{aq}^{\ominus} \rightleftharpoons [Q^{\oplus}OR^{\ominus}]_{org} + Na^{\oplus}X_{aq}^{\ominus} \qquad \text{(both phases)}$$

In addition, the following equilibria may be of importance:

$$[Q^{\oplus}X^{\ominus}]_{org} + Na^{\oplus}OH_{aq}^{\ominus} \rightleftharpoons [Q^{\oplus}OH^{\ominus}]_{org} + Na^{\oplus}X_{aq}^{\ominus} \qquad \text{(both phases)}$$

$$[Q^{\oplus}OH^{\ominus}]_{org} + ROH_{org} \rightleftharpoons [Q^{\oplus}OR^{\ominus}]_{org} + H_2O \qquad \text{(organic phase)}$$

Finally, ether formation occurs irreversibly:

$$[Q^{\oplus}OR^{\ominus}]_{org} + R'X \longrightarrow R'OR + [Q^{\oplus}X^{\ominus}]_{org} \qquad \text{(organic phase)}$$

We have seen in Chapter 1, Section 1.3.4, that very hydrophilic $OH^{\ominus}$ ions are difficult to extract in competition with halide ions. Since $OR^{\ominus}$ is more lipophilic than $OH^{\ominus}$, the two equilibria involving $[Q^{\oplus}OH^{\ominus}]$ are probably only of marginal importance. Still, the interplay of equilibria is obviously much more complicated here than in simple displacements.

Freedman and Dubois [7] prepared unsymmetrical ethers from primary alkyl chlorides and alcohols by stirring a fivefold excess of 50% aqueous sodium hydroxide over alcohol at 25–70 °C, using an excess of alkylchloride as solvent and 3–5 mol-% NBu_4HSO_4 as catalyst. In a detailed mechanistic investigation, one equivalent of benzyl chloride in THF was reacted with an excess of *n*-butanol and sodium hydroxide, saturated with sodium chloride. Under these conditions the reaction was first-order in catalyst and benzyl chloride. Using equivalent amounts of reactants, clear second-order kinetics were observed through 80% of the reaction. The activation energy and entropy were found to be 13.9 ± 0.5 kcal/mol and −25.5 ± 1.6 e.u., respectively. As expected, the initial counterion of the catalyst had a pronounced effect on the rate because of competitive extraction. Rates were largest with $HSO_4^{\ominus}$,*) very small with $I^{\ominus}$, and even smaller with $ClO_4^{\ominus}$.

---

*) Transformed in NaOH into $SO_4^{2\ominus}$, which cannot be extracted, *cf.*, Chapter 1.

These results are fully consistent with normal PTC where extraction of $^{\ominus}OR$ and possibly—to a minor extent—its formation in the organic phase are fast processes. Concerning the relative extractability of $OH^{\ominus}$, these findings are revealing: if 10^{-2} M NBu_4HSO_4, 50% NaOH saturated with NaCl, and THF are equilibrated, 86% of the cation goes into the organic phase. Of this only 10.7% is in the $OH^{\ominus}$ form, the rest is present as chloride. Starting with a 1 M solution of butanol in THF all the $NBu_4^{\oplus}$ is found in the organic phase, and only 5.4% of it is in the $Cl^{\ominus}$ form, the rest is presumably present as $NBu_4^{\oplus}{}^{\ominus}OC_4H_9$.

During the last 20% of the reaction, when the mixture becomes depleted of ROH, the excess alkyl halide is increasingly hydrolyzed, and this new alcohol is transformed into a symmetrical ether [7]:

$$R'Cl \longrightarrow R'OH \longrightarrow R'O{-}R'$$

This type of reaction was first discovered by Herriott and Picker [3, 23]. These authors also investigated the reaction between secondary alkyl bromides, aqueous NaOH, and PTC catalysts. A high percentage (81–94%) of the substrate underwent elimination, the remainder hydrolysis, which is typical for reactions in the organic phase as distinct from conversions in the aqueous phase.

The results of Freedman and Dubois [7], mentioned above, have been subjected to computational scrutiny by Gordon and Kutina [10]. This indicated that the original authors had worked within the experimental limits predicted for a linear second-order plot, assuming that the above equilibria applied. Freedman and Dubois used a rather unusual system, THF/concentrated sodium hydroxide saturated with NaCl, to demonstrate the extraction of the anion of a fairly weak acid, *n*-butanol. The present authors [25] choose a more typical PTC system: 0.01 mol TEBA (benzyltriethylammonium chloride) in 100 ml benzyl cyanide were equilibrated with 100 ml concentrated sodium hydroxide.*) Of the theoretical amount of basicity 63% was found in the organic phase. Thus, only about $\frac{1}{3}$ of the catalyst is in the chloride form under these conditions, and a normal PTC reaction must be possible. The basicity of the organic phase must be due to the ion pair $[PhCH_2N^{\oplus}Et_3{}^{\ominus}CH(CN)Ph]$. The alternative ion pair $[PhCH_2N^{\oplus}Et_3OH^{\ominus}]$ would either undergo rapid acid-base reaction with the phenyl acetonitrile excess or remain in the aqueous phase since it is not lipophilic enough to be extracted to any appreciable extent. It was shown that in the equilibration of a benzene solution of $[NBu_4^{\oplus}Cl^{\ominus}]$†) with NaOH, about 50% of the cation is found in the concentrated NaOH. Only 4% of the catalyst, in the organic phase, is in the $OH^{\ominus}$ form [30]. Although the solubility of $[NR_4^{\oplus}OH^{\ominus}]$ is better in a medium like phenyl acetonitrile than in benzene, the difference in the $OH^{\ominus}$ and $Cl^{\ominus}$ extraction constants does not permit the presence of much hydroxide in the organic phase (*cf.*, Section 1.3). Furthermore, as previously stated above, this minute amount would react with the solvent phenyl acetonitrile.

---

*) Although the exact pK_a of benzyl cyanide in an aqueous environment is not known, it is estimated to be of the order of 15–16; phenylacetic ester has a pK_a of 17 [24].
†) TEBA is even more hydrophilic and virtually insoluble in benzene.

Let us now consider the much weaker CH-, and NH-acids. It has been demonstrated that PTC alkylations can be performed in the presence of concentrated aqueous sodium hydroxide as the deprotonating agent, even with acids as weak as pK_a 22–25 (depending on the reference for pK_a values). This is rather surprising since the intermediate anions are more basic than water by many powers of ten.

In earlier work on PTC reactions in the presence of strong inorganic bases, the mechanism was believed to involve extraction of $[Q^{\oplus}OH^{\ominus}]$ into the organic medium, followed by deprotonation and alkylation of the substrate:

$$[Q^{\oplus}X^{\ominus}]_{org} + HO_{aq}^{\ominus} \rightleftharpoons [Q^{\oplus}OH^{\ominus}]_{org} + X_{aq}^{\ominus} \quad \text{(both phases)}$$

$$[Q^{\oplus}OH^{\ominus}]_{org} + \text{H Sub} \rightleftharpoons [Q^{\oplus}Sub^{\ominus}]_{org} + H_2O \quad \text{(organic phase)}$$

$$[Q^{\oplus}Sub^{\ominus}]_{org} + RX \longrightarrow [Q^{\oplus}X^{\ominus}]_{org} + \text{R-Sub} \quad \text{(organic phase)}$$

Many arguments have been advanced against this mechanistic scheme. We have seen in Chapters 1 and 2 that the extraction of hydroxide is difficult *per se* because of both its solubility and its small extraction constant. It becomes even more unlikely if more lipophilic anions such as the halides compete for the catalytic amount of quaternary ammonium cation present. As more and more halide is liberated in the course of the reaction, inhibition should increase; this, however, is not the case, at least for chloride. Furthermore, in an homogeneous phase the deprotonation equilibrium between the quaternary ammonium hydroxide and the H-substrate should lie far to the left because of the difference in acidities.

A more recent hypothesis proposed by Mąkosza [26, 27] makes the assumption that substrate deprotonation occurs at the phase boundary. If no catalyst is present, a bilayer-like arrangement of alkali metal cations on the aqueous side and deprotonated substrate anions on the organic side of the phase boundary results. Because of mutual insolubilities at the counter phases, the ions are immobilized and rather unreactive. This situation resembles that of normal adsorptions on a surface.

Mąkosza has shown that some alkylation reactions are possible without the presence of PT catalyst when the interphase is the obvious reaction site [26, 27, 31]. For example, when phenyl acetonitrile, *n*-butyl iodide and 50% aqueous sodium hydroxide are stirred at 80 °C for 5 hours, 70% alkylation occurs [31]. It was shown that the concentration of the nitrile in the aqueous phase was below 2 ppm, and the concentration of the sodium salt of the substrate in the organic phase was about 5 ppm. Thus, the only reaction possible is an interfacial one [31]. Although under these conditions, alkyl bromides R′Br alone do not react with phenylacetonitrile, the use of an $R'Br/C_4H_9I$ mixture leads to two alkylated products, Ph—CH(R′)—CN and Ph—$CH(C_4H_9)$—CN. The latter was obviously formed *via* R′I. Since a separate experiment showed no halide exchange between R′Br and aqueous sodium iodide, halide exchange appears to occur on the organic side of the phase boundary between iodide liberated from C_4H_9I during alkylation and R′Br [31]. From these and other results [26, 27] of interphase reactions, the importance of the phase boundary is emphasized. The higher stirring rates necessary for constant, reproducible results in

PTC reactions using aqueous sodium hydroxide (see above) may point in the same direction although the higher viscosity of concentrated NaOH is an additional factor undoubtedly.

Deprotonation at the interphase also seems to occur in normal PTC reactions. Here the action of the catalyst appears to be twofold:

(a) It aids in deprotonation. This effect is demonstrable in the acceleration of H/D exchange by PT catalysts. It would appear that the positive charge of the cation opposite the reacting center facilitates proton removal.
(b) It aids in detaching the "anchored"*) substrate anion from the interphase. Of greater importance than (a) is this liberation of the deprotonated substrate from the interphase attended by transfer of the original counterion of the catalyst to the aqueous phase.

In general this equilibrium is favorable because typical organic substrate anions are more lipophilic than halide ions. Occasionally, however, "catalyst poisoning" by lipophilic inorganic anions has been described (*e.g.*, perchlorate, picrate,†) iodide) if the intermediate organic anion is of medium lipophilicity. Thus, we arrive at the following very satisfactory mechanistic picture for the PTC alkylation of weak acids:

$$Na_{aq}{}^{\oplus}OH_{aq}{}^{\ominus} + HSub_{org} \rightleftharpoons Na^{\oplus}_{\text{anchored at interphase}} + Sub^{\ominus}_{\text{anchored at interphase}} + H_2O \quad \text{(interphase)}$$

$$Sub^{\ominus}_{\text{anchored at interphase}} + [Q^{\oplus}X^{\ominus}]_{org} \rightleftharpoons [Q^{\oplus}Sub^{\ominus}]_{org} + X_{aq}{}^{\ominus} \quad \text{(both phases)}$$

$$Q^{\oplus}Sub^{\ominus}]_{org} + RX \longrightarrow R\text{-}Sub + [Q^{\oplus}X^{\ominus}]_{org} \quad \text{(organic phase)}$$

Contrary to when $[Q^{\oplus}X^{\ominus}]_{org}$ is extracted from neutral aqueous solutions, here ion pairs $[Q^{\oplus}Sub^{\ominus}]_{org}$ cannot be hydrated to any great extent because of two factors: concentrated NaOH is a strong desiccative, and H_2O being a stronger acid would decompose the ion pair.

It might be suggested from the above experiments in the absence of catalysts that only small concentrations of $[Q^{\oplus}Sub^{\ominus}]_{org}$ are present and that the final reaction might take place at the interphase. This in fact is not the case.

Using concentrated sodium hydroxide and a PTC catalyst, stable solutions of ion pairs containing the anion of a weak acid such as fluorene‡) can even be prepared. As mentioned in Chapter 1, a 0.1 M solution of a lipophilic salt such as tetra-*n*-hexylammonium chloride in benzene will exchange chloride for hydroxide up to a

---

*) The expression is not meant to imply that one and the same anion is kept "anchored" for long, but rather that a dynamic process of exchange may occur.
†) A previous statement from our laboratory [28], that the rate of alkylation of benzyl cyanide is as fast with NBu_4 picrate as with NBu_4Cl was in error. Picrate performs very poorly. In the generation of dihalocarbenes, however, picrates are good catalysts. It has been found that the picrate ion is destroyed in the course of this reaction, being exchanged for chloride eventually [25].
‡) pK_a 20.5–22.8 depending on the reference compound and method of determination [24].

28:72 $OH^{\ominus}:Cl^{\ominus}$ ratio, if it is equilibrated with an equal volume of 50% NaOH [30]. A similar experiment with a benzene solution, 1 M in fluorene and additionally 0.1 M in $NHex_4Cl$ shows, by titration, that 81% of the possible basicity is in the benzene layer [25]. The large increase in the titration value in the presence of fluorene, the color of the organic phase, and the position of the equilibrium with ammonium hydroxide (*vide supra*) show that the "basicity" is present as $[NHex_4^{\oplus}C_{13}H_9^{\ominus}]_{org}$.

Having clarified the PTC alkylation mechanisms we shall now consider the mechanism of dihalocarbene generation. We have thoroughly investigated the situation for dichlorocarbene, but the arguments we shall now present apply to all carbene PTC reactions. Competition reactions have shown that the dichlorocarbene, generated in PTC reactions, is identical with that generated by other methods [1, 29] and is not a carbenoid. Furthermore, it can be demonstrated that, under PTC conditions, carbenes CXY can undergo halide exchange giving CX_2 and CY_2.*) In addition, and in contrast to all other methods of dihalocarbene generation, the reaction at room temperature is not an irreversible, fast, one-step process. Whereas potassium *tert*-butoxide/chloroform reacts at −20 °C whether or not a substrate is present and $LiCCl_3$ decomposes irreversibly at temperatures as low as −72 °C, the PTC mixture, chloroform/concentrated NaOH/catalyst, retains its ability to produce CCl_2, even over a period of several days at room temperature, if no reactant is present. Since H/D exchange between chloroform and concentrated $NaOD/D_2O$ is very fast, even in the absence of a catalyst, the first step must be deprotonation at the phase boundary. Presumably this involves the same kind of bilayer discussed above.

$$HCCl_3 \underset{\text{fast}}{\rightleftharpoons} CCl_3^{\ominus}{}_{\text{anchored}} \quad \text{(phase boundary)}$$

In the absence of a catalyst hydrolysis is very slow although the rate increases somewhat on the addition of a quaternary ammonium salt (*cf.*, Table 3–12, p. 179). When olefins are added, the rate of chloroform consumption increases and concurrently the hydrolysis is suppressed. The extent of this effect depends on the nucleophilicity of the olefin. In other words, the olefin and water compete for the CCl_2, and the olefin wins since it is in the same phase as the carbene precursor.

Analogous to the alkylations of weak CH- and NH-acids and the fluorene experiment discussed above, one would expect the catalyst to detach $CCl_3^{\ominus}$ from the interphase to form $[NR_4^{\oplus}CCl_3^{\ominus}]_{org}$. To test this, a 0.1 M solution of $[NR_4^{\oplus}Cl^{\ominus}]$ in chloroform was equilibrated with concentrated NaOH. After separation and drying, one aliquot of the organic phase was stirred with an olefin, a second was titrated with HCl, and a third was carefully concentrated. No evidence of $[NR_4^{\oplus}CCl_3^{\ominus}]$ or its derived products was found [9, 29, 32].

Further evidence indicates that $[NR_4^{\oplus}CCl_3^{\ominus}]$ is unstable at room temperature. Solid sodium trichloroacetate can be partially solubilized into chloroform with the aid of $[NR_4^{\oplus}Cl^{\ominus}]$ forming $[NR_4^{\oplus\ominus}O_2CCCl_3]$. This salt decarboxylates slowly at room temperature, and the liberated dichlorocarbene can be trapped by olefins. In the

*) For a detailed discussion see Section 3.3.13.

absence of a trapping agent $[NR_4^{\oplus}CCl_3^{\ominus}]$ is not isolable, instead intensive tarring occurs [33]. Although it is unstable in solution, $[NR_4^{\oplus}CCl_3^{\ominus}]$ must be the intermediate for CCl_2 generation. A mechanism of PTC carbene reactions, which fits all available evidence, is shown below. The first step involves deprotonation at the interphase and formation of "anchored" trihalomethylide ions, H_2O stays in the aqueous phase:

$$HCCl_{3\,org} + Na_{aq}^{\oplus} + OH_{aq}^{\ominus} \rightleftharpoons CCl^{\ominus}_{3\,\text{anchored to interphase}} + Na^{\oplus}_{\text{anchored to interphase}} + H_2O \quad \text{(interphase)}$$

This is followed by "detaching" of the anion. The equilibrium, however, must lie far to the left, thus stabilizing all subsequent steps:

$$CCl^{\ominus}_{3\,\text{anchored to interphase}} + [NR_4^{\oplus}X^{\ominus}]_{org} \rightleftharpoons [NR_4^{\oplus}CCl_3^{\ominus}]_{org} + X_{aq}^{\ominus} \quad \text{(both phases)}$$

It is known that the formation of the carbene is reversible:

$$[NR_4^{\oplus}CCl_3^{\ominus}]_{org} \rightleftharpoons [NR_4^{\oplus}Cl^{\ominus}]_{org} + [CCl_2]_{org} \quad \text{(organic phase)}$$

Adduct formation is irreversible:

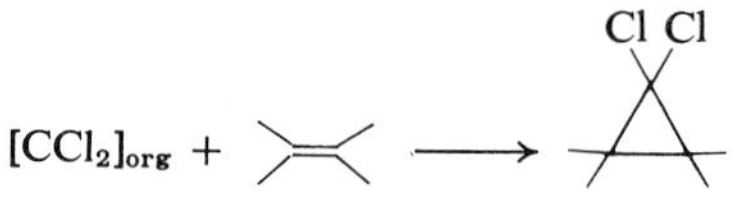

$$[CCl_2]_{org} + \text{>C=C<} \longrightarrow \text{1,1-dichlorocyclopropane (Cl Cl)} \quad \text{(organic phase)}$$

Scheme 2–3

There is competition with hydrolysis which is irreversible and slow because of the phase boundary:

$$[CCl_2]_{org} + OH_{aq}^{\ominus} \longrightarrow [HOC^{\ominus}Cl_2] \longrightarrow \text{formate + CO (both phases)}$$

$$[CCl_2]_{org} + H_2O_{aq} \longrightarrow [HOCHCl_2] \longrightarrow \text{formate + CO (both phases)}$$

Scheme 2–4

In addition, exchange processes can occur:

$$[CCl_2]_{org} + [NR_4^{\oplus}X^{\ominus}]_{org} \rightleftharpoons [NR_4^{\oplus}CXCl_2^{\ominus}]_{org} \quad \text{(organic phase)}$$

$$[NR_4^{\oplus}CXCl_2^{\ominus}]_{org} + X_{aq}^{\ominus} \rightleftharpoons CXCl^{\ominus}_{2\,\text{anchored to interphase}} + [NR_4^{\oplus}X^{\ominus}]_{org} \quad \text{(both phases)}$$

$$CXCl^{\ominus}_{2\,\text{anchored to interphase}} \longrightarrow HCXCl_2$$

Scheme 2–5

Other slow side processes involve attack of $CCl_3^{\ominus}$ and CCl_2 on $HCCl_3$ (*cf.*, Section 3.3.13).

Once more it should be stressed that the important feature of this mechanistic proposal is the stabilization of the carbene precursor in the form of a trihalomethylide anion "dynamically anchored" to the phase boundary. The kinetics of such reactions and of alkylations of weak acids have not been investigated. Such an investigation would not be easy in view of the heterogeneity, competition reactions, complicated equilibria, and because of general restrictions by the latter on the linearity of second-order rate constants (*cf.*, [10]). In spite of this lack the known facts appear to fit the above mechanism.

We have just seen how PTC alkylations and carbene generations in the presence of concentrated sodium hydroxide can follow three slightly different mechanisms. Phase transfer catalysis, therefore, clearly involves a whole range of mechanisms.

The simplest PTC mechanism in the presence of strong base (*e.g.*, mechanisms of H/D exchange and isomerization reactions) probably involves reaction at the interphase. Many other reaction mechanisms have not been investigated in depth. PTC mechanisms may vary depending on the substrate and reaction conditions involved. β-Eliminations, for example, may occur at the interphase where deprotonation would be facilitated by the catalyst. Alternatively, the presence of a small amount of quaternary ammonium hydroxide ion in the organic phase could affect the deprotonation. There is, however, a third mechanism known. This has been observed in the absence of base at elevated temperatures. In nonpolar aprotic media relatively unsolvated halide ions behave as bases (*cf.*, Chapter 1); thus pentachloroethane, for example, is dehydrochlorinated by ammonium halides in a patented technical process:

$$Cl_3C{-}CHCl_2 \xrightarrow[\approx 200°C]{NR_4X} Cl_2C{=}CCl_2 + HCl(gas)$$

One may expect, therefore, that some elimination reactions involve formation of the salt $[NR_4^{\oplus}HX_2^{\ominus}]$. Subsequently, this diffuses to the interphase where neutralization occurs.

Other PTC reactions obviously require the extraction of hydroxide ion. Notable among these are hydrolysis and saponification reactions. As has been shown, using dichloromethane as a model substrate and NBu_4HSO_4 as catalyst, fairly rapid hydrolysis occurs as long as $[NBu_4^{\oplus}OH^{\ominus}]_{org}$ can be extracted into the organic phase in the absence of a more lipophilic anion [30]. Due to increasing pairing of the liberated chloride ion with the catalyst anion in the organic phase, the reaction eventually becomes very slow.

PTC saponifications are extremely sensitive to many factors including solvent, catalyst, and substrate structure [34]; hence several mechanisms may be possible. If the carboxylic acid liberated is very hydrophilic (*e.g.*, a low molecular weight dicarboxylic acid) PTC acceleration of hydrolysis is very effective. In contrast, PTC saponification of lipophilic acid esters is sluggish. Variation of the original catalyst anion (cation = $NBu_4^{\oplus}$) in the saponification of diethyl adipate under fixed conditions led to the following order of reaction rates: $HSO_4^{\ominus} \gg Cl^{\ominus} > Br^{\ominus} > I^{\ominus} \gg$

$ClO_4^{\ominus}$. In the reaction with NBu_4HSO_4, the ammonium salt in the organic phase was $[NBu_4^{\oplus\ominus}O_2C—(CH_2)_4—COOH]$ [34]. This is apparently easily exchanged at the interphase for $OH^{\ominus}$. With other anions, the extraction of $OH^{\ominus}$ or extraction and transport of $^{\ominus}O_2C—(CH_2)_4—COOH$ is partially inhibited by more lipophilic inorganic ions.

In the saponification of diethyl adipate, neutral surfactants such as

$$C_{16}H_{33}—(O—CH_2—CH_2)_n—R \ (R = OH, OCOCH_3, OCH_3, Cl)$$

were at least as effective as the best onium salts [34]. The soap sodium stearate was also very effective, thus indicating the possibility of surface reactions (*cf.*, Table 2–7, page 41). Although the polyethers could complex sodium hydroxide in a way similar to the crown ethers, the effect of such reagents has often been described as that of a surfactant or micelle. Further details of the mechanism are yet to be clarified. The rate of ester hydrolysis in a water-toluene mixture in the absence of catalyst was determined largely by the rate of diffusion. Vigorous stirring brought about an enormous rate acceleration. Addition of the anionic surfactant, sodium dodecyl sulfate, however, had a slightly retarding influence [41].

Preliminary experiments have shown that the catalytic effectiveness of onium salts in saponifications does not parallel their influence on surface tension, at least under typical PTC conditions [35]. On the other hand, many hydrolysis reactions in micellar and reversed micellar systems are reported, and some other types of reactions considered in this book could occur in a micellar environment [11].

In the reaction between alkyl halides and base the possibility of a micellar phase as site of reaction is excluded, *inter alia*, on the basis that effective PT catalysts are large, lipophilic onium salts which are often symmetrical [23]. Typical micelle forming agents have a small polar head group (*e.g.*, $—NMe_3^{\oplus}$) and a long lipophilic tail. Although some symmetrical tetraalkylammonium salts can aggregate to some extent in water [36], it is open to question whether or not this is due to micelle formation [37]. Furthermore, symmetrical onium ions have lower aggregation numbers than typical micelle-forming agents and are worse solubilizers for organic substrates [38]. Even more important, however, is the fact that typical micellar reactions are carried out in "homogeneous" aqueous or organic solutions. In PTC reactions, the organic substrate and the bulk of the catalyst are normally in the organic phase. Transport and diffusion processes become very cumbersome if substrate, products, and organic solvent have to travel to and from the aqueous phase micelles. Both micelles and inverted micelles have been excluded as participating in some PTC reactions. In fact it is open to question whether or not such aggregates, whose association numbers in organic solvents are usually less than 10, can even be considered as micelles.

Under certain reaction conditions, however, micellar catalysis could be involved. "Aliquat 336" (methyl-trioctyl-ammonium chloride), for example, is a very effective lipophilic PTC catalyst (*vide infra*). Alone it cannot form a micelle. In aqueous solution, in the absence of organic solvents, it is present as an oily suspension. If, however, a nonionic micelle forming agent (*e.g.*, polyoxyethyleneoleyl alcohol) is added, Aliquat is trapped in or on the nonionic micelles. The catalyst thus produced proves very effective in many processes [39]. In water at very low concentrations (10^{-4}–10^{-5} M), Aliquat 336 forms aggregates by itself. These appear to be much smaller than conventional globular micelles and catalyze nucleophilic hydrolysis reactions [40]. *Clearly, hydrolysis must be scrutinized further.*

The kinetics of a PTC process involving crown ethers has been investigated recently [55]. The basic mechanisms are the same as with onium salts.

2.2. Empirical Catalyst Evaluations

In this section, we shall review some of the comparative data from the literature concerning catalysts. We will first consider performance of different classes of onium salt and crown complexed cation catalysts. Secondly, the reaction selectivity achieved through the use of certain types of catalysts will be discussed.

The main factors governing catalyst behavior have been described in Sections 1.3.1 to 1.3.4, and 2.1. The catalyst cation must be sufficiently lipophilic to guarantee both sufficient solubility of the extracted ion pair and a high extraction coefficient in the organic phase. Both the original catalyst anion and that liberated in the course of the reaction should be as hydrophilic as possible in order that the ratio of extraction constants is favorable. Preparative aspects, such as the separation of products from the used catalyst, may require the use of a less lipophilic catalyst which can be removed with water in the workup. Therefore it is sometimes necessary for a compromise to be made concerning the choice of onium salt (*cf.*, Section 3.1),

However, it must be remembered that in nonpolar media, even with the most lipophilic onium cations, solubility and extractability are only prerequisites of a chemical reaction. As was pointed out in Chapter 1, when comparing various catalysts a general direct relationship between their solubilizing or extracting effects, and "anion activation" or rates, cannot be made. Other factors besides the precursory equilibria must be considered. Ion-pair anion-cation interactions and the amount of water of hydration coextracted into the organic medium are especially important. It is not surprising, therefore, that the overall catalytic performance of a homologous series of catalyst cations, *e.g.*, the symmetrical tetraalkylammonium salts, will show a dramatic increase in activity ranging from very hydrophilic to lipophilic ions and then, when these other factors gain importance, a slight decrease.

Although many authors have compared the performance of a few catalysts for the reaction under consideration, these values are not generally useful because of the different reaction conditions. It has been pointed out that even in parallel experiments misinterpretations of comparisons result if the concentrations of catalyst cation and inorganic cation vary [10]. We shall discuss work involving comparisons under essentially constant reaction conditions. In most cases yields only were calculated, but in some cases rate constants were also determined.

Hydrophilic catalysts are useful in basic reactions only because of salting-out effects. so we shall consider these reactions separately as in the previous section.

Table 2–2. Second-Order Rate Constants for the Reaction of Thiophenoxide with 1-Bromooctane in Benzene/Water [4].

Catalyst	Rate Constant $k \times 10^3$, $M^{-1}s^{-1}$*)
NMe_4Br	<0.0016
$NProp_4Br$	0.0056
NBu_4Br	5.2
NBu_4I	7.4
$NMeOct_3Cl$(Aliquat 336)	31
$N(PhCH_2)Et_3Br$ (TEBA-bromide)	<0.0016
Pyridyl-BuBr	<0.0016
Pyridyl-HepBr	0.023
Pyridyl-$C_{12}H_{25}Br$	0.092
$NHexEt_3Br$	0.015
$NOctEt_3Br$	0.16
$N(C_{10}H_{21})Et_3Br$	0.24
$N(C_{12}H_{25})Et_3Br$	0.28
$N(C_{16}H_{33})Et_3Br$	0.48
$N(C_{16}H_{33})Me_3Br$	0.15
PPh_4Br	2.5
PPh_4Cl	2.7
PPh_3MeBr	1.7
PBu_4Cl	37
$POct_3EtBr$	37
$P(C_{16}H_{33})Et_3Br$	1.8
$AsPh_4Cl$	1.4
Dicyclohexano-18-crown-6	41

*) Observed rate constants extrapolated to 0.00137 mole of catalyst.

Rate constants for the reaction between thiophenoxide and 1-bromooctane in benzene/water [4] are given in Table 2–2.

As expected, the larger more lipophilic compounds are the most effective and the more symmetrical ions are better than those with only one long chain. Substituted phosphonium ions are somewhat more effective than similarly substituted ammonium ions. As was pointed out already in Chapter 1, benzyl and aryl groups are not catalytically equivalent to C_7 or C_6 alkyl groups. In heptane, rates are more than a hundredfold slower and in the more polar solvent *o*-dichlorobenzene they increase 5–300-fold, especially for the weaker catalysts. The maximum rate constant was ten times faster for the PTC reaction than the homogeneous reaction in ethanol and about 300 times slower than in absolute DMF [4]. It should, of course, be remembered that the actual experimental rate of the PTC reaction can be regulated by the amount of catalyst used.

Yields and reaction times for the displacement reaction between *n*-octyl bromide and inorganic iodide were determined, with various crown ethers and Tributyl-*n*-hexadecylammonium bromide serving as catalysts (*cf.*, Table 3–1, p. 58). Only certain cryptates with a side chain performed better than the onium salt.

Table 2–3. Extraction of 1.0 mmole of $MnO_4^{\ominus}$ from 0.02 M Aqueous Solution into Benzene by Quaternary Salts [42].

Catalyst	mmole	mmole $MnO_4^{\ominus}$ in benzene
NMe_4Cl	1.51	0.0
TEBA	1.63	0.74
NBu_4Br	1.54	0.95
PBu_4Cl	1.54	0.96
$N(C_{16}H_{33})Me_3Br$	1.53	0.86
Aliquat 336	1.52	0.93

The extractive power of the catalyst is probably the most important factor in permanganate oxidations. It was shown that a moderate excess of lipophilic ammonium salts can extract $MnO_4^{\ominus}$ almost quantitatively ([42], Table 2–3).

In borohydride reductions, reactions are much faster with quaternary ammonium catalysts carrying a hydroxyl group β to the nitrogen, *e.g.*, (-)-N-dodecyl-N-methylephedrinium bromide, than with either Aliquat 336 or tributylhexadecylphosphonium bromide [43]. This particular catalyst, however, was less effective than more normal ones in other typical PTC reactions. One possible explanation for its unique effect in borohydride reductions is that the carbonyl group is activated towards $BH_4^{\ominus}$ attack through hydrogen bonding. Some authors claim to have found slight asymmetric induction in PTC borohydride reductions with optically active hydroxyl group carrying catalysts (*cf.*, Section 3.4.1).

For solid-liquid PTC, a comparison of the catalytic abilities between a typical quaternary ammonium salt, Aliquat 336, 18-crown-6, and tetramethylethylenediamine was made [65]. The ammonium catalyst turned out to be equivalent or superior to the others in displacement reactions with acetate, fluoride, and adeninyl anions. However, the cyanide anion reacted at least 100 times faster when catalyzed by the crown relative to Aliquat 336 (*cf.*, Section 1.5).

Let us now consider alkylation reactions in the presence of concentrated sodium hydroxide. Mąkosza [44] compared a number of catalysts in the ethylation of benzyl cyanide by ethyl chloride under standard conditions. The results are shown in Table 2–4.

These early results showed that TEBA is a more or less optimal catalyst. Any other compounds in Table 2–4 approaching the performance of TEBA are either hygroscopic or more expensive to prepare. Interestingly, yields are lower with the higher homologues of TEBA. This is apparently because of specific anion-cation interactions in the ion pairs [$NR_4^{\oplus}$ $PhCH^{\ominus}$–CN]. This steep rise of yields to a maximum and subsequent falling off with increasing lipophilicity of the catalyst cation has been repeatedly observed (*cf.*, Tables 2–6 to 2–9).

Mąkosza also found that many inorganic foreign ions in the aqueous medium competitively suppress the alkylation. Among those with the most strongly adverse

Table 2–4. Influence of Catalyst Structure on the Yield of the Reaction between Benzyl Cyanide and Ethyl Chloride [44].

Catalyst	Yield (%)	Catalyst	Yield (%)
$N(PhCH_2)Me_3Cl$	32	p-H_3CO—$PhCH_2$—NEt_3Cl	54
$N(PhCH_2)Me_2EtCl$	40	p-Cl—$PhCH_2$—NEt_3Cl	38
$N(PhCH_2)MeEt_2Cl$	45	$N(PhCH_2)Et_2(CH_2$—$CH{=}CH_2)Cl$	8
$N(PhCH_2)Et_3Cl$(TEBA)	50	N-allyl-N-benzyl-piperidinium chloride	5
$N(PhCH_2)Et_2PropCl$	47	$H_2C{=}CH$—CH_2—NEt_3Cl	23
$N(PhCH_2)EtProp_2Cl$	44	$H_2C{=}CH$—CH_2—$NProp_3Cl$	6
$N(PhCH_2)MeEtPropCl$	43	$N(PhCH_2)Et_3Br$	35
$N(PhCH_2)Et_2BuCl$	45	NEt_4Cl	51
$N(PhCH_2)Prop_3Cl$	43		
$N(PhCH_2)_2Et_2Cl$	15		

effects were $I^{\ominus}$ and $ClO_4^{\ominus}$; $Br^{\ominus}$, $NO_2^{\ominus}$, $CO_3^{2\ominus}$, $CN^{\ominus}$ were less severe. Fluoride, chloride, and sulfate showed no effect [44]. The benzylation of benzyl cyanide was investigated under PTC conditions, where only a small amount of foreign ion originally brought in by the catalyst could disturb. Even here very large effects can be observed [45] (*cf.*, Table 2–5). Although there are some differences not easily explained, the trend follows the extraction coefficient differences on the whole, that is, very lipophilic anions are not easily exchanged for deprotonated substrate molecules at the interphase.

Table 2–5. Yields of Products from Benzyl Cyanide and Benzyl Chloride with Tetrabutylammonium Catalysts*) [45, 46].

Catalyst Anion	% Monobenzylated	% Bisbenzylated
$HSO_4^{\ominus}$	42.8	11.8
$Cl^{\ominus}$	40.6	9.3
$Br^{\ominus}$	35.1	7.4
$I^{\ominus}$	8.6	0.5
$ClO_4^{\ominus}$	24.1	3.5
β-naphthalenesulfonate	40.2	12.8
benzoate	45.3	12.2
p-nitrobenzoate	44.1	12.2
picrate	15.1	3.7

*) Conditions: 0.05 moles each of $PhCH_2CN + PhCH_2Cl$, 0.001 mole catalyst, 0.1 mole conc. NaOH, 1 h, 0 °C.

An even larger variety of catalyst cations and crown ethers were compared for the same reaction [45], but no clearcut trends emerged.

Dockx investigated the influence of catalyst chain length on styrene formation in the reaction of phenethyl bromide and concentrated sodium hydroxide [47] (*cf.*, Table 2–6).

Table 2–6. Yields of Styrene from Phenethyl Bromide and Concentrated Sodium Hydroxide with Different Catalysts [47].

Catalyst	Yield (%)
NEt_4Br	3
$NPropEt_3Br$	7
$NBuEt_3Br$	12
$NPentEt_3Br$	50
$NHexEt_3Br$	53
$NHepEt_3Br$	45.5
$NOctEt_3Br$	44
$N(C_{10}H_{21})Et_3Br$	42
$N(C_{12}H_{25})Et_3Br$	38

A comparison of catalysts for a saponification reaction was also carried out [34, 35]. But it must be stressed that these results are valid only for an ester that has a very hydrophilic acid anion (*cf.*, Table 2–7). The strong effect of the soap, sodium stearate, shows that various mechanisms are possible (*cf.*, Section 2.1). It is noteworthy that open chain polyethers are as effective here as tetrabutylammonium hydrogensulfate, while crowns are not useful.

Table 2–7. Yields in Saponification of Diethyl Adipate by Concentrated Sodium Hydroxide and Various Catalysts*) [34].

Catalyst	Yield (%)	Catalyst	Yield (%)
none	11	18-crown-6	11
NBu_4HSO_4	93	dibenzo-18-crown-6	12
NBu_4Cl	32	NBu_4Br	18
NBu_4Br	18	$NPent_4Br$	20
NBu_4I	15	$NHex_4Br$	45
NBu_4ClO_4	14	$NHep_4Br$	46
NBu_4 benzoate	32	$NOct_4Br$	39
Aliquat 336	50	sodium stearate	60
$NOct_4Br$	40	$NEt_4^{\oplus}C_8F_{17}SO_3^{\ominus}$	39
$N(C_{16}H_{33})Me_3Br$	39	$C_{16}H_{33}(O—CH_2—CH_2)_{29}—OCOCH_3$	95
TEBA	18	$C_{16}H_{33}(O—CH_2—CH_2)_{30}—OCH_3$	84
cetyl pyridinium Cl	20	$C_{16}H_{33}(O—CH_2—CH_2)_{30}—Cl$	76
15-crown-5	23	$C_{16}H_{33}(O—CH_2—CH_2)_{30}—OH$	74

*) Conditions: 0.01 mole substrate, 0.05 mole 50% NaOH, 5 ml petroleum ether, 0.0001 mole catalyst, 1 h, room temperature.

Industrial chemists were the first to make rough comparisons for dichlorocarbene reaction catalysts [48]. Tables 2–8 and 2–9 give more recent results on dichlorocarbene and dibromocarbene generations, respectively.

Table 2–8. Catalyst Comparison in Dichlorocarbene Generations. Yield of Dichloronorcarane from Cyclohexene*) [9, 29].

Catalyst	Yield (%)	Catalyst	Yield (%)
NMe_4I	1	$N(C_{16}H_{33})Prop_3Br$	39.0
NEt_4Br	44.2	$N(C_{16}H_{33})Bu_3Br$	35.5
$NProp_4Cl$	34.2	$N(C_{16}H_{33})Me_2(PhCH_2)Cl$	48.2
$NProp_4Br$	26.1	$NEt_4^{\oplus}C_8H_{17}SO_3^{\ominus}$	4.0
NBu_4HSO_4	45.7	$PProp_4Cl$	35.4
NBu_4Cl	38.8	PBu_4Cl	38.3
NBu_4Br	29.1	$P(C_{16}H_{33})Bu_3Br$	39.6
NBu_4I	22.6	PPh_3furfuryl Br	1
NBu_4ClO_4	18.6	$PPh_3(PhCH_2)Cl$	1
NBu_4benzoate	24.5	PPh_4Cl	1
NBu_4*p*-nitrobenzoate	16.3	$AsPh_4Cl$	62.0
NBu_4naphthalenesulfonate	8.6	$C_{16}H_{33}$pyridinium Cl	1
NBu_4picrate†)	36.6	18-crown-6	54.9
$NPent_4Br$	32.4	15-crown-5	55.0
$NHex_4Br$	34.8	dicyclohexano-18-crown-6	41.7
$NHep_4Br$	28.4	dibenzo-18-crown-6	18.2
$NOct_4Br$	23.4	$NProp_3$	74.0
Aliquat 336	41.9	$C_{16}H_{33}(OCH_2—CH_2)_{30}OH$	7.9
TEBA	51.0	$C_{16}H_{33}(O—CH_2—CH_2)_{29}OCOCH_3$	7.0
TEBA-bromide	35.2	$C_{16}H_{33}(O—CH_2—CH_2)_{40}OCH_3$	7.6
$N(PhCH_2)Prop_3Cl$	48.3	$C_{16}H_{33}(OCH_2—CH_2)_{40}Cl$	6.3
$N(PhCH_2)Bu_3Cl$	48.8	$C_{12}H_{25}NO(Me)_2$	8.5
$N(C_{16}H_{33})Me_3Br$	42.6	$C_{16}H_{33}NO(Me)_2$	8.3
$N(C_{16}H_{33})Et_3Br$	39.7		

*) Conditions: 0.1 mole cyclohexene, 0.4 mole $CHCl_3$, 0.2 mole 50% NaOH, 0.001 mole catalyst, 4 h, 23 °C.
†) Picrate ion is destroyed under the reaction conditions.

An inspection of Tables 2–8 and 2–9 reveals a number of interesting effects. Catalyst performance in dichlorocarbene reactions does not parallel those of dibromocarbene. Adverse effects of lipophilic anions brought in by the catalyst are much more pronounced in the CCl_2 series. This is in perfect agreement with the mechanism proposed in Section 2.1: $CBr_3^{\ominus}$ is much more lipophilic than $CCl_3^{\ominus}$, and its detachment from the interphase cannot be as easily suppressed. Similarly, different catalyst cations do not influence yields as much in the case of CBr_2 as with CCl_2. However, it is remarkable that tertiary amines, which are transformed into the actual catalyst $R_3NCHX_2^{\oplus}X^{\ominus}$ under the reaction conditions (see Section 3.3.13) are the best catalysts with CCl_2 but rather poor with CBr_2. The very poor catalytic action of phenyl substituted phosphonium salts contrasts sharply with the excellent performance of tetraphenylarsonium chloride. The former are decomposed by hydroxide yielding R_3PO. Finally, in agreement with the mechanistic concepts developed above, neutral surfactants, amine oxides, and open chain polyethers, as well as anionic surfactants, are very weak catalysts.

Table 2–9. Catalyst Comparisons in Dibromocarbene Generation [49]. Yields of Dibromonorcarane from Cyclohexene.*)

Catalyst	Yield (%)	Catalyst	Yield (%)
TEBA	44.8	$N(C_{16}H_{33})Me_2(PhCH_2)Cl$	49.0
Aliquat 336	52.6	$C_{16}H_{33}$pyridiniumCl	38.5
$NProp_3$	33.0	18-crown-6	47.5
NBu_3	31.4	Dibenzo-18-crown-6	48.0
$NProp_4Br$	40.0	NBu_4HSO_4	44.5 ‡)
NBu_4Br	48.7	NBu_4Cl	38.2 ‡)
$NPen_4Br$	48.0	NBu_4Br	39.0 ‡)
$NHex_4Br$	55.0	NBu_4I	38.0 ‡)
$NHep_4Br$	50.2	NBu_4ClO_4	32.5 ‡)
$NOct_4Br$	43.0	NBu_4benzoate	35.5 ‡)
PPh_4Cl	0.7	NBu_4*p*-nitrobenzoate	29.7 ‡)
$AsPh_4Cl$	41.0	NBu_4picrate †)	37.0 ‡)
$N(C_{16}H_{33})Me_3Br$	48.3	$NBu_4\beta$-naphthalenesulfonate	44.0 ‡)
$N(C_{16}H_{33})Bu_3Br$	51.5		

*) Conditions: 0.1 mole cyclohexene, 0.4 mole 50% NaOH, 0.4 mole $CHBr_3$, 0.001 mole catalyst, 4 h, room temperature.
†) Picrate ion is destroyed under the reaction conditions.
‡) Only 0.2 mole NaOH in the reaction mixture.

For practical comparative purposes, changes in reaction times and other variables are kept to a minimum. Thus, most of the catalysts mentioned above can in fact be applied to carbene generation reactions.

Japanese authors have claimed that dichlorocarbene generated using different catalysts may differ in "selectivity" defined by the ratio of mono to bis addition to polyolefins. Thus, whereas β-hydroxylalkyltrialkylammonium catalysts were described as giving almost exclusively monoadducts, cetyl-trimethylammonium gives polyadducts [50, 51]. At the same time optically active dichlorocarbene adducts were said to be formed using optically active hydroxyl group carrying catalysts [51]. This was later shown to be in error [52]. Attempts at repeating the selectivity experiments also failed [53].

More recently, results on the catalytic dependence of selective addition to one or two double bonds of conjugated or nonconjugated di- and trienes have been published [54]. Pseudo-first-order rate constants for mono- and bisaddition of dichlorocarbene to isoprene were reported. Their ratio was said to increase sharply in the order: cetyl-trimethylammonium > quinuclidine > tributylamine > 1,4-diazabicyclo[2.2.2]-octane. In other words, the amount of bisadduct decreases on going from cetyl-trimethylammonium to 1,4-diazabicyclo[2.2.2]octane: the former gives 88% yield (mono- : bis- = 2:1) under certain conditions, the latter a 23% yield of pure mono adduct. These results are difficult to reconcile with other information available on catalytic behavior and of dichlorocarbene reactions. The results in Table 2–8 show that tertiary amine is one of the most efficient catalysts even though the authors of Ref. [54] have found them much less active than quaternary ammonium salts. Conditions were, however, slightly different (less concentrated NaOH!). It was assumed that nitrogen ylides acted as carbene carriers; this was subsequently disproved by Mąkosza (see Section 3.3.13). Claimed "selectivities" must obviously await further confirmation.

2.3. Unusual and Polymer Supported Catalysts

Apart from onium salt and crown ether catalysts, other compounds have proven useful. Among these are open chain polyethers of ethylene glycol and molecules that contain a number of such side-chains bound to a central aromatic nucleus (so-called "polypod" molecules). Their catalytic action is mechanistically similar to that of crown ethers. Although they are sometimes less effective as extractants, this disadvantage often can be overcome by using higher concentrations (*cf.*, Chapter 3). Even more unusual types of catalysts have been used sometimes. Exchange reactions were performed in the presence of 10–15 mol% of phosphoramides (**A**). In these R was any lipophilic group but not methyl [16]. The compounds **B** to **F** were also used

$[(H_3C)_2N]_2PON(CH_3)R$ **A**

$(RO)_2PO—CH_2—SOR^1$ **B**

$(RO)_2PO—CR_2^1—SO_2R^2$ **C**

$(EtO)_2PO—CH_2SO—CH_2—PO(OEt)_2$ **D**

$PhSO—CH_2—SOPh$ **E**

$(EtO)_2PO—CH_2—PO(OEt)_2$ **F**

in the alkylation of benzyl methyl ketone by alkyl halides and aqueous sodium hydroxide at room temperature [17]. Mąkosza frequently uses DMSO as "co-catalyst" or in rare cases as the only catalyst in similar alkylations.*) It seems likely that the catalytic action of compounds **A** to **F** and of DMSO is in making the salts or the sodium hydroxide soluble in the organic medium through specific solvation and/or chelation.

Although pyridinium salts are usually unsuitable as PTC agents, it has recently been found that 2-dialkylaminopyridinium salts (**G**) can be used as catalysts for alkylation in the presence of concentrated sodium hydroxide [18].

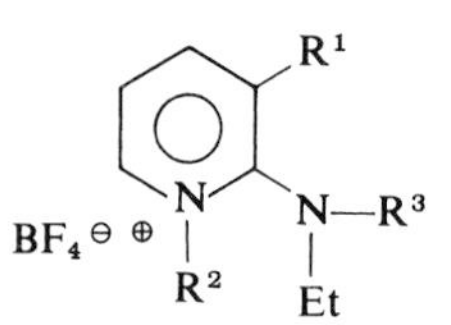

R^1, $BF_4^{\ominus}$ $\oplus$ N, R^2, N—R^3, Et

(R^1 = H, CH_3; R^2 = Me, Et; R^3 = long chain)

G

Scheme 2–6

*) For examples, see Chapter 3.

A number of authors have described PT catalysts chemically fixed on polymer supports. These are potentially of great interest in technical processes since the catalysts are easy to remove after reaction or can be used in continuous processes. This catalytic technique has been called "tri-phase-catalysis" [19, 21, 22]. In a 1-bromooctane displacement reaction using supported ammonium salts, first-order kinetics were found. The activity of the resin is directly proportional to the number of $—CH_2NR_3^{\oplus}$ groups present when 1–21% of the phenyl rings in a polystyrene contain such groups. Increasing ring substitution in microporous polystyrene with $—CH_2—NMe_3^{\oplus}$ to 46–76%, resulted in a sharp drop in catalytic activity. Commercial anion exchange resins are normally less suitable as PT catalysts [19]. Further results for immobilized onium salts, crown ethers, and cryptants [20] indicate that the mechanism is basically similar to "normal" PTC reactions.

Recently, it has been demonstrated that the hydrolysis of 1-bromoadamantane in toluene/water using a resin consisting of polystyrene with polyether side-chains gives 1-hydroxyadamantane [21]. This has been termed a "solid-phase cosolvent" effect and has been further studied using microtechniques with spin tracers [22]. Polymer supported phosphoric acid triamides are being used as PT catalysts too [57]. In general, catalytic effects of polymer supported catalysts appear to be smaller than those of free catalysts. However, they may in the long run prove more favorable if suitable polymer-bound catalysts become commercially available. The potential of this area is still being actively developed. Practical aspects including the types of polymer-bound catalysts so far investigated are found in Chapter 3.

It should be mentioned here that triphase catalysts of the cosolvent type show a selectivity in the reaction between phenoxide and a homologous series of alkyl bromides which is not found in simple PTC. Thus, the conversion of 1-bromobutane is much faster than that of 1-bromooctane [58]. (For a review on triphase catalysis *cf.* [56].)

Very recent papers report on model reactions for liquid-liquid PTC with unusual cyclic phosphonium and arsonium salts [59] and for solid-liquid PTC with a special "octopus molecule" [60] with glycol ethers [61], polyethyleneamines [62], tetramethylethylenediamine [63], and substituted β-amino phosphoramides [64] as catalysts.

3. Practical Applications of Phase Transfer Catalysis

3.1. General Experimental Procedures

3.1.1. The Catalysts

In patents, numerous quaternary ammonium, phosphonium, arsonium, antimonium, bismuthonium, and tertiary sulfonium salts have been claimed as catalysts. In practice, however, only a limited number of ammonium and phosphonium salts are widely used. In general the more common catalysts are commercially available. Here are some current prices of typical fine chemicals: many onium salts cost between U.S.$10–60/100 g, some even as much as U.S.$150/100 g, and crown ethers are even more expensive. The industrial preparation of these compounds, however, is not a difficult process, so that prices will undoubtedly decrease if there is a market for larger quantities. The cheapest catalyst available in 1977 was "tricaprylylmethylammonium chloride," a technical mixture containing C_8–C_{10} alkyl groups. It is sold under the trade names "Aliquat 336" (Fluka AG, Buchs, Switzerland; General Mills Co., Kankasee, Illinois, USA) or "Adogen 464"*) (Aldrich Chemical Comp., Milwaukee, Wisconsin, USA). Compounds like benzyl triethylammonium chloride, ("TEBA") or bromide ("TEBA-Br"), benzyl trimethylammonium chloride, bromide, or hydroxide ("Triton B," in methanol or in water), tetra-*n*-butylammonium chloride, bromide, iodide or hydroxide (in water, methanol, and methanol/toluene), and cetyltrimethylammonium bromide or chloride, are available from all the major chemical supply houses. Eastman Organic Chemicals†) (Rochester, New York, 14650, USA) for instance also offer tetra-*n*-pentyl-, tetra-*n*-hexyl-, and trioctylpropylammonium chlorides and bromides. The bromides are generally cheaper. Fluka AG†) (Buchs, Switzerland) also has "15-crown-5," "18-crown-6," "dibenzo-18-crown-6," "dicyclohexano-18-crown-6," tributylhexadecyclophosphonium bromide, ethyltriphenylphosphonium bromide, trimethyloctadecylammonium bromide, tetrabutylammonium hydrogen sulfate, and tetraphenyl-phosphonium chloride.

Among those available from the Aldrich Chemical Comp.†) are: benzyltributylammonium chloride, benzyltriphenylphosphonium iodide, "dibenzo-" and "dicyclohexano-18-crown-6," triphenyl-*n*-propylphosphonium bromide, tetrabutylammonium hydrogen sulfate, and tetrabutylphosphonium chloride. In addition to other compounds Merck-Schuchardt†) (Darmstadt, Germany) offer 4,7,13,16,21-pentaoxa-1,10-diazabicyclo[8.8.5]-tricosane ("Kryptofix 221"®), 4,7,13,18-tetraoxa-1,10-diazabicyclo[8.5.5]-eicosane ("Kryptofix 211"®) and the more frequently used "[2.2.2]", "Kryptofix 222"® (4,7,13,16,21,24-hexaoxa-1,10-diazabicyclo[8.8.8]-hexacosane) and its benzoderivative ("Kryptofix 222 B"®).

The laboratory preparation of most of these catalysts is easy. Since the quaternization

---

*) Registered trade name of Ashland Chemical Co., USA.
†) Mentioning these firms does not imply that some of these chemicals cannot be bought from other suppliers or that these companies are necessarily the cheapest purveyors.

of tertiary amines is especially fast in acetonitrile, this solvent has been extensively used in recent years. Some examples are presented as follows. Benzyltributylammonium chloride is obtained when tri-*n*-butylamine and benzyl chloride are refluxed in acetonitrile for 1 week, the yield is 86% [1]. Benzyltriethylammonium chloride is prepared in almost quantitative yield by boiling benzyl chloride and triethylamine in toluene for 4–5 days [2]. If a good yield is not required then the components may be kept overnight at room temperature in acetonitrile [2] or acetone [3]. Tributylhexadecylphosphonium bromide is obtained by heating 1-bromohexadecane and tributylphosphine at 65 °C for 3 days [4]. Tetrahexylammonium bromide in 59% yield can be made by boiling the pertinent amine and bromide in acetonitrile for 48 hours. Tri-*n*-butylethylammonium hydrogen sulfate is formed by refluxing tributylamine with diethyl sulfate in acetonitrile for 5 hours, removing the solvent, and refluxing the residue for 48 hours in water containing a little concentrated sulfuric acid. This is to hydrolyze the ethyl sulfate formed to hydrogen sulfate [5].

From the above examples it is seen that quaternizations are mainly performed using alkyl bromides, iodides, and sulfates. Of the chlorides, only methyl, benzyl, and allyl chloride are suitable. Others react too slowly with tertiary amines to be of practical use. In quaternizations with dialkyl sulfates, tetraalkylammonium alkyl sulfates are formed: $NR_4^{\oplus}{}^{\ominus}OSO_2OR$. As the example above shows, these can be hydrolyzed to hydrogen sulfates.

These salts are useful as PT catalysts only if the anion accompanying the catalyst (or in Brändström's words the preparative ion pair extractive) is not distributed in the organic phase to a much larger extent than the anion to be reacted. Of the more common catalyst anions, hydrogen sulfate and chloride are the most desirable. Bromide is reasonably good in many cases, and only iodide may lead to serious difficulties, especially if the reaction is performed with an alkyl iodide which liberates more iodide ion during the reaction. In such a case "catalyst poisoning" by preferential extraction of iodide to the organic phase might bring the conversion to a standstill.

In this case, and with reactions involving isolated ammonium salts in homogeneous media, anion exchange could be of interest. It should be stressed, however, that in the majority of typical PTC reactions no such exchange is necessary.

Older methods still in use for anion exchange are summarized in Houben-Weyl [6] but can often be replaced by solvent extraction procedures. Applied techniques are:

(a) Reaction of $NR_4^{\oplus}$ Halide$^{\ominus}$ with the silver salt of the desired anion (an old expensive method).

(b) Passing either an aqueous or methanol/water solution of the quaternary onium salt through an exchange resin containing the desired anion. The following anions are, from left to right, decreasingly tightly bound to the polymer: $SO_4^{2\ominus} > I^{\ominus} > NO_3^{\ominus} > Br^{\ominus} > Cl^{\ominus} > OH^{\ominus} > F^{\ominus}$ [7]. A detailed description of the preparation of quaternary ammonium hydroxides in methanolic solution using a strongly basic polystyrene alkyl quaternary ammonium hydroxide resin is given in Organic Syntheses [8].

(c) Double comportion with an alkali metal salt in a solvent in which the desired quaternary ammonium or phosphonium salt is soluble, while the alkali metal salt of the unwanted anion crystallizes.
(d) Reaction of an ammonium iodide compound with an acid HX in the presence of an epoxide [6]:

$$NR_4^{\oplus}I^{\ominus} + H_2C\!-\!CH\!-\!R \text{ (epoxide, bridged by O)} + HX \longrightarrow NR_4^{\oplus}X^{\ominus} + IH_2C\!-\!CH(OH)R$$

Scheme 3–1

(e) Reaction of quaternary ammonium bromides and iodides with dimethyl sulfate in relatively high boiling solvents (*e.g.*, chlorobenzene). Methyl bromide or iodide is evolved, and the methyl sulfate formed is later hydrolyzed to the hydrogen sulfate [9]. Expensive CH_3I is regenerated in this way.

$$NR_4^{\oplus}X^{\ominus} + (CH_3)_2SO_4 \longrightarrow NR_4^{\oplus\ominus}OSO_2OCH_3 + CH_3X$$

(f) Oxidation of an iodide salt with hydrogen peroxide in the presence of an acid HX. The quaternary ammonium triiodide which is insoluble in water precipitates out [6, 33]:

$$3NR_4^{\oplus}I^{\ominus} + H_2O_2 + 2HX \longrightarrow NR_4^{\oplus}I_3^{\ominus} + 2NR_4^{\oplus}X^{\ominus} + 2H_2O$$

(g) Anion exchange using solvent extraction. This is often the easiest method.

No complications arise if the anion to be introduced is more lipophilic than the one to be exchanged (*e.g.*, iodide exchanged for chloride, chloride for hydrogen sulfate, azide for neutral sulfate, cyanide for chloride).*) The compound is dissolved in methylene chloride and shaken with an excess of a concentrated†) aqueous solution of the alkali metal salt of the anion to be introduced, washing once or twice with fresh solutions. Tetrabutylammonium hydrogen sulfate is often a good starting material, since the sulfate formed after neutralization does not disturb. Among the salts prepared in crystalline form using this method [5, 894] are halides, cyanides, azides, nitrites, benzoates, phenolates, and the enolates of β-diketones, β-cyanoesters, β-ketosulfones, and dimethyl benzoylmalonate.

Difficulties may arise if the quaternary onium salt formed is hydrophilic and is lost to some extent in the aqueous phase or if the extraction constants of the two anions are similar. Brändström has solved some of these problems [10, 37]. Another possible solution is to suppress the solubility of the $NR_4^{\oplus}$ salts in water by using concentrated sodium hydroxide as the aqueous phase. Mąkosza [11] recommends anion exchange in concentrated aqueous sodium hydroxide using benzene, or *o*-dichlorobenzene, or mixtures of these solvents with acetonitrile as the organic phase. Usually it is un-

*) For the complete "anion extraction" series, *cf.*, Section 1.3.4.
†) Highest possible concentrations give best results.

necessary that the desired sodium salt and the starting quaternary ammonium salt are fully dissolved in the two phase mixture because in many cases the exchange reaction leads to more soluble compounds. In this way starting from chlorides the following salts can be prepared: bromides, iodides, tetrahydridoborates, rhodanides, perchlorates, hexacyanoferrates(III), cyanides, permanganates, nitrites, nitrates, and azides. Starting from sulfates or fluorides hydroxides can be prepared although their solubility is rather limited in nonpolar organic media, unless the cation employed is very large [11, 12].

In another procedure, quaternary ammonium halides are dissolved in methanol in the presence of the acid whose anion is to be introduced. On concentration, the original anion is removed as methyl halide [1017], though the generality of this simple method remains to be demonstrated. Other workers react the original ammonium halide with boron trifluoride etherate, whereupon the tetrafluoroborate is formed. This anion can be exchanged easily [1018].

The preparation of fluorides presents a special problem since this anion is very hydrophilic and therefore difficult to extract and obtain free of water of hydration. Tetrabutylammonium fluoride is made by the neutralization of aqueous quaternary ammonium hydroxide with hydrofluoric acid followed by crystallization of a clathrate ($NBu_4^{\oplus}F^{\ominus} \cdot 32.8\ H_2O$) at 0–10 °C. Dehydration at 30–50 °C/15 Torr and further drying of the paste obtained at 30–40 °C/0.5 Torr over phosphorus pentoxide for 15–20 hours, gives the anhydrous compound, which is extremely hygroscopic [24, 25, 26]. Evaporation of aqueous tetraethylammonium fluoride solutions results in crystalline $NEt_4^{\oplus}F^{\ominus} \cdot 2H_2O$. This partially decomposes on further drying to give:

$$N(C_2H_5)_4^{\oplus}F^{\ominus} \longrightarrow N(C_2H_5)_3H^{\oplus}F^{\ominus} + H_2C{=}CH_2 \text{ [36]}$$

A very useful form of dry $NR_4^{\oplus}\ F^{\ominus}$ catalyst is made by evaporating an aqueous solution of the salt in the presence of a three-fold excess of silica gel, adding methanol, evaporating again, and drying at 100 °C. This material is not hygroscopic and remains an active catalyst after being left in air for long periods [1019].

Even simple crown ethers are quite expensive but their laboratory preparation is not difficult. 1,4,7,10,13,16-Hexaoxacyclooctadecane (18-crown-6) (**1**) can be prepared from triethylene glycol and its ditosylate by several routes: with potassium *tert*-butoxide in benzene [13] (33% yield), in THF [14] (60% yield), or from the same glycol plus the commercially available dichloride (1,8-dichloro-3,6-dioxaoctane) with potassium hydroxide in aqueous THF [15] (40% yield). Finally, with potassium hydroxide in THF bis(2-chloroethyl)ether and tetraethylene glycol give 18-crown-6 in 30% yield [20]. The raw material is purified *via* an acetonitrile complex. (For an Organic Syntheses procedure see [1006])

Dibenzo-18-crown-6 (**2**) is made similarly using an Organic Syntheses procedure [16] or in a slightly modified form which gives improved yields [17]. Catalytic hydrogenation of this compound over ruthenium-on-alumina results in the formation of dicyclohexano-18-crown-6 (**3**) [16].

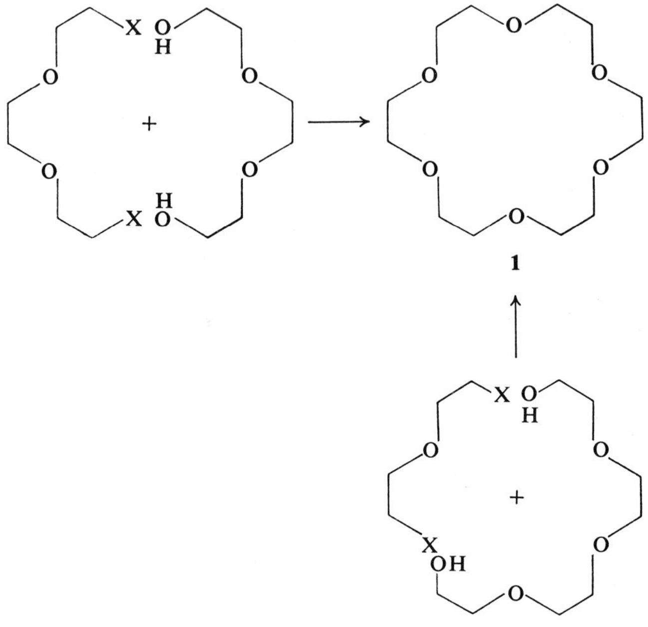

1

Scheme 3–2

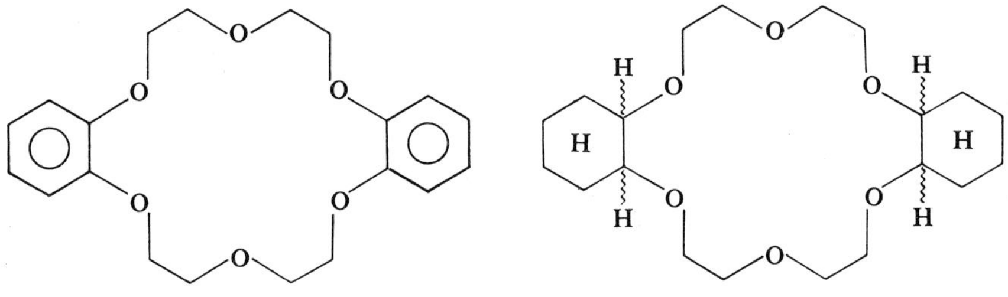

2

3

Scheme 3–3

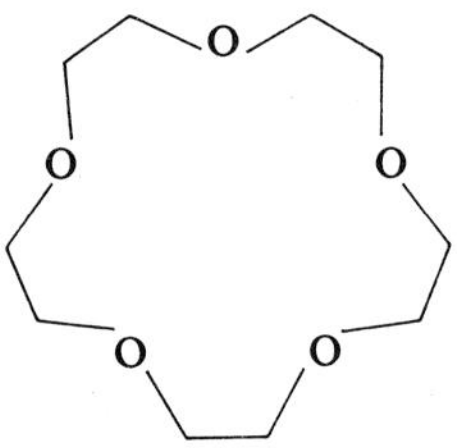

4

Scheme 3–4

Five diastereomers are possible, and the hydrogenation leads to a mixture of the *cis-syn-cis* and *cis-anti-cis* compounds. Specific processes leading to the *trans-anti-trans* and *trans-syn-trans* isomers have been developed [18], and the optically active (+)-*trans-anti-trans* compound has also been prepared [19].

1,4,7,10,13-Pentaoxacyclopentadecane (15-crown-5) (**4**) is obtained in 38% yield by the reaction of 0.5 mole triethylene glycol with 1.25 mole bis(2-chloroethyl)ether and 2.3 mole sodium hydroxide in boiling dioxane for 24 hours [20]. An alternative synthesis starting from 1,8-dichloro-3,6-dioxaoctane and diethylene glycol gave only 14% yield [21]. Selective oligomerization of ethylene oxide in the presence of lithium, cupric, or zinc tetrafluoroborates gives mainly 15-crown-5 complexes. Preparative details of this very simple process have not yet been published [22].

Lehn's cryptate [222] (Kryptofix 222®, 1,10-diaza-4,7,13,16,21,24-hexaoxa-bicyclo-[8.8.8]-hexacosane) (**5**) was prepared as shown below using high dilution techniques in the two ring forming steps [23].

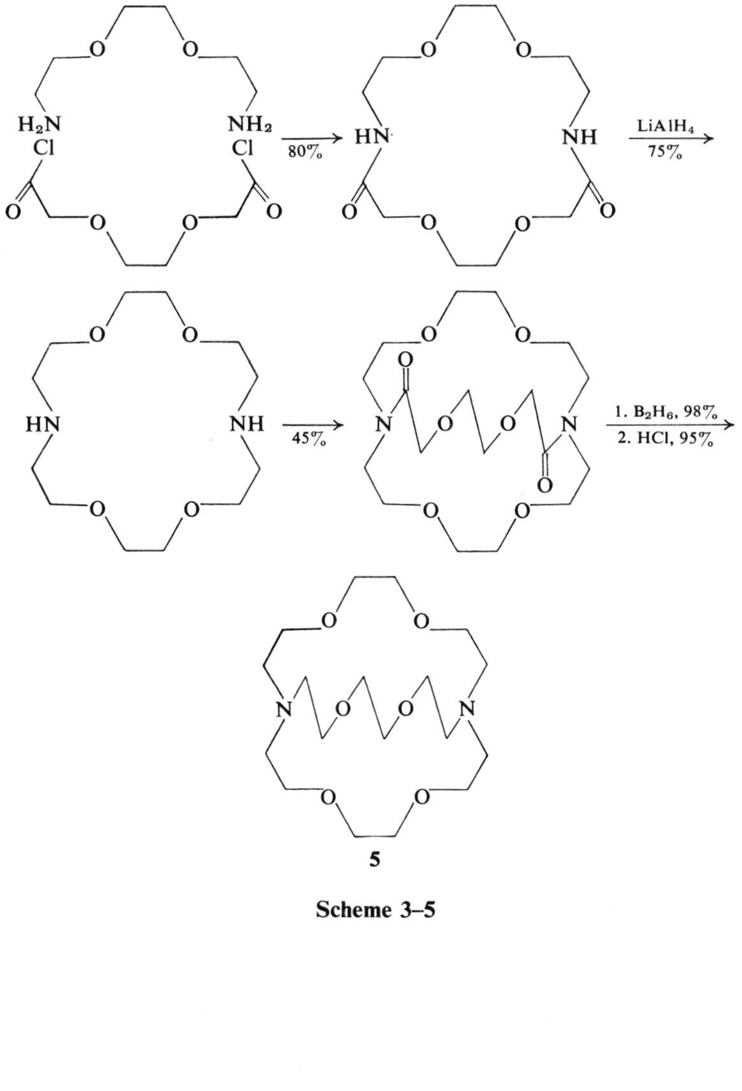

5

Scheme 3–5

The preparation of numerous other complexing agents of this general type has been described, but the majority of these have not found many practical applications. Some of these ligands are mentioned in Chapter 1, Section 4.

3.1.2. Reaction Conditions

Typical reaction conditions are described in later sections although these are not always optimal. The influence of the catalyst structure on the rate and yield, for instance, has until recently not been well understood [27, 28]. The influence of the solvent, catalyst, and mixing are discussed in Chapters 1 and 2, but here we present some practical hints.

3.1.2.1. Solvent

Liquid substrates are often used neat as the organic phase. In principle, many organic solvents can be used. They must not be even partially miscible with water, however, to protect the ion pairs against extensive hydration. It must be kept in mind that in very nonpolar solvents such as heptane or benzene, ion pairs are partitioned to a minimal extent only from the aqueous to the organic phase unless the cation anion-combination is very lipophilic. TEBA, for instance, is very ineffective with benzene/water [28] and even with dichloromethane/water [2]. Tetra-*n*-butylammonium or larger ions such as tetra-*n*-pentylammonium, tetra-*n*-hexylammonium, or Aliquat 336 are recommended for use with these solvents. In general, more favorable partitioning of the ion pairs into the organic phase and faster rates are observed in solvents like dichloromethane, 1,2-dichloroethane, and chloroform. Although many alkylations are much faster than carbene reactions, if the latter are undesired chloroform should not be used in the presence of aqueous sodium hydroxide. With chlorinated solvents, the choice of catalyst is less critical. Strong nucleophiles, however, may attack these solvents. This reaction is particularly well known for dichloromethane solutions of good nucleophiles in the presence of less reactive alkylating agents, when unwanted side products may become abundant [29, 30]:

$$2NR_4^{\oplus}X^{\ominus} + CH_2Cl_2 \longrightarrow X\text{—}CH_2\text{—}X + 2NR_4^{\oplus}Cl^{\ominus}$$
$$X = \text{O-Aryl, O-Alkyl, SR, } O_2C\text{—R, CH(Acceptor)}_2$$

Under suitably modified conditions production of these compounds can be achieved. In cases were such solvent intervention is to be avoided, however, *o*-dichlorobenzene has been advocated as an inert chlorinated medium in spite of its rather poor extracting capacity for ion pairs.

3.1.2.2. Stirring

Most laboratory PTC preparations may be performed using magnetic stirring. It must be mentioned, however, that results are sometimes not reproducible, especially in the presence of viscous 50% sodium hydroxide, when stirring is too slow. Recommended stirring rates are: >200 rpm for essentially neutral PTC reactions in water/

organic medium [27], 750–800 rpm for solid-liquid reactions and reactions in the presence of sodium hydroxide [31, 32]. For some solid-liquid reactions high-shear stirring may be necessary.

3.1.2.3. Amount of Catalyst

In the literature, quantities varying from between a few mole-percent and several moles have been used with various systems. Since reaction rates are dependent on catalyst concentration, a very small amount of catalyst seems desirable only if the conversion is highly exothermic or the catalyst is very expensive. In most cases 1–3 mole-% is normal.

In some cases a molar amount of catalyst is desirable:

(a) if iodide ion is set free in the course of the reaction and tends to tie up the onium salt in the organic phase;
(b) if the alkylating agent is very unreactive;
(c) if the alkylating agent is prone to side reactions (*e.g.*, hydrolysis with the water/alkali metal hydroxide present);
(d) if a selective reaction in a polyfunctional molecule is desired.

3.1.2.4. Stability of the Catalyst

Although catalysts normally function quite well a few points should be kept in mind. The ammonium and phosphonium salts may be subject to destruction under the reaction conditions. Quaternary hydroxides for instance are known to undergo Hofmann elimination, not only at elevated temperatures, but even (depending on the structure) on standing in solution at room temperature. In addition, depending on the groups present, the following process is possible:

$$R_3NR'^{\oplus}OH^{\ominus} \longrightarrow R_3N + R'OH$$

The ease with which a particular residue is separated from a quaternary ammonium hydroxide is as follows [34]: β-phenylethyl > allyl > benzyl > ethyl > propyl > cyclohexyl > methyl > isobutyl > phenyl.

Under certain conditions, strong nucleophiles like phenolates or thiolates might give benzyl ethers or sulfides with benzyl substituted quaternary ammonium salts [211, 226, 231, 933]. Many other dealkylations have been described. Recently, such reactions have proven useful preparatively especially in removing benzyl and methyl groups under very mild conditions (*e.g.*, with thiolates or other nucleophiles in HMPT at room temperature [848–850]). It was also reported that dequaternizations of heterocycles are possible with thiolates under PTC [1020, 1021].

It is reassuring, however, that under most conditions, the commonly used catalysts are reasonably stable for days at room temperature. Elevated temperatures lead to decomposition: after 7 hours at 60 °C or 100 °C, tetrabutylammonium salts gave

52 and 92% tributylamine in the presence of concentrated aqueous caustic soda, respectively. TEBA gives diethylbenzylamine in high yield, while benzyltrimethylammonium chloride gives about equal amounts of dibenzyl ether and dimethylbenzylamine [12]. The former compound is formed in a way similar to the phenolate reaction in these consecutive steps:

$$C_6H_5CH_2{-}NMe_3^{\oplus}OH^{\ominus} \longrightarrow C_6H_5CH_2{-}OH + NMe_3$$

$$C_6H_5CH_2{-}NMe_3^{\oplus} + {}^{\ominus}O{-}CH_2C_6H_5 \longrightarrow C_6H_5CH_2{-}O{-}CH_2C_6H_5 + NMe_3$$

Similarly, benzylhexadecyldimethylammonium bromide gives the following decomposition products with NaOH: dimethylhexadecylamine, dimethylbenzylamine, dibenzyl ether, and 1-hexadecene [12].

Whereas these decompositions occur at elevated temperatures, the degradation of benzyldimethylphenylammonium hydroxide while concentrating an alcoholic solution even at room temperature is reported [225]. The ethoxide, propoxide, and hydroxide of *d*-allylbenzylmethylphenylammonium decomposed in ethanol or water when concentrated or dried [1007].

Phosphonium salt catalysts are more stable in the presence of concentrated aqueous base than the corresponding ammonium salts unless they are activated for Wittig-type reactions or carry phenyl substituents, in which case they decompose yielding triphenylphosphine oxide [1027]. Thus, triphenylalkylphosphonium salts undergo Wittig-PTC conversions at room temperature, but tributylhexadecylphosphonium bromide was unchanged after 16 hours at 100 °C [12].

Benzyl substituted catalysts may be particularly prone to attack by oxidizing agents [2]. In the absence of aqueous sodium hydroxide, phosphonium salt catalysts are believed to be more stable (up to 200 °C) than the corresponding ammonium salts [4]. On the other hand, it was observed that during the preparation of thioethers some phosphonium ions were rather rapidly consumed even under mild conditions [27].

3.1.2.5. Choice of Catalyst

In neutral media a good catalyst should have 15 or more C-atoms. In a preliminary study of a new PTC reaction, tetrabutylammonium salts, in particular hydrogen sulfate, or Aliquat 336, are recommended in neutral or acidic media. In the presence of concentrated aqueous base, TEBA and Aliquat 336 should be tried first. But, as stated in other sections, numerous other catalysts such as onium salts, crown ethers, and cryptants may prove effective. Older references implied that "solid-liquid PTC" required a crown ether catalyst whereas in fact onium salts are equally good [624, 953]. It should be mentioned that occasionally unorthodox catalysts are advocated: polyamines [117, 118], open-chain polyethers [47], 2-dialkylamino pyridinium salts [312], phosphoramides [851], and others [46]. Benzyltributylammonium salts are conventional catalysts that can be easily prepared in the laboratory and are often very useful.

3.1.2.6. Separation and Regeneration of the Catalyst

Separating the catalyst from the product after synthesis does not, as a rule, meet with difficulties since in most cases only the catalyst is soluble in water. Sometimes it is sufficient to wash the reaction mixture repeatedly with water. In other cases, the original solvent is evaporated, the residue is treated with water and reextracted with a solvent such as ether. To regenerate the catalyst salt, tetrabutylammonium salts from various sources should be collected and dissolved in dichloromethane. After shaking with an excess of aqueous concentrated sodium iodide solution at $pH < 3$ and removal of the solvent, the iodide thus formed can be crystallized from butanone [33]. This is converted into tetrabutylammonium hydrogen sulfate by one of the following three processes [33]:

(1) $$NBu_4^{\oplus}I^{\ominus} + H_2SO_4 + (n\text{-}C_8H_{17})_3N \xrightarrow[\text{water}]{\text{toluene}} NBu_4^{\oplus}HSO_4^{\ominus} + (n\text{-}C_8H_{17})_3NH^{\oplus}I^{\ominus}$$

(2) $$NBu_4^{\oplus}I^{\ominus} + (CH_3)_2SO_4 \longrightarrow NBu_4^{\oplus\ominus}OSO_2OCH_3 + CH_3I$$
$$NBu_4^{\oplus\ominus}OSO_2OCH_3 + H_2O \xrightarrow{H^{\oplus}} NBu_4^{\oplus}HSO_4^{\ominus} + CH_3OH$$

(3) $$3\,NBu_4^{\oplus}I^{\ominus} + 2H_2SO_4 + H_2O_2 \xrightarrow[\text{water}]{Cl_2C{=}CHCl} NBu_4^{\oplus}I_3^{\ominus} + 2NBu_4^{\oplus}HSO_4^{\ominus} + 2H_2O$$

Using the following procedure the recovery of 18-crown-6 is often possible: The reaction mixture is repeatedly washed with an acidified saturated potassium chloride solution. The combined washes of several reaction runs are evaporated using a rotary evaporator, and the solid residue is extracted repeatedly with methylene chloride. The combined extracts are dried over magnesium sulfate, filtered, and evaporated. The resulting solid contains potassium chloride and can be sublimed at 0.05 Torr/130–140 °C or recrystallized from acetonitrile to give a product reactive enough for most purposes in spite of containing traces of potassium chloride [35].

3.1.2.7. Polymer-Bound Catalysts

Although the catalyst is not usually recovered, under certain conditions, for large scale or continuous processes for example, recovery becomes a necessity. Here insoluble, polymer-bound catalysts ("triphase catalysis") have a wide potential. As detailed in other sections, such catalysts may be quaternary ammonium salts [62, 64, 68, 775, 860, 902], phosphonium salts [775, 858], phosphoric acid triamides [1025], crown ethers [775, 858, 1023, 1024], polyethers [316,1022], and cryptants [775] bound to a cross-linked polystyrene matrix. Occasionally, commercial anion exchange resins are used for PTC reactions [299, 616] but are less suited because of the high proportion of $—NR_3^{\oplus}$ groups in the polymer [64]. The kinetic activity of specially prepared polymer-bound catalysts may well approach that of soluble PTC agents. Recently a facile economical synthesis of widely porous resins and supported catalysts prepared therefrom, has been described [1008]. It is to be expected, therefore, that once such polymer-bound catalysts are commercially available they will find a wide application in industry, although their use in the laboratory may be limited. Their long-term stability under the reaction conditions must be tested [933].

In more detailed studies it was found that polymer-supported catalysts show the highest pseudo-first-order rates when the catalytic site is connected by a long chain "spacer" of 30 to 40 atoms to the matrix. Most notably in the solvent *n*-heptane, a parallel rise in catalytic activity with chain length was observed with ammonium [860], phosphonium [858], and crown ether [858] head groups. This is believed to bring about a better protruding, in a sense a "solubilization," of the catalytic center into the organic solvent.

Similar quaternary phosphonium catalysts bound to silica gel have been made and found active in exchange reactions, alkylations, and borohydride reductions [859]. Compound **A**, shown below was more active than **B**.

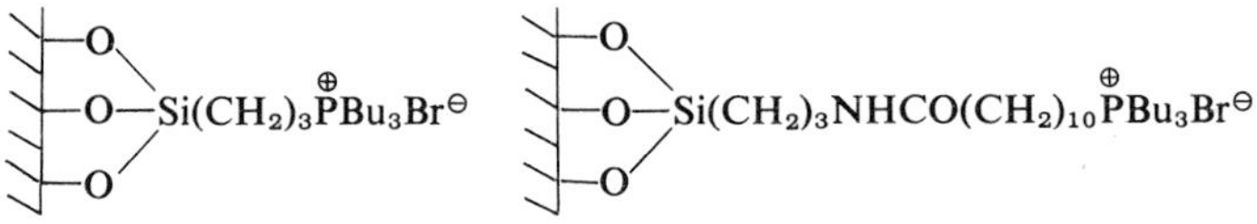

$$(\equiv\text{O})_3\text{Si(CH}_2)_3\overset{\oplus}{\text{P}}\text{Bu}_3\text{Br}^{\ominus} \qquad (\equiv\text{O})_3\text{Si(CH}_2)_3\text{NHCO(CH}_2)_{10}\overset{\oplus}{\text{P}}\text{Bu}_3\text{Br}^{\ominus}$$

A **B**

Scheme 3–6

Onium salts bound to Aerosil-200 or aminoalkyl Corning Glass are active catalysts too [1026].

3.1.2.8. Miscellaneous

In organic/aqueous two-phase systems, the aqueous solution should be rather concentrated in order to aid extraction through ion pair association in the organic phase. Since quite a few PTC reactions are exothermic, it is wise to start an unknown conversion with a cooling bath at hand or to add the reagent gradually. Persistent emulsions are frequent with reactions in the presence of concentrated sodium hydroxide. Neutralization or centrifugation often aid in the workup, but it is often convenient to wash the mixture repeatedly with water in order to remove any excess base. This helps in the separation of the phases. It should be remembered that $NR_4^{\oplus}HSO_4^{\ominus}$ needs an extra equivalent of base for the neutralization.

3.2. Displacement Reactions without Extra Base

In this section simple substitutions of the general type $RX + Y^{\ominus} \rightarrow RY + X^{\ominus}$ will be considered. Alkylations in the presence of added aqueous base are discussed in Section 3.3.

3.2.1. Formation of Alkyl Halides

Phase transfer catalytic methods have been used for the following types of substitutions:

alkyl (aryl, acyl) halide ⟶ different halide (halide exchange reactions)

alkyl sulfonate ⟶ alkyl halide

primary alcohol + HCl ⟶ alkyl chloride

diaryldiazomethane ⟶ diarylmethyl fluoride

nucleophilic aromatic substitution

3.2.1.1. Halide Exchange

Preparatively, halide exchange is most important for the synthesis of iodides and fluorides. But PTC techniques can also be used in the synthesis of isotopically labeled chlorides, bromides, and iodides. Starks found complete $^{35}Cl/^{36}Cl$ equilibration between 1-chlorooctane and aqueous $Na^{36}Cl$ after 5 hours of boiling with a quaternary salt catalyst. The analogous iodide-radioiodide process was complete after 5 minutes at 100 °C [4]. For the exchange of chemically nonequivalent groups, $X^{\ominus}$ and $Y^{\ominus}$, both the extraction equilibrium of the two ion pairs $Q^{\oplus}X^{\ominus}$ and $Q^{\oplus}Y^{\ominus}$ and the chemical equilibrium of the reaction $RX + Y^{\ominus} \rightarrow RY + X^{\ominus}$ can limit the conversion. If, for instance, an alkyl bromide is to be transformed into a chloride, an excess of aqueous sodium chloride must be used. Once the PTC equilibrium is reached, the aqueous phase has to be replaced with fresh NaCl solution. This may have to be repeated several times [4] (*cf.*, Chapter 2). An easier method, even for a simple I → Cl exchange, therefore, is to mix equimolar amounts of the alkyl iodide and anhydrous NBu_4Cl (which is easy to prepare) and distill off the alkyl chloride as it is formed [899].

$$RX + NaI \longrightarrow RI + NaX$$

For primary alkyl iodide preparations, the exchange of chloride or bromide with aqueous NaI can be accomplished in either 5 or 2 hours, respectively, at 108 °C in the presence of a few mol-% quaternary ammonium salt [38, 39]. Reactions are much faster with molar amounts of catalyst; Brändström transformed chloroaceto-2,6-xylidide into the corresponding iodo compound by refluxing only 90 minutes (92% yield) [899]. Under these conditions secondary bromides give predominantly olefins. Besides quaternary onium salts, crown ethers have also been used as catalysts [43–45, 48]. A survey of the catalytic efficiency of various catalysts used for the conversion of a 5 molar excess of aqueous saturated alkali metal iodide with *n*-octyl bromide using 0.05 molar equivalents of catalyst is presented in Table 3–1.

Table 3–1. Catalytic Efficiency of Various Catalysts for the Conversion K(Na)I + *n*-Octyl Bromide (5 molar excess of aqueous saturated alkali metal salt, 0.05 molar equivalents of catalyst).

Salt	Catalyst	Temperature (°C)	Time (h)	Yield (%)	Reference
KI	none	80	24	<4	44
KI	dibenzo-18-crown-6 (**2**)	80	40	80	44
KI	dicyclohexano-18-crown-6 (**3**)	80	3	100	44
KI	[2.2.2] (**5**)	60	14	90	48
KI	$n\text{-}C_{16}H_{33}PBu_3^{\oplus}Br^{\ominus}$	60	1	93	48
NaI	benzo-15-crown-5	80	21	80	44
KI	**6a**	60	0.2	100	48
KI	**6b**	60	0.5	92	48

As can be seen from these results, the onium salt is a better catalyst than most crowns, exceeded only by the modified aza-macrobicyclic polyethers **6a** and **6b**.*) These, however, are not commercially available. Rate data for ammonium, phosphonium, crown ether, and cryptate catalysts bound to insoluble polymers [775, 1023, 1026], and for crown ethers with lipophilic sidechains [776], show normal rates. The three-phase catalysts are easy to remove, but again have to be specially prepared.

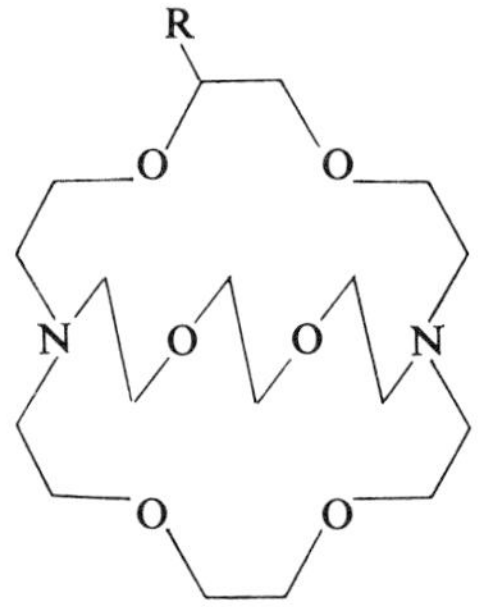

6 **a**: R = $n\text{-}C_{14}H_{29}$
b: R = $n\text{-}C_{11}H_{23}$

Scheme 3–7

Instead of expensive crown ethers, so-called "polypod" molecules and technical mixtures of open-chain polyethers have been advocated as catalysts [46, 47]. The most efficient polypod molecule of those tested, was the substituted triazene **7**. Its catalytic ability was similar to that of tributylhexadecylphosphonium bromide and a little less than that of alkyl substituted aza-macrobicyclic polyethers **6**. Thus, with 1 mole-% of **7**, the *n*-octyl bromide → iodide conversion (5-fold excess KI), gave 85% yield in 3 hours at 60 °C [46]. 2 mole-% of technical polyethylene glycol

*) Even higher rates are found with tricyclohexano[2.2.2] (perhydrotribenzohexaoxadiaza-[8.8.8]eicosane) [992].

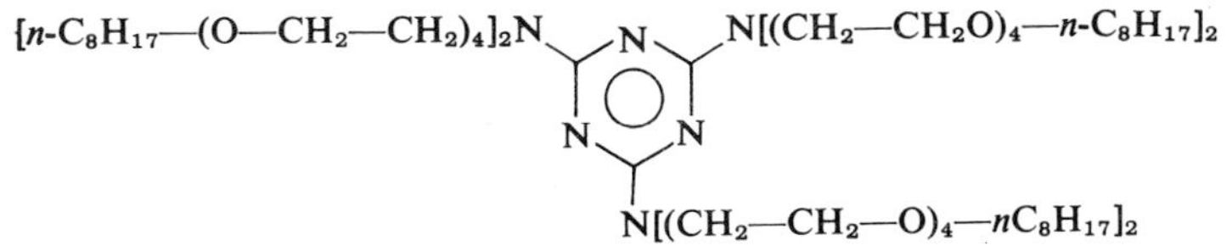

$[n\text{-}C_8H_{17}\text{—}(O\text{—}CH_2\text{—}CH_2)_4]_2N$–(1,3,5-triazine)–$N[(CH_2\text{—}CH_2O)_4\text{—}n\text{-}C_8H_{17}]_2$, with $N[(CH_2\text{—}CH_2\text{—}O)_4\text{—}nC_8H_{17}]_2$ on the third ring carbon

7

$$H_3C\text{—}O(CH_2\text{—}CH_2\text{—}O)_n\text{—}CH_3$$

8

Scheme 3–8

dimethyl ether (n = 8.39) with **8** as catalyst yielded 44% for the same reaction after 3 hours at 100 °C [47]. Other catalysts tested for the Finkelstein reaction included cyclic phosphonium and arsonium salts [1029], special crowns and cyclic aminoethers [1023, 1028], and hexaalkyl phosphoramides [851].

$$CH_2Cl_2 \xrightarrow{NaI} CH_2Cl\ I + CH_2I_2$$

Using tributylhexadecylphosphonium bromide [40] or one of several other catalysts [41], methylene chloride was transformed into the products shown above. The reaction was carried out in a shaking autoclave at 100–110 °C, for 10 to 20 hours, with a molar ratio CH_2Cl_2/NaI = 2.4 and chloroiodomethane and diiodomethane were obtained in 67% and 14% yield respectively. At 150 °C such exchange is possible without catalyst [1030]. In chloroform or bromoform there is no such exchange, presumably because of steric hindrance to the approach of the nucleophile. However, the exchange is possible (although preparatively unattractive) *via* a carbene mechanism in the presence of aqueous base and PTC catalyst [42].

PTC displacement reactions proceed much more rapidly if methanesulfonate is used as the leaving group. The order of reactivities for the following conversion is

$$n\text{-}C_8H_{17}OSO_2Me + KX \xrightarrow[\text{dicyclohexano-18-crown-6}]{H_2O\text{/no extra solvent}} n\text{-}C_8H_{17}X + K^{\oplus\ominus}OSO_2Me$$

$X = I^{\ominus} > Br^{\ominus} > Cl^{\ominus} \gg F^{\ominus}$*) [44]. The fastest reaction is over after 7 minutes at 100 °C, the slowest is 65% complete after 42 hours at 115 °C. Compound **9**, which is easily hydrolyzed, was prepared as shown below [49]. **10** was made from the corresponding tosylate in homogeneous acetonitrile solution by refluxing 6 hours with anhydrous $NBu_4^{\oplus}F^{\ominus}$ [26].

Although, as we have seen above, secondary bromides give mainly alkenes under PTC substitution conditions, more reactive mesylates give secondary halides in reasonably good yields. Starting with optically active 2-octyl mesylate, optically active chloride (89% optical purity, 83% yield), and bromide (82% optical purity, 78% yield) were obtained after 1.5 and 0.5 hours respectively at 100 °C in the presence

*) With tri-n-butylhexadecylphosphonium salt catalysts in chlorobenzene/water, the reactivity ratio for the same reaction was $I^{\ominus}:Br^{\ominus}:Cl^{\ominus}$ = 1:1.1:0.60 [50].

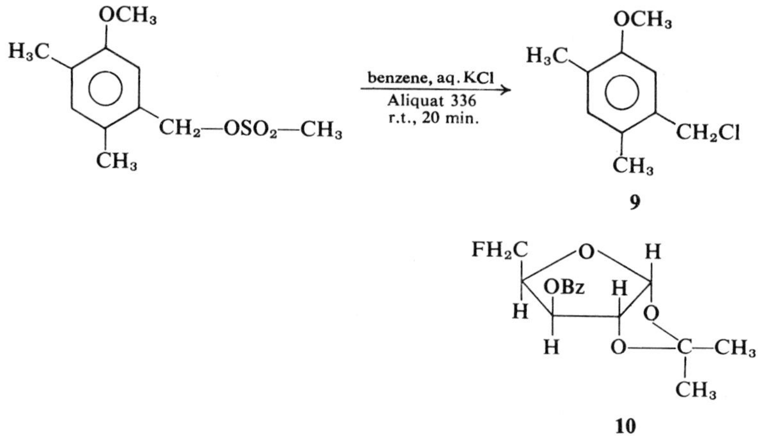

Scheme 3–9

of 5 mole-% Aliquat 336 or tributylhexadecylammonium bromide. In order to minimize racemization through repeated exchange, equimolar amounts of inorganic salt were used in the case of fluorides which reacted too sluggishly. Even then the iodide racemized [51]. It is very likely that the amount of catalyst used influences the optical yield, but this was not tested.

Quite good yields for various halogen exchange processes have been reported with the "triphase" catalysts after boiling for 24 to 100 hours [62]. The catalysts used were cross-linked polystyrene-bound trialkylbenzylammonium salts. They can be removed by simple filtration after the reaction. Most of the more active catalysts are not yet commercially available, and normal anion exchange resins are rarely useful in this case. It has been suggested that alternatively the anion exchange resin can be loaded with the required ion and boiled with the substrate in an inert solvent [777]. This method appears much less attrative than PTC!

3.2.1.2. Exchange for Fluoride

Turning now to the more specific preparation of fluorides, it must be remembered that the relatively unsolvated fluoride ion pairs in PTC reactions behave both as nucleophiles and as bases, so that competition between substitution, hydrolysis, and elimination arises. Montanari and co-workers [52] stirred primary or secondary alkyl bromides, chlorides, or mesylates at 100–160 °C with a saturated KF solution and catalytic amounts of tributylhexadecylphosphonium bromide for 7 to 16 hours. Primary and benzylic compounds gave fluoride yields of 70 to 90%: secondary substrates gave much lower yields. Cyclohexyl chloride and 2-octyl bromide gave olefin and alcohol only. 2-Octylmesylate, however, gave 54% fluoride.

It is of no advantage to use crown ethers as catalysts in the aqueous/organic two-phase system [44, 45], but Liotta advocates the use of 18-crown-6 for solid/liquid PTC (excess solid KF/benzene or acetonitrile) [53]. For most primary and benzylic halides, reaction times are similar to the liquid/liquid onium salt catalyzed processes

(24 to 120 hours). Cl/F exchange in aryl chloromethyl sulfides, however, requires 4–5 days of refluxing in acetonitrile with KF/crown [895]. Contrary to observations reported for the aqueous/organic reaction, 2-octyl bromide can be transformed into the fluoride (32%) by this method. Elimination, however, is again prominent. Chloro-2,4-dinitrobenzene and acetyl chloride were converted into the respective fluorides under these conditions [53]. For some of these solid/liquid processes, Kryptofix [2.2.2] seems to be the best of the crown catalysts [60]. Again, open-chain polyethylene glycol derivatives such as **8** can be used for the same purpose more economically. These catalysts are less reactive than the crowns, but this can be compensated for easily by using a higher concentration of the cheap technical product. Solvents recommended here are benzene, acetonitrile or the polyethylene-glycol ether itself [61, 1015]. The exchange of a chlorine α- to a carbonyl group is possible as shown in these examples (KF, CH_3CN, 18-crown-6):

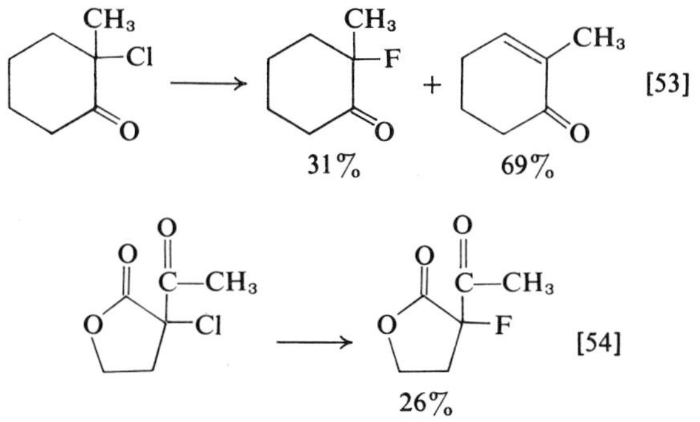

CH_3 –Cl ⟶ CH_3 –F (31%) + CH_3 (69%) [53]

C–CH_3 –Cl ⟶ C–CH_3 –F (26%) [54]

Scheme 3–10

Further examples of Cl/F exchange are presented in Scheme 3–11:

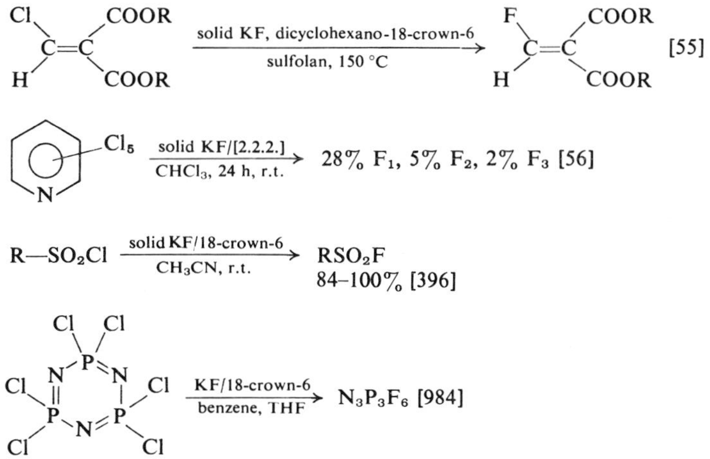

Cl(H)C=C(COOR)(COOR) —solid KF, dicyclohexano-18-crown-6 / sulfolan, 150 °C→ F(H)C=C(COOR)(COOR) [55]

Cl_5 (pyridine) —solid KF/[2.2.2.] / $CHCl_3$, 24 h, r.t.→ 28% F_1, 5% F_2, 2% F_3 [56]

$R—SO_2Cl$ —solid KF/18-crown-6 / CH_3CN, r.t.→ RSO_2F 84–100% [396]

$N_3P_3Cl_6$ —KF/18-crown-6 / benzene, THF→ $N_3P_3F_6$ [984]

Scheme 3–11

Other applications of fluoride (eliminations, additions) will be discussed in other sections.

3.2.1.3. Nucleophilic Aromatic Substitution

The displacement of halogen from an aromatic ring in chloro-2,4-dinitrobenzene and perchloropyridine was mentioned above. Making use of the very strong electron acceptor properties of the diazonium group, Gokel and co-workers [80] reacted 0.1 M of 4-bromobenzenediazonium tetrafluoroborate in methylene chloride or chloroform with a five molar excess of a quaternary ammonium chloride (TEBA or tetrabutylammonium chloride). Although the tetrafluoroborate itself is insoluble, a double comportion (gegenion exchange reaction) with the ammonium chloride brings the aryldiazonium moiety into solution [81]. After 24 hours at 30 °C, about 70% of the bromide was displaced by chloride. Subsequent reduction with H_3PO_2 and catalytic amounts of cuprous oxide gave the following product mixture [80]:

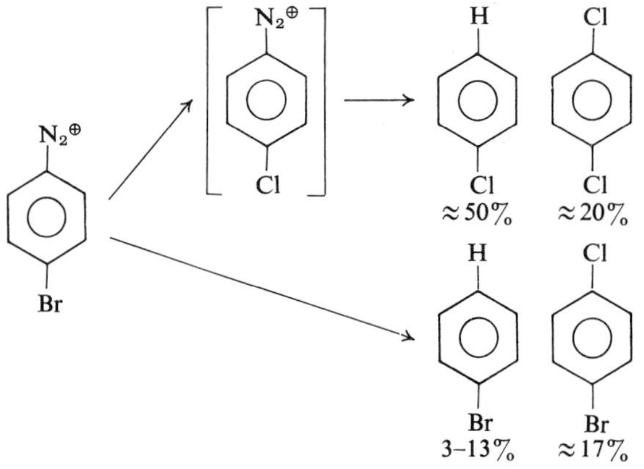

Scheme 3–12

A more dilute solution gave an even higher exchange if a 5-molar excess of chloride was used. Interestingly enough, the exchange was retarded by the addition of 18-crown-6 which complexes the diazonium group thus dispersing the positive charge and weakening the tendency towards nucleophilic displacement.

4-Nitrophthalic anhydride was transformed into the 4-fluoro compound in 42% yield by boiling with KF/dibenzo-18-crown-6 in acetonitrile for 43 h [1031].

3.2.1.4. Related Radical Reactions of Diazonium Compounds

An alternative to the Sandmeyer reaction is made possible when *p*-bromobenzene-diazonium tetrafluoroborate is treated with potassium acetate and 18-crown-6 at room temperature [81]. It is believed that the crown-solubilized potassium acetate undergoes a metathetical counterion exchange process with the originally undissolved diazonium salt. Transient diazoacetate is thought to give the diazoanhydride which leads eventually to aryl radicals that attack the solvent [81, 227]:

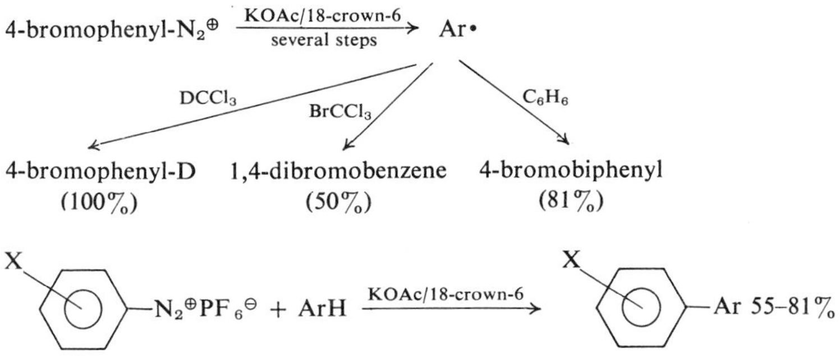

$$\text{4-bromophenyl-}N_2^{\oplus} \xrightarrow[\text{several steps}]{\text{KOAc/18-crown-6}} Ar\bullet$$

$Ar\bullet$ + $DCCl_3$ → 4-bromophenyl-D (100%)

$Ar\bullet$ + $BrCCl_3$ → 1,4-dibromobenzene (50%)

$Ar\bullet$ + C_6H_6 → 4-bromobiphenyl (81%)

$$X\text{-}C_6H_4\text{-}N_2^{\oplus}PF_6^{\ominus} + ArH \xrightarrow{\text{KOAc/18-crown-6}} X\text{-}C_6H_4\text{-}Ar \quad 55\text{–}81\%$$

Scheme 3–13

As shown above, this method has been developed for the synthesis of mixed biaryls from stable, easy-to-handle aryldiazonium tetrafluoroborates or hexafluorophosphates [227]. Much higher yields compared to conventional procedures were obtained. Reduction of the diazonium group (formally a substitution of hydrogen) also occurs if the compound is heated in methylene chloride with catalytic amounts of dicyclohexano-18-crown-6 and traces of copper powder [93] ($\frac{1}{2}$ to 4 hours, 40 °C), alternatively, it can be stirred with hypophosphorous acid in chloroform in the presence of a little cuprous oxide and (not always necessary) 18-crown-6 [94].

Based on this crown-cation complexation and subsequent aryl radical generation, high yield syntheses for aryl bromides and iodides were developed [855]. The halogenations are carried out in chloroform solution from stable, safe aryldiazonium tetrafluoroborates in the presence or catalytic amounts of 18-crown-6 and a small excess of bromotrichloromethane for the bromide preparation, and iodomethane or molecular iodine for the iodide preparation. Some reduction and chlorination products (0–8%) are formed. Hexachloroethane from the solvent bromotrichloromethane is another byproduct.

3.2.1.5. Halides from Alcohols, Ethers, and Diazomethanes

$$ROH \xrightarrow[\text{catalyst}]{HCl} RCl$$

The PTC extraction of hydrogen halides is a less common process. For example, in the equilibration of methylene chloride with concentrated hydrochloric acid only a small amount of acid is transferred into the organic phase. In the presence of tetrabutylammonium chloride ($\sim$0.1 M in CH_2Cl_2), however, a little more than an equimolar amount of hydrochloric acid can be titrated in the methylene chloride layer [57]. It would appear that the hydrogen bond between chloride and hydrochloride is strong enough to permit the extraction of $NBu_4^{\oplus}HCl_2^{\ominus}$.*) More lipophilic catalysts

*) Species such as $NR_3 \cdot 2HCl$ have been assumed in benzene solution [879]; $NEt_3H^{\oplus}HCl_2^{\ominus}$ is a known compound both in the solid state and in solution [880].

(*e.g.*, $NHep_4Cl$) permit extraction of hydrochloric acid even into benzene. Similarly, $NBu_4^{\oplus}HF_2^{\ominus}$ can be extracted from concentrated aqueous potassium hydrogen difluoride. Although there remains some chloride ion associated with the quaternary cation, titration of the acid in the organic layer is again higher than theoretical. With hydrobromic acid/$NBu_4^{\oplus}Br^{\ominus}$ on the other hand, less than the equivalent quantity of HBr corresponding to the salt is extracted [57].

To date, the effective transfer of hydrohalides into nonpolar organic solvents has been exploited in only a few studies. Landini and co-workers [58] transformed *n*-alkanols into the corresponding chlorides by heating the alcohols with 5 moles HCl and 0.1 mole tributylhexadecylphosphonium bromide for 30 to 45 hours at 105 °C. Yields for the C_6 to C_{16} alcohols were greater than 90%. Reaction rates and yields were lower in the case of partially water miscible alcohols. When alcohols were boiled with concentrated hydrobromic acid, the presence of a catalyst barely accelerated the formation of the bromides [2]; however, the cleavage of dialkyl and alkylaryl ethers is strongly accelerated by boiling with 47% HBr in the presence of tributylhexadecylammonium bromide [1032].

$$Ph_2CHN_2 \longrightarrow Ph_2CHF$$

The reaction of diphenyldiazomethane in methylene chloride with aqueous potassium hydrogen difluoride in the presence of tetrabutylammonium perchlorate for 48 hours in the dark, gave 50% diphenylmethyl fluoride, together with 10% benzhydrol and 35% of a polymer. Other diaryldiazomethanes behaved similarly, but 9-diazofluorene was stable under the reaction conditions, and phenyldiazomethane gave only 14% of the fluoride [59].

Another possibility for transforming alcohols into chlorides under basic conditions is offered by the PTC dichlorocarbene reaction. This will be discussed in Section 3.3.13.2.

Additional patent references pertinent to this section can be found in Table 3–2.

Some typical preparative procedures are presented below.

1-Fluorooctane [52]: 0.1 mole 1-chlorooctane, 0.5 mole $KF \cdot 2H_2O$, 30 ml water, and 0.01 mole tributylhexadecylphosphonium bromide were stirred in an autoclave for 7 hours at 160 °C. After cooling, the organic phase was separated, washed with water, concentrated H_2SO_4, and water, dried over $CaCl_2$, and distilled to yield 77% (10.2 g), b.p. 142–144°, of the desired product.

(+)-(S)-2-Chlorooctane [51]: 0.05 mole freshly prepared optically active (R)-2-octyl methanesulfonate, 0.05 mole KCl, 0.0025 mole Aliquat 336, and 5 ml water were stirred at 100 °C for 90 minutes. The organic layer was separated, and the aqueous phase extracted with pentane. The combined organic solutions were washed with water, concentrated H_2SO_4, and again with water, then dried over $CaCl_2$. After evaporation of the solvent and distillation of the residue *in vacuo*, the product was obtained: 74% (5.5 g) b.p. 69–70 °C/20 torr, $[\alpha]_D^{20}$ + 33.4 ° (neat, l = 1) (89% optically pure).

Table 3–2. Further Patent and Rare Journal References on Displacements of Sections 3.2.1, 3.2.2, and 3.2.4.

Type of Reaction	Conditions/Remarks	Reference
$ArCOCl \rightarrow ArCOF$	KF/H_2O ammonium salts	[130]
$ArSO_2Cl \rightarrow ArSO_2F$	KF/H_2O ammonium salts	[131]
$(Me_2N)_2POCl \rightarrow (Me_2N)_2POF$	KF/H_2O ammonium salts	[132]
$BrCH_2{-}CH_2Br \rightarrow ClCH_2{-}CH_2Cl$	$NaCl/H_2O$ phosphonium salts	[133]
$ROH + HCl \rightarrow RCl$	onium salts	[134]
$HO(CH_2)_4OH \rightarrow Cl{-}(CH_2)_4{-}Cl$	HCl/H_2O ammonium salts	[135]
$Cl(CH_2)_3{-}Br \rightarrow Cl(CH_2)_3{-}CN$	$NaCN/H_2O$ ammonium salts	[136, 137, 138]
$RX \rightarrow RCN$	$(K)NaCN/H_2O$ onium salts	[139, 140, 141]
$RX \rightarrow RNO_2$	$NaNO_2$ onium salts	[141]
$CH_2Br_2 \rightarrow CH_2(SCN)_2$	$NaSCN/H_2O$ onium salts	[142]
$RX \rightarrow RSCN$		[143]
$Me_3Sn(CH_2)_2X \rightarrow Me_3Sn(CH_2)_2{-}SO_2{-}Ar$ $\rightarrow Me_3Sn(CH_2)_2{-}OAr$ $\rightarrow Me_3Sn(CH_2)_2{-}SR$	metal$^{\oplus}$ $^{\ominus}O_2SAr$ onium salts $^{\ominus}OAr$ onium salts $^{\ominus}SR$ onium salts	[144]
$RCl \rightarrow RSO_3Na$	$Na_2\ SO_3/H_2O$, NR_4X	[1033]
polymer—$CO_2{-}CH_2{-}HC{-}CH_2$ (epoxide, O) $\longrightarrow$ polymer—$CO_2{-}CH_2{-}CH(OH){-}CH_2{-}SO_3^{\ominus}$	Na_2SO_3/H_2O, NBu_4X	[145]
$R{-}CO{-}Cl \rightarrow R{-}CO{-}N_3 \rightarrow R{-}N{=}C{=}O$	NaN_3/H_2O, Aliquat 336, NR_4X	[146, 147, 946]
$RCl + KOCN \longrightarrow$ (R—N, N—R, N—R ring with three C=O: O, O, O)	(K)NaOCN dry, ketone solvent, onium salt	[148, 149]
$CH_2{=}CH{-}CH_2Cl + EtOH + KOCN \rightarrow CH_2{=}CH{-}CH_2{-}NH_2{-}COOEt$	CH_3CN, phosphonium salts	[150]
$RCl + KOCN \rightarrow RN{=}C{=}O$	KOCN, trace H_2O, ammonium salts	[151]

3.2.2. Preparation of Nitriles

Among the most thoroughly studied PTC reactions, the earliest applications dating back to the 1950s and early 1960s, is the substitution R—X → RCN [69, 73]. The fundamental mechanism of PTC [4, 63]. and the kinetics of "triphase" catalysis [64] have been investigated in this exemplary system. Many catalysts of widely varying structure have been tested, and the general pattern observable in the literature is similar to that discussed for halide exchange. Normally, quaternary ammonium or phosphonium salts are used [4, 38, 39], and the use of crown ethers [44, 77] or polypod molecules such as **7** [46] does not seem to bring about important preparative improvements*) except in special cases [1035], such as when the substrate is water sensitive and solid KCN is employed. With allylic and benzylic halides for instance, hydrolysis may under certain conditions, compete unfavorably with substitution when aqueous cyanides are used. Nevertheless, in the literature one finds instructions for running such reactions using aqueous KCN [70, 71, 82, 897]. When it is desirable to avoid the presence of water, then solid/liquid PTC with 18-crown-6 catalysts has been found useful in the preparation of substituted benzyl cyanides as well as trimethylsilyl cyanide [65] and cyanoformates (R = phenyl, benzyl, alkyl, 62–94%) [66].

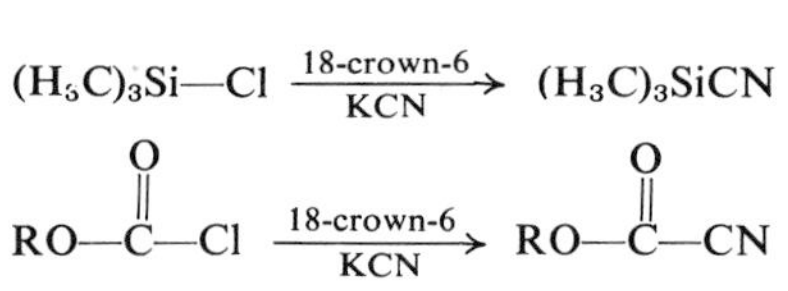

$$(H_3C)_3Si{-}Cl \xrightarrow[KCN]{\text{18-crown-6}} (H_3C)_3SiCN$$

$$RO{-}\overset{\overset{\displaystyle O}{\|}}{C}{-}Cl \xrightarrow[KCN]{\text{18-crown-6}} RO{-}\overset{\overset{\displaystyle O}{\|}}{C}{-}CN$$

Scheme 3–14

Replacing crown ethers by onium salt catalysts in solid/liquid PTC has not yet been tried, but probably these would work as well with dry inorganic cyanides as they do with other salts. On the other hand, Simchen and Kobler advocate the use of previously prepared, isolated quaternary ammonium cyanide in nonprotic solvents like DMSO, acetonitrile or methylene chloride with hydrolysis sensitive compounds [67]. Polyethyleneglycol ethers **8** are again cheaper than crown ethers although less reactive [47, 61] as catalysts. In addition to these "triphase" catalysis has been recommended lately [62, 64, 68, 775, 860]. This technique is in principle very attractive as outlined in the preceding section. In fact, Amberlite IRA-400 ion exchange resin has been patented as a catalyst for the synthesis of adiponitrile from 1,4-dichlorobutane [69]. It was shown recently, however, that the catalytic activity of the "triphase" catalysts fixed to cross-linked polystyrene is related to the number of $NR_4^{\oplus}$ groups present. A high level of ring substitution—as is usual with commercial ion exchange resins—is detrimental to the PTC reaction [64].

Finally, even primary, secondary, or tertiary amines can function as catalysts for cyanide formation [70, 72] as long as they are not too small and therefore extremely

---

*) Although reaction rates are increased in homogeneous solution, even in HMPT [978].

water soluble. Apparently, the amines are alkylated *in situ* to form quaternary ammonium salts, the real catalysts.

Cyanide displacement, catalyzed by onium salts, is much faster with primary (2–8 hours at 100 °C, >95% yield) than with secondary substrates. Cyclohexyl and tertiary alkyl halides give mainly elimination products. 2-Chlorooctane leads to 85–90% displacement along with 10–15% olefin formation. Substitution of optically active 2-octyl methanesulfonate gave the inversion product with a maximum of 30% racemization [4]. Benzylic substrates can be reacted at room temperature (17 hours). Leaving group effects are as expected, *i.e.*, OMes > Br > Cl.

With iodide and *p*-toluenesulfonate the reaction is quenched after a short time due to the pairing of the catalyst cation with these anions [4]. This can be overcome either by using more catalyst*) or by replacing the aqueous phase repeatedly by fresh cyanide solution, if necessary. Besides simple alkyl and benzyl cyanides, some more complicated compounds have been prepared [74, 75, 967]; of these the formation of aroyl cyanides is worth noting [76]:

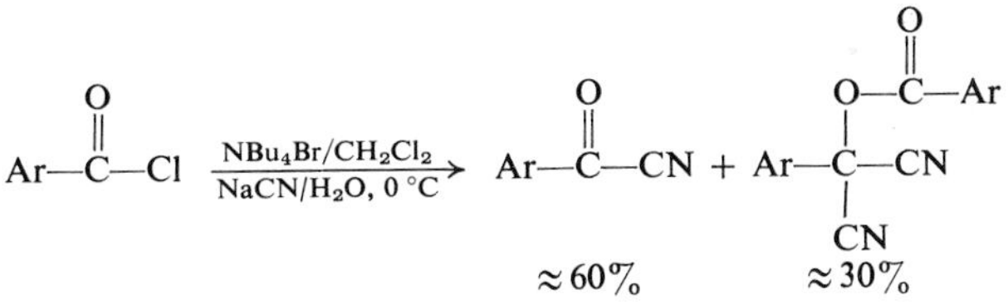

$$Ar{-}C({=}O){-}Cl \xrightarrow[NaCN/H_2O,\ 0\ ^\circ C]{NBu_4Br/CH_2Cl_2} \underset{\approx 60\%}{Ar{-}C({=}O){-}CN} + \underset{\approx 30\%}{Ar{-}C(CN)_2{-}O{-}C({=}O){-}Ar}$$

Scheme 3–15A

Hardly accessible vinylnitriles can be made conveniently in excellent yields, ≈90%, and in a stereospecific manner by treating vinyl halides with two equivalents of KCN in benzene in the presence of a little $Pd(PPh_3)_4$ and 18-crown-6 at 75 °C [893].

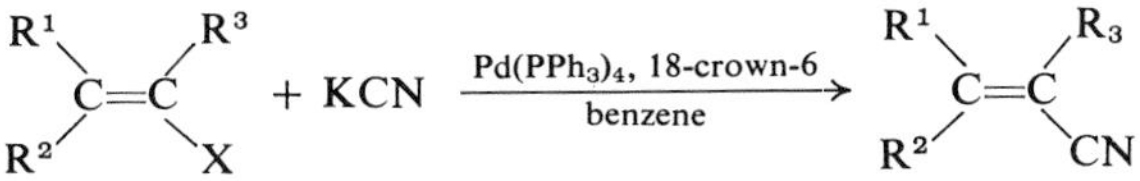

$$R^1R^2C{=}CR^3X + KCN \xrightarrow[benzene]{Pd(PPh_3)_4,\ 18\text{-crown-6}} R^1R^2C{=}CR_3CN$$

Scheme 3–15B

α-Iminonitriles (imidoyl cyanides) are made by stirring imidoyl chlorides in chloroform with aqueous potassium cyanide in the presence of hexadecyltrimethylammonium chloride or TEBA at room temperature [1036]:

$$Ar{-}C(Cl){=}N{-}C_6H_5 \rightarrow Ar{-}C(CN){=}N{-}C_6H_5$$

In all the $CN^\ominus$ displacements considered so far only cyanides—and no isocyanides—were formed. It has been found, however, that homogeneous solutions of preformed

---

*) With mesylates or bromides the reaction could proceed too vigorously and become uncontrollable if too much catalyst is added [4].

tetramethylammonium dicyanoargentate in acetonitrile react with activated halides exclusively to give the isonitriles [78]:

$$2RX + NMe_4^{\oplus}Ag(CN)_2^{\ominus} \longrightarrow 2R{-}NC + NMe_4^{\oplus}AgX_2^{\ominus}$$

With triphenylmethyl and benzhydryl halides, this reaction proceeded at room temperature within 2–15 hours with over 90% yields. 4-Nitro-substituted compounds required refluxing for a few hours. Methyl bromide was very slow, methyl iodide had to be refluxed with the reagent for 30 minutes, and no reaction could be observed with phenyl iodide and benzoyl chloride [78]. Preparation of isonitriles by the PTC carbylamine reaction will be discussed in Section 3.3.13.2.

Finally, an oxidative cyanation and a photocyanation, both somewhat related to PTC, will be mentioned. Anodic cyanation of naphthalene, stilbenes, anisol, and other aromatic ethers and amines has been performed in methylene chloride/aqueous NaCN with tetrabutylammonium sulfate or Aliquat 336 as the supporting electrolyte/transfer agent [79, 863, 864, 1034]. Under these conditions, 9,10-dialkylanthracenes

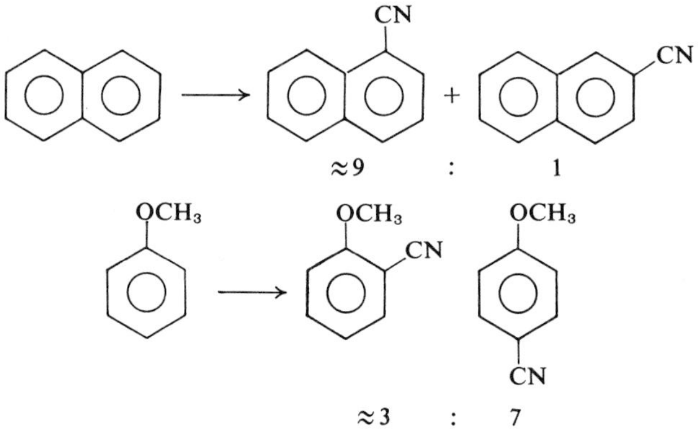

≈9 : 1

≈3 : 7

Scheme 3–16

add both nitrile and isonitrile groups in the 9- and 10-positions [95]. Anodic acetoxylations have been similarly performed [862].

A photocyanation of aromatic hydrocarbons was carried out in anhydrous acetonitrile or dichloromethane using KCN/18-crown-6 [883], NBu_4CN [916] or aqueous $KCN/NBu_4CN/CH_2Cl_2$ [916].

Additional references for patents relevant to the main subject of this section can be found in Table 3–2.

Some typical experimental procedures are presented below.

Crown ether catalysis: Trimethylsilyl cyanide [65]: 0.1 mole dry KCN, 0.11 mole trimethylsilyl chloride, and 0.4 mole 18-crown-6 were refluxed in the absence of O_2 for 24 hours. The mixture was distilled directly from the reaction vessel through a Vigreux column. 36% yield, b.p. 114–117 °C.

Onium salt catalysis: 1-Cyanooctane [4]: 0.67 mole 1-chlorooctane, 2 moles NaCN, 25 ml water, and 0.01 mole tributylhexadecylphosphonium bromide were refluxed at 105 °C for 2 hours. Workup of the organic phase gave a 94% yield in 97% purity.

3.2.3. Ester Formation

Under normal displacement conditions in protic solvents, carboxylate anions are among the weakest nucleophiles. Among the major factors contributing to their poor performance must be the strong solvation of the anion. Ion pairs in nonpolar aprotic solvents, as produced in PTC reactions, should and do show enhanced reactivity.

Early references to the uses of onium salts relevant to our subject are mainly from the patent literature. There are, in fact, numerous processes patented for making polymers from alkali metal salts of mono- or dicarboxylic acids with dihaloalkanes or epichlorohydrin, as well as from alkali metal salts of phenols with dicarboxylic acid chlorides, disulfonic acid chlorides, phosgen, and thiophosgen, catalyzed either in a paste or in an organic solvent/water mixture at elevated temperatures. Possible catalysts include quaternary ammonium, phosphonium, arsonium, and tertiary sulfonium salts. Benzyltriethyl-, and -trimethylammonium salts, however, are frequently used in the examples described. Table 3–3 sums up these references. Onium salts are also used as catalysts for reactions between acids and epoxides. Most of these patented processes do not appear to be PTC-like in character (*cf.*, [965]).

Turning now to substrates of lower molecular weight, there are a few variants of reactions between carboxylate salts and alkylating agent with or without the use of additional solvent [96, 97]. Sometimes, free acids are used together with tertiary amines. It is not clear in some cases whether the trialkylhydrogenammonium salt solubilizes the carboxylate or whether a tetraalkylammonium halide is first quickly formed with the alkylating agent and then acts as the true catalyst [98, 99, 100]. For instance, the preparation of di-*n*-octyl- and dibenzyl phthalates from sodium phthalate (both with 91% yield) is described in patents [101]. Using trimethyloctylammonium bromide or benzyltrimethylammonium chloride catalysts, the aqueous/organic mixture containing alkylating agent in excess is distilled azeotropically (water/alkylating agent coming off) and finally refluxed at 150–160 °C. The reaction conditions used here may not have been optimal because the catalysts were not very lipophilic. Thus, when methyltrioctylammonium chloride (Aliquat 336) is the catalyst, 1-bromodecane was transformed into its acetate using an eightfold excess of aqueous sodium acetate at 105 °C. The reaction was essentially complete in one hour [38, 39].

$$n\text{-}C_{10}H_{21}\text{—Br} \longrightarrow n\text{-}C_{10}H_{21}\text{—OAc}$$

cis-3,5-Dibromo-1-cyclopentene **A** could be transformed into the *cis*-diacetoxy compound **B** in carbon tetrachloride/aqueous potassium acetate again using Aliquat 336 [102, *cf.*, 427]. Conversions of 68% after 5 hours at 60 °C and 75% after 9 hours

Table 3–3. Further Patent and Rare Journal References on Formation of Esters (aqueous solution of (or solid) salts, onium salt catalysts).

Types of Reaction	Reference
$CH_2{=}C(R){-}CO_2^{\ominus}K^{\oplus} + H_2C{-}CH{-}CH_2Cl$ (epoxide, bridging O)	[152 to 161]
$RCO_2^{\ominus}metal^{\oplus}$ / $^{\ominus}O_2C{-}R{-}CO_2^{\ominus}metal^{\oplus\oplus}$ $+ H_2C{-}CH{-}CH_2Cl$ (epoxide, bridging O)	[162 to 168, 276]
$R{-}CO_2^{\ominus}metal^{\oplus}$ / $^{\ominus}O_2C{-}R{-}CO_2^{\ominus}metal^{\oplus\oplus}$ $+ ClCH_2{-}CH{=}CR_2$	[169 to 173]
$2R{-}CO_2^{\ominus}metal^{\oplus}$ / $^{\ominus}O_2C{-}R{-}CO_2^{\ominus}metal^{\oplus\oplus}$ $+ ClCH_2{-}CH{=}CH{-}CH_2Cl$	[174, 175]
$^{\ominus}O_2C{-}R{-}CO_2^{\ominus}metal^{\oplus\oplus}$ + Cl-polymer → cross-linking	[176]
$^{\ominus}O_2C{-}R{-}CO_2^{\ominus}metal^{\oplus\oplus}$ / $R{-}CO_2^{\ominus}metal^{\oplus}$ $+ R'X$	[177 to 180]
$C_nF_{2n+1}{-}CH_2CH_2X + K^{\oplus\ominus}O_2C{-}R$	[181]
$CH_3(CH{=}CH)_2{-}CO_2^{\ominus}K^{\oplus} + Cl{-}CH_2{-}CH_2{-}O{-}CH{=}CH_2$	[182]
$CH_2{=}C(R){-}CO_2^{\ominus}metal^{\oplus} + Cl{-}CH_2{-}SR'$	[183]
$CH_2{=}C(R){-}CO_2^{\ominus}metal^{\oplus} + Cl{-}CH_2{-}COOEt$	[184]
$XCH_2{-}CH(OCOR){-}CH_2X + metal^{\oplus\ominus}O_2CR'$ → glyceride	[185]
cyanuric acid, Na salt + $ClCH_2{-}HC{-}CH_2$ (epoxide, bridging O)	[186]
bisphenol-A Na-salt + $ClCH_2{-}HC{-}CH_2$ (epoxide, bridging O)	[186]
$+ Cl_2CO$	[187 to 189, 277, 917]
$+ Cl{-}CO{-}R{-}COCl$	[190 to 194]
$+ ClO_2S{-}R{-}SO_2Cl$	[195, 196]
$+ Cl_2CS$	[196]
HO–C₆H₄–C(=CCl₂)–C₆H₄–OH $+ ClCO{-}C_6H_4{-}COCl$	[919]
3,5,6-trichloro-2-pyridinyl $O^{\ominus}Na^{\oplus}$ (pyridine ring with Cl, Cl, Cl, N, $O^{\ominus}Na^{\oplus}$) $+ (RO)_2PSCl$	[197 to 199]
$+ R^1(R^2O)PSCl$	[200]
penam (R^1, S, CH_3, CH_3, O, N, $CO_2^{\ominus}Na^{\oplus}$) $+ BrR^2$	[839]
benzene-1,3-di($CO_2^{\ominus}Na^{\oplus}$) $+ ClCO_2Et$	[926]
$BrCH_2{-}C_6H_4{-}CH_2Br + K_2CO_3$ (crown ether) → polycarbonate	[1037]

at 42 °C were reported. The product mixture contained only 3–5% of the allylic-rearranged product, 3,4-diacetoxy-1-cyclopentene. Similarly, cyclohexenone was converted into 4-acetoxy-2-cyclohexen-1-one by N-bromosuccinimide bromination and consecutive PTC displacement (7 hours at 33 °C).

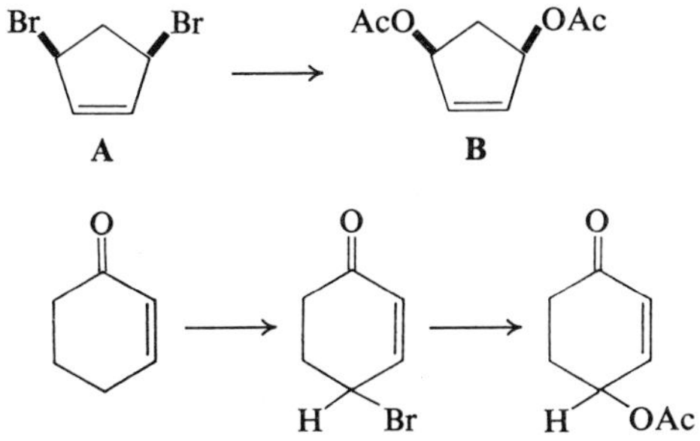

Br Br → AcO OAc

A B

O O O

→ →

H Br H OAc

Scheme 3–17

Ion pairs extracted from an aqueous medium are always accompanied by some water molecules and hence are somewhat less reactive than they would be under strictly anhydrous conditions. It is sometimes preferable, therefore, to isolate and dry the quaternary ammonium salts before the alkylation [104]. The latter step is often performed in methylene chloride. Except for reactions with some sluggishly reacting alkyl chlorides, this solvent can be used safely because it is itself a very slow alkylating agent. Boiling dry sodium benzoate with TEBA for 24 hours in dichloromethane gave only 15% dibenzoyloxymethane [103]. Various (including sterically hindered) methylene diesters were prepared by refluxing the isolated tetrabutylammonium carboxylates in methylene chloride for 2–3 days [104].

$$2RCO_2^{\ominus} \longrightarrow RCOO{-}CH_2{-}OCOR$$

Brändström's standard procedure [105] for esterification consists of neutralizing equivalent amounts of the acid and tetrabutylammonium hydrogen sulfate with a 2 N sodium hydroxide solution, adding an excess of alkylating agent dissolved in dichloromethane, and refluxing for up to 30 minutes. Separation of the ammonium salt in the case of iodides is achieved by evaporating the methylene chloride and dissolving the residual ester with ether. If a tetrabutylammonium bromide has been formed, this can be extracted from the etheral solution with water. When dimethyl sulfate is the alkylating agent, only catalytic amounts of tetrabutylammonium hydrogen sulfate are recommended, because the methyl sulfate anion formed is distributed preferentially into the aqueous phase to such an extent that it does not hinder the extraction of the carboxylate ion.

Using this method, Brändström was able to obtain yields of up to 90%, even in the case of sterically hindered *o,o'*-dimethyl or dimethoxy substituted benzoic acids. Most dicarboxylic acids did not present any problem. Only very hydrophilic acids gave lower yields (*e.g.*, tartaric acid, 40%). Amino acids, however, could not be esterified.

Using only 1 mole of tetrabutylammonium salt, a hydroxybenzoic acid can be selectively alkylated at the carboxylic group [106]. Substituted β-hydroxy dithiocinnamic acids **C** were similarly esterified at room temperature [107]. H. Ehrsson [108] developed a method for the gas chromatographic determination of carboxylic acids and

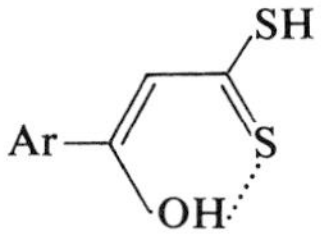

C

Scheme 3–18

phenols. This method involves extracting the acid as an ion pair into methylene chloride and forming its pentafluorobenzyl bromide derivative. Here the reaction rate was enhanced both by the structure of the counterion and by increasing its concentration. For a rapid reaction, the more lipophilic tetra-*n*-pentylammonium compound was again much better than tetrabutylammonium. Extractive alkylations for the analysis of pharmaceutic agents have been reviewed [1052], and in a recent solid/liquid PTC micromethod sodium carbonate was the base [1053].

Solubilization, extraction, and "anion activation" of alkali metal carboxylates by various complexants has been investigated extensively. Knöchel [109] compared many ligands (crown ethers, aminopolyethers, nonactin, polypods, and open-chain polyethers) as PT catalysts in the reaction between solid potassium acetate and benzyl chloride in acetonitrile. The results showed that there is no simple correlation between the extent of solubilization and the reaction rate.

The following factors influence the overall performance:

(i) stability of the complex formed;
(ii) lipophilicity of the ligand;
(iii) rigidity of the ligand.

Dicyclohexano-18-crown-6 and Kryptofix [221] (**11**) were the most active catalysts. In another study it was found, however, that Aliquat 336 was as good a catalyst as crown ethers in solid/liquid PTC [1041].

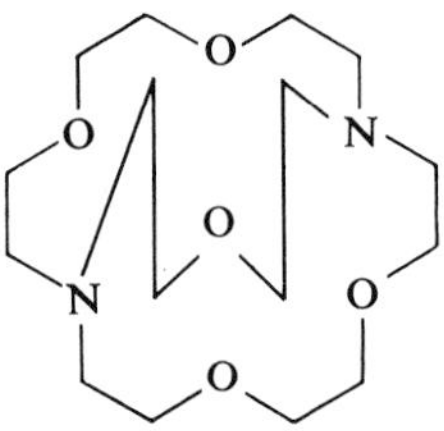

11

Scheme 3–19

Liotta and co-workers [110] showed that high concentrations of 18-crown-6-complexed potassium acetate can be achieved in benzene (up to 0.8 M) and acetonitrile. At least 80% of the crown ether present is complexed. It is more economical, however, to use only 3 to 10 mole-% of the catalyst together with excess solid potassium acetate in acetonitrile, where reaction is faster, or in benzene. The acetate reagent so used is termed "naked" acetate by Liotta, but as shown previously in this book the term is rather misleading (see p. 19). Acetate is still a very powerful nucleophile but a very weak base under these conditions. Reactions with primary alkyl halides require several days at room temperature or a few hours at reflux. Activated bromides (*e.g.*, benzyl bromide) require only a few hours at room temperature. Virtually no elimination products are formed from primary halides, thus once more reflecting the low basic character of the "naked" acetate. The reaction rates for different leaving groups were approximately Br:OTs:Cl = 4:2:1, as tested with 1-substituted hexanes. The two secondary halides investigated gave only 10% olefinic elimination products each.

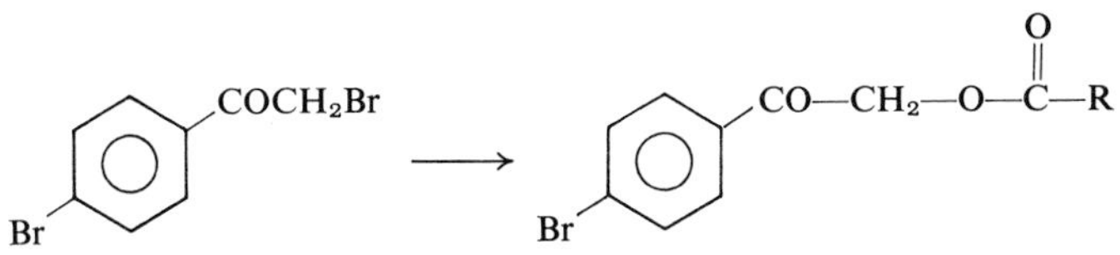

$$Br{-}C_6H_4{-}COCH_2Br \longrightarrow Br{-}C_6H_4{-}CO{-}CH_2{-}O{-}\overset{\displaystyle O}{\overset{\|}{C}}{-}R$$

Scheme 3–20

p-Bromophenacyl esters were obtained similarly by refluxing the potassium salts of carboxylic acids with *p*-bromophenacyl bromide in acetonitrile, or in benzene which is slower, for 15 minutes in the presence of 5 mole-% dicyclohexano-18-crown-6 or 18-crown-6 [111]. Isolated yields from between 90 and 100% were obtained, even with sterically hindered carboxylates like 2-iodobenzoic acid or mesitoic acid. An alternative to refluxing (up to 30 minutes in some cases), several hours stirring at room temperature is possible. It is claimed that this method is superior in yield and ease of manipulation to the previous synthesis of phenacylesters [111].

Again yields between 85 and 99% were realized by boiling potassium salts of simple carboxylic acids with alkyl bromides in benzene in the presence of a small amount of Kryptofix [222]® (**5**) for 3 hours [112]. Methylthiomethyl esters have been proposed as carboxylic acid protection groups. They are formed by boiling the potassium carboxylate with chloromethyl methyl sulfide and catalytic amounts of sodium iodide and 18-crown-6 in benzene. Deprotection is effected by consecutive treatment with $HgCl_2$ in refluxing acetonitrile/water and hydrogen sulfide [1042]:

$$R{-}COOH + ClCH_2SCH_3 \rightarrow R{-}CO{-}OCH_2SCH_3$$

Analytical applications of crown ether catalyzed esterifications have also been developed [113, 114, 1038]. The limit of UV-detectability of phenacylesters after liquid chromatographic separation was 1 ng for C_2 and 50 ng for C_{20} acid derivatives [114]. Below is a more complicated application of PTC techniques [115]:

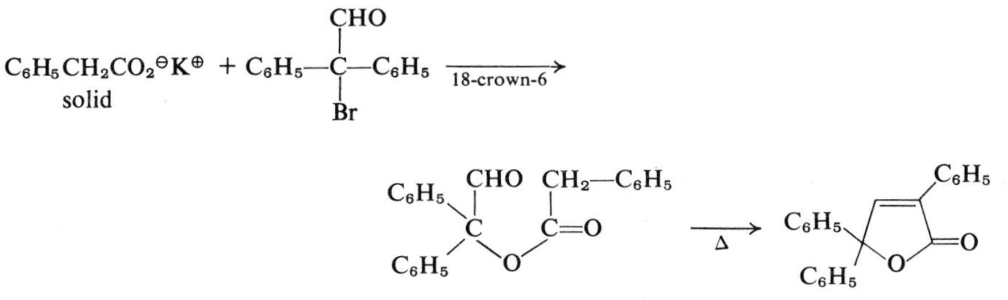

$$C_6H_5CH_2CO_2^{\ominus}K^{\oplus} \text{ (solid)} + C_6H_5-\underset{Br}{\overset{CHO}{C}}-C_6H_5 \xrightarrow{\text{18-crown-6}}$$

$$(C_6H_5)_2C(CHO)-O-C(=O)-CH_2-C_6H_5 \xrightarrow{\Delta} \text{2,2,4-triphenyl-furan-5(2H)-one ring } (C_6H_5, C_6H_5, C_6H_5, =O)$$

Scheme 3–21

The first step of the Merrifield peptide synthesis is the attachment of an N-protected amino acid to a chloromethyl polystyrene copolymer by ester formation. It can be accelerated by using the K-salts of Boc-amino acids with molar amounts of 18-crown-6 in DMF [972].

Apart from crown ethers, other complex ligands have been recommended: technical polyethyleneglycol ethers (**8**, $n \approx 8$) [61, 1004], and various substituted phosphoric acid triamides [117] and diamines [117], of which N,N′-tetramethylethylenediamine (**12**) is the most simple [117, 118]. It was shown, however, that **12** is converted to the monobenzyl chloride—the actual catalyst—in the course of the reaction [1013].

$$H_3CO-(CH_2-CH_2-O)_n-CH_3$$
8

$$Me_2N-CH_2-CH_2-NMe_2$$
12

When solid salts were reacted with bromides or benzyl chloride as the alkylating agent and a catalyst in benzene (in the case of **8**) or acetonitrile (in the case of **12**), yields were almost quantitative even with secondary alkylating agents and sterically hindered acids [118].

In principle, the present section is dedicated to reactions under neutral conditions, but in order to keep all the ester-forming reactions together, a few conversions in the presence of sodium hydroxide will be dealt with.

An interesting question is whether or not ester formation from acid chloride and alcohol is possible under PTC conditions. This type of reaction is often carried out with sulfonic and phosphoric acid chlorides (*vide infra*). Schotten-Baumann reactions were attempted using benzoyl chloride and various alcohols in the presence of a PT catalyst and aqueous sodium hydroxide. No standard conditions were found for all substrates; although the reaction is faster than when uncatalyzed, a part of the chloride is hydrolyzed and the rate is dependent on the lipophilicity of the alcohol [602]. Synthesis of N,N-dialkylaminosulfenyl carbamate insecticides was achieved in 38–75% isolated yields from carbamoyl fluorides and alcohols in aqueous NaOH/toluene (NBu_4HSO_4 as catalyst) or in NaOH/dichloromethane (TEBA as catalyst) [1040]:

$$R^1R^2N—S—N(CH_3)—COF \rightarrow R^1R^2N—S—N(CH_3)—COOR^3$$

The monotosylation of partially protected sugar diols by tosyl chloride has been described using dichloromethane/aqueous sodium hydroxide and an excess of tetrabutylammonium hydrogen sulfate [119] (30 min, room temperature). Depending on the stereochemistry of the starting material, varying amounts of derivatives were formed: *e.g.*, methyl 4,6-O-benzylidene-β-D-glucopyranoside gave 55% 2-tosylate, 31% 3-tosylate, and 7% 2,3-ditosylate, while methyl 4,6-O-benzylidene-α-D-mannopyranoside yielded 95% 2-tosylate.

Dialkyl sulfaminates of steroidal phenols (**D**) were obtained in benzene/aqueous sodium hydroxide with 10 mole-% TEBA as catalyst (1.5 h at 60 °C). The second

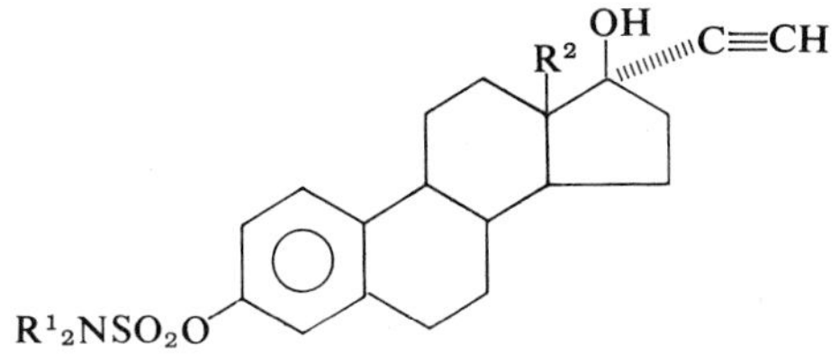

OH, R^2, C≡CH, $R^1{}_2NSO_2O$

D

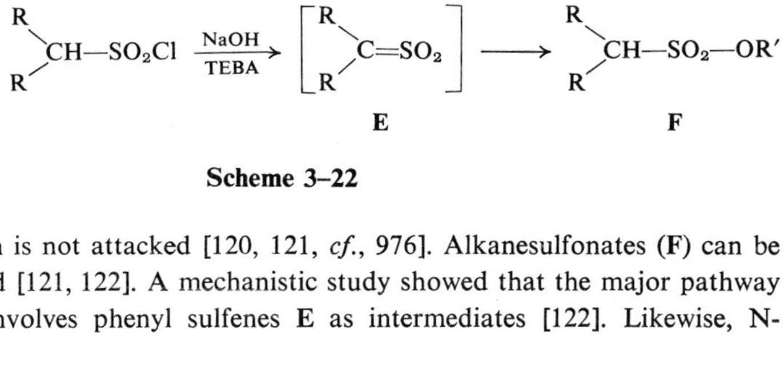

$$R_2CH—SO_2Cl \xrightarrow[\text{TEBA}]{\text{NaOH}} [R_2C{=}SO_2] \longrightarrow R_2CH—SO_2—OR'$$

E **F**

Scheme 3–22

hydroxyl function is not attacked [120, 121, *cf.*, 976]. Alkanesulfonates (**F**) can be similarly prepared [121, 122]. A mechanistic study showed that the major pathway to these esters involves phenyl sulfenes **E** as intermediates [122]. Likewise, N-

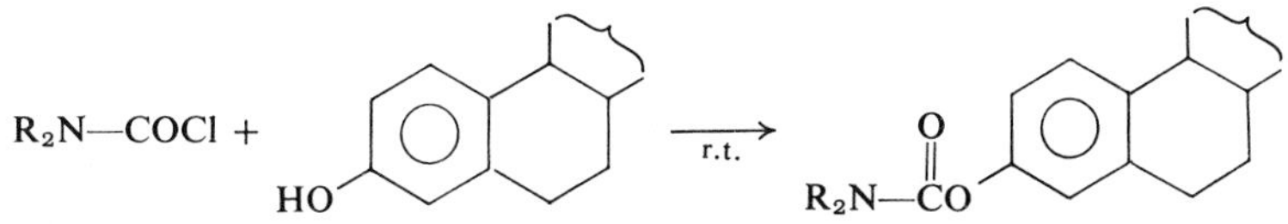

$R_2N—COCl$ + HO– (steroid phenol) $\xrightarrow{\text{r.t.}}$ $R_2N—\overset{O}{\overset{\|}{C}}O$– (steroid)

Scheme 3–23

disubstituted carbamoyl chlorides react with steroid phenolic hydroxyl radicals dissolved in chlorinated hydrocarbons/aqueous sodium hydroxide in the presence of various onium salt and crown either catalysts [123]. In the first reaction of Scheme 3–24 the monoalkyl phosphate was neutralized with tetramethylammonium hydroxide in methanol, the solvent removed, and the residue refluxed with 1 mole of an alkyl halide [124]. It is quite likely that this conversion can be performed in a two-phase system using a sufficiently lipophilic catalyst. Similarly, mixed dialkyl phosphites were made by extraction of tetrabutylammonium monoalkyl phosphites into CH_2Cl_2

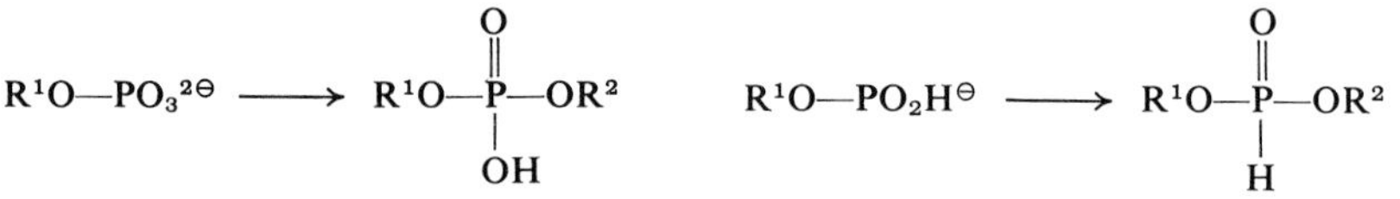

$$R^1O{-}PO_3^{2\ominus} \longrightarrow R^1O{-}P(=O)(OH){-}OR^2 \qquad R^1O{-}PO_2H^{\ominus} \longrightarrow R^1O{-}P(=O)(H){-}OR^2$$

Scheme 3–24

and alkylation in acetonitrile at 50 °C [957] (Scheme 3–24), and this was extended to the preparation of mixed trialkyl phosphates [1044].

$$R^1X + K^{\oplus}\ {}^{\ominus}S_2C{-}OR^2 \rightarrow R^1S{-}CS{-}OR^2 \rightarrow R^1SR^2$$

$$RS{-}CS{-}OC(CH_3)_3 \rightarrow RSH$$

Various O,S-dialkyl dithiocarbonates were prepared from alkyl halides or mesylates and potassium O-alkyldithiocarbonates in the presence of Aliquat 336 at 0 °C or room temperature without any extra solvent [915, 1009, 1043]. By heating the raw reaction mixture with potassium hydroxide at 80 °C, a simple one-pot synthesis of unsymmetrical sulfides (25–85% yield, depending on the structures) became possible [1043]. If R^2 in the dithiocarbonate is *tert*-butyl, heating of the weakly alkaline reaction mixture gave alkanethiols in 60–91% yield [915].

An interesting example of co-catalysis was recently developed by Freedman and co-workers [125]. If no catalyst is used, the reaction between a phenolate in water and a dialkyl thiophosphoryl chloride in methylene chloride leads mainly to hydrolysis products.

$$ArO^{\ominus}Na^{\oplus} + (MeO)_2P(S)Cl \longrightarrow ArO{-}PS(OMe)_2$$

Activation of the phenolate by a PT catalyst, *e.g.*, tetrabutylammonium hydrogen sulfate, accelerates the reaction considerably and increases the yield of the desired product. On the other hand, although the nucleophilic catalyst N-methylimidazole activates the acylating agent, increasing the rate even more, hydrolysis is also increased. Using both catalysts together results in a unique effect in which the combined rate enhancements of both catalysts are surpassed and only 6% of the phosphoryl chloride is hydrolyzed.

Zwierzak *et al.* have developed a method for PTC phosphorylations of amines, hydroxylamines, hydrazine, and alcohols. Applying Atherton and Todd's procedure, intermediate dialkyl phosphorohalidates (**H**) can be made *in situ* from dialkyl phosphites (**G**) and tetrahalomethanes:

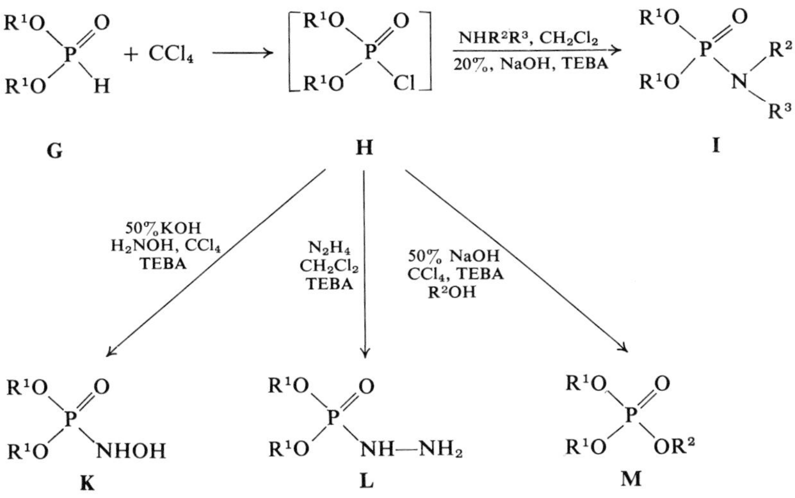

$(R^1O)_2P(O)H$ + CCl_4 ⟶ [$(R^1O)_2P(O)Cl$] —(NHR^2R^3, CH_2Cl_2; 20%, NaOH, TEBA)→ $(R^1O)_2P(O)NR^2R^3$

G **H** **I**

H —(50% KOH, H_2NOH, CCl_4, TEBA)→ $(R^1O)_2P(O)NHOH$ **K**

H —(N_2H_4, CH_2Cl_2, TEBA)→ $(R^1O)_2P(O)NH{-}NH_2$ **L**

H —(50% NaOH, CCl_4, TEBA, R^2OH)→ $(R^1O)_2P(O)OR^2$ **M**

Scheme 3–25

In one-pot reactions, compounds **I** [126], **K** [127], **L** [128], and **M** [129] can be prepared in very satisfactory yields.

Some typical experimental procedures are presented below.

Crown ether catalysis. p-Bromophenacyl acetate [111]: 1 mmole KOAc, 1 mmole ω,*p*-dibromoacetophenone, and 15 mg dicyclohexano-18-crown-6 were suspended in 10 ml acetonitrile. After 15 minutes of refluxing the solvent was removed under vacuum and the residue redissolved in benzene and washed through a short column of dry silica gel to remove the crown ether. Evaporation gave 252 mg (98%), m.p. 85–86 °C of the desired product.

Quaternary ammonium salt catalysis (*small amount*). 1-Decyl acetate [38]: 0.25 mole 1-bromodecane, 2 mole sodium acetate trihydrate, and 10 g Aliquat 336 were stirred at 105 °C for 2 hours. Water was added, and the mixture extracted with an organic solvent. Normal workup and distillation gave an almost quantitative yield.

Catalysis using molar amounts of tetrabutylammonium ion. Methyl ester (general procedure) [105]: 0.1 mole carboxylic acid and 0.1 mole NBu_4HSO_4 were dissolved in 100 ml 2M NaOH. 0.12 mole CH_3I in 100 ml of dichloromethane were added and the mixture was stirred at reflux temperature for 15 minutes. After separation of the phases the organic solvent was removed. The residue was treated with 100 ml of ether, whereupon the NBu_4I precipitated. It (100% yield) was filtered off, and the ether was evaporated. The residue was recrystallized or distilled. Yields were at least 90%; methyl 2,4,6-trimethylbenzoate: 94%.

3.2.4. Other Displacements

Besides those leading to halogenides, nitriles, and esters, many other substitution reactions involving PTC are known. In general, both the catalyst and reaction conditions are similar to those discussed in the preceding sections. It should be remembered that ion pairs present in relatively nonpolar, aprotic solvents show a different order of reactivities than the more usual reactions in aqueous ethanol or similar media. Hence, some "poor" nucleophiles perform very well in PTC reactions.

$$R{-}X \xrightarrow{\text{aq. SCN}^{\ominus}} R{-}SCN$$

When the above reaction is carried out with primary and secondary substrates, refluxing for 1–24 hours, yields of 90% are obtained [4, 38, 39, 73, 82, 1045]. Besides quaternary ammonium salts and crown ethers [45], many primary, secondary, and tertiary amine catalysts were tested. The latter must be quaternized in the reaction mixture thus leading to somewhat longer reaction times [82]. In homogeneous acetonitrile solution, 18-crown-6-complexed potassium thiocyanate is a comparatively weak nucleophile, reacting 32 times slower with benzyl tosylate than potassium acetate [83]. Six SCN groups can be introduced by substitution for chlorine into hexachlorocyclotriphosphazene [984].

$$R{-}X \longrightarrow R{-}ONO_2$$

It has been reported that nitrite displacements are possible in aqueous/organic mixtures using PTC conditions, but no details were published [4]. It has been claimed that the process is not very efficient [65]. With 18-crown-6 as catalyst, solid nitrite was reacted in acetonitrile at 25–40 °C giving nitro compounds as the major products and nitrite esters as side products. The yields and product distributions are similar, although slightly lower, to nitrite displacements in DMSO [65]: 50–70% nitro compounds with primary halides. Cyclohexyl bromide gives almost exclusively the elimination products.

Although strictly speaking it does not embody a PTC process, another procedure should be mentioned here. First, a commercially available strongly basic anion-exchange resin (*e.g.*, Amberlite IRA 900) is converted into the nitrite form, washed with water, ethanol, and benzene, and dried *in vacuo*. Then, a twofold excess is stirred and heated with a bromocarboxylic ester or an alkyl bromide giving the nitro compound. For example: 1-nitropropane (25 °C, 15 h, 47%), phenylnitromethane (25 °C, 4 h, 87%), ethyl 2-methyl-2-nitropropionate (50 °C, 24 h, 60%) [116]. A number of other reactions on anion-exchange resins loaded with specific reagent-ions are known (*cf.*, [116]).

$$R{-}X \longrightarrow R{-}NO_2 + R{-}ONO$$

Nitrate displacements under PTC conditions have not yet been carried out. It was shown, however, that nitrate can be a very active nucleophile in homogeneous solution in the presence of a cation complexing agent. Depending on the solvent, acetobromoglucose and silver nitrate/Kryptofix [222] gave mixtures of solvolysis product **A** and nitrate ester **B** in ratios ranging from **A**:**B** = 98:1 (methanol) to 0:100 (diglyme) [84].

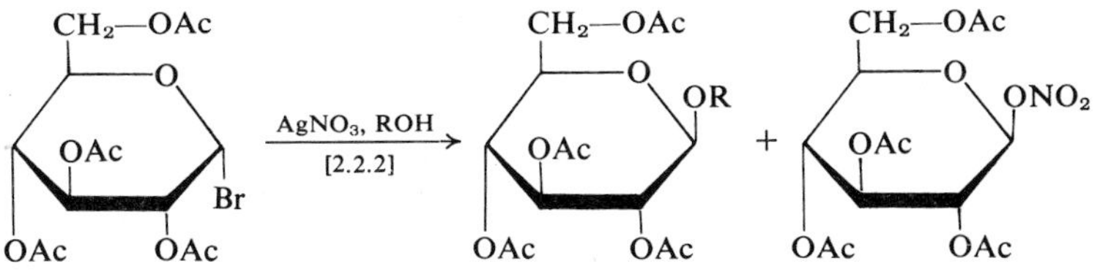

CH_2—OAc, OAc, Br, OAc, OAc $\xrightarrow[\text{[2.2.2]}]{AgNO_3,\ ROH}$ CH_2—OAc, OR, OAc, OAc, OAc (**A**) + CH_2—OAc, ONO_2, OAc, OAc, OAc (**B**)

Scheme 3–26A

$RX \rightarrow RN_3$

When *n*-alkyl bromides are reacted with aqueous 25% NaN_3 using Aliquat 336 as catalyst, almost quantitative yields of azides are obtained after only 6 hours at 100 °C [85]. Chlorides react somewhat slower, iodides faster. Cyclohexyl halides give about 25% cyclohexene.

As depicted below, satisfactory yields of amino acids are obtained from α-bromoesters either in water with NBu_4Br as catalyst or in benzene with 18-crown-6. Glycidate is not an intermediate in the reaction with α-bromo-β-hydroxypropionate (R = $HOCH_2$) since the ring opening of the epoxide is in the opposite direction [865].

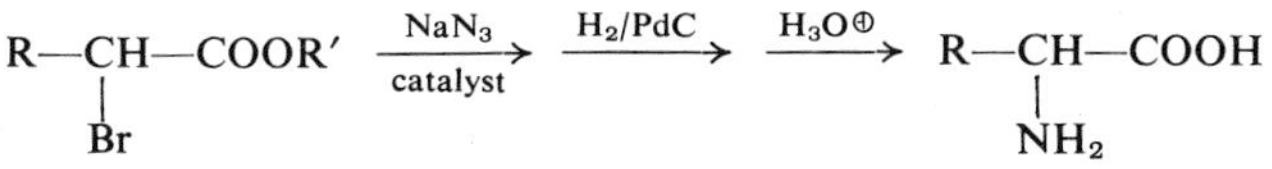

$$R\text{—}CH(Br)\text{—}COOR' \xrightarrow[\text{catalyst}]{NaN_3} \xrightarrow{H_2/PdC} \xrightarrow{H_3O^{\oplus}} R\text{—}CH(NH_2)\text{—}COOH$$

R = H, Me, Et, $HOCH_2$, $MeOCH_2$

$$\text{(epoxide)—COOR} \xrightarrow[\text{catalyst}]{NaN_3} N_3\text{—}CH_2\text{—}CH(OH)\text{—}COOR$$

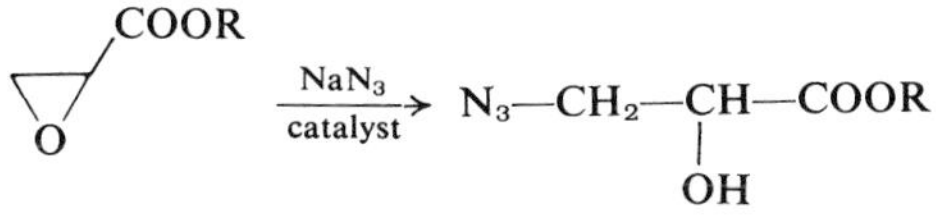

Scheme 3–26B

More recent investigations on the reaction of *erythro* and *threo* methyl β-hydroxy- (and β-methoxy-) β-phenyl-α-bromopropionates showed a dependence of regio- and stereochemistry both from the catalyst used and the relative configuration of the reactant [1046].

The reaction between solid potassium azide and benzyl bromide in benzene and acetronile with technical polyethylene glycol dimethyl ethers (**8**) as catalysts went to completion after 8 and 2 hours respectively [61].

The substitution of azide for chloride in polyvinyl chloride (PVC) has also been investigated [88, 89]. The heterogeneous mixture of PVC powder and aqueous sodium azide reacted at 80 °C in the presence of quaternary ammonium salts. Tetrabutylammonium chloride and bromide were better catalysts than the more usual surfactants [88]. In solvents such as THF, where PVC is soluble but the sodium salt is not, tetrabutylammonium chloride is again the best catalyst [88].

The authors discuss the kinetic phenomena observed for long chain surfactants in terms of the adsorption of the ammonium salt/catalyst on the polymer both with solid and dissolved PVC [88, 89].

$$\mathrm{RCO{-}X \longrightarrow RCO{-}N_3 \longrightarrow R{-}N{=}C{=}O}$$

Tetrabutylammonium azide can be made in pure crystalline form by direct extraction from an aqueous mixture of tetrabutylammonium hydrogen sulfate, sodium azide, and excess sodium hydroxide [86]. This in turn can be transformed into acyl azides using acid chlorides in toluene solution at 25 °C. Higher temperatures, 50–90 °C, result in Curtius rearrangement yielding various isocyanates, 52–89% [86]. Using PTC, this reaction is performed in the presence of quaternary ammonium chloride by reacting aqueous sodium azide directly with acyl chlorides, followed by pyrolysis. This procedure proved effective even for relatively insoluble acid chlorides, and has been patented [87].

$$\mathrm{R{-}X + OCN^{\ominus} \longrightarrow R{-}N{=}C{=}O}$$

Direct displacement to give isocyanates is also possible. In 1958 the reaction of benzyl chloride with sodium cyanate either without any solvent or in DMF, catalyzed by tetraethylammonium iodide (170 °C, very fast) was patented [90]. The same catalyst was used for the preparation of silyl compounds in DMF (145 °C, 30 min ≈89% yield) [91]. There are some indications that the ammonium iodide first displaces the chloride in the starting material. This is supported by the observation that KI also catalyzes the reaction, although less efficiently [91].

$$\mathrm{R_3Si(CH_2)_nCl \longrightarrow R_3Si(CH_2)_n{-}N{=}C{=}O}$$

$$\mathrm{(Cl(CH_2)_n{-}SiRR')_2O \longrightarrow (O{=}C{=}N{-}(CH_2)_n{-}SiRR')_2O} \quad [91]$$

$$\mathrm{RBr + KOCN(H_2O) \xrightarrow[\text{Aliquat 336}]{} R{-}NH{-}CO{-}NH{-}R} \quad [38, 39]$$

If the substitution is carried out in the presence of water, urea derivatives are of course isolated [38, 39]. Allyl chloride and bromide have also been reacted with solid potassium cyanate in N-methylpyrrolidone or *o*-dichlorobenzene in the presence of crown

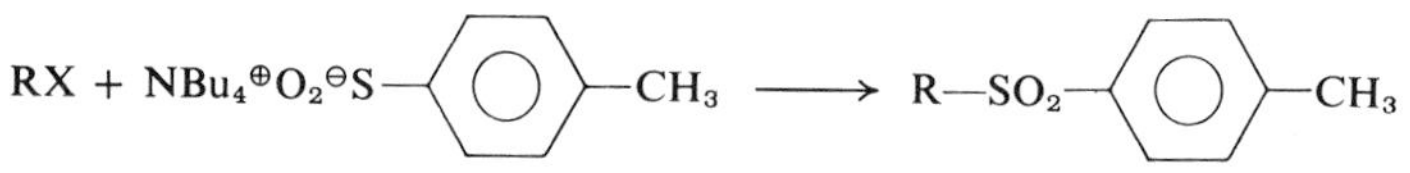

$$\mathrm{RX + NBu_4^{\oplus}O_2^{\ominus}S{-}C_6H_4{-}CH_3 \longrightarrow R{-}SO_2{-}C_6H_4{-}CH_3}$$

Scheme 3–27

ethers [60]. Here, as in the cases considered above, dimerization and trimerization of the isocyanates may interfere in the reaction.

Tetrabutylammonium *p*-toluenesulfinate was isolated from sodium *p*-toluenesulfinate and $NBu_4^{\oplus}Cl^{\ominus}$ by ion pair extraction with CH_2Cl_2. The reagent reacted smoothly in THF with primary alkyl, benzyl, and allyl halides at temperatures ranging from 10–40 °C to give the corresponding sulfones in acceptably high yields [92]. Other compounds converted into aryl sulfones were isopropyl bromide, chloroacetonitrile, ethyl bromoacetate, phenacyl chloride, 1,3-dichloroacetone (mono- or bissubstitution is possible depending on the reaction conditions), methylene bromide (mono product). 1,2-Dibromoethane underwent partial substitution, partial dehydrobromination [92] (*cf.*, Scheme 3–27).

Table 3–2 gives further patent references for displacement reactions.

3.2.5. Thiols and Sulfides

Sulfide and thiolate anions are known to be strong nucleophiles. In addition they can be partitioned easily from an aqueous to an organic phase with a quaternary onium counterion. Thus, they should be ideal substrates for PTC reactions. In fact, basic work on the mechanism PTC and the evaluation of different catalysts using the system thiophenol/alkylating agent was done by Herriott and Picker [28, 201].

$$PhCH_2Br + KHS \longrightarrow PhCH_2SH$$
$$RX + Na_2S \longrightarrow R\text{—}S\text{—}R$$

Benzyl thiol was made from benzyl bromide and potassium hydrogen sulfide in benzene at room temperature within 0.5 hours using polyethylene glycol dimethyl ether (**8**) as the complexant/catalyst [61]. Almost quantitative yields of symmetrical sulfides were obtained by heating primary alkyl bromides or chlorides with aqueous sodium sulfide in the presence of tributylhexadecylphosphonium bromide or tetrabutylammonium chloride with vigorous stirring at 70 °C for up to 40 minutes [202, 1010]. The benzyl chloride was consumed in less than 10 minutes. Secondary alkyl halides reacted more slowly, at 110 °C 2 hours were needed. Sterically hindered neopentyl bromide gave dineopentyl sulfide after 500 minutes at 70 °C without any rearrangement [202]. Dilauryldimethylammonium chloride can also be used as a catalyst although a longer reaction time (2 hours at 105 °C with *n*-octylchloride) is required [203]. With Aliquat 336, *p*-nitro-chlorobenzene yields 42% di-(*p*-nitrophenyl) sulfide under the same conditions [204]. The use of dicyclohexano-18-crown-6 with K_2S does not seem to present any experimental advantages [45]. PTC is also useful in the preparation of dialkyl polysulfides [203].

$$RS^{\ominus}Na^{\oplus} + R'X \longrightarrow R\text{—}S\text{—}R'$$

One of the earliest reactions reported was the alkylation of thiophenol and ethanethiol using substituted allyl chlorides in the presence of 10 N KOH and catalytic amounts of catalysts such as dibenzyldimethylammonium chloride [211]. Unsymmetrical sulfides can be prepared in the same way as the symmetrical analogues except for shorter reaction times or lower temperatures. Landini and Rolla [202] used tributylhexadecylphosphonium bromide in the absence of solvent. Under these conditions even neopentyl bromide and sodium thiophenolate give 78–85% of the unsymmetrical sulfide [856]. Herriott and Picker prefer benzene/Aliquat 336 [205]. With aryl thiolates and a twofold excess of the alkylating agent >90% yields can be obtained at room temperature within less than one hour, even with secondary halides [205]. The reaction with alkyl mercaptides requires some refluxing [202, 203]. As in other displacement reactions, polypod ligands have also been tested as catalysts. They did not prove sufficiently efficient for routine use [46]. In the reaction of sodium *p*-methylthiophenol with (R)-(–)-2-chlorooctane 98% inversion to (S)-(+)-2-octyl-*p*-tolyl sulfide was observed [202].

$$CH_2X_2 + 2RS^{\ominus}Na^{\oplus} \longrightarrow (RS)_2CH_2$$

In this reaction symmetrical formaldehyde dithioacetals are formed as byproducts if dichloromethane is the solvent. In the absence of other alkylating agents >90% yields are obtained within 15 minutes at room temperature (catalyst: Aliquat 336) [205]. The dithioacetals can be readily alkylated at the central carbon atom, leading to aldehydes after hydrolysis [205]. Using CH_2BrCl, compounds of the type Ar—S—CH_2Cl were prepared [994]. Substituted 1,3-benzoxathioles (**A**) were prepared from the aqueous disodium salt, methylene bromide, and Adogen 464, which is the same as Aliquat 336, by refluxing for 9.5 hours [206].

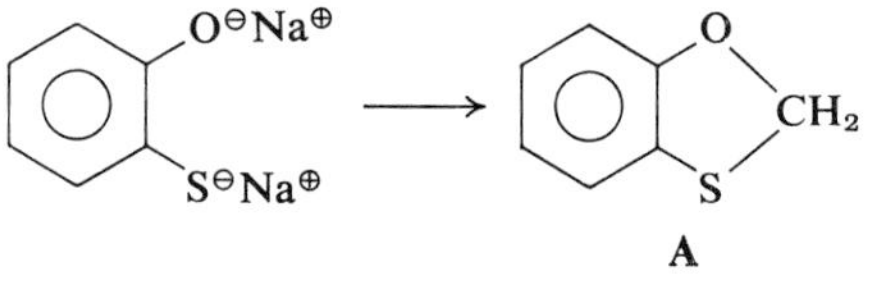

$$C_6H_4(O^{\ominus}Na^{\oplus})(S^{\ominus}Na^{\oplus}) \longrightarrow C_6H_4(O)(S)CH_2 \quad (\mathbf{A})$$

Scheme 3–28

Reduction of compound **B** by tetrabutylammonium tetrahydridoboranate in CH_2Cl_2 at $-30\ ^{\circ}C$ rather surprisingly yields **D** [207]. Its formation may be explained by the alkylation of the primary reduction product **C** by an excess of **B**:

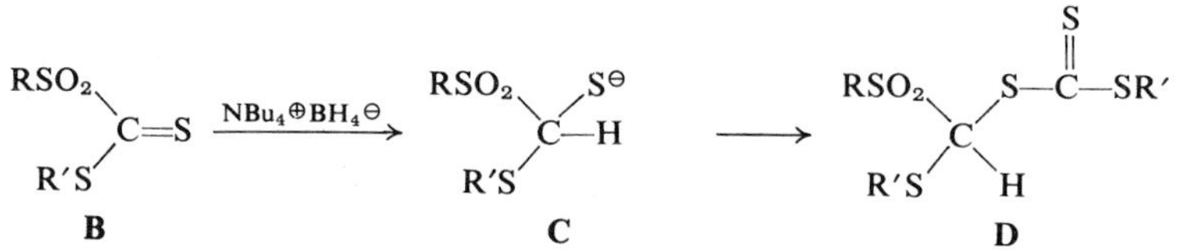

$$\underset{\mathbf{B}}{(RSO_2)(R'S)C{=}S} \xrightarrow{NBu_4^{\oplus}BH_4^{\ominus}} \underset{\mathbf{C}}{(RSO_2)(R'S)C(H)S^{\ominus}} \longrightarrow \underset{\mathbf{D}}{(RSO_2)(R'S)C(H)S{-}C({=}S){-}SR'}$$

Scheme 3–29

The alkylation of thiocompounds using the "extractive alkylation" procedure will be reviewed in the section on the alkylation of ambident anions. Deprotonated compounds **E** and **G**, however, behave chemically as true thiolate anions. Their alkylation, therefore, is discussed here. Reacting **E** with propargyl bromide in methylene chloride in the presence of 2 moles of aqueous sodium hydroxide and 1 mole of tetrabutylammonium hydrogen sulfate for 18 hours at room temperature yielded the allenyl sulfide **F** [208].

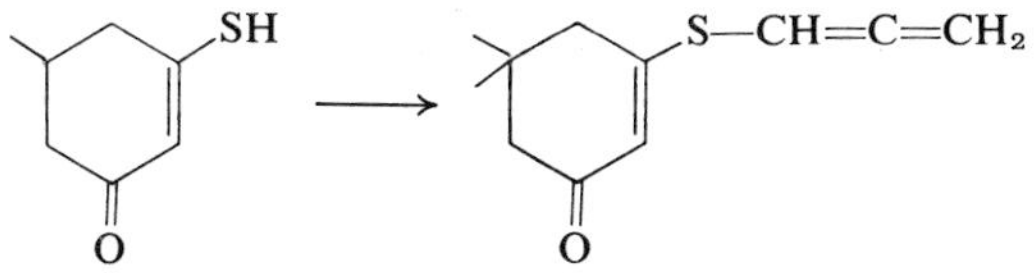

SH → S—CH=C=CH₂

E **F**

R(HS)C=C(H)COOEt ⟶ R(R′S)C=C(H)COOEt + R(R′S)C=C(COOEt)H

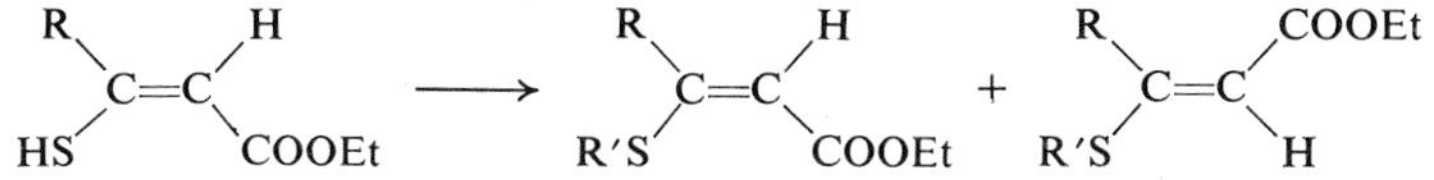

G (*Z*)-**H** (*E*)-**H**

Scheme 3–30

Compound **G** (R = CH_3, C_6H_5) was alkylated in either CH_2Cl_2 or $CHCl_3$ at room temperature for 1.5 to 5 hours with allyl, crotyl, and propargyl bromides [209]. Tetrabutylammonium hydrogen sulfate was used as the extractant. For R = CH_3, both the *Z* and *E* isomers of **H** were formed while with R = C_6H_5 only the *Z* isomer was observed. Starting from 2-mercaptocinnamic acids (**G**-acids, R = C_6H_5) and molar amounts of alkylating agent, first S-, then ester alkylation can be performed successively [907].

Dicyclohexano-18-crown-6 catalyzes the formation of sulfides from thiols and chloro-2,4-dinitrobenzene in acetonitrile, even if no base is present and the thiol is not in the thiolate form [210]. This reaction, however, which involves the formation of hydrochloric acid, is much slower than that of a thiolate.

An unusual way of preparing certain sulfides, and mechanistically quite different from the thiolate reactions, consists of reacting aliphatic or aromatic thiocyanates with C—H-acidic compounds in the presence of concentrated NaOH/TEBA [300]:

$$\text{R—S—CN} + \text{HCCl}_3 \xrightarrow[\text{TEBA}]{\text{NaOH}} \text{R—S—CCl}_3 + \text{NaCN}$$

$$\text{R—S—CN} + \text{Ph—CH(R')—CN} \xrightarrow[\text{TEBA}]{\text{NaOH}} \text{Ph—C(SR)(R')—CN} + \text{NaCN}$$

Scheme 3–31A

Phenyl ethinyl sulfides are formed from phenylacetylene and alkyl thiocyanates using concentrated NaOH/TEBA [769]:

$$\text{Ph—C}\equiv\text{CH} + \text{RSCN} \xrightarrow[\text{TEBA}]{\text{NaOH}} \text{Ph—C}\equiv\text{C—SR} + \text{CN}^{\ominus}$$

Thioalkylations of arylacetonitriles are possible with concentrated NaOH/TEBA/ trace DMSO, and polysulfides are formed as inexplicable by-products [1047]:

$$PhCH(R^1) - CN + S + R^2X \rightarrow PhCR^1(SR^2) - CN$$

$$R^2X + S \rightarrow R^2 - (S)_n - R^2.$$

Again a neutral type of PTC reaction is the conversion of 2-chlorocyclohexanone with aqueous sodium sulfide in the presence of NBu_4Cl yielding—depending on the conditions—either of the products shown [995]:

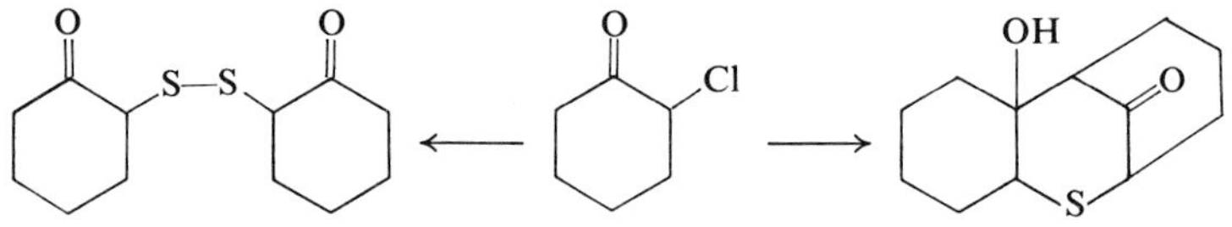

Scheme 3–31B

3.3. Reactions in the Presence of Additional Bases

In this section we shall first discuss alkylations in the presence of bases (3.3.1 to 3.3.5) and then other reactions with bases.

3.3.1. General Remarks

Although the ultimate step of most base catalyzed displacement reactions is mechanistically identical to that in the substitution reactions, weakly acidic substances require deprotonation first. Thus, at least one more equilibrium is involved.

The most frequently used inorganic bases are concentrated aqueous or solid alkali metal hydroxides. Here mixing may be critical since solutions tend to be viscous and form emulsions. Yields and reaction rates may not be reproducible if stirring is not adequate. Sometimes mechanical stirring or vibromixing is better than magnetic stirring.

In general, aqueous sodium hydroxide/ammonium salt catalysts are able to deprotonate substrates with a pK_a value down to 22–25, depending on the reference for pK_a values used [212, 213]. Fluorene can be alkylated under these conditions, acetonitrile cannot [214]. However, acetonitrile undergoes aldol reactions with powdered potassium hydroxide/18-crown-6 [215]. The potential of solid powdered KOH/catalyst (possibly in the presence of an additional inert drying agent) has not yet been fully explored. Recently, solid alkali metal carbonates were shown to be powerful bases under PTC [128, 1048]. Dietrich and Lehn [359] showed that the combination Kryptofix[222] (**5**)/$NaNH_2$/THF is capable of deprotonating diphenyl- and triphenylmethane. Combinations such as solid potassium *tert*-butoxide/crown ether/suitable solvent, seem promising. We will now consider some of the better studied homogeneous systems.

It is known that the rate of deprotonation by potassium *tert*-butoxide is both solvent and concentration dependent. The complexity of the system can be explained not only by the complex series of equilibria which exists between monomers and aggregates of the potassium *tert*-butoxide ion pairs, but also by the solvent complexes [216, 217]. Only the monomeric form of the base is catalytically active in isomerizations, racemizations, and H/D-exchange. The relative amount of monomer, of course will be largest at low base concentrations. Benzyltrimethylammonium *tert*-butoxide in *tert*-butanol (10^{-3} M) was shown to be at least 1000 times more basic than K-*tert*-butoxide at the same concentration [218]. Addition of polar solvents such as DMSO to a solution of $KOC(CH_3)_3$ in *tert*-butanol enhances the basicity in a number of ways: it increases the dielectric constant, specifically solvates the potassium cations (thereby shifting the monomer/aggregate-equilibrium towards the monomers), and establishes a new equilibrium with its conjugate base [218]. Potassium *tert*-butoxide in pure DMSO, therefore, must be a much stronger base than in *tert*-butanol: H/D-exchange rates are estimated to be larger by a factor of 10^5 [216]. Even in this solvent, however, the basicity is improved through the addition of crown ethers. At concentrations between 0.01 M and 0.33 M, a linear dependence of the isomerization rate from $KOC(CH_3)_3$/crown-complex concentration was observed, thus showing that the association of ion pairs can in fact be effectively suppressed by the crown ether [219]. It was shown quite recently that crowns enhance the nucleophilicity of potassium *tert*-butoxide more than its basicity in tetrahydrofurane or benzene: benzyl chloride gave 77–83% benzyl *tert*-butoxide ether and only small amounts of stilbene. No carbene adducts were formed. Diphenylmethane could be partly deprotonated [931].

Conversion of ion pair aggregates or of contact ion pairs into solvent separated ion pairs by crown ethers always leads to more reactive anions. This effect has been utilized in a great number of instances, some of those applying to PTC are described in other parts of this book. Concerning base catalyzed reactions, however, it should be mentioned in passing, that in homogeneous media not only rates [434, 918] but even the *syn*/*anti* ratio and regioselectivity of HX eliminations is influenced by the presence of crown ethers [220, 221, 222].

18-Crown-6 is even capable of partially solubilizing potassium hydride in THF. This combination was shown to metalate triphenylmethane (pK_a 31.5), diphenylmethane (pK_a 33.1), di-*p*-tolylmethane (pK_a 35.1), and even to some extent di-2,4-xylylmethane (pK_a 36.3) [430]. Reaction rates were rather slow, but improvement should be possible by varying experimental conditions.

Benzyltrimethylammonium *tert*-butoxide in *tert*-butanol is prepared by simple mixing of potassium *tert*-butoxide with the ammonium halide in *tert*-butanol and centrifuging off the precipitated KX [218]. Application of this reagent and other onium salts to other solvent systems and substrates did not prove valuable; extensive decomposition occurred [1060]. Better results were experienced in the case of crown ethers combined with various strong solid bases. The use of expensive crown ethers for large-scale reactions would be justified if savings on expensive anhydrous solvents or organometallic compounds could be made.

In concluding the general remarks on base catalyzed reactions, one more point should be mentioned. Quite a number of reactions are known to be catalyzed by "Triton B" (benzyltrimethylammonium hydroxide) in various homogeneous solutions. These include: Michael additions, aldol condensations, and autoxidations of carbanions [223]. Presumably, many of them could be performed in a PT catalyzed two-phase reaction using the same or another ammonium salt. Only relatively few applications, however, are so far known (*vide infra*).

3.3.2. Preparation of Ethers

The first ether preparations using PTC were accidental discoveries. In 1951 Jarrousse found that secondary alcohols and β-chloroethyldimethylamine in the presence of aqueous sodium hydroxide gave ethers:

$$ROH + ClCH_2{-}CH_2{-}N(CH_3)_2 \xrightarrow[\text{aq. NaOH}]{\text{PTC}} R{-}O{-}CH_2{-}CH_2{-}N(CH_3)_2$$

Scheme 3–32

He suggested that trace amounts of N,N,N′,N′-tetramethylpiperazinium dichloride formed during the reaction acted as a catalyst [224]. Furthermore, he showed that a mixed ether is formed from benzyl chloride and cyclohexanol with aqueous sodium hydroxide catalyzed by TEBA. Jarrousse's work, however, seems to have attracted little attention. Similarly, early reports on PTC alkylations of phenol and benzyl alcohol by substituted allyl chlorides in the presence of KOH/quaternary ammonium chlorides, remained lost in the literature [211, 225, 226]. At about the same time, some processes were described in the patent literature that could be classed broadly as PTC reactions, *e.g.*, preparation of epoxy resins from biphenols [186, 228], or cyanuric acid [186] and epichlorohydrin in the presence of inorganic base and an ammonium salt.

$$2RX \xrightarrow[\text{aq. NaOH}]{\text{PTC}} R{-}O{-}R$$

It was only in 1972 that Herriott and Picker [28] reported that the reaction of 1-bromooctane with 2N NaOH in the presence of a quaternary ammonium salt gave ether together with 20% 1-octanol and 5% 1-octene. Addition of the alcohol to the reaction medium raised both the yield and the rate of ether formation, thus indicating that hydrolysis of the alkyl halide is the first step. One may ask then why ethers are formed in preference to the normal hydrolysis products, the alcohols. In the presence of concentrated sodium hydroxide, there is an equilibrium-controlled deprotonation of the alcohols. Irrespective of whether these are hydrophilic and found mainly in the aqueous phase or are in the organic phase from the beginning, the alkoxide anions are more lipophilic than the hydroxyl ions and are therefore extracted preferentially. Since they are also the stronger nucleophiles, the result is ether formation.

Further examples for the preparation of symmetrical ethers from halides or sulphates using Aliquat 336 as the preferred catalyst are found in patents [229]. Using $(C_{12}H_{25})_3N^{\oplus}CH_2—CO_2CH_3\ Br^{\ominus}$ as the catalyst, 1,4-dichlorobutane is transformed into tetrahydrofurane [230].

$$RX + R'OH \xrightarrow[\text{aq. NaOH}]{\text{PTC}} R—O—R'$$

If 2 moles of primary alcohol per mole of primary bromide were used yields of mixed simple ethers were reported to be about 60% [28]. The symmetrical ether, however, was still to some extent formed. In some cases this can give rise to separation difficulties. Simple secondary alkyl halides could not be transformed into ethers by this method because the competing elimination reaction (giving olefins) and hydrolysis disturbed.

Optimum PTC conditions for the formation of unsymmetrical ethers are reported to consist of a fivefold excess of 50% aqueous sodium hydroxide to alcohol, an excess of alkyl chloride (preferably used as solvent), and 3–5 mole-% of tetrabutylammonium hydrogen sulfate, with stirring between 25–70 °C [232, 1051]. Primary alcohols gave yields of 80–95% after 3–4 hours but secondary alcohols required longer reaction times or larger catalyst concentrations. In a standardized reaction, a pronounced effect of the original catalyst counterion was observed: hydrogen sulfate (sulfate is formed in the medium) is better than iodide which is better than perchlorate [232]. Less than 10% of the symmetrical ether was formed and then largely only in the last 20% of the reaction when the relative concentrations become unfavorable. Such difficulties cannot occur in the preparation of methyl esters with dimethyl sulfate. Almost quantitative yields were observed in the presence of tetrabutylammonium iodide [233], but a more hydrophilic catalyst anion could prove advantageous. It was found that the reaction can also be run in the alcohol itself or in various solvents such as petroleum ether, benzene, ether, or dichloromethane. Smooth reaction occurs with saturated primary alcohols as well as with "activated" primary, secondary, and tertiary alcohols which contain unsaturated or electron withdrawing groupings. Saturated secondary alcohols are consumed more sluggishly, and tertiary alcohols do not react at all. Sterical hindrance does not appear to present problems. Water soluble compounds, however (*e.g.*, carbohydrates), cannot be reacted because their ion pairs are not present in the organic phase. 1,4-Dibromobutane can be transformed into 1-benzoxy-4-bromobutane by this process in 88% yield [874]. The adducts formed by the addition of chloroform across the double bonds of carbonyl compounds (**A**) can be O-methylated under these conditions, although many alternative reaction pathways exist, under various reaction conditions (*cf.*, Section 3.3.13.2). Merz and

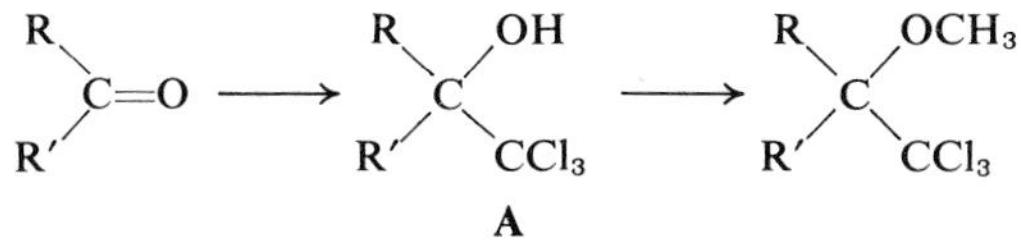

$$RR'C{=}O \longrightarrow RR'C(OH)CCl_3\ (\mathbf{A}) \longrightarrow RR'C(OCH_3)CCl_3$$

Scheme 3–33

Tomahogh [235], working with 1 mole carbonyl substrate, 1.2 moles dimethyl sulfate, 1 mole-% TEBA in 3 moles chloroform and 2.4 moles 50% aqueous sodium hydroxide at 30–35 °C did not isolate the intermediate compounds **A**.

Other authors [236] use methyl iodide with tetrabutylammonium iodide as catalyst in benzene/12 M NaOH. Yields are often as high as 80–95%, sometimes much lower.

A more difficult etherification is that between the endiolate of benzoin and the bis-tosylate of diethylene glycol [428]:

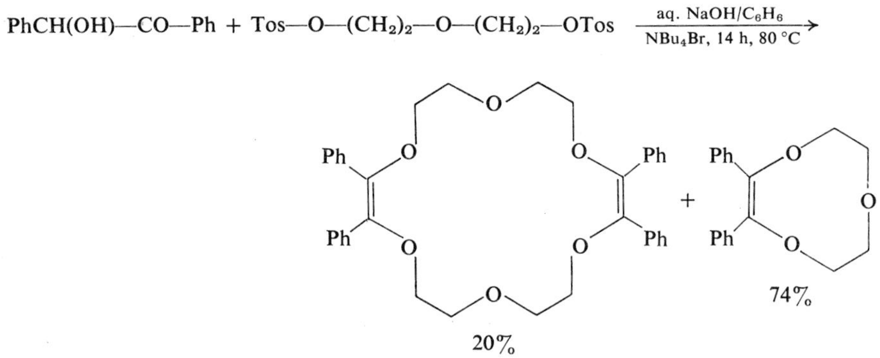

$$PhCH(OH)\text{—}CO\text{—}Ph + Tos\text{—}O\text{—}(CH_2)_2\text{—}O\text{—}(CH_2)_2\text{—}OTos \xrightarrow[NBu_4Br,\ 14\ h,\ 80\ °C]{aq.\ NaOH/C_6H_6}$$

20% + 74%

Scheme 3–34

1,2-O-isopropylidene-glycerol was benzylated after stirring several hours at 100 °C with excess benzyl chloride/50% aqueous NaOH and catalytic amounts of benzyl-tri-*n*-butylammonium bromide [399].

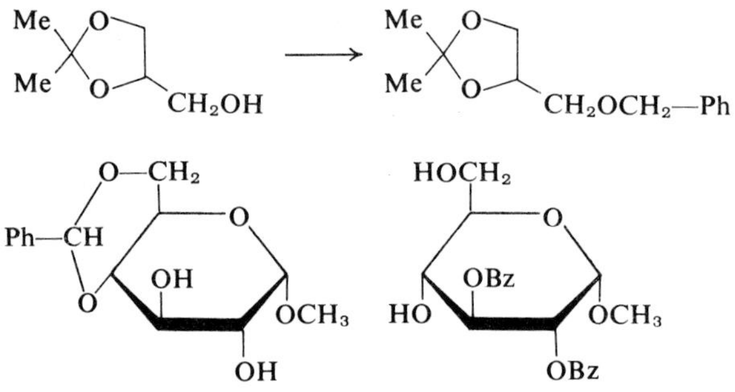

Me, Me, O, O, CH_2OH ⟶ Me, Me, O, O, CH_2OCH_2—Ph

O—CH_2, Ph—CH, O, OH, OCH_3, OH — **B**

$HOCH_2$, O, OBz, HO, OCH_3, OBz — **C**

Scheme 3–35

Monobenzylation of monosaccharide-diols was similarly achieved by boiling compounds such as methyl 4,6-O-benzylidene-α-D-glucopyranoside (**B**) or methyl 2,3-di-O-benzyl-α-D-glucopyranoside (**C**) in methylene chloride with 1.7 moles benzyl bromide, 20 mole-% $NBu_4^{\oplus}\ HSO_4^{\ominus}$, and a 5% solution of sodium hydroxide for 48 hours [239]. Total yields were ≈80%, and the products formed from **B** were:

6% 2,3-dibenzyl ether, 54% 2-benzyl ether, and 20% 3-benzyl ether. With **C**, 18% methyl 2,3,4-tribenzyl-α-D-glucopyranoside and 68% 2,3,6-tribenzyl ether were obtained. The PTC preparation of 1,2,5,6-tetraoctadecyl-D-mannitol *via* a lipophilic 3,4-derivative has been described [1005]. Other O-alkylations of isopropylidene and benzylidene monosaccharide derivatives were carried out at room temperature with methyl bromide, benzyl chloride, or 1,2-dichloroethane in benzene/50% aqueous sodium hydroxide [240, 1049]. Methyl iodide was a slower alkylation agent because the iodide set free during the reaction poisoned the catalyst. Of the two catalysts (TEBA and NBu_4Br) the latter was much more efficient. Using dichloromethane both as the alkylating agent and as the solvent, formaldehyde acetals were obtained (*cf.*, below) [240].

The use of NaOH/Aliquat 336 also permits the preparation of α-(allyloxy)-acetonitriles (**E**) from aldehyde cyanohydrins (**D**) and allyl bromides in dichloromethane [237]. In the strongly basic medium part of the cyanohydrin anions decompose giving aldehyde and cyanide. These ions compete for the allyl halide, so that compounds such as **F** are formed as byproducts. Compound **H** has been made similarly [238]. Although the yields are not too good (27–46%) compounds such as **E** are useful intermediates that can be transformed into β,γ-unsaturated ketones such as **G** with lithium diisopropylamide in THF at −78 °C. [2,3]-Sigmatropic rearrangement followed by the elimination of LiCN gives **G**.

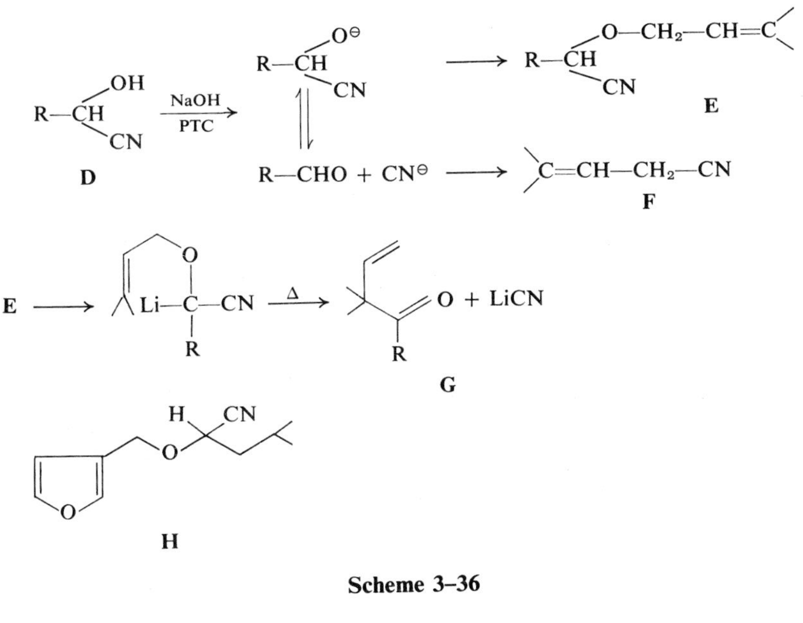

Scheme 3–36

Chloromethylated polystyrene can be transformed into an ether derivative by PTC also [1050].

Using PTC, phenolate anions can be O-alkylated even more easily than aliphatic alcohols. Ugelstad and co-workers carried out kinetic measurements in homogeneous dioxane solution. Tetrabutylammonium phenolate ion pairs reacted with butyl bromide $3 \cdot 10^4$ times faster than the normal potassium salt, and a further 2–3 times faster, if potassium was complexed with dicyclohexano-18-crown-6 [249].

In true PTC reactions, yields (mostly in the range of 80–95%) compare favorably with more conventional methods, especially in critical cases where C-alkylation intervenes [29, 278]. C-Alkylation was always less than 5% [29, 380]. The following alkylating agents were tested: primary and secondary alkyl halides, allyl, and benzyl halides, and epichlorohydrin and dialkyl sulfates [29, 966]. Benzyl-tri-*n*-butylammonium salts were used as catalysts, and the reactions were done in a methylene chloride/aqueous sodium hydroxide mixture, with very efficient stirring (vibromixer) at ambident temperature. Steric hindrance did not interfere with methylation. Thus, even 2,4,6-tri-*tert*-butylphenol with dimethyl sulfate gave a yield of 93%. No reaction was observed in the absence of the catalyst. With ethyl or isopropyl iodide, however, virtually no transposition of the highly hindered phenols occurred.

Here, as in many other PTC reactions, although catalytic amounts of the quaternary ammonium salt were sufficient with many alkylating agents, this was not the case with iodides. Iodide is extracted in preference to phenolate into the organic phase, thus poisoning the catalyst. Benzyl chloride, on the other hand, is so reactive that it is hydrolyzed before all the phenolate has been transferred to the organic phase by the limited amount of catalyst. Both with iodides and benzyl halides, therefore, stoichiometric amounts of phenol/catalyst salt are recommended.*)

When an excess of a reactive alkylating agent is used, O-alkylation of 2′-hydroxychalcones occurs in excellent yields [875]. Otherwise ring closure to flavanones partially intervenes.

In the absence of an alkylating agent, the solvent dichloromethane competes for the phenolate anion and diaryloxymethanes are formed as side products [29]. This reaction can be modified slightly to give the latter in very high yield [234]. The phenol is stirred in methylene chloride with solid powdered potassium hydroxide in the presence of 5–10 mole-% Aliquat 336 for 8–16 hours at room temperature. The reaction rate is markedly slower if aqueous alkali metal hydroxide or the less lipophilic catalyst TEBA is used. With methylene bromide, aqueous sodium hydroxide, and Adogen 464 catalyst, methylene ethers of catechols can be prepared [244, 245]. Methylene chloride can also be used [873].

$$ArOH + CH_2Cl_2 \longrightarrow ArO{-}CH_2{-}OAr$$
$$R{-}OH + CH_2Cl_2 \longrightarrow RO{-}CH_2{-}O{-}R$$

Since alkoxylates are less nucleophilic than phenolates, the preparation of formaldehyde acetals by the solid KOH process requires about 15 hours for 65–80% yield [234]. Methylenation of 1,2-diols including carbohydrates is also possible [909, 954].

*) A different explanation [29] has been forwarded in the case of benzyl chloride, which was considered as very unreactive under the reaction conditions.

The more critical conversion shown below was accomplished at room temperature by stirring overnight [429]:

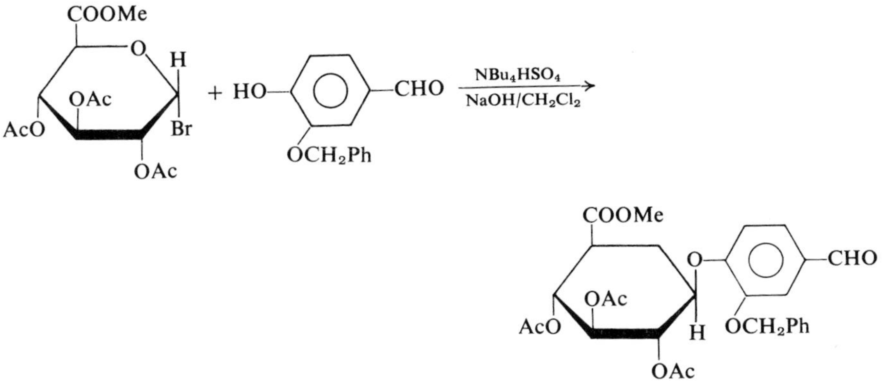

COOMe, O, H, OAc, AcO, Br, OAc + HO—C₆H₃(OCH₂Ph)—CHO $\xrightarrow[NaOH/CH_2Cl_2]{NBu_4HSO_4}$ COOMe, O—C₆H₃(OCH₂Ph)—CHO, OAc, AcO, H, OAc

Scheme 3–37

Here, as in other cases, tetrabutylammonium hydrogen sulfate was used as the catalyst for the preparation of phenol ethers [243]. TEBA was used with the heterocyclic analogue **I** [247]. *p*-Nitro substituted phenolethers **K** were obtained in nucleo-

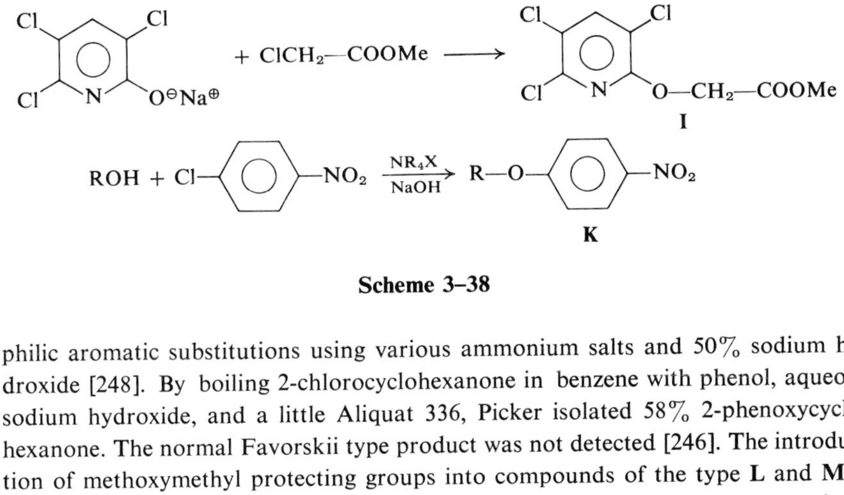

Cl, Cl, Cl, N, $O^{\ominus}Na^{\oplus}$ + $ClCH_2$—COOMe ⟶ Cl, Cl, Cl, N, O—CH_2—COOMe

I

ROH + Cl—C₆H₄—NO_2 $\xrightarrow[NaOH]{NR_4X}$ R—O—C₆H₄—NO_2

K

Scheme 3–38

philic aromatic substitutions using various ammonium salts and 50% sodium hydroxide [248]. By boiling 2-chlorocyclohexanone in benzene with phenol, aqueous sodium hydroxide, and a little Aliquat 336, Picker isolated 58% 2-phenoxycyclohexanone. The normal Favorskii type product was not detected [246]. The introduction of methoxymethyl protecting groups into compounds of the type **L** and **M** is possible when the potassium salt of the corresponding phenol is stirred in acetonitrile with chlorodimethyl ether and 10 mole-% 18-crown-6 at room temperature [241].

In an improved procedure, the phenol in CH_2Cl_2, a small amount of Adogen 464, and sodium hydroxide in water are stirred, while a three- to fivefold excess of the chloroether is dropped in slowly (79–91% yield) [950]. Recent publications report on PTC alkylations of *o*-acylphenols [1056] and reactions between phenolates and epichlorohydrin [1057] and cyanuric chloride [1058].

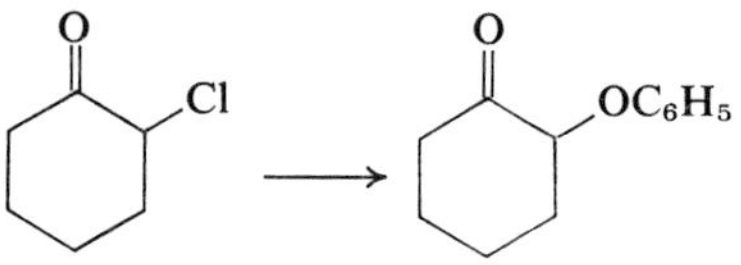

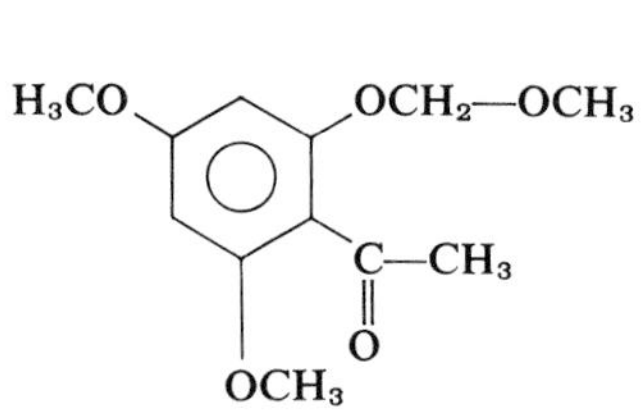

Scheme 3–39

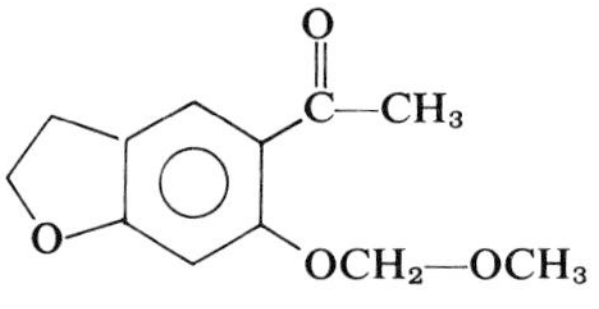

What has been called "triphase catalysis" has been used for the synthesis of phenol and, with less success, aliphatic ethers [62, 316]. A solid polystyrene resin cross-linked by 2% divinylbenzene and carrying 12% ring substitution with $—CH_2N(Me)_2(n\text{-}C_4H_9)^{\oplus}\ Cl^{\ominus}$ groups was used. Ninety-seven percent *n*-butyl phenyl ether was isolated from phenolate in water, and alkyl bromide in toluene, after 10 hours at 90 °C [62]. Using what has been called the "solid-phase cosolvent," where the resin carries no ionic sidegroups, the catalyst was prepared by reacting chloromethylated polystyrene (11% ring substitution, 2% cross-linked with divinyl benzene) with polyethylene glycol monomethyl ether and sodium hydride. When sodium phenoxide in water was heated with 1-bromobutane in toluene, in the presence of this polymer for 12 hours at 110 °C, 60% of the ether was formed. In the absence of catalyst less than 5% was found [313]. Although in this procedure separation from the solid catalyst is easy, it remains to be seen whether or not this is a preparatively competitive alternative. As has been mentioned in other sections, normal anion exchange resins are not useful as catalysts in "triphase catalysis."

Quantitative gas-chromatographic determinations of phenol containing analgesic pentazocin (**N**) and compound **O** by extractive alkylation have been preformed by analytical chemists using **Brändström's** technique. The first manipulations involved concentration by benzene extraction from human plasma at pH 10.5, reextraction into aqueous dilute acid, and addition of aqueous sodium hydroxide. Then derivatization was accomplished by stirring with pentafluorobenzyl bromide and tetrabutyl-

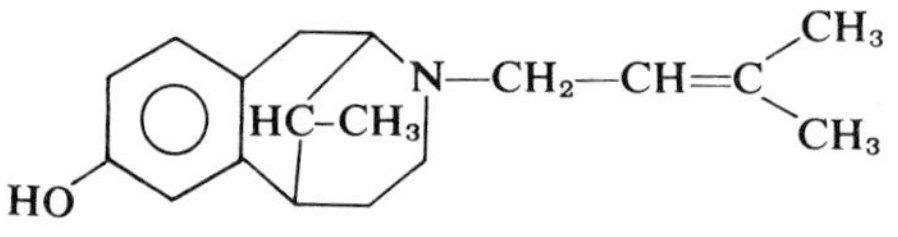

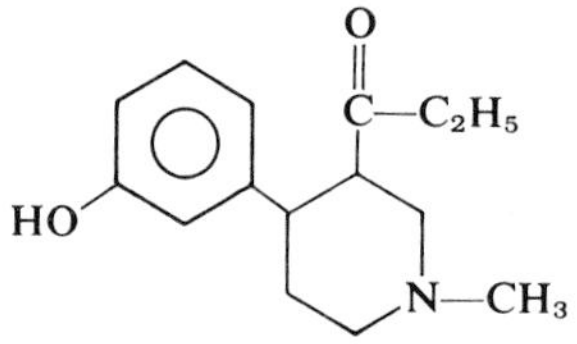

Scheme 3–40

ammonium hydrogen sulfate in dichloromethane at room temperature [250]. In another analytical application, estrogen-type compounds were extracted as ion pairs with tetrahexyl ammonium ions into methylene chloride and alkylated by methyl iodide prior to gas chromatography. The very lipophilic cation ensures rapid extraction and methylation at room temperature [242]. Other analytical applications [1054] and a review on analytical extractive alkylations [1052] have appeared. Pentafluorobenzyl bromide reacts in two-phase alkylations with phenolates, but not with carboxylates, *without PTC catalyst*, thus providing an easy differentiation [1055].

The formation of enol ethers of ketones, O- *vs.* C-alkylation of β-diketones, β-ketoesters, and heterocycles will be discussed in Section 3.3.5. The preparation of epoxides by the Darzens reaction is described in Section 3.3.8, and sulfoniumylide reactions to form epoxides can be found in Section 3.3.11.

Typical experimental procedures are presented below.

Preparation of methyl ethers [*cf.*, 233]. A solution of the alcohol (0.5 mole) and 1 g tetrabutylammonium halide in 200 ml petroleum ether (b.p. 50–70 °C) is equilibrated with 1.3 moles of 50% aqueous sodium hydroxide by vigorous stirring for 15–30 minutes. There is a slight evolution of heat. 0.6 to 0.75 moles of dimethyl sulfate are added dropwise while cooling over a period of 1 hour at such a rate that the temperature does not exceed 45 °C. The reaction mixture is stirred for 2–3 hours and after the addition of 10 ml conc. aq. NH_3 for a further 30 minutes at room temperature. Finally, the mixture is poured into water and the organic phase separated, washed with water, dried over Na_2SO_4, and worked up in the usual way. Yields depend to some extent on efficient stirring. For semimicro batches with magnetic stirring a 1.5-fold excess of $(CH_2)_3SO_4$ is recommended.

Preparation of phenol ethers [29]. A mixture of 50 ml CH_2Cl_2, 50 ml of water, 10 mmoles of the phenol, 15 mmoles of NaOH, 20–30 mmoles of alkylating agent and 0.1–1 mmoles of benzyltri-*n*-butylammonium bromide*) is stirred with a vibromixer at room temperature for 2–12 hours. The organic layer is separated, and the aqueous layer extracted twice with 20 ml portions of CH_2Cl_2. The combined organic extracts are evaporated, water added, and the mixture extracted with ether or pentane. The organic phase is washed twice with 2N NH_3 (to remove the dimethyl sulfate) or when necessary treated with methanolic NH_3 (to destroy the diethyl sulfate), then finally with a saturated NaCl solution. After drying over Na_2SO_4, the solvent is evaporated and the residual ether purified by distillation or crystallization.

3.3.3. N-Alkylations

Classically, N-alkylations are carried out in a two-phase system containing sodium carbonate or alkali hydroxide. The secondary or tertiary ammonium salt formed must diffuse to the phase boundary in order to be converted to the amine. The rate

*) When CH_3I or benzyl chloride is the alkylating agent, stoichiometric amounts of the catalyst are more satisfactory.

of reaction will be determined by the nucleophilicity of the amine, and it is obvious that a PT catalyst will not greatly influence the reaction for normal amines. Aqueous sodium hydroxide is not a strong enough base to deprotonate unactivated amines. If, however, the NH-groups are acidified by neighboring electron withdrawing groups, deprotonation becomes possible.

$$pK_a\ NH_3\ 34\text{–}36, \qquad ArCONH_2 \approx 25, \qquad PhSO_2NH_2 \approx 10$$

Thus, the catalyst can serve in the generation of the very nucleophilic amide anion in one of two ways: (1) to transport hydroxide ions for deprotonation into the organic layer, or (2) to detach deprotonated molecules from the phase boundary.

Most of the work in this field is in accordance with these predictions. In fact, there are only three reports in the literature where the alkylation of nonactivated NH-bonds is done in the presence of a quaternary ammonium compound catalyst. Aromatic amines were alkylated with 1,3-dichloro-2-butene and benzyl chloride/aqueous potassium or sodium hydroxide. Yields were 2–3 times better if bis[3-chloro-2-butenyl]dimethylammonium chloride was used as catalyst [225, 226, 251]. Similarly, cyclic bisalkylations of primary amines **A** (R = CH_3, $(CH_3)_2CH$, *tert*-butyl) with compound **B** was successful in the solvent mixture chloroform/ethanol/water with benzyltrimethylammonium methoxide present. Yields were of the order 65–90% [252]. Finally, the reaction of anthranilonitrile with l-chloro-triacetyl-*D*-ribopyranose (10% yield) was possibly accelerated by PTC [1059].

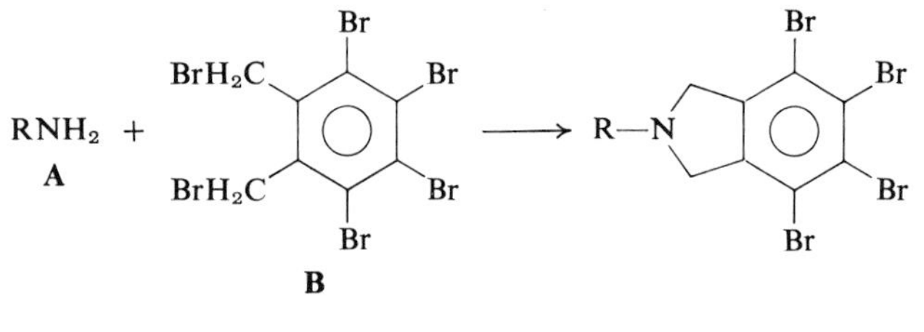

RNH_2 (**A**) + **B** (1,2-bis(bromomethyl)-3,4,5,6-tetrabromobenzene: BrH_2C, BrH_2C, Br, Br, Br, Br) ⟶ R—N (tetrabromoisoindoline: Br, Br, Br, Br)

Scheme 3–41

In hydrazobenzene, the NH-bonds are somewhat more strongly acidified, hence a small equilibrium concentration of the anion might be present. In the presence of potassium carbonate, in ethanol or DMF monoalkylation did not give any good results. Applying the extractive alkylation technique with methylene chloride as solvent in the presence of 0.5–1 mole tetrabutylammonium hydroxide and 0.5–1 mole aqueous sodium hydroxide, boiling 1–24 hours until neutrality, remarkably improved the yield. Primary iodides, benzyl bromide, and allyl bromide gave good yields, secondary iodides were less satisfactory [253] (Scheme 3–42). Similarly, phenylhydrazones **C** can be alkylated to give **D**, yields between 43 and 98%, by stirring with 50% aqueous sodium hydroxide, excess alkylating agent, and 5 mole-% tetrabutylammo-

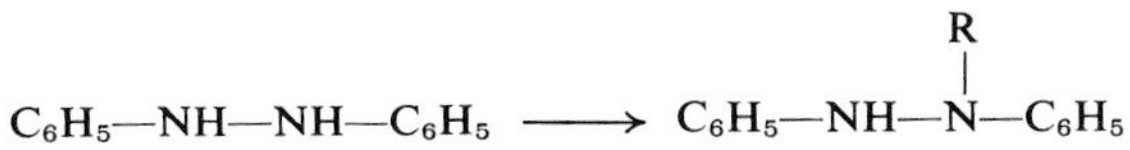

$$C_6H_5\text{—NH—NH—}C_6H_5 \longrightarrow C_6H_5\text{—NH—N(R)—}C_6H_5$$

R	Yield (%)
CH_3	79
C_2H_5	42
$i\text{-}C_3H_7$	13
$C_6H_5CH_2$	79
$CH_2\text{=CH—}CH_2$	79

Scheme 3–42

nium chloride for (0.5 h, 30–60 °C) [254]. Hydrolysis with dilute sulfuric acid gives 1-alkyl-1-phenylhydrazines **E**. Compounds such as **F** were alkylated in dichloromethane/15% aqueous sodium hydroxide with benzyltrimethylammonium catalyst (8 h, room temperature) [273]. Without alkylating agent, using conc. $NaOH/NBu_4Cl$, compounds such as **F** decomposed to give diazo compounds [908]. Diazoamino compounds (1,3-diaryltriazenes) were also PTC alkylated [1062].

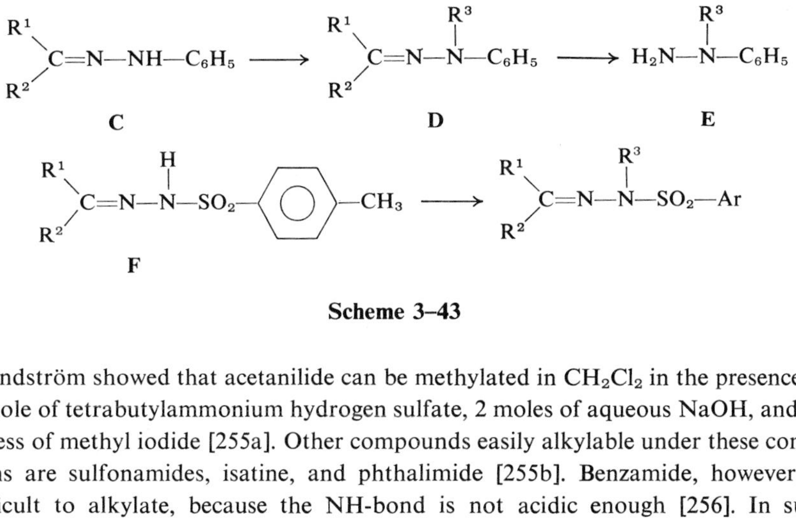

$$R^1R^2C\text{=N—NH—}C_6H_5 \longrightarrow R^1R^2C\text{=N—N(}R^3\text{)—}C_6H_5 \longrightarrow H_2N\text{—N(}R^3\text{)—}C_6H_5$$

C D E

$$R^1R^2C\text{=N—N(H)—}SO_2\text{—}C_6H_4\text{—}CH_3 \longrightarrow R^1R^2C\text{=N—N(}R^3\text{)—}SO_2\text{—Ar}$$

F

Scheme 3–43

Brändström showed that acetanilide can be methylated in CH_2Cl_2 in the presence of 1 mole of tetrabutylammonium hydrogen sulfate, 2 moles of aqueous NaOH, and an excess of methyl iodide [255a]. Other compounds easily alkylable under these conditions are sulfonamides, isatine, and phthalimide [255b]. Benzamide, however, is difficult to alkylate, because the NH-bond is not acidic enough [256]. In such cases the amide-NH can be deprotonated by alkali metal hydrides in inert solvents, and alkylation is performed in the presence of 18-crown-6 [941] or Aliquat 336 [942].

Alkylations in the presence of NaOH have also been performed with only 1–10 mole-% catalyst, reaction times and reaction temperatures varied with the alkylating agent, *e.g.*, for compounds of the type **G**: $(CH_3)_2SO_4$, 45 min, room temperature and $n\text{-}C_4H_9I$: 22 h, 80 °C. Yields were of the order 80 to 95%. In the absence of catalyst when R^1 = alkyl, the reactions were extremely slow. Formanilides, however, reacted in a two-phase system without TEBA [257]. The rate of competing amide hydrolysis was very low in all cases. Cyclic amides such as pyrrolidone, ϵ-caprolactame, and ω-caprylolactame, can also be alkylated at 40–50 °C in the presence of concentrated sodium hydroxide/benzene/TEBA [847]. Higher members require longer reaction times. Even β-lactames were alkylated [1063] or formed by intramolecular cyclization [1064] in high yields.

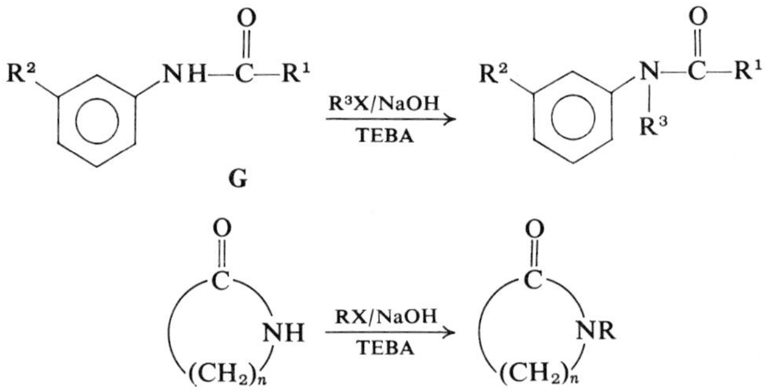

$R^2C_6H_4$–NH–CO–R^1 (G) $\xrightarrow[\text{TEBA}]{R^3X/NaOH}$ $R^2C_6H_4$–N(R^3)–CO–R^1

cyclic amide C(=O)–NH–$(CH_2)_n$ $\xrightarrow[\text{TEBA}]{RX/NaOH}$ C(=O)–NR–$(CH_2)_n$

Scheme 3–44

Solid potassium phthalimide was alkylated in toluene in the presence of 10 mole-% tributylhexadecylphosphonium bromide at 100 °C: Benzyl chloride (20 min); cyclohexyl or neopentyl bromide (40 h), yield 85–100%. Optically pure alkyl methanesulfonates afforded products with about 90% inversion. For some secondary substrates and neopentyl bromide, competing elimination reactions were observed. Reaction rates were much lower in the solvent amyl alcohol. This observation agrees with the assumption that unsolvated ion pairs are necessary for fast PTC reactions [258]. (For reactions in the presence of crown ethers, *cf.* [1108].) Optically active 2-phthalimido esters of low to medium optical purity were obtained in a solid/liquid PTC alkylation with 2-bromoalkanoates using (−)- or (+)-benzyl-cinchonidinium chloride as catalysts [940]. Alkylation of cyanamides with excess RX in the presence of 50% aqueous sodium hydroxide and an onium salt gave dialkylated products only. These could be cleaved to secondary amines easily [1065]. Activation of NH-bonds

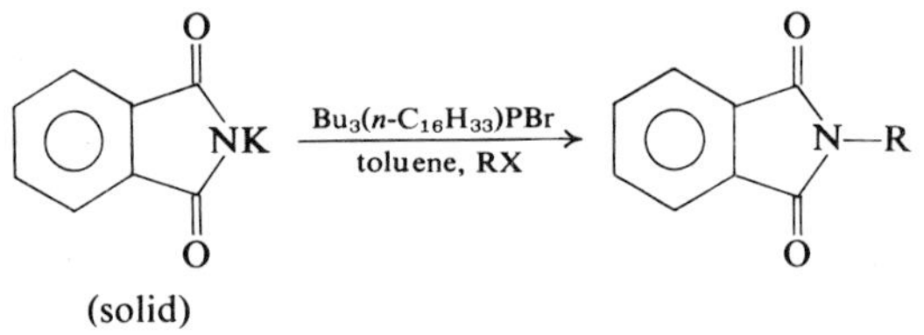

Phthalimide–NK (solid) $\xrightarrow[\text{toluene, RX}]{Bu_3(n\text{-}C_{16}H_{33})PBr}$ Phthalimide–N–R

Scheme 3–45

can also be achieved by phosphoric acid groupings. Zwierzak [272] transformed primary into secondary amines with diethyl phosphoramidates as intermediates.

These in turn were alkylated and subsequently cleaved. Steps (1) and (2) are PTC processes:

$$\text{Step (1)}\quad (R^1O)_2\overset{O}{\overset{\|}{P}}H + CX_4 + R^2NH_2 \xrightarrow[\text{TEBA}]{\text{NaOH}} (R^1O)_2\overset{O}{\overset{\|}{P}}—NH—R^2 + HCX_3 \quad [126]$$

$$\text{Step (2)}\quad (R^1O)_2\overset{O}{\overset{\|}{P}}—NH—R^2 \xrightarrow[NBu_4HSO_4]{\text{toluene, NaOH}} (R^1O)_2\overset{O}{\overset{\|}{P}}—\overset{R^3}{\overset{|}{N}}—R^2 \quad [272]$$

$$\text{Step (3)}\quad (R^1O)_2\overset{O}{\overset{\|}{P}}—\overset{R^3}{\overset{|}{N}}—R^2 \xrightarrow{\text{HCl}} H\overset{R^3}{\overset{|}{N}}—R^2$$

Scheme 3–46

Diethyl chlorophosphite is an intermediate in the first reaction (*cf.*, Section 3.2.3). In the first step the amine and dialkyl phosphite in methylene chloride are added to the two-phase system consisting of dichloromethane and carbon tetrachloride or -bromide, 20% aqueous sodium hydroxide, and TEBA at room temperature or lower. Yields are usually around 80–90%. The alkylation step (2) is carried out in toluene/ 50% NaOH with tetrabutylammonium hydrogen sulfate as the catalyst (4 h, reflux). Yields are mainly between 67 and 98%. Hydrolysis (3) finally is carried out by dissolving the amide in THF, saturating it with gaseous HCl and keeping the mixture at room temperature for 12 hours [272]. Alternatively, easily accessible diphenylphosphine amide, Ph_2PONH_2, can be mono- or bisalkylated in benzene/conc. NaOH using NBu_4HSO_4 as catalyst. On hydrolysis, primary or secondary amines are obtained [885]. Diphenylphosphinic hydrazides were PTC alkylated using solid $NaOH/Na_2CO_3$ as base combination [1061].

As a weak base and nucleophile, aziridine is difficult to alkylate under normal conditions because of easy rearrangement to open chain products such as **H**. When it and benzyl chloride or *n*-butyl chloride are stirred with TEBA in 30% aq. NaOH, for 6 hours at room temperature and 8 days at 45 °C, respectively, quantitative yields of **I** were obtained [259]. The N-inversion of aziridine is accelerated by traces of water in the medium. As no inversion is observed in the system aziridine/CH_2Cl_2/TEBA/ concentrated aqueous NaOH, the organic phase must be anhydrous [943].

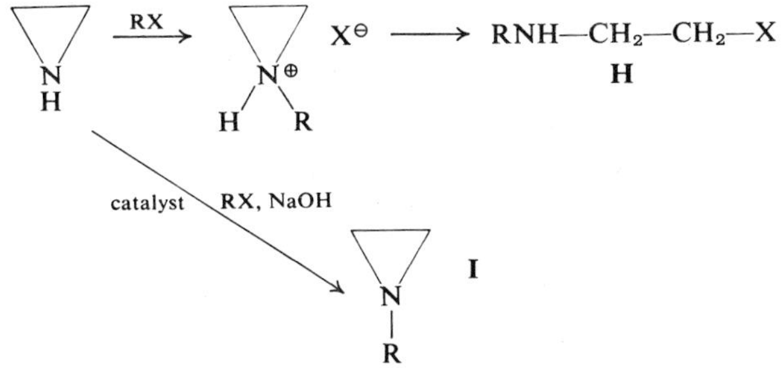

Aziridine (NH) $\xrightarrow{RX}$ aziridinium ($N^{\oplus}$, H, R) $X^{\ominus}$ $\longrightarrow$ $RNH—CH_2—CH_2—X$ **H**

Aziridine (NH) $\xrightarrow[\text{catalyst}]{RX,\ NaOH}$ N–R aziridine **I**

Scheme 3–47A

3,5-Dicyano-1,4-dihydropyridines **K** (R^1, R^2, R^3 = alkyl or aryl) can be alkylated with alkyl iodides or benzyl bromides in toluene or methylene chloride using benzyldodecyldimethylammonium bromide with yields of between 52 and 86% [260]. Without a catalyst such reactions are possibly only in an homogeneous KOH/DMSO mixture.

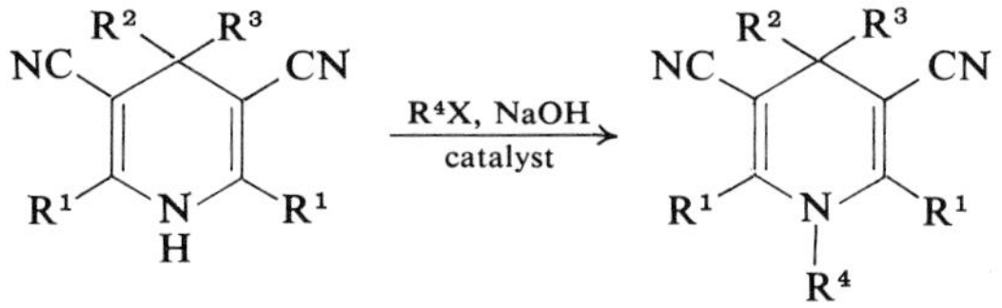

R^4X, NaOH, catalyst

K

Scheme 3–47B

Much has been published on the alkylation of indoles [261–264, 934, 935]. Working without solvent [261, 262] or in benzene [263], with concentrated sodium hydroxide as base, reaction times were as follows: at room temperature 6–22 hours, at 50 °C about 1 hour. Yields of **L** were 80–90%. The following catalysts were used: TEBA [261], Aliquat 336 [262], or NBu_4HSO_4 [263]. Whereas the more reactive halides and alkyl sulfates required a few mole-% catalyst, secondary bromides, methyl iodide, and *p*-nitrophenyl bromide needed molar amounts. With allyl and benzyl halide, the side products **M** and **N** became more dominant. On closer inspection it was seen that

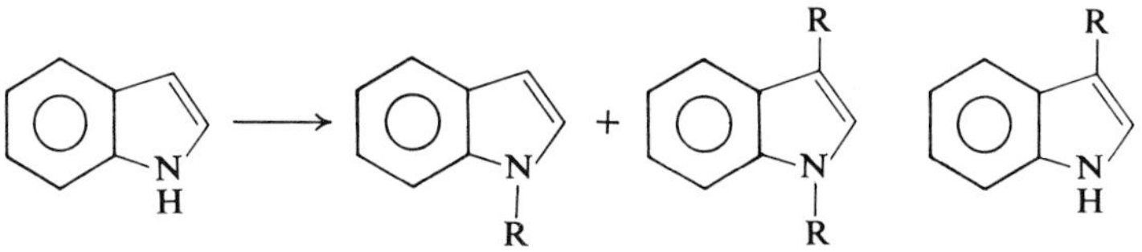

L **M** **N**

Scheme 3–48

by using molar amounts of catalyst even reactive reagents such as these gave a smaller proportion of **M** and **N** [262]. Alternatively, compounds such as **L** can be prepared in the absence of catalyst using two-phase solvent systems containing moderate amounts of DMSO or HMPT. Reactions occur more vigorously and less **M** and **N** are formed [261]. PTC N-sulfonations [936] and N-acylations [1107] were described also. KF/18-crown-6 was used to put Z-protecting groups onto tryptophan [921].

Similarly, carbazole, pyrrole, benzotriazole, and diphenylamine can be N-alkylated in the presence of TEBA [261, 937, 938]. N-Alkyl-2-chlorophenothiazine derivatives (**O**), which are important therapeutic agents, can be obtained in medium to low yields by stirring or refluxing in benzene or xylene in the presence of various ammonium salts and 6M NaOH [275, 279]. On the other hand, tetrabutylammonium bromide

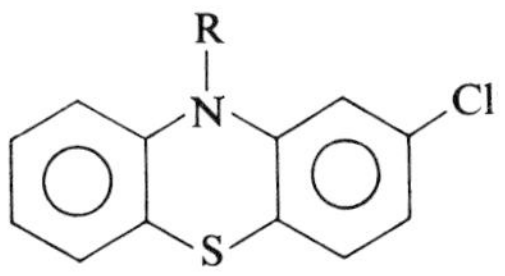

R N Cl S

O

Scheme 3–49

and concentrated sodium hydroxide was the preferred combination for alkylations of the unsubstituted imidazole (**P**) and pyrazole (**Q**) [274]. Maximum yields from **P** using 1-bromobutane and benzyl chloride in benzene were 71% and 76%, respectively (3 h, reflux). Bromocyclopentane was also used as the alkylating reagent. It is interesting to note that both **P** and **Q** are almost quantitatively present in the concentrated sodium hydroxide at the start of the reaction, but during the course of the reaction the catalyst extracts small amounts of deprotonated substrate for alkylation.

Sometimes intramolecular alkylation is so favorable that the reaction proceeds without added catalyst at the phase boundary. Thus, aziridine formation from **R** and **S** was observed by simply stirring the benzene solution with 50% aqueous sodium hydroxide [265]. This is preparatively interesting since attempted eliminations in homogeneous pyridine or potassium hydroxide/methanol solution lead to double bond formation or subsequent undesired cleavage of the aziridine ring.

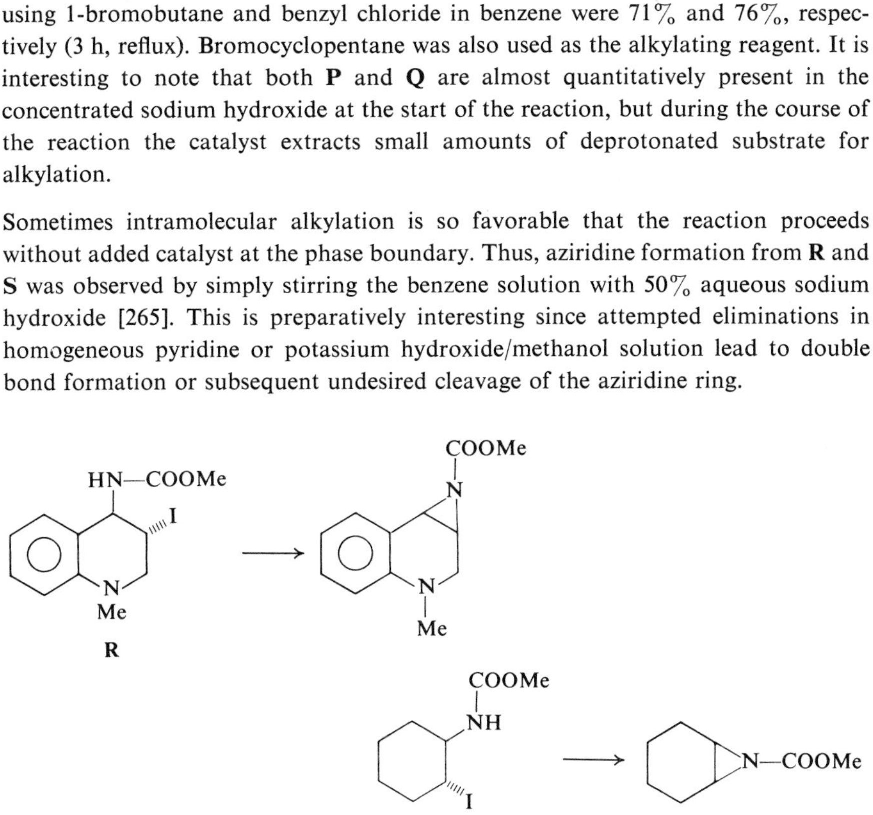

HN—COOMe I N Me → COOMe N N Me

R

COOMe NH I → N—COOMe

S

Scheme 3–50

Analytical chemists have made use of this extractive alkylation procedure to methylate minute quantities of compounds isolated from biological samples prior to gas chromatography or mass spectrometry studies: for example, nitrazepam (**T**) [266], chlorothalidone (**U**) [267], barbiturates (**V**) phenobarbitone and pentobarbitone [268], and substituted 5-aryl-5-phenylhydantoins (**W**) [269]. Mono- to poly-methyl derivatives were formed with methyl iodide as the alkylating agent and tetraalkylammonium hydroxides as the extracting agent/counterion. The influence of the lipophilic character of both the anion and cation on the reaction rate is especially important. Since the alkylation step in the organic layer is fast, overall reaction rates depend mainly on the equilibrium of the ion pair extraction process. With phenobarbitone (**V**, R = C_6H_5) in a phosphate buffer of pH 10/CS_2 at 25 °C using tetrabutylammonium there was only 10% reaction after 60 minutes, whereas, after the same time using tetrapentylammonium the derivatization was complete. Tetrahexylammonium, however, needed only 5 minutes for quantitative reaction under these conditions. Changing the anion to the more lipophilic pentobarbitone (**V**, R = $CH(CH_3)—CH_2—C_2H_5$), it is now sufficient to use tetrapentylammonium for 100% derivatization after only 15 minutes [268]. Similar ion pair extraction techniques were

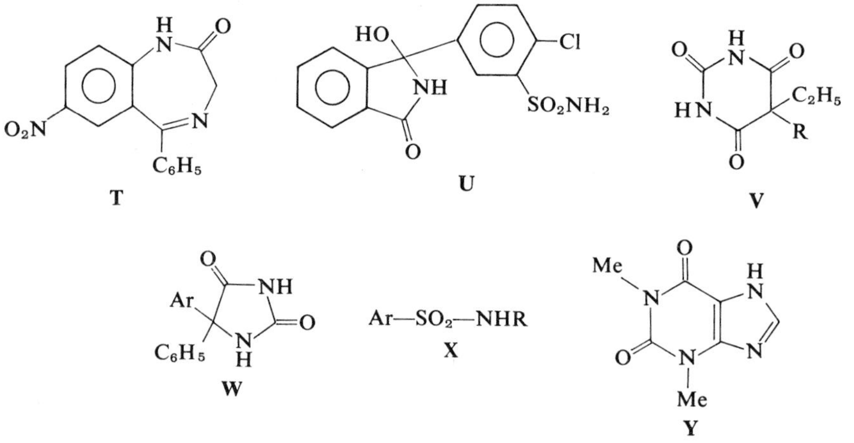

Scheme 3–51

used for analytical determination of sulfonamides **X** [270] and theophyllin **Y** [271] as pentafluorobenzyl derivates. For further extractive alkylations of pharmaceutical agents *cf.*, [939, 983, 1052].

Formal "alkylation" of secondary and primary amines as well as hydrazine by dichlorocarbene leading to formamides, isonitriles, and diazomethane, respectively, are discussed in Section 3.3.13.2, N- *vs.* O- or S-alkylation is covered in Section 3.3.5.

Typical experimental procedures are presented below.

Alkylation of acetanilides.

(a) Molar amount of catalyst: 0.1 mole tetrabutylammonium hydrogen sulfate was dissolved in 100 ml 2M NaOH. To this solution were added 0.1 mole acetanilide and 0.18 mole methyl iodide in 100 ml CH_2Cl_2. The mixture was refluxed for 30 minutes. After separation of the two phases, the organic layer was evaporated, and 100 ml of

ether was added to the residue leaving tetrabutylammonium iodide which was filtered off. Removal of the solvent gave the N-acetyl-N-methylaniline, yield 90% [255a].
(b) Small amount of catalyst: A mixture of 0.02 mole acetanilide, 0.001 mole TEBA, 50 ml benzene, 0.1 mole NaOH, and 4 ml H_2O is vigorously stirred at room temperature, whereupon a paste is formed. At 60–70 °C, 0.04 mole of *n*-propyl bromide are added dropwise, and the mixture is heated and stirred for 200 minutes, after which the substance dissolves. The organic phase is separated, washed with 2N HCl, water, and dried. After removal of the solvent the product is crystallized, m.p. 45–48 °C, 82% yield [257].

3.3.4. C-Alkylation of Activated CH-Bonds

Because of its great synthetic interest the alkylation of carbanion is among the most extensively studied PTC reactions. In 1951 Jarrousse [224] found that phenylacetonitrile can be alkylated by ethyl and benzyl chlorides in the presence of concentrated aqueous base and TEBA. This reaction was developed and optimized by Mąkosza, who also established its generality. It was shown that the PTC method has many advantages over more conventional procedures which require not only expensive bases (sodium amide, potassium *tert*-butoxide, metal hydrides, triphenyl methide, *etc.*) but also anhydrous solvents (absolute ether, benzene, dimethylsulfoxide, dimethylformamide, *etc.*). In many cases PTC is simpler than other procedures and because it is highly selective it gives good yields of purer products. TEBA is the most commonly used catalyst*) for such transformations.

3.3.4.1. Alkylation of Arylacetonitriles

The alkylation of phenylacetonitrile with ethyl bromide/concentrated aqueous sodium hydroxide/1 mole-% of TEBA is described in Organic Syntheses [280]. The reaction is mildly exothermic and takes 3–5 hours at 28–35 °C, yield 78–84%. For easier separation of the product from unreacted starting material, the latter is converted into α-phenyl cinnamonitrile by the addition of benzaldehyde to the reaction mixture when the main reaction is completed [280]. Using this general procedure a large variety of substituted arylacetonitriles was prepared (*cf.*, Table 3–4 and the Scheme 3–52 below). The reaction mechanism has been investigated extensively. Using conventional procedures, active methylene compounds usually give mixtures of mono- and bisalkylation products. However, in PTC the concentration of ion paired carbanions in the organic phase is low because of the low concentration of catalyst. When $PhCH_2CN$ is a stronger acid than Ph—CH(R)—CN, which is very

*) Many other onium salts can be used too and reaction variables were investigated extensively [997]. Commercial anion exchange resins gave markedly slower rates in triphase catalysis [958].

Table 3–4. Catalytic C-Alkylations of Arylacetonitriles.

Substrate	Alkylating Agent	Product	Yield (%)*)	Reference
$PhCH_2CN$	C_2H_5Br	Ph—CH(C_2H_5)—CN	78–84	[280]
ArCH(R)—CN (R = H, alkyl)	R′X (R′ = alkyl, allyl, benzyl, subst. benzyl)	Ar—CRR′—CN	*), †)	[281–83, 285, 308, 314, 315]
Ph_2CH—CN	RX	Ph_2C(R)—CN	>90	[284]
$PhCH_2CN$	X—$(CH_2)_n$—X	Ph—CH(CN)—$(CH_2)_n$—X	*)	[286]
		Ph(CN)C< cyclic $(CH_2)_n$	*)	
		Ph—CH(CN)—$(CH_2)_n$—CH(CN)—Ph	*)	
$PhCH_2CN$	$PhCHCl_2$	Ph—CH(CN)—CH(Ph)—CH(CN)—Ph	93	[286]
PhCH(R)CN	X—$(CH_2)_n$—X	Ph—C(R)(CN)—$(CH_2)_n$—X	*)	[287]
		Ph—C(R)(CN)—$(CH_2)_n$—C(R)(CN)—Ph	*)	
Ph—CH(R)—CN	Cl—$(CH_2)_n$CN	Ph—CR(CN)—$(CH_2)_n$—CN	70–90	[288, 774]
		Ph—C(CN)[$(CH_2)_n$—CN]—$(CH_2)_n$—CN ‡)	*)	

PhCH(R)—CN	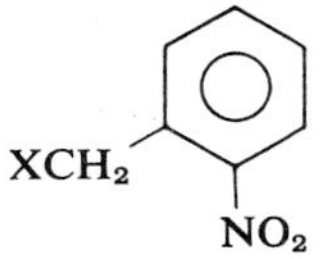 2-nitrobenzyl: $XCH_2-C_6H_4-NO_2$ (o)	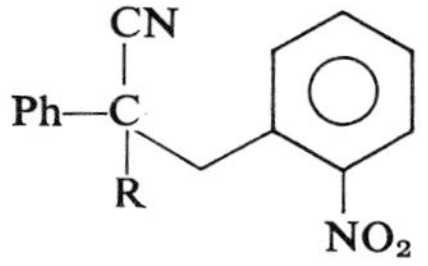 $Ph-C(CN)(R)-CH_2-C_6H_4-NO_2$ (o)	56–98	[289]
$PhCH_2CN$	$XCH_2-CR=CR'R''$	$Ph-CH(CN)-CH_2-CR=CR'R''$	25–60	[290]
PhCH(R)—CN	ClCH(R′)—OR″	$Ph-CR(CN)-CH(R')-OR''$	7–95*)	[291]
$PhCH_2CN$	ClCH(R′)—OR″	$Ph-CH(CN)-CH(R')-CH(CN)-Ph$	68–88	[291]
$PhCH_2CN$	$X-CH_2-COOR'$§)	$Ph-CH(CN)-CH_2-COOR'$ + $Ph-C(CN)(CH_2COOR')-CH_2-COOR''$	*)	[294]
PhCH(R)CN	$X-CH_2-COOR'$§)	$Ph-CR(CN)-CH_2COOR'$	26–95*)	[292]
$Ph-CH(CN)-(CH_2)_nNR_2$	R′X	$Ph-CR(CN)-(CH_2)_nNR_2$	34–88*)	[293]
$PhCH_2CN$	$R-N(CH_2CH_2Cl)_2$ (R = alkyl, tosyl)	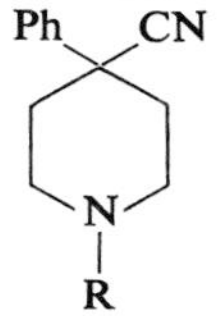 4-Ph-4-CN-piperidine, N—R	30–68	[297, 317]
PhCH(R)CN	$Br(CH_2)_2C(CH_3)_2NO_2$	$Ph-CR(CN)-(CH_2)_2C(CH_3)_2-NO_2$	20–80	[309]
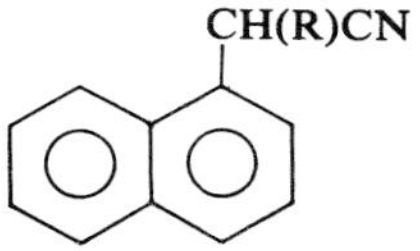 1-naphthyl—CH(R)CN	R′X	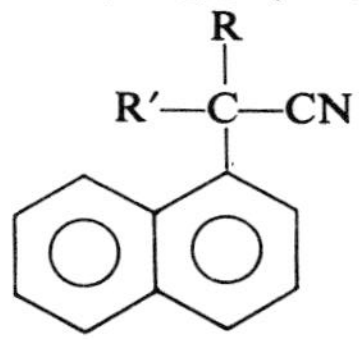 1-naphthyl—C(R)(R′)—CN	*), †)	[301]

(Continued)

Table 3-4. (Continued)

Substrate	Alkylating Agent	Product	Yield (%)*)	Reference
Ph—CH(—O—C(CH$_3$)(R″)—OR′)—CN	RX	Ph—CR(—O—C(CH$_3$)(R″)—OR′)—CN	71–81	[295, 320]
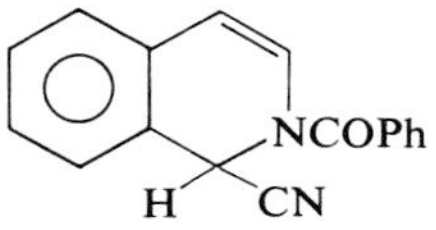 1-H-1-CN-2-NCOPh-1,2-dihydroisoquinoline	RX	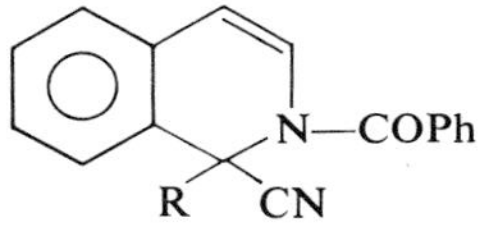 1-R-1-CN-2-(N—COPh)-1,2-dihydroisoquinoline	76–82	[310, 1068]
PhCH(R)CN	R′SCN	Ph—CR(SR′)—CN	45–77	[300]
PhCH(R)CN	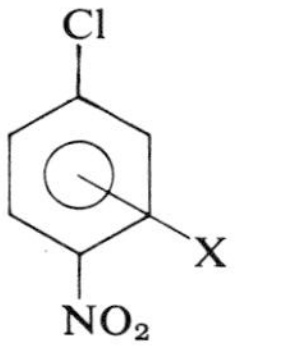 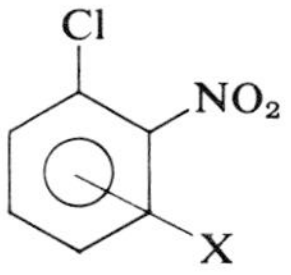 Cl-, NO$_2$-, X-substituted benzene (Cl / NO$_2$ / X)	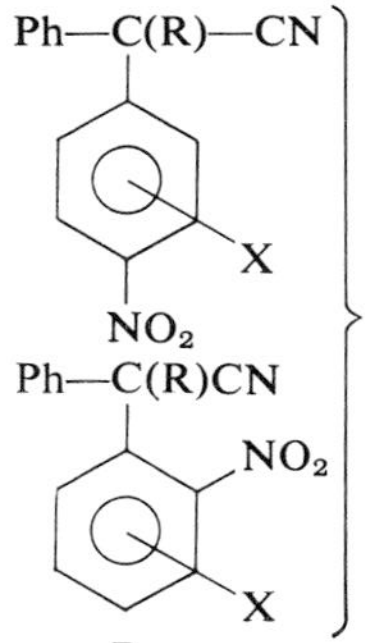 Ph—C(R)—CN on NO$_2$-, X-substituted ring	59–92	[303–307, 313]
	Cl-, NO$_2$-, X-substituted benzene (Cl / NO$_2$ / X)	Ph—C(R)CN on NO$_2$-, X-substituted ring		
Ar—CH(R)—CN	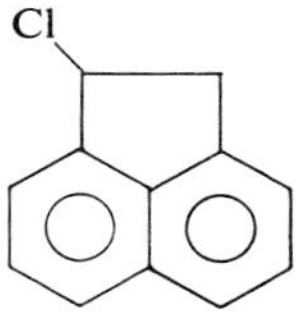 1-chloroacenaphthene (Cl)	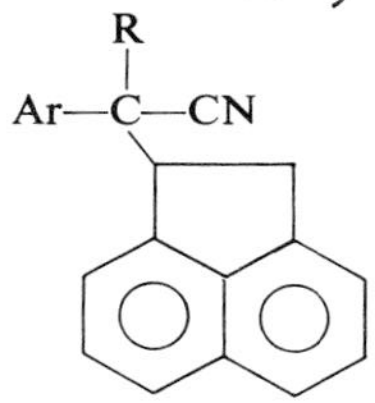 Ar—C(R)—CN on acenaphthene (+ bisprod.)	58–67 (7–11)	[330]

*) Depending on substituents and conditions.
†) Generally high.
‡) For R = H, n = 1, 2.
§) R′ = tert or sec alkyl.

often the case, bisalkylation is disfavored, and the monoalkylation step is highly selective. From the experimental results however, it is noticeable that very reactive alkylating agents (*e.g.*, allyl and benzyl halides) tend to give relatively more bis-alkylation. In cases where this is desirable, higher temperatures are recommended [283]. At 0 °C, benzylation of benzyl cyanide gave mono- to bisproduct ratios of 3 to 6 with various catalysts after 1 hour (40–60% conversion) [852].

In general, alkyl bromides are better alkylating agents than chlorides if concentrated sodium hydroxide is used in considerable excess [281, 282, 285]. Iodides and to some extent bromides (with a lower excess of NaOH) have an adverse effect on alkylation [281], because iodide is extracted in preference to the phenylacetonitrile anion into the organic phase. The use of iodides, therefore, is not recommended when using catalytic amounts of TEBA. Instead, Brändström used molar quantities of tetra-butylammonium salts with alkyl iodides [321]. As expected, in the more usual TEBA reactions, secondary halides reacted more sluggishly and gave lower yields than primary halides [283, 285]. Solvents had a retarding influence; alcohols more or less inhibited the reaction whereas if calcium hydroxide was substituted for sodium hydroxide, no reaction occurred [281]. In spite of the fact that some reactions were carried out at 60–80 °C, hydrolysis of nitrile and ester groupings was almost nonexistent. Only at temperatures of about 90 °C did this become noticeable.

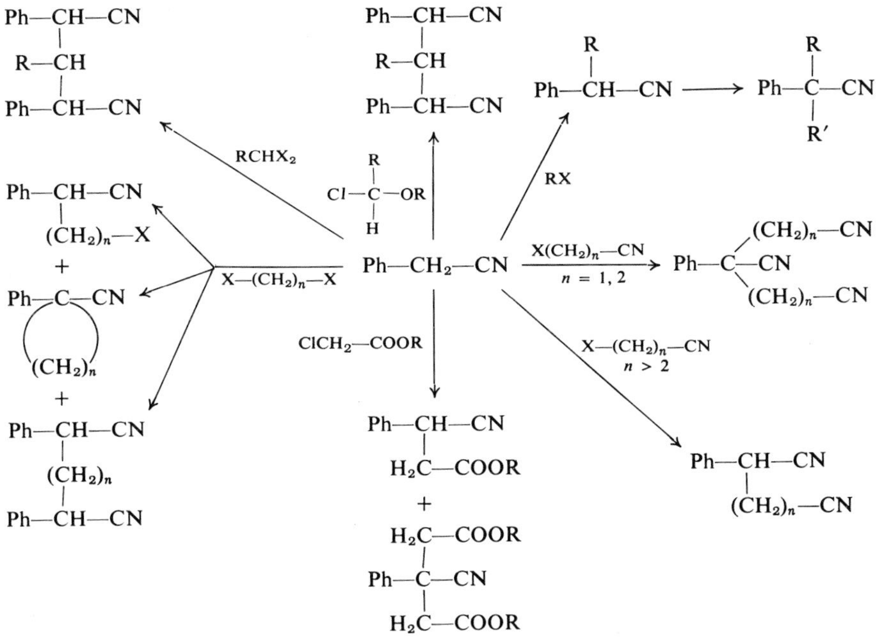

Ph—CH—CN | R—CH | Ph—CH—CN ← ($RCHX_2$) — Ph—CH_2—CN

Ph—CH—CN | R—CH | Ph—CH—CN ← (Cl—C(R)(H)—OR) — Ph—CH_2—CN

Ph—CH_2—CN —(RX)→ Ph—CH(R)—CN ⟶ Ph—C(R)(R′)—CN

Ph—CH_2—CN —(X—$(CH_2)_n$—X)→ Ph—CH($(CH_2)_n$—X)—CN + Ph—C($(CH_2)_n$, ring)—CN + Ph—CH—CN | $(CH_2)_n$ | Ph—CH—CN

Ph—CH_2—CN —(X$(CH_2)_n$—CN, $n = 1, 2$)→ Ph—C($(CH_2)_n$—CN)($(CH_2)_n$—CN)—CN

Ph—CH_2—CN —($ClCH_2$—COOR)→ Ph—CH(H_2C—COOR)—CN + Ph—C(H_2C—COOR)(H_2C—COOR)—CN

Ph—CH_2—CN —(X—$(CH_2)_n$—CN, $n > 2$)→ Ph—CH($(CH_2)_n$—CN)—CN

Scheme 3–52

Recently, alkylations and acylations were performed in the presence of solid K_2CO_3/dibenzo-18-crown-6 in benzene solutions [1048].

The preceding scheme surveys the different types of compounds accessible by PTC. Many reactions have been performed using simple alkyl, allyl, and benzyl halides [280–285, 290] and phenylacetonitrile, alkylphenyl- or diphenylacetonitriles. Substituted phenyl rings can also be used [301, 314, 315, 318]. Reactions with dihalogen compounds may result in the formation of homocyclic [286] and heterocyclic [297, 317] rings. Alkylation with *p*-nitrobenzyl chloride in the presence of relatively dilute sodium hydroxide and quaternary ammonium catalyst is at least partly a radical process because radical intermediates have been trapped [319]. When concentrated sodium hydroxide and TEBA are used, substitution is the main process for *o*-, *m*-, and *p*-nitrobenzylchlorides although dinitrostilbenes and, in some cases typical radical coupling products are also formed [289, 308]. Halonitriles [287, 288], *tert* or *sec* alkyl haloacetates*) [292, 294], many other substituted haloalkanes [297, 309], 9-chlorofluorene [1066], and even α-chloroethers [291] have also been used as alkylating agents. The latter compounds give simple substitution products with alkylphenylacetonitriles, and substituted glutaronitriles with phenylacetonitrile [291]. α-Dialkylamino- [293] and α-alkoxyphenylacetonitriles react similarly [295, 320], and

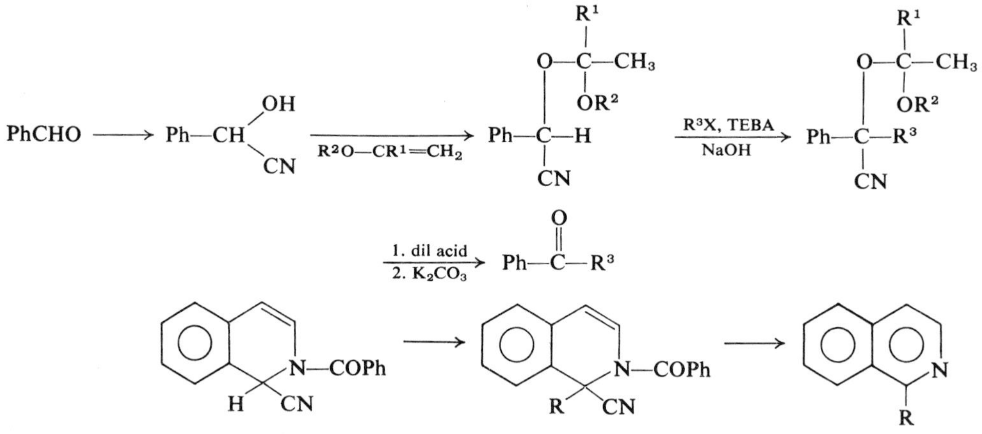

Scheme 3–53

the reactions are facilitated by the addition of DMSO. Since the starting materials can be easily prepared from benzaldehyde cyanohydrins, the sequence shown above constitutes a synthesis of arylketones from benzaldehyde [295, 320] or other aromatic aldehydes. Reissert compounds were found to react exothermically with alkyl bromides and benzyl chloride [310, 1068] and with 9-chloroacridine [870] in the two-phase system, the adducts being further transformed into isoquinolines.

---

*) Methyl and ethyl esters are hydrolyzed too fast.

Because carbanions are soft bases, they preferentially attack the sulfur atom of thiocyanates. Thus, it is reasonable, although surprising at first sight, that thioalkylation of phenylacetonitriles by alkylthiocyanates has in fact been observed under PTC conditions [300]:

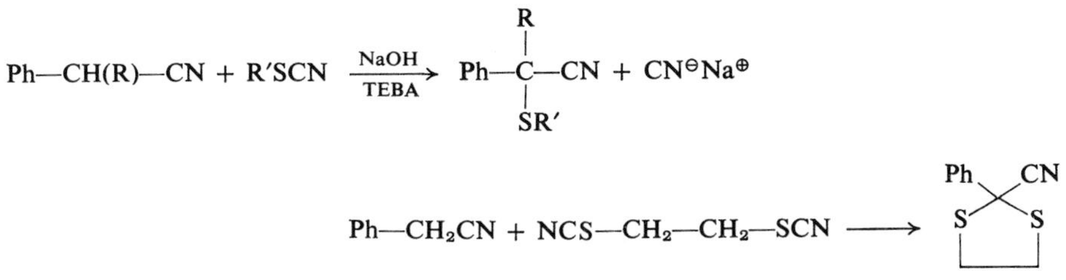

$$Ph{-}CH(R){-}CN + R'SCN \xrightarrow[TEBA]{NaOH} Ph{-}C(R)(SR'){-}CN + CN^{\ominus}Na^{\oplus}$$

$$Ph{-}CH_2CN + NCS{-}CH_2{-}CH_2{-}SCN \longrightarrow \text{2-Ph-2-CN-1,3-dithiolane}$$

Scheme 3–54

PTC appears to be the most convenient method for the nitroarylation of alkylphenyl- and diphenylacetonitriles using both *o*- and *p*- halonitrobenzenes and their substituted derivatives [303–305, 313] and chloronitropyridines [1067]. Phenylacetonitrile itself does not react because the traces of nitroarylation product formed poison the catalyst. X can be halogen, CN, COPh, phenyl, alkyl, NO_2 or COO-*tert*-Bu.

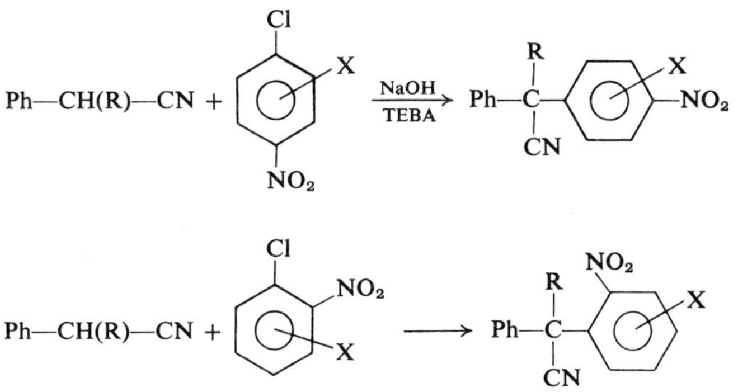

$$Ph{-}CH(R){-}CN + Cl{-}C_6H_3(X){-}NO_2 \xrightarrow[TEBA]{NaOH} Ph{-}C(R)(CN){-}C_6H_3(X){-}NO_2$$

$$Ph{-}CH(R){-}CN + Cl{-}C_6H_3(NO_2)(X) \longrightarrow Ph{-}C(R)(CN){-}C_6H_3(NO_2)(X)$$

Scheme 3–55

With 2,4-dichloronitrobenzene more displacement of the 4-chloroatom is observed [304, 306]. When nitro groups are located *ortho* or *para* to strongly electron drawing substituents even they can be displaced [305, 307, 313]:

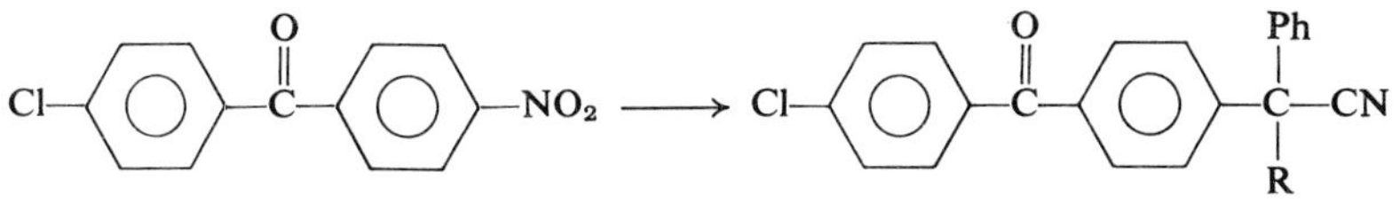

$$Cl{-}C_6H_4{-}C({=}O){-}C_6H_4{-}NO_2 \longrightarrow Cl{-}C_6H_4{-}C({=}O){-}C_6H_4{-}C(Ph)(R){-}CN \quad [305, 307]$$

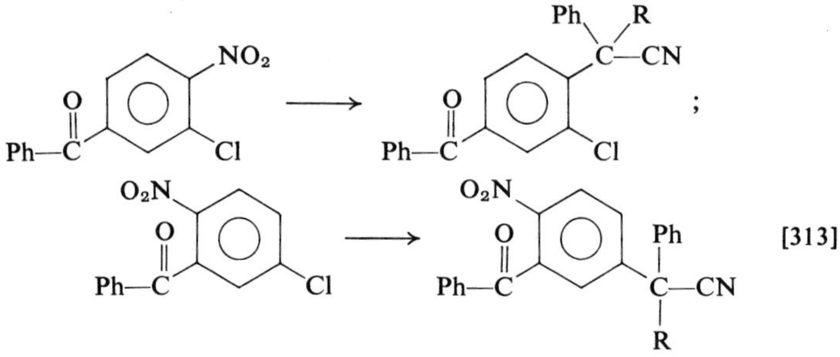

[313]

Scheme 3–56A

Similarly, nucleophilic substitution of chlorine in 9-chloroacridine by phenylacetonitrile derivatives was performed in benzene, conc. NaOH and a little DMSO at 50 °C in the presence of NBu_4Cl [870]. It should be mentioned that diphenyl-

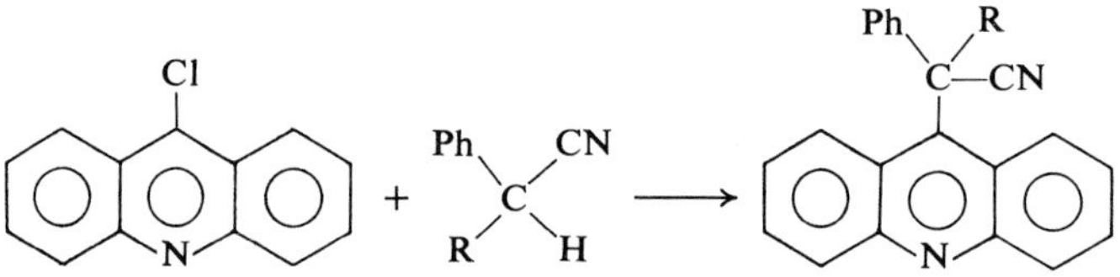

Scheme 3–56B

acetonitrile and sterically hindered alkylphenylacetonitriles can undergo various redox processes with aromatic nitrocompounds in the strongly alkaline PTC medium [307]. In some cases these become dominant. TEBA was the usual catalyst-employed for these reactions. It has been pointed out that dibenzo-[18]-crown-6 [302] could also be used, although it has no apparent advantages. A comparison of several catalysts was made and it was found that TEBA and tetrabutylammonium bromide were more suitable than hexadecyltrimethylammonium bromide, tributylhexadecylphosphonium bromide, or Dowex 1 resin (triphase catalysis) [299]. Occasionally, specially prepared solid polymer-supported catalysts were also used [902]. 2-Dialkylaminopyridinium tetrafluoroborates (**A**) have recently been found to be effective catalysts [312]. Even trialkylamines can be catalysts in a one-pot dialkylation because excess RX generates the true catalysts [1069].

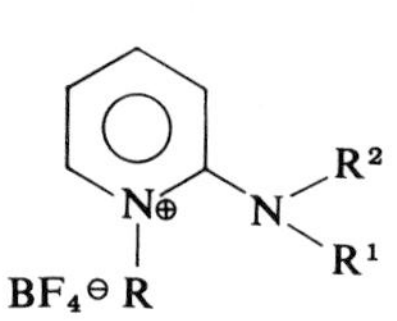

A

Scheme 3–57

A mechanistically very important fact is that very active alkylating agents (*e.g.*, benzyl chloride) react with phenylacetonitrile even in the absence of catalyst, although the reaction is much slower than in PTC. At elevated temperatures (80 °C), alkyl iodides also react rather fast without catalyst [298]. Such observations as well as the competitive alkylation results stress the importance of the interphase in alkylations [298].

3.3.4.2. Alkylation of other Activated Cyanides and Isocyanides

Aliphatic nitriles which do not have activating substituents cannot be alkylated under PTC conditions. However, a phenylthio group such as in S-phenylthioglycolonitrile **B** [296], the analogue selenocompound [876], or a dithiocarbamate group such as in **C** [322], is enough to permit alkylation at room temperature. Compounds such as **C** are intermediates of a new ketone synthesis [322], and the selenium compounds can be transformed into α,β-unsaturated nitriles by treatment with N-chlorosuccinimide in moist acetonitrile or with H_2O_2 in THF/water [876].

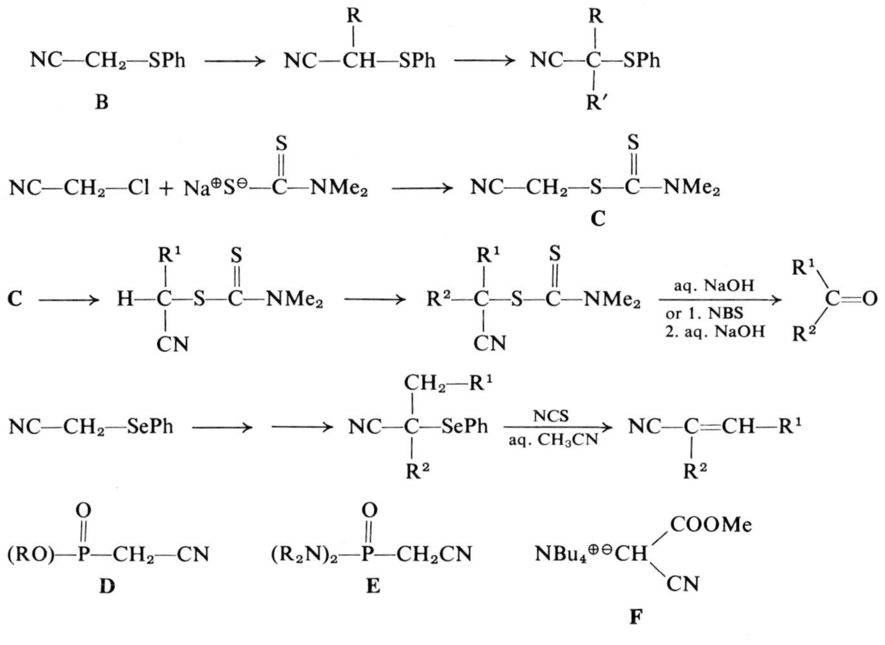

$$\underset{\mathbf{B}}{NC{-}CH_2{-}SPh} \longrightarrow NC{-}CH(R){-}SPh \longrightarrow NC{-}C(R)(R'){-}SPh$$

$$NC{-}CH_2{-}Cl + Na^{\oplus}S^{\ominus}{-}C({=}S){-}NMe_2 \longrightarrow \underset{\mathbf{C}}{NC{-}CH_2{-}S{-}C({=}S){-}NMe_2}$$

$$\mathbf{C} \longrightarrow H{-}C(R^1)(CN){-}S{-}C({=}S){-}NMe_2 \longrightarrow R^2{-}C(R^1)(CN){-}S{-}C({=}S){-}NMe_2 \xrightarrow[\text{or 1. NBS, 2. aq. NaOH}]{\text{aq. NaOH}} R^1R^2C{=}O$$

$$NC{-}CH_2{-}SePh \longrightarrow \longrightarrow NC{-}C(CH_2{-}R^1)(R^2){-}SePh \xrightarrow[\text{aq. } CH_3CN]{\text{NCS}} NC{-}C(R^2){=}CH{-}R^1$$

$$\underset{\mathbf{D}}{(RO){-}P({=}O){-}CH_2{-}CN} \qquad \underset{\mathbf{E}}{(R_2N)_2{-}P({=}O){-}CH_2CN} \qquad \underset{\mathbf{F}}{NBu_4^{\oplus\ominus}CH(COOMe)(CN)}$$

Scheme 3–58

Nitriles carrying phosphonic ester (**D**) or -amide (**E**) substituents are even more activated. Catalytic alkylation of these compounds was performed with molar amounts of tetra-*n*-butylammonium bromide in dichloromethane at 45 °C (1 hour, 30–80% yield) [323–325] with iodides as the alkylating agent, or with 5–10 mole-% TEBA with alkyl bromide [325]. Cyanoacetic ester can be susbstituted with nitroaryl groups (TEBA, *p*-chloronitroaryl compounds) [303]. Brändström isolated the tetrabutylammonium salt of methyl cyanoacetate (**F**) by solvent extraction and then alkylated the salt in chloroform. The reactions were finished after 10 to 120 minutes at room temperature. Contrary to the examples given above, a mixture of starting material and bisalkylated and monoalkylated product was isolated [326]. Malodinitrile appears to give exclusively dialkylation products, even with catalytic amounts of ammonium salt [4, 38]. Tosylacetonitrile also leads to a mixture of mono- and bisalkylation products under PTC conditions. With 1,8-diazabicyclo[5.4.0]undecene in homogeneous benzene solution only monoadduct is found [888]. Alkylation of both malodinitrile and cyanoacetate with 1,2-dibromoethane in concentrated NaOH using molar amounts of TEBA, gave directly cyanocyclopropane-1-carboxylic acid [329].

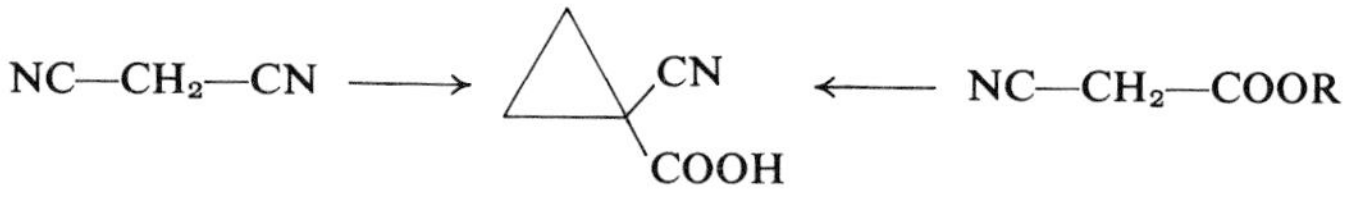

NC—CH$_2$—CN ⟶ (cyclopropane: CN, COOH) ⟵ NC—CH$_2$—COOR

Schiff bases Ph$_2$C=N—CH$_2$COOR and Ph$_2$C=N—CH$_2$CN can be made easily. They are alkylated in a simple PTC process (10% NaOH, NBu$_4$HSO$_4$ for the ester, 50% NaOH, TEBA for the nitrile) and then hydrolyzed to yield amino acids [1070]. A convenient synthesis of α-mercaptoalkanoic acids involves PTC alkylation of S-cyanomethyl thiocarbamates or dithiocarbamates (NC—CH$_2$S—CX—NR$_2$, X=O, S) and ensuing hydrolysis [1071].

Chemical modifications of chloromethylated and cyanomethylated polystyrene resins were achieved by basic PTC reactions with malodinitrile or chloroacetonitrile respectively [1072].

Alkylation α to isonitrile groups has only recently been studied. Activating groups are necessary. Working with **G**/TEBA/conc. NaOH, only allyl and benzyl halogenides

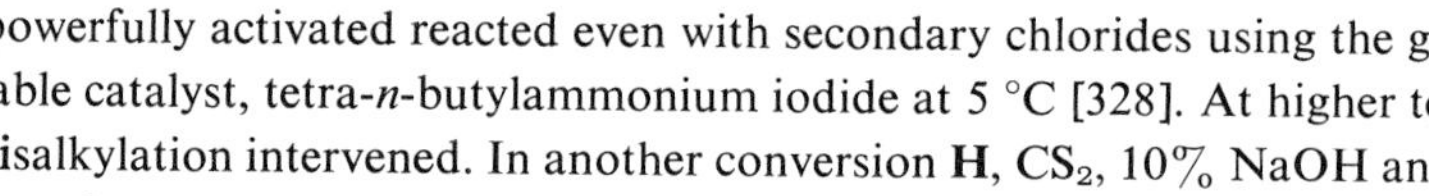

C=N—CH$_2$—COO—*tert*-butyl

G

H$_3$C—C$_6$H$_4$—SO$_2$—CH$_2$—N=C

H

Scheme 3–59

reacted, the latter yielding a bisbenzyl compound [327]. **H**, on the other hand, being more powerfully activated reacted even with secondary chlorides using the generally unsuitable catalyst, tetra-*n*-butylammonium iodide at 5 °C [328]. At higher temperatures bisalkylation intervened. In another conversion **H**, CS$_2$, 10% NaOH and molar amounts of tetrabutylammonium bromide gave the salt shown below, which can be further alkylated to yield 5-alkylthio-4-tosyl-1,3-thiazoles in excellent yields [431]:

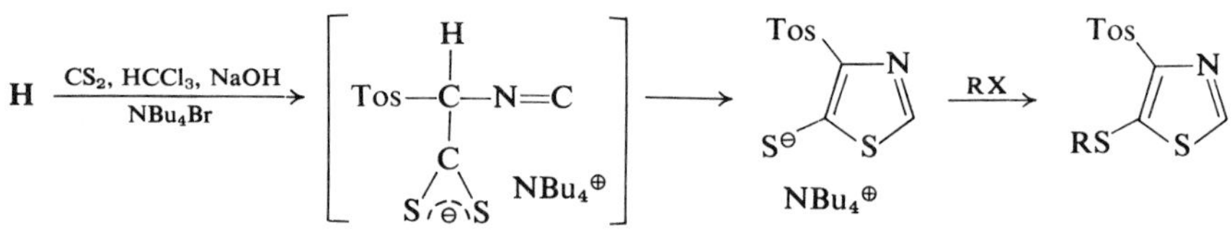

H $\xrightarrow[\text{NBu}_4\text{Br}]{\text{CS}_2,\ \text{HCCl}_3,\ \text{NaOH}}$ [Tos—C(H)(CS$_2^{\ominus}$)—N=C NBu$_4^{\oplus}$] ⟶ (thiazole: Tos, N, S$^{\ominus}$, S) NBu$_4^{\oplus}$ $\xrightarrow{\text{RX}}$ (thiazole: Tos, N, RS, S)

Scheme 3–60

3.3.4.3. Alkylation of Malonic Ester

As early as 1954 Babayan alkylated ethyl malonate with substituted alkyl chlorides in the presence of 10% quaternary ammonium compounds/40% aqueous potassium hydroxide [340]. The potentiality of these experiments remained undiscovered for many years. Later, Brändström applied his extractive alkylation procedure (molar amount or slight excess of tetra-*n*-butylammonium hydrogen sulfate, 2 mole 2M NaOH, CH_2Cl_2) to the alkylation of malonic esters with alkyl iodides [321]. Except in the case of methyl iodide, only monoalkylation is normally observed.

With Mąkosza's method (TEBA/conc. NaOH) exothermic reactions between *tert*-butyl malonate and ethyl bromide, benzyl, and allyl chloride are observed [338]. Mixtures of mono- and bissubstitution products were obtained when equimolar amounts of the reactants were used. Less active alkylating agents, *e.g.*, butyl bromide, gave better yields if some DMSO was added to the reaction mixture. When excess reagent was used double alkylations were possible. With α,ω-dibromoalkanes, both three- and five-membered rings were formed in 90 and 75% yields, respectively [338]. Ethyl malonate and 1,2-dibromoethane gave cyclopropane-1,1-dicarboxylic acid. In a one-pot process double alkylation and hydrolysis occurred in the presence of a molar amount of TEBA [329]. Nitroarylation with substituted *p*-nitrochlorobenzenes (TEBA/conc.NaOH) was analogous to the reaction of phenylacetonitrile considered above [303].

Recent additions to the literature include alkylations using solid alkali metal carbonates with onium salt catalysts [1048] and the application of the extractive alkylation method to a complex malonic ester [1073].

The alkylation of acetoacetic esters and related compounds is considered in Section 3.3.5.

3.3.4.4. Alkylation of Benzyl Ketones and Arylacetic Esters and Amides

These reactions are summarized in Table 3–5.

Although enolate anions can in principle undergo both O- and C-alkylation, C-alkylation is the usual process. Thus, Brändström's technique (1 mole catalyst) leads to 92% mono-C-product with phenylacetone and methyl iodide [321, 1075]. Although Mąkosza's method with TEBA and various alkyl, allyl, and benzyl halides gives high yields of monoproduct if the alkylating agent is present in slight excess, with a greater excess (2.5 to 3 moles) disubstitution is observed. It is mechanistically interesting that even in the absence of TEBA slow reaction with alkyl and benzyl halides occurs [331]. This is a strong indication that the interface is the site of reaction. Furthermore, it was found that many primary, secondary, and tertiary amines catalyze the alkylation of phenylacetone by butyl bromide and other alkyl halides in the presence of 50% aqueous potassium hydroxide or solid KOH, provided that the reaction conditions permit the formation of the quaternary ammonium salt, which is the true catalyst, *in situ* [432]. Several amines showed no catalytic effect, *e.g.*, those

Table 3–5. PTC-Alkylations of Benzyl Ketones, Arylacetic Esters, and Amides.

Substrate	Alkylating Agent	Product	Yield (%)	Reference
Ph—CH$_2$—COCH$_3$	CH$_3$I	Ph—CH(CH$_3$)—COCH$_3$	92	[321]
Ph—CH$_2$—COR	1-chloroacenaphthene (H, Cl on the five-membered ring)	Ph—CH—COR, with the CH bonded to C-1 (bearing H) of an acenaphthene ring	44–72	[330]
Ph—CH$_2$—COR′	RX (1 mole)	Ph—CH(R)—COR′	77–94	[331, 333, 334]
	RX (excess)	Ph—CH$_2$—COR	*)	
Ph—CH$_2$—COR	X—(CH$_2$)$_n$—X, n = 1–4	Ph—C—COR, with (CH$_2$)$_n$ closing a ring on C Ph—CH(COR)—(CH$_2$)$_n$—CH(COR)—Ph Ph—CH(COR)—(CH$_2$)$_n$—X	*)	[332–334, 903]
Ph—CHR—COR	Cl—(CH$_2$)$_2$—COOCH(CH$_3$)$_2$	Ph—CR(COR)—(CH$_2$)$_2$—COOCH(CH$_3$)$_2$	68–92	[333]

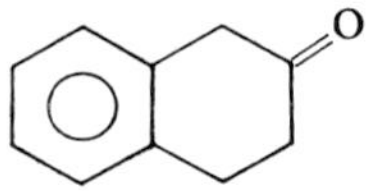	RX	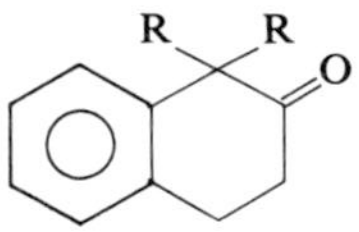	30 (R = alkyl) 80–90 (R = benzyl, allyl)	[334, 335]
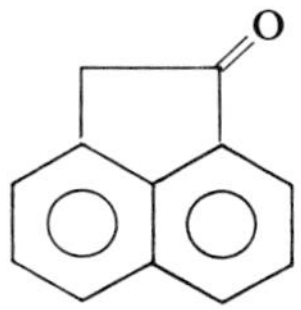	RX	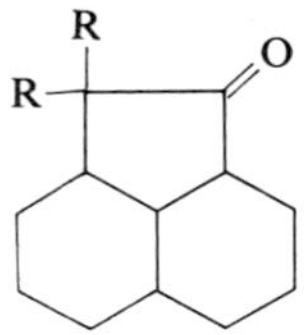	†)	[334, 336]
Ph—CH$_2$—COR	ClCH$_2$OR′	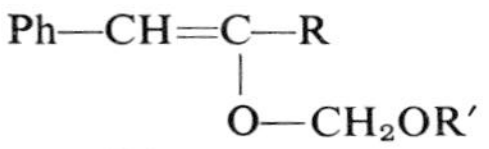 Ph—CH=C(—R)—O—CH$_2$OR′	25–60	[337]
 Ph—CH(R)—COOtBu	R′X	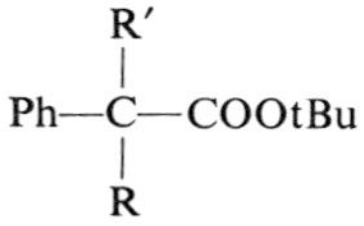 Ph—C(R′)(R)—COOtBu	45–70†)	[338]
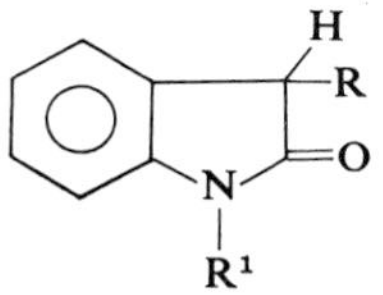	R^{2}X	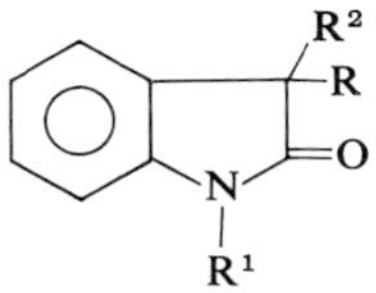	50–89	[339]

(*Continued*)

Table 3–5. (Continued)

Substrate	Alkylating Agent	Product	Yield (%)	Reference
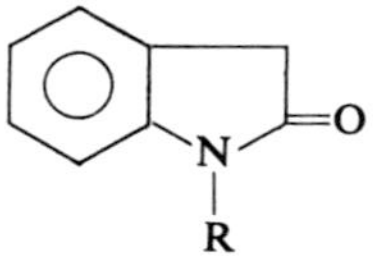 N–R, =O	X—$(CH_2)_n$—X	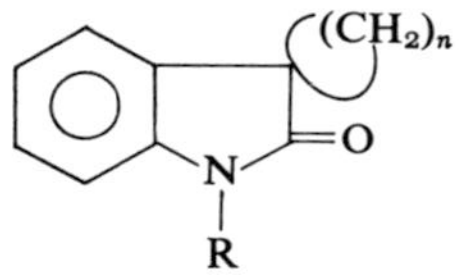 $(CH_2)_n$, N–R, =O	54–72†)	[339]
	$PhCH_2X$	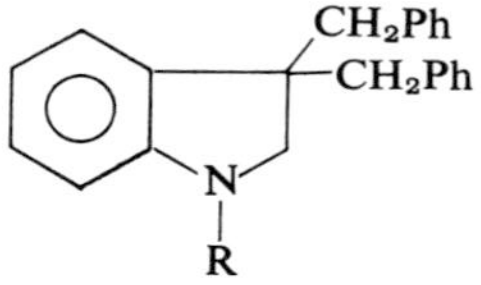 CH_2Ph, CH_2Ph, N–R	73	
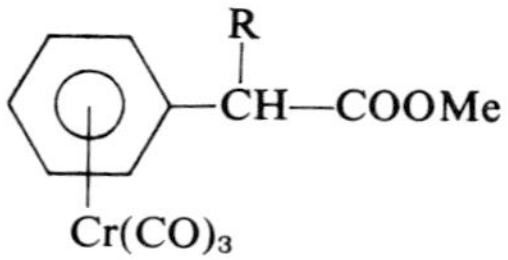 R, CH—COOMe, $Cr(CO)_3$	R′X	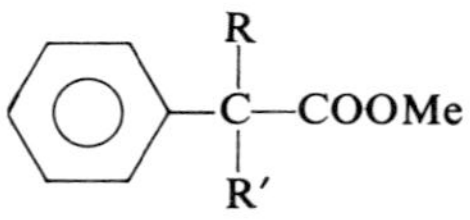 R, C—COOMe, R′	100‡)	[341, 930]
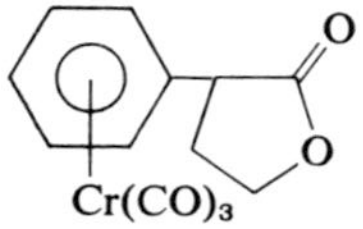 O, O, $Cr(CO)_3$	R′X	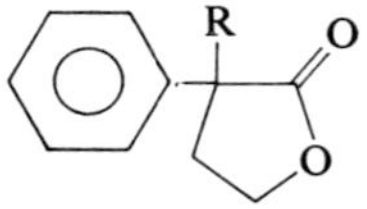 R, O, O	100‡)	

*) Yields depend on conditions.
†) For good yields a little DMSO is added to the reaction mixture.
‡) Complex destroyed with ceric salt.

with sterically hindered nitrogen or with aniline or pyridine structures. The failure of tributylamine to act as a catalyst in the butylation of phenylacetonitrile [433] was due to excessively mild reaction conditions. Quaternization to give tetrabutylammonium bromide, the "active" catalyst, was not possible under the reaction conditions. At 100 °C the reaction does in fact occur [432].

1,2-dibromoethane and 1,4-dibromobutane give compounds **I** or **K** and **L**, respectively, as the main products; with 1,3-dibromopropane, however, the mixture of products shown below is formed [332]:

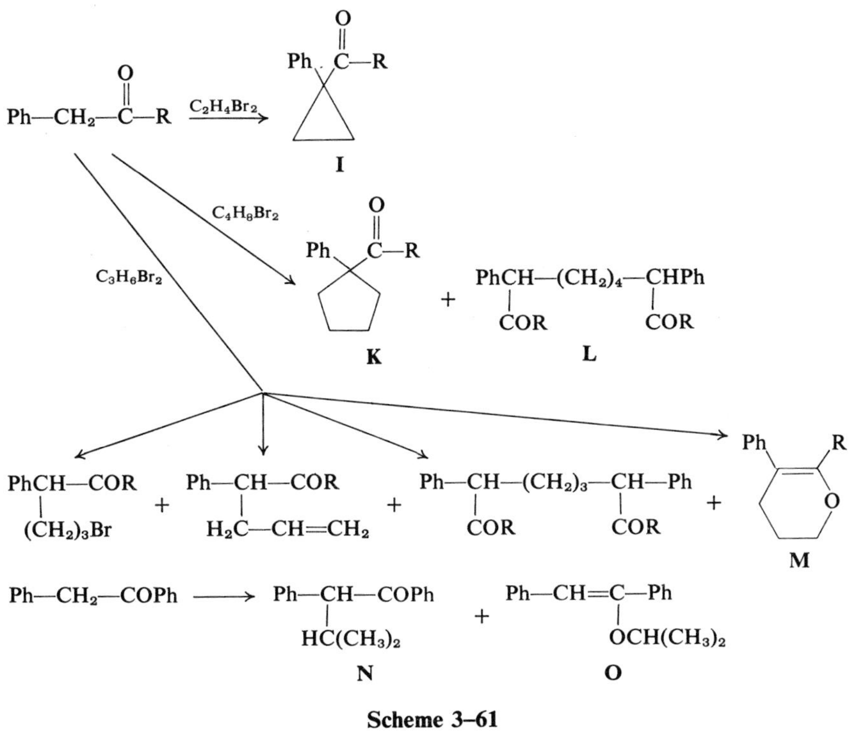

Ph—CH$_2$—C(=O)—R $\xrightarrow{C_2H_4Br_2}$ 1-Ph-1-(C(=O)—R)cyclopropane **I**

Ph—CH$_2$—C(=O)—R $\xrightarrow{C_4H_8Br_2}$ 1-Ph-1-(C(=O)—R)cyclopentane **K** + PhCH(COR)—(CH$_2$)$_4$—CH(COR)Ph **L**

Ph—CH$_2$—C(=O)—R $\xrightarrow{C_3H_6Br_2}$ PhCH(COR)—(CH$_2$)$_3$Br + Ph—CH(COR)—H$_2$C—CH=CH$_2$ + Ph—CH(COR)—(CH$_2$)$_3$—CH(COR)—Ph + 6-R-5-Ph-3,4-dihydro-2H-pyran **M**

Ph—CH$_2$—COPh ⟶ Ph—CH(HC(CH$_3$)$_2$)—COPh **N** + Ph—CH=C(OCH(CH$_3$)$_2$)—Ph **O**

Scheme 3–61

Compound **M** is a result of O-alkylation. With isopropyl bromide and desoxybenzoin both C- and O-alkylation products **N** and **P** were also obtained in the ratio 2.3:1 [333]. Similarly, isopropylation of tetralone gave about equal amounts of O- and C-products [335], and diphenylacetaldehyde lead exclusively to O-derivatives [1048]. In the TEBA catalyzed reaction between benzyl ketones and α-chloroethers no simple C-alkylation products were isolated, O-reaction was the main process [337]. *Tert*-butyl arylacetates were less reactive in alkylations than the corresponding arylacetonitriles. For good yields it was necessary to add small amounts of DMSO [338]. The bulky *tert*-butyl groups are necessary to prevent hydrolysis of the esters. Methyl esters can be used if the active methylene group is acidified. Thus, arene Cr(CO)$_3$-π complexes are suitable because of the electron withdrawing effects of the metal atom [341, 930]. They are prepared by refluxing with chromium hexacarbonyl. After almost quantitative alkylation (CH_2Cl_2 or benzene solution, cetyltrimethylammonium bromide/50% NaOH, 1.5 to 3 hours at room temperature) the complexes can easily be destroyed with ceric salts [341, 930].

For the synthesis of 3-alkyl, 3,3-dialkyl, or nitrophenyl products from N-substituted oxindoles, TEBA plus DMSO as cocatalyst were used [339, 1074]. The PTC alkylation of a complex heteroaryl-acetic ester is part of a recent elegant alkaloid synthesis [1076].

All the reactions discussed so far have been catalyzed by ammonium salts, usually TEBA. 2-Dialkylaminopyridinium salts have recently been found to be active catalysts, although not better than TEBA [312]. Catalysts of the crown ether type [45] and substituted azamacrobicyclic polyethers [48a] have also been applied, but their advantages are outweighed by their price.

3.3.4.5. Alkylation of other Doubly Activated CH-Bonds

β-Ketosulfones such as those pictured below are readily extractable as ion pairs with counterions, for example tetrabutylammonium. They were monoalkylated either after isolation and replacement of the original extracting solvent dichloromethane by ethyl acetate, or by direct extractive alkylation (Brändström method). Yields were generally better than 70% [355].

$$ArCO{-}CH_2{-}SO_2R \qquad RCH_2{-}CO{-}CH_2{-}SO_2Ph$$

In a three step synthesis of 1,4-diketones, intermediate β-ketosulfones were alkylated with substituted phenacyl bromides as shown [342]:

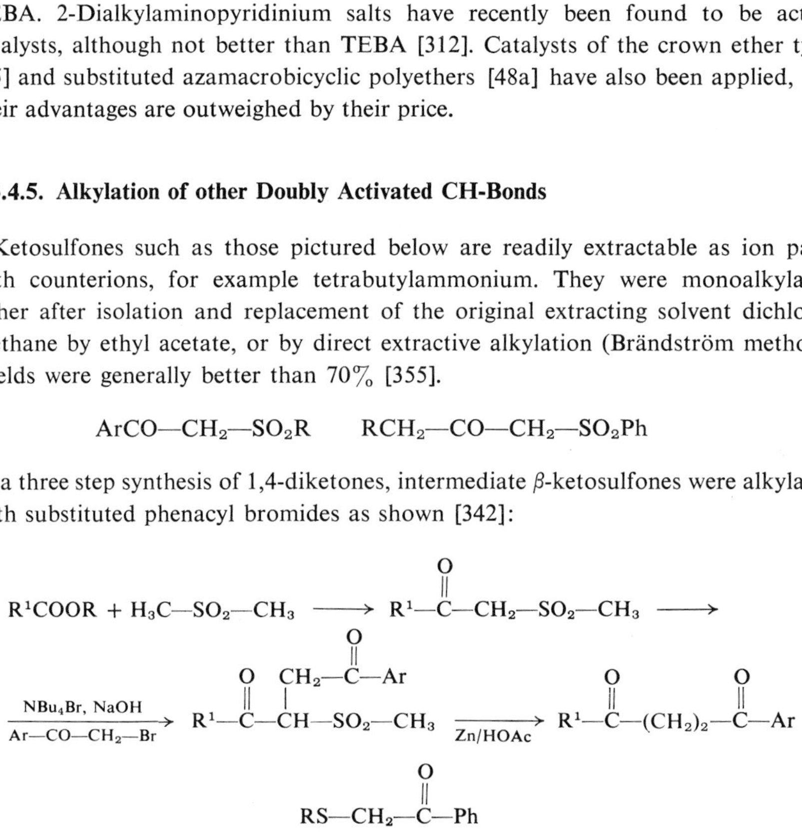

$$R^1COOR + H_3C{-}SO_2{-}CH_3 \longrightarrow R^1{-}C(=O){-}CH_2{-}SO_2{-}CH_3 \longrightarrow$$

$$\xrightarrow[Ar-CO-CH_2-Br]{NBu_4Br,\ NaOH} R^1{-}C(=O){-}CH(CH_2{-}C(=O){-}Ar){-}SO_2{-}CH_3 \xrightarrow[Zn/HOAc]{} R^1{-}C(=O){-}(CH_2)_2{-}C(=O){-}Ar$$

$$RS{-}CH_2{-}C(=O){-}Ph$$

P

Scheme 3–62 A

The β-ketosulfone anion/tetrabutylammonium ion pairs were extracted into chloroform and dried prior to alkylation. It was not tested whether or not this precaution is really necessary [342]. Yields of the alkylation were 63–80%.

Acetophenone ω-sulfides (**P**) are another class of compounds that can be alkylated easily in the presence of TEBA [343]. Depending on the alkylating agent and the radical R, varying amounts of O-alkylation were observed; with less reactive alkyl halides complex mixtures were obtained. In the latter case the cocatalyst was DMSO [343].

$$\underset{\mathbf{Q}}{\underset{(X = Cl, Br)}{Ar{-}SO_2{-}CH_2X}} \qquad \underset{\mathbf{R}}{(EtO)_2PO{-}CH_2{-}COOEt}$$

Halomethyl aryl sulfones (**Q**) as well as compounds such as **R** can be similarly alkylated with TEBA/conc. NaOH [344, 323, 1077]. The alkylation of ethyl 3-methyl-4-benzenesulphonylbut-2-enoate with 1-bromo-3-methyl-2-butene or 1-bromo-3,7-dimethyl-2,6-octadiene is of preparative interst [906] (CH_2Cl_2/NaOH/TEBA). Reductive desulfuration provides an attractive route to terpenes:

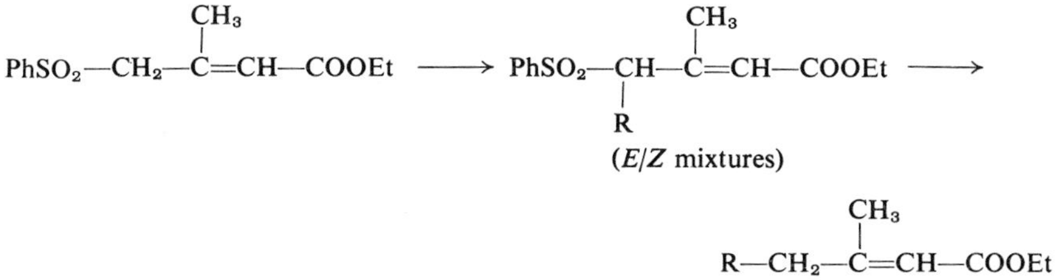

$$PhSO_2{-}CH_2{-}\overset{\overset{CH_3}{|}}{C}{=}CH{-}COOEt \longrightarrow PhSO_2{-}\underset{\underset{R}{|}}{CH}{-}\overset{\overset{CH_3}{|}}{C}{=}CH{-}COOEt \longrightarrow$$

(*E*/*Z* mixtures)

$$R{-}CH_2{-}\overset{\overset{CH_3}{|}}{C}{=}CH{-}COOEt$$

Scheme 3–62B

Similar PTC alkylations of β-phenylsulfonylesters proceed almost quantitatively [1078]. The alkylation of chloromethanesulfonamides gave 55–88% yield when a small amount of HMPT was present beside NBu_4Br/50% NaOH [1087]. The activated CH-bonds of α- and γ-methyls in pyridinium salts can be alkylated by PTC also. As an example, 1,2,4,6-tetramethylpyridinium iodide was transformed into a 20% yield of 4-*tert*-butyl-2,6-diisopropyl-1-methylpyridinium iodide with methyl iodide [1079].

3.3.4.6. Alkylation of Simple Carbonyl and Carboxyl Derivatives

This type of reaction is much more difficult to perform under PTC conditions than those previously considered: first, the CH-bonds are much less acidic, thus longer reaction times or higher temperatures are needed; second, competing aldol type reactions may dominate. In general, therefore, the alkylation of simple aliphatic aldehydes and ketones is not synthetically useful although several examples are known [334]. In patents it is described how acetone can be alkylated in the presence of sodium hydroxide and a phosphonium or ammonium salt catalyst with very reactive substituted allyl chlorides in good yields [345–348, 1081]:

$$CH_3{-}CO{-}CH_3 + Cl{-}CH_2{-}CH{=}C(CH_3)_2 \longrightarrow CH_3{-}CO{-}CH_2{-}CH_2{-}CH{=}C(CH_3)_2$$

Alkylation of cyclohexanone, 2-methylcyclohexanone, 3-methylcyclohexanone, and cyclopentanone with the same prenyl chloride (conc. NaOH/TEBA) gave yields of 20–50% along with many side products [352]. Interestingly enough, very similar yields were observed when solid potassium hydroxide was the condensing agent in the absence of any catalyst.

With acetophenones, however, the combination aqueous sodium hydroxide/TEBA (50 °C, 3 h) was more effective than solid KOH in this same type of conversion. Again yields were not very good [352]. A mixture of mono- and dimethylation products of acetophenone was obtained using the Brändström extractive alkylation technique [356]. This synthetically poor performance can be explained by the limited acidity of the aryl ketones. Earlier kinetic studies showed that in ether the ion pairs with quaternary ammonium cations once formed are in fact quickly alkylated

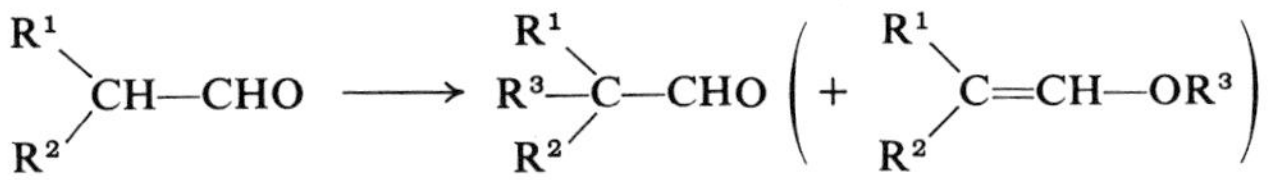

$$R^1R^2CH{-}CHO \longrightarrow R^3{-}CR^1R^2{-}CHO \; \left(+ \; R^1R^2C{=}CH{-}OR^3 \right)$$

Scheme 3–63

[353]. Aldehydes containing only one α-hydrogen atom (such as isobutyraldehyde) can be alkylated by active reagents such as methyl iodide, allyl, and benzyl halides in benzene/concentrated sodium hydroxide using tetrabutylammonium salts as catalysts [349, 350, 1080]. To minimize self-condensation, alkylating agent and aldehyde are added dropwise to the stirred reaction mixture at temperatures between 20–70 °C, but in spite of this yields are often low. Isobutyraldehyde gave only C-alkylation products. But 2-ethylhexanal, having a more sterically hindered α-C-atom leads to mixtures of both C- and O-alkylation products [349]. As expected, isopropyl bromide gave only O-alkylation in 21% yield, as did chloroacetic esters in poor yield [351].

The platinum-salt-complex **S**, on the other hand, was alkylated next to the carboxylate group by either dimethyl sulfate or benzyl chloride (NBu_4HSO_4, $NaOH/H_2O$/acetone) in excellent yields [354].

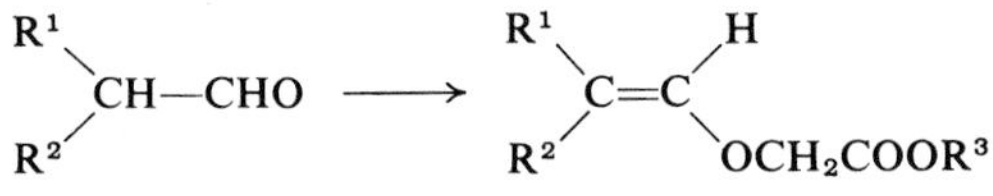

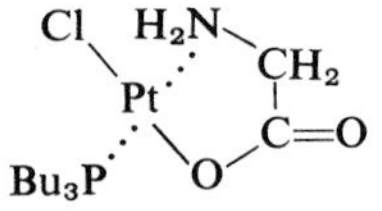

$$R^1R^2CH{-}CHO \longrightarrow R^1R^2C{=}C(H)OCH_2COOR^3$$

S: Cl, H_2N, CH_2, Pt, C=O, Bu_3P, O

Scheme 3–64

3.3.4.7. Alkylation of Hydrocarbons

The system aqueous sodium hydroxide/TEBA is basic enough to deprotonate both indene and fluorene. A variety of indene derivatives was obtained from primary alkyl, benzyl, and allyl halides in satisfactory yields of 50–73% [358]. Fluorene

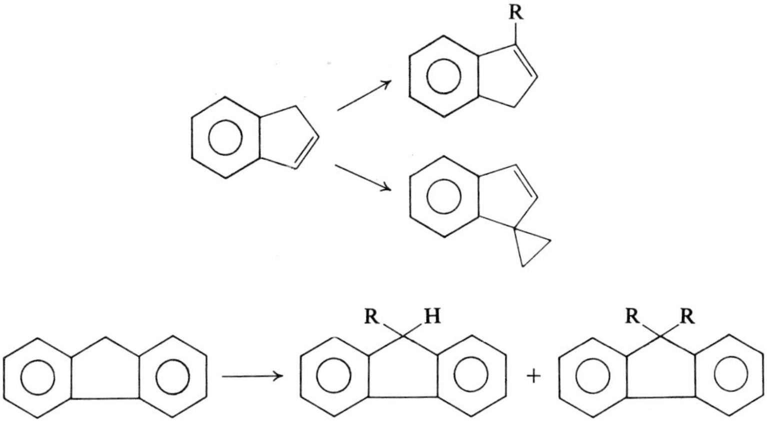

R

R H R R

+

Scheme 3–65

proved to be somewhat more difficult to alkylate as it is a much weaker acid. Best results, as in other cases of weaker acids, required the addition of a little DMSO to the reaction mixture, which contained saturated bromides. At 80–100 °C mixtures of mono- and bisalkylation products were obtained [357]. The alkylation of cyclopentadiene itself should be easy and has been mentioned in the literature without experimental details [214, 360], but presumably it will lead to an unpleasant mixture. Crown ether catalysis has also been applied to indene alkylations [45]. It should be emphasized that complexing agents may have wider application in very basic media in the absence of water than onium salts which decompose too easily.

Using the azamacrobicyclic polyether Kryptofix [2.2.2] (**5**) and solid potassium hydroxide/THF or sodium amide/Kryptofix[2.2.2]/THF, Dietrich and Lehn were able to deprotonate compounds with very high pK_a values [359]. With the latter system the colored anions of triphenylmethane and diphenylmethane were generated and trapped as the benzylated products [359].

Typical experimental procedures are presented below.

Arylacetonitrile alkylation, catalytic amount of TEBA. 4-Bromo-2,2-diphenylbutyronitrile: When 0.025 mole diphenylacetonitrile, 0.06 mole 1,2-dibromoethene, 0.1 g TEBA, and 10 ml 50% aqueous sodium hydroxide were stirred together a mildly exothermic reaction occurred. After 5 hours the phases were separated, and the excess 1,2-dibromoethane removed by steam distillation. The residue was extracted and crystallized from ligroin, m.p. 68–69 °C, yield 94% [287, 360].

Alkylation of tert-butyl phenylacetate, catalytic amounts of TEBA + DMSO. tert-Butyl 2-phenylhexanoate: 0.025 mole *tert*-butyl phenylacetate, 0.038 mole 1-bromobutane, 12 ml 50% aqueous sodium hydroxide, 0.1 g TEBA, and 5 ml DMSO were stirred at 60° C for 3 hours. The mixture was diluted with water and extracted with benzene. After drying and removing the solvent, the product was purified by vaccum distillation, b.p. 129–131 °C (8 torr), yield 45% [338].

Alkylation of malonic ester, using molar amounts of catalysts. Diethyl ethylmalonate: 0.24 mole NaOH in 50 ml water are added to 0.12 mole tetrabutylammonium hydrogen sulfate in 50 ml water. A solution of 0.1 mole of diethyl malonate and 0.2 mole of ethyl iodide in 75 ml dichloromethane is added, and the mixture is stirred and refluxed for 20 minutes. The phases are separated, and the organic solvent removed on a rotary evaporator. 200 ml ether are added to the residue and the undissolved NBu_4I (100% yield) is filtered off. After the solvent is removed an almost quantitative yield is obtained [321].

3.3.5. Alkylation of Ambident Anions

In the preceding section reactions leading to C-alkylation products were considered. However, in a study of simple carbonyl enolates a certain tendency towards O-alkylation was encountered. In this section we shall deal with reactions of β-dicarbonyl compounds and related classes where the question of C- *vs.* heteroatom-alkylation is more complex.

A summary of the factors influencing the structure of the anion and the direction of alkylation and/or acylation will be given first. In this field, the application of PTC centers around the question as to how the mode of attack (*e.g.*, C- *vs.* O-) can be changed, in addition to the experimental advantages of PTC compared to older procedures. Consequently, we shall discuss PTC processes, homogeneous quaternary ammonium counterion and crown ether mediated processes together with "classical" methods. It will be shown that the direction of C- *vs.* heteroatom alkylation depends on many factors (substrate structure, alkylating agent, solvent, concentration, the nature of cation, presence of crown ether, temperature, *etc.*) and that a full, quantitative understanding of the interaction of these factors has not yet been reached. Extensive discussions are found in references [361, 368, 896].

Nonrigid β-dicarbonyl anions can exist in principle in three conformations: U-, W-, and S- (for sickle)-shapes [361]:

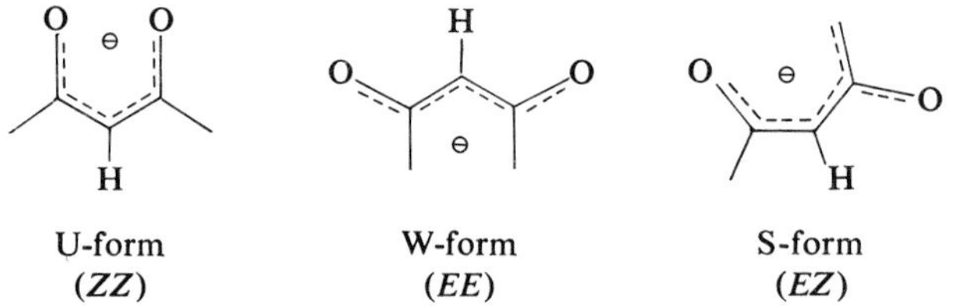

U-form
(*ZZ*)

W-form
(*EE*)

S-form
(*EZ*)

Scheme 3–66

In general in solution there is an equilibrium between these forms that depends on various factors (*vide infra*), but the introduction of a rigid skeleton may restrict it to one form. The properties of these three forms, in particular the acidity and the tendency towards ion pair association and dissociation, are very different. It has been shown that in nonpolar solvents the alkali metal enolates of acylic β-keto compounds are present preferentially in the U-form and that a strong "association" between anion and cation exists [362]. Even in aqueous solution this association remains strong

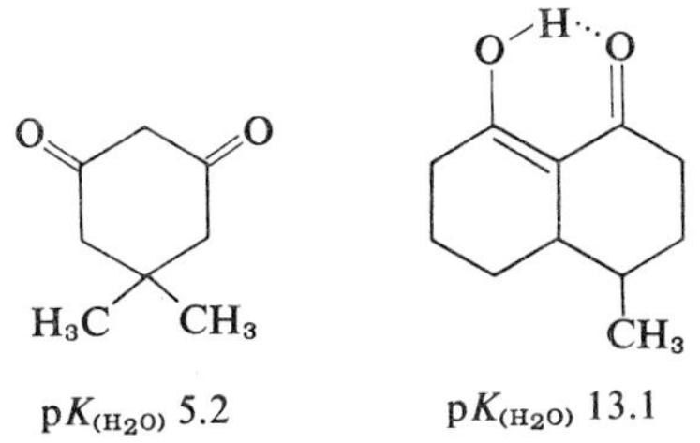

$pK_{(H_2O)}$ 5.2 $pK_{(H_2O)}$ 13.1

Scheme 3–67

[362]. Under similar conditions enolates favoring the W-form are dissociated to a higher extent [362]. Two factors determine this behavior: (a) the associated U-enolate is stabilized, relative to the associated W-enolate, by chelation of the metal, and (b) the dissociated U-enolate is destabilized, relative to the dissociated W-form enolate, by the electrostatic repulsion of the partly charged oxygens. These factors play a more decisive role in aprotic than protic solvents where H-bridges decrease intramolecular electrostatic interaction [361]. It should be added that the associated U-shaped salts tend to form higher aggregates in nonpolar solvents, and that, depending on solvent and counterion, contact or solvent separated ion pairs as well as dissociated ions can be present.

Recent physicochemical studies (far infrared, NMR spectroscopy, conductivity measurements) in THF and DMSO confirm the picture of associated U-shaped ion pairs as the predominant species. It is especially noteworthy that tetrabutylammonium salts behave similarly to the alkali metal salts, thus showing that the latter are ionic and—even more important—that there is no strict requirement for the cation to bind a particular region of the anion. While smaller alkali metals might be best accommodated in the plane between the O-atoms (true chelate) the ammonium ion could be above the plane of the U-shaped anion [363].

The X-ray crystal structure of the 18-crown-6 complexed potassium salt of ethyl acetoacetate has the $K^{\oplus}$ coordinated to the six O-atoms of the crown ether and the two O-atoms of the planar U-shaped enolate [364].

Previously many authors considered only the U- and W-shaped enolates, but recently it has been shown that what had been taken for the symmetrical W-form in methanol solution is in fact the sterically unrestricted S-form. The second NMR methyl signal for acetylacetonate (W-form) was in fact overlapping with another signal [365, 366]. Low temperature NMR studies in CD_3OD indicated that 23% of the U-shaped chelate is present while the nonchelated S-form accounts for most of the anion. In methanol the solvent molecules compete as a ligand for sodium, so that chelation is relatively unimportant [365]. In pyridine even if an excess of 18-crown-6 was present a greater preference for the U-form was found [366]. An increase in crown ether concentration brought about a decrease in the concentration of the U-form. As a result of a careful study in methanol, it was concluded that with sodium the U-shaped form is chelated and associated, whereas the S-shaped anion is largely dissociated. The lithium enolate of acetylacetonate is practically undissociated, while the potassium salt is in the dissociated S-form more so than the sodium salt [366].

Alkylation and acylation of β-dicarbonyl compounds can lead to three primary products, C-, O_{cis}-(Z), and O_{trans}-(E), alkyl or acyl products. Gelin and co-workers [367] using acetoacetic ester as an example, gave a survey of these pathways as shown below:

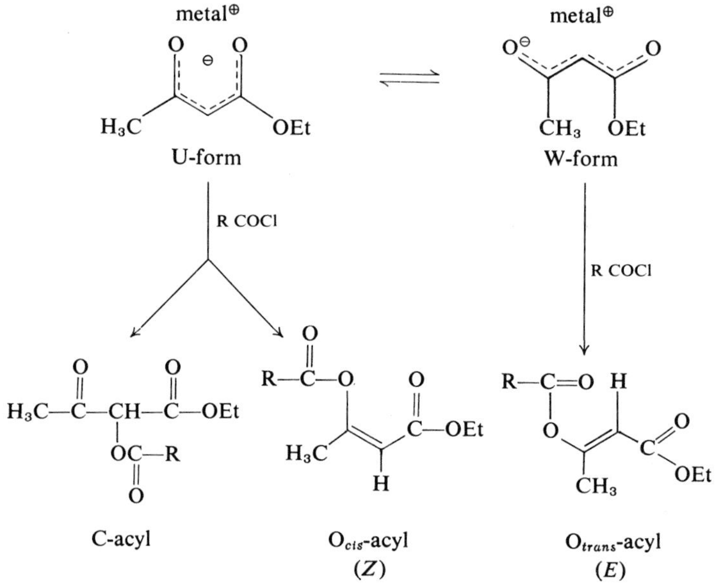

metal$^{\oplus}$ U-form

metal$^{\oplus}$ W-form

R COCl

C-acyl

O_{cis}-acyl (Z)

O_{trans}-acyl (E)

Scheme 3–68

In the above scheme the U-form leads to a mixture of O_{cis}- and C-products, while the W-form gives only O_{trans}-products. Although this type of reasoning is widespread in the literature, the various enolate forms are in fact reflected in the products only if alkylation is faster than equilibration. Although this has been tacitly assumed by many authors, it was shown in at least one case that conclusions about product formation from enolate structures are not possible [889]. With symmetrical β-diketones, moreover, the S-shaped form, which has often been neglected [364], can give rise to both O_{cis}- and O_{trans}- products while the W-form of the enolate leads only to O_{trans}-alkyl product. It is quite possible that some of the discrepancies in the literature are due to the neglect of the S-shaped enolate. Further complications arise if it is uncertain whether kinetic or thermodynamic control is operative. Noble [368] summarized the rules of the alkylation of ambident anions: "The freer the anion, the greater the tendency for alkylation at the most electronegative site." Thus, O-alkylation is promoted by these factors:

(1) a polar, aprotic solvent
(2) a large counterion
(3) a low concentration of enolate anion

(4) a "hard" leaving group in the alkylating agent, or a reagent with low S_N2 reactivity
(5) a sterically hindered alkylating agent.

Classical alkylations in ethanol using alkyl halides tend towards the least electronegative site, thus giving predominant C-alkylation. Hydrogen bonding with the solvent makes the U-shaped enolate more stable and shields the O-atoms against attack. In general, ion pair association leads to a decrease in the percentage of alkylation at the more electronegative site. Thus, reaction at the least electronegative site, *i.e.*, at C, is observed in aprotic solvents of low polarity (THF, ethers, $CHCl_3$, CH_2Cl_2, C_6H_6). In contrast, in dipolar aprotic solvents (DMSO, HMPT, DMF) O-alkylation predominates [361, 367]. The dissociated W-form of the tetrabutylammonium salt of ethyl acetoacetate is the main species in very dilute solution, thus O-alkylation prevails. However, the C-directing effect of less polar solvents is still apparent: O/C = 7.7 in DMF and 3.9 in DME [361].

Brändström and Junggren investigated the isopropylation of tetrabutylammonium acetylacetonate in various solvents [369]. It should be remembered, however, that in this case steric factors favor O-alkylation strongly. As can be seen from Table 3–6, the C/O ratio increases with falling dielectric constant ϵ and Taft parameter P (which is a measure of the solvating properties of a solvent). Acetone seems to be the only exception.

Table 3–6. Isopropylation of NBu_4 Acetylacetonate (0.05 mole salt in 50 ml solvent and 0.1 mole isopropyl iodide).

	Taft Parameter P	Dielectric Constant ϵ	C (%)	O (%)	C/O
DMSO	2.60	48.9	42	58	0.72
acetone	1.95	20.5	42	58	0.72
acetonitrile	2.30	37.5	48	52	0.92
$CHCl_3$	2.30	4.8	51	49	1.04
dioxane	1.25	2.2	63	33	1.91
toluene	0.55	2.4	69	5	13.8

For general PTC work therefore, it might be inferred that toluene is the best solvent for obtaining a high percentage of C-alkylation. It should be noted that the most frequently used PTC solvent, chloroform, shows no significant effect in this admittedly critical case. In a similar study no solvent influence (DMF *vs.* $CHCl_3$) was found for the alkylation of dialkyl oxaloacetates and alkyl 2-cyano-2-phenyl-pyruvates [370].

O_{cis}- *vs.* O_{trans}-acylation or -alkylation is another important subject. To demonstrate the complexity of the various factors involved, we shall consider the reaction between the sodium salt of ethyl acetoacetate and O,O-diethyl phosphorochloridothioate [371]:

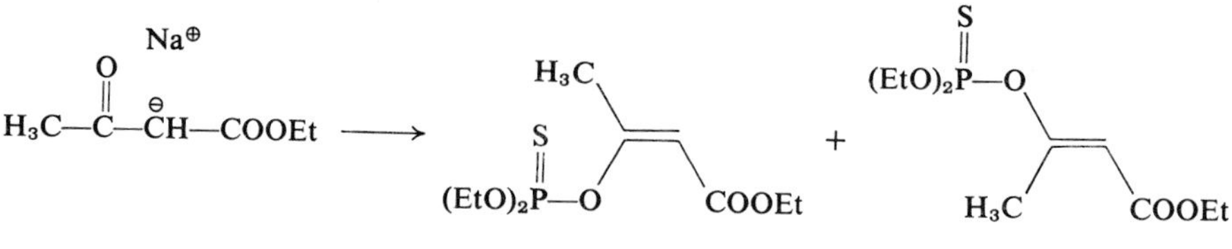

$$H_3C-\overset{O}{\overset{\|}{C}}-\overset{\ominus}{C}H-COOEt\ (Na^{\oplus}) \longrightarrow O_{cis}(Z) + O_{trans}(E)$$

$O_{cis}(Z)$: $H_3C(\text{(EtO)}_2P(=S)-O)C=CH-COOEt$, $O_{trans}(E)$: $(\text{EtO})_2P(=S)-O(H_3C)C=CH-COOEt$

Scheme 3–69

The acid chloride directs the reaction towards the electronegative O, and polar solvents favor the dissociated W-shaped enolate form. In such media, therefore, the $O_{trans}(E)$ product is almost exclusively formed, whereas in benzene, the $O_{cis}(Z)$ product (formed from the associated U-form enolate) predominates.

Kolind-Andersen and Lawesson [373] applied this reaction to many derivatives of the type **A**. They gave the Z/E ratio for the reaction with the sodium salt in refluxing ethanol and with the (previously isolated) tetrabutylammonium salt in dichloromethane.

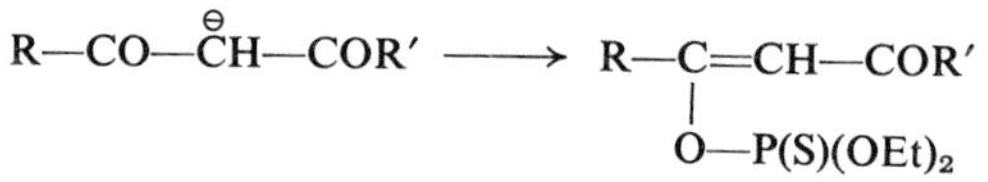

$$\underset{\mathbf{A}}{R-CO-\overset{\ominus}{C}H-COR'} \longrightarrow \underset{Z \text{ and } E}{R-\underset{O-P(S)(OEt)_2}{\underset{|}{C}}=CH-COR'}$$

$R = CH_3$, Ph
$R' = OMe$, OEt, CH_3, Ph

Scheme 3–70

As expected, the sodium salt in ethanol (U-shape enolate) gave almost exclusively Z product; the tetrabutylammonium salt in dichloromethane (dissociated W shape enolate) gave principally the E (O_{trans}) product.

The influence of the cation is also seen in the alkylation of ethyl acetoacetate with ethyl tosylate in dimethoxyethane [372]:

	$Li^{\oplus}$	$Na^{\oplus}$	$K^{\oplus}$	$NBu_4^{\oplus}$
O_{cis}/O_{trans}	≈ 20	6	0.9	traces O_{cis}
C/O	≈ 2.2	6	4.7	0.26

Here the "hard" leaving group and an increasing size of the cation favor O-alkylation. O_{trans} is favored by the larger cations and by the W-form enolate. Under similar conditions the NBu_4-salt together with various other alkylating agents leads to more C-alkylation [372] (*cf.*, [382] for another example):

(0.05 M $NBu_4^{\oplus}$ $CH_3COCH^{\ominus}-COOEt$, DME solvent)

	0.1 M EtI	0.3 M EtBr	0.2 M EtOTs
C/O	8.4	2.9	0.26

A word of caution is however warranted: As already stated, in principle low concentration favors O-alkylation (particularly O_{trans}) [368, 371]. In certain cases, however a concentration independent product ratio has been observed [370, 1083], while in other cases an O/C ratio which is independent of the alkylating agent leaving group (benzyl chloride, bromide, tosylate) was found [380]. The correlation between S_N2 reactivity of the alkylating agent and the O/C ratio has been questioned too [370]. It is clear that the various factors involved are very complex.

Cambillau *et al.* [377] studied the alkylation of the sodium enolates of ethyl acetoacetate both with and without 18-crown-6 and cryptate [222] in THF with ethyl tosylate and iodide. Although the reaction rate was higher in the presence of crown ether, it was even higher with the cryptate. In general the percentage of O-alkylation, O_{trans}-alkylation in particular, was also higher. Crown complexation breaks up the ion pair aggregates with a consequent increase in reaction rates. Furthermore, the equilibrium is displaced towards the dissociated W-form of the enolate. Other extensive investigations using various crowns, alkali cations, and tetraphenylarsonium ion in different solvents lead to qualitatively similar conclusions [378, 379, 1082]. The influence of the crown ether in shifting the ratio from relatively more C- to more O-alkylation, was strongest in the less polar solvents [379]. The influence of crown ether and cryptate [222] on the C/O alkylation of methoxycarbonylcyclohexanone in DMSO/CH_3OH was also investigated [382].

We will now consider some applications. Two procedures related to PTC are possible: (a) the isolation of enolate quaternary ammonium salts by solvent extraction and subsequent reaction in the same or a new solvent, and (b) "true" PTC reactions in a two phase medium. Both have been extensively used.

In some cases, the use of a quaternary ammonium salt of an ambident anion is the only possible method for alkylation. Dialkyl acylmalonates, for example, are rather strong acids, and the corresponding anions are weak nucleophiles. Conventional alkylation methods fail for this reason and because after alkylation the compound can split off the acyl group very easily. Using excess alkyl iodide, Brändström alkylated solutions of the tetrabutylammonium salts of dimethyl benzoylmalonate in dichloromethane with the following results [374]:

	C (%)	O (%)
CH_3I	100	0
EtI	54	46
n-BuI	47	53
iso-PropI	14	86

As has been suggested previously, toluene as solvent might increase the percentage of C-alkylation [369].

Returning once more to acetylacetone, this compound presents problems under conventional reaction conditions in protic solvents since it is alkylated much slower than ethyl acetoacetate or diethyl malonate. Tetrabutylammonium salts (previously isolated by solvent extraction) were alkylated in chloroform with the results shown below [375] (*cf.* also [1083]):

Alkylation of Acetylacetone (10–30 min. reflux).

	mono-C	di-C	O
CH_3I	98.5		1.5
EtI	72	16	12
iso-PropI	50.5	—	49.5
n-BuI	87	—	13

R. A. Jones *et al.* investigated the acylation of acetylacetone and ethyl acetoacetate in dichloromethane/2N NaOH in the presence of molar amounts of tetrabutylammonium hydrogen sulfate [837]. As in the case of O,O-diethyl phosphorochloridothioate, considered above, both acetyl and benzoyl chloride gave exclusively O-acylation. The *E*-isomers were more abundant with the pentanedione reactions and predominated with acetoacetic ester [93–96%*E*:7–4%*Z*). (In Jones's paper [837] the product formulas were inadvertently interchanged, but the text is clear and consistent.)

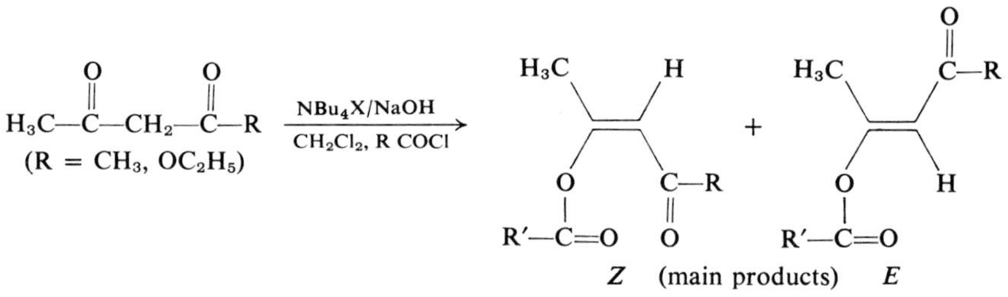

$H_3C—C(=O)—CH_2—C(=O)—R$ (R = CH_3, OC_2H_5) $\xrightarrow[CH_2Cl_2,\ R\ COCl]{NBu_4X/NaOH}$ $(H_3C)(R'C(=O)O)C=CH(C(=O)R)$ + $(H_3C)(R'C(=O)O)C=CH(C(=O)R)$

Z (main products) *E*

Scheme 3–71

Using only catalytic amounts of quaternary ammonium salts, the author [854] observed only very slow reaction in the ethylation of acetylacetone. The O/C alkylation ratio was unaffected by both the catalyst concentration and the size of the catalyst cation (within experimental variation). Working with previously isolated tetra-*n*-pentylammonium acetylacetonate in methylene chloride (20 °C, 15 min) the following O/C ratios were observed:

C_2H_5I 1:5.7, C_2H_5Br 1:2, $(C_2H_5O)_2SO_2$ 2:1

The benzylation of acetylacetone in aqueous/organic media with tetrabutylammonium bromide catalyst in molar quantity was also studied [380]. The ratio of O/C products varied between 1:3 and 1:1 with various leaving groups and conditions.

True PTC alkylations of ethyl acetoacetate and acetylacetone were first carried out using benzyl chloride and 1,3-dichloro-2-butene, aqueous potassium hydroxide, and dibenzyldimethylammonium chloride or di-(3-chlorobut-2-enyl)dimethylammonium chloride in equimolar amounts [226, 340].

Methyl acetoacetate was alkylated in chloroform/aqueous sodium hydroxide, in the presence of molar amounts of tetrabutylammonium hydrogen sulfate [376]. With methyl, ethyl, and butyl iodides, only C-alkylation (mainly mono- with a little bis-product) was observed. Isopropyl iodide gave C/O products in the ratio 3:1. Under the same conditions with a 2.4 molar amount of CH_3I, benzyl acetoacetate gave 66% mono-C- and 33% di-C-product [398].

In another approach the acetoacetic ester dissolved in benzene was first reacted with sodium hydride [397]. After addition of the alkylating agent and 10 mole-% of Aliquat 336, the reaction mixture was refluxed for 8 hours. Very good yields of mono-C-products are obtained with allyl and benzyl halides. Simple aliphatic halides require longer reaction times and give some O-alkylation and thus there is little advantage in using PTC [397].

An unusual method of preparing mono-C-alkylation products of alkyl acetoacetates and 1,3-diketones was introduced by Clark and Miller [381]. In aqueous solution tetraethylammonium fluoride and the β-diketone in THF form strong H-bonded monosolvate complexes. After removal of the solvent and reaction with excess alkyl iodide in chloroform at room temperature excellent yields of mono-C-products are obtained.

If the chiral PT catalyst, (—)-N-benzyl-N-methylephedrinium bromide (**B**), in chloroform solution is used, asymmetric alkylation of compound **A** has been claimed by Fiaud [383].

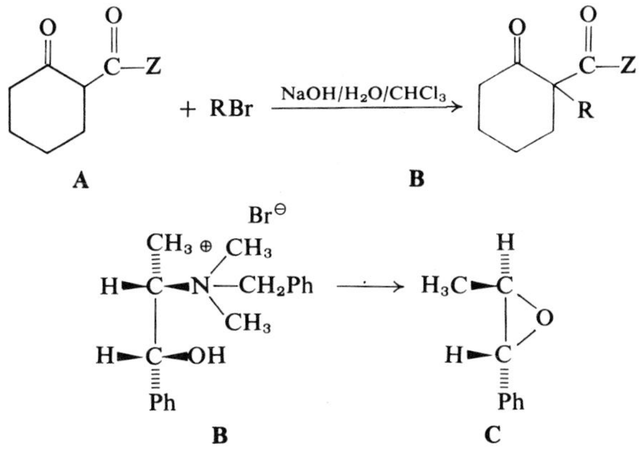

A + RBr $\xrightarrow{NaOH/H_2O/CHCl_3}$ (**B**)

B ⟶ **C**

Scheme 3–72A

Six percent maximum asymmetric induction; no optical activity in hexane solution, and lower optical yields with a catalyst carrying a third N-methyl group instead of the benzyl radical were reported. This is the first asymmetric PTC alkylation reported in the literature. It should be noted though that compounds of the type **B** decompose easily under PTC conditions to give asymmetric epoxides **C** [384]. Indeed, a re-investigation of Fiaud's supposedly highest optical induction showed that at least 98%—perhaps all—of the rotation observed in the raw reaction product was due to impurities [949]. Thus, as was the case in another supposedly asymmetric PTC reaction, the observed rotation could result from **C** or compounds derived therefrom [384]. Compound **A** can also be reacted first with sodium hydroxide, then alkylated in the presence of quaternary ammonium salt to give 84% C-alkylation product [385].

Another recently proposed alkylation method is related to PTC [116]: The OH-form of an anion exchange resin (*e.g.*, Amberlite IRA 900) is treated with an ethanolic solution of a cyclic β-dicarbonyl compound. The β-diketonate resin is then shaken with the alkylating agent in ethanol (giving almost exclusive C-alkylation) or in toluene (mainly C- with some O-alkylation).

Using NBu_4Br as catalyst, various benzoins and desoxybenzoin were alkylated [900]. Dimethyl sulfate yielded almost exclusively an E/Z mixture of α,β-dimethoxystilbene with the E isomer in slight excess. Allyl and benzyl derivatives, on the other hand, gave mainly α-allyl- or benzylbenzoin and allyl- or benzyl ethers respectively. This is in agreement with the hypothesis that soft alkylating agents give mainly C-alkylation products. For ethylation in benzene at 60 °C with a variety of leaving groups the following results were obtained:

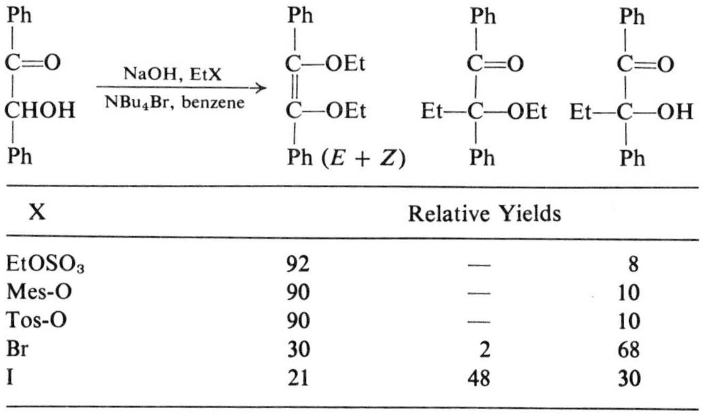

Ph–C(=O)–CHOH–Ph $\xrightarrow[\text{NBu}_4\text{Br, benzene}]{\text{NaOH, EtX}}$ Ph–C(OEt)=C(OEt)–Ph ($E + Z$) + Ph–C(=O)–C(Et)(OEt)–Ph + Ph–C(=O)–C(Et)(OH)–Ph

X	Ph–C(OEt)=C(OEt)–Ph ($E + Z$)	Ph–C(=O)–C(Et)(OEt)–Ph	Ph–C(=O)–C(Et)(OH)–Ph
	Relative Yields		
$EtOSO_3$	92	—	8
Mes-O	90	—	10
Tos-O	90	—	10
Br	30	2	68
I	21	48	30

Scheme 3–72B

The O/C alkylation ratio in the ethylation of desoxybenzoin for various leaving groups was as follows: $OSO_2OC_2H_5$ 2.22, OMes 3.0, OTos 2.7, Br 0.39, I 0.03 [900].

Activated methylene compounds such as **D** can be transformed into their anions and extracted with tetrabutylammonium ion into an organic medium. Reaction with carbon disulfide gives tautomeric anions **E** and **F** of dithioacids and of gem. mercapto-thiolates respectively, which in turn can be deprotonated by base to give dianions **G**.

Examples of **D** so reacted were diethylmalonate [391], methylcyanoacetate [391], malodinitrile [391], acetylacetone [30], benzoylacetone [30], dibenzoylmethane [30], methyl and ethyl acetoacetate [392], ethyl benzoylacetate [392], and cyanoacetone [392]. Some of the salts **E**/**F** were isolated in crystalline form. Stirring in chloroform at room temperature for up to 18 hours with an alkylating agent leads to dithioesters **H** and ketene thioacetals **I**. Since the anions **F** are very similar in basicity to the anions derived from the dithioesters **H**, deprotonation of the compounds **H** occurs to some extent even if in the beginning only monosalts are present. Consequently, bisalkylation products **I** are always formed to some extent [30, 391, 392]. For example, if X = Y = COOEt, and R = ethyl or allyl, **H** and **I** are formed in a 2:1 ratio. Dialkylation by methylene chloride gives **K**. Various rearrangements of these sulfur compounds and intramolecular condensations have been described [30, 391, 392; 107, 209].

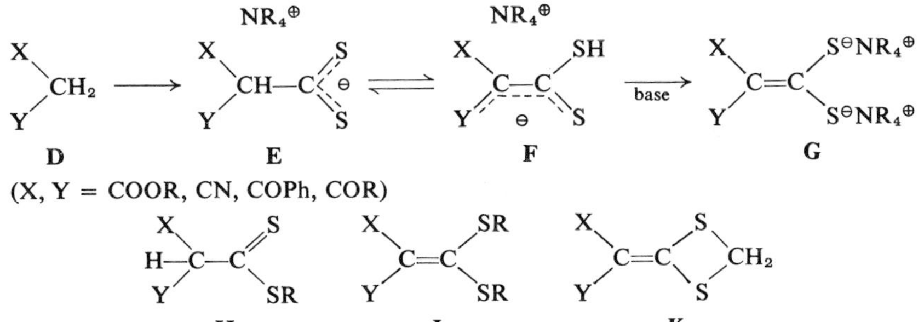

Scheme 3–73

The PTC reaction between 2-nitropropane and *p*-nitrobenzyl chloride in benzene/aqueous NaOH/NBu_4OH gave only C-alkylation in a radical anion chain process, while benzyl chloride gave S_N2 O-alkylation leading ultimately to the aldehyde [838]:

$$p\text{-}NO_2\text{—}C_6H_4\text{—}CH_2Cl \longrightarrow p\text{-}NO_2\text{—}C_6H_4\text{—}CH_2\text{—}CMe_2\text{—}NO_2$$

$$C_6H_5\text{—}CH_2Cl \longrightarrow C_6H_5\text{—}CH_2\text{—}O\text{—}N(O){=}CMe_2 \longrightarrow \longrightarrow C_6H_5CHO$$

As mentioned previously, phenols do not normally exhibit any complications in C-alkylations. 2,4,6-Trihydroxyphenylketones, however, give only C-alkyl compounds in PTC-reactions with prenyl bromide [1084]. Anthrone and acridone yield O-methyl

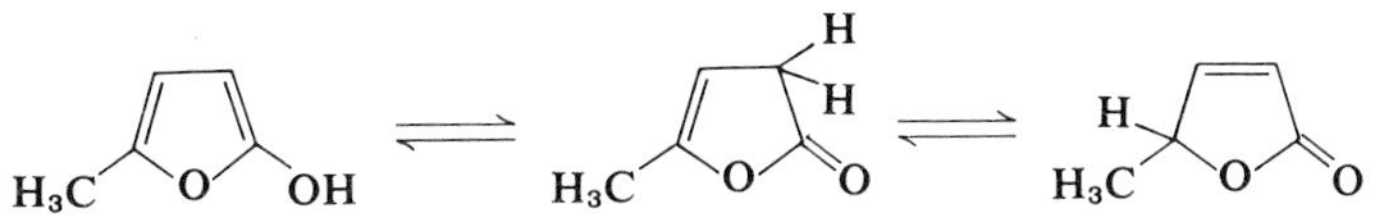

Scheme 3–74

products in contrast [1085]. Related heterocyclic tautomeric systems have been investigated extensively. In the furane series, extractive alkylation (molar amounts of

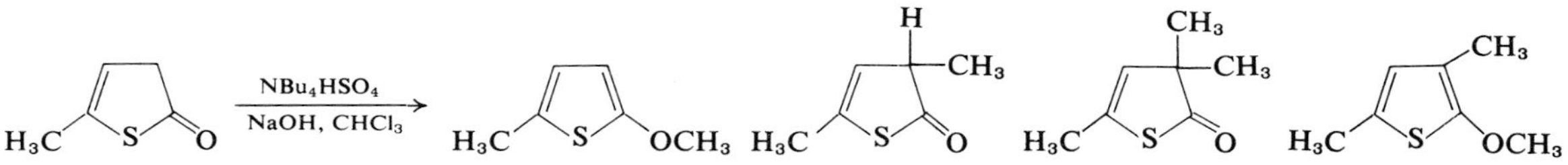

H_3C S O $\xrightarrow[\text{NaOH, CHCl}_3]{\text{NBu}_4\text{HSO}_4}$ H_3C S OCH_3 H_3C S O (H, CH_3) H_3C S O (CH_3, CH_3) H_3C S OCH_3 (CH_3)

Reagent	Relative Yields (%)			
CH_3I	12	86	2	—
$(CH_3O)_2SO_2$	90	4	2	4

Scheme 3–75A

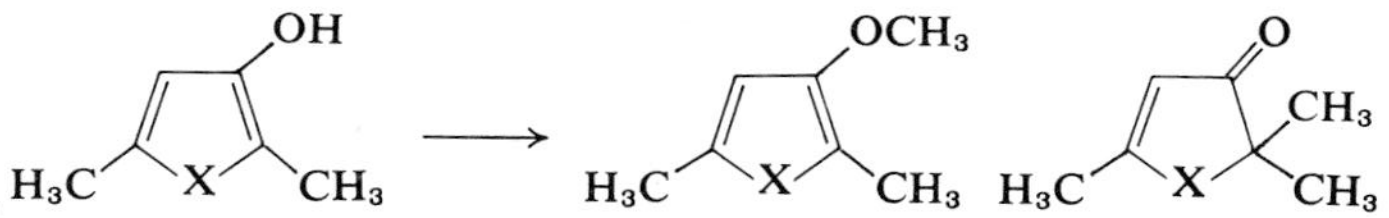

H_3C X CH_3 (OH) $\longrightarrow$ H_3C X CH_3 (OCH_3) H_3C X CH_3 (O, CH_3)

X	Reagent	Relative Yields (%)	
O	CH_3I	2	98
	$(CH_3O)_2SO_2$	54	46
S	CH_3I	35	65
	$(CH_3O)_2SO_2$	90	10
Se	$(CH_3O)_2SO_2$	93	7

Scheme 3–75B

tetrabutyl ammonium hydrogen sulfate, excess CH_3I) gave only C^3-alkylation in low yield due to competing ring opening [387]. In the case of 2-hydroxythiophenone and its tautomers [386] a number of products were obtained. Not unexpectedly, regio-selectivity, depending on the leaving group of the alkylating agent, was observed: the better leaving group methoxysulfonyl gives a higher yield of O-alkylation. The C/O ratio is similar to that found with thallium salts. The ion pair extraction method, however, is easier and safer since it occurs at room temperature. A considerable number of other 2-hydroxythiophene and -selenophene derivatives carrying substituents in various positions were reacted similarly. The trends observed were analogous to the example given above [386–390]. 3-Hydroxyfuranes, -thiophenes, and -selenophenes have also been investigated (Scheme 3–75B [904]).

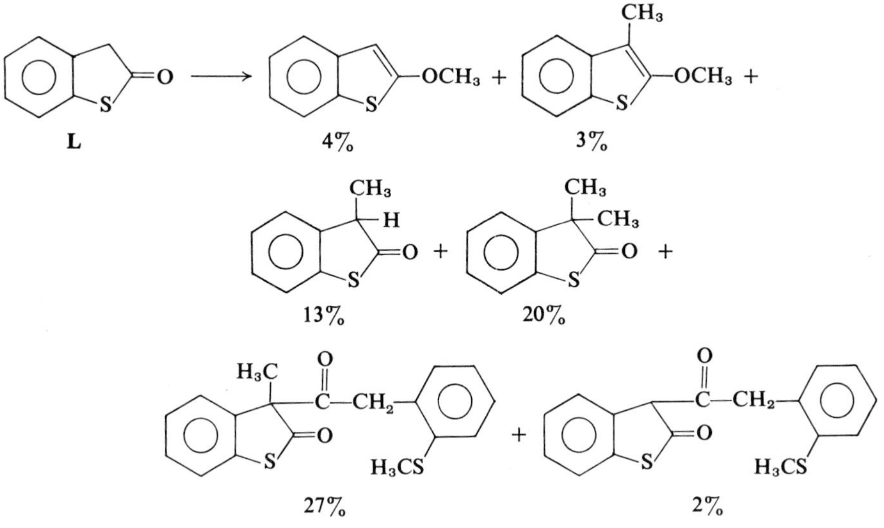

L 4% 3% 13% 20% 27% 2%

Scheme 3–76

Benzo[b]thiophene-2(3H)one (**L**) was methylated after extraction of the tetrabutylammonium salt into chloroform and drying of the solution. The mixture shown below was formed. Alternative reactions of the sodium or the thallium salts also gave complex mixtures which were less easy to separate than that obtained by the solvent extraction process [395].

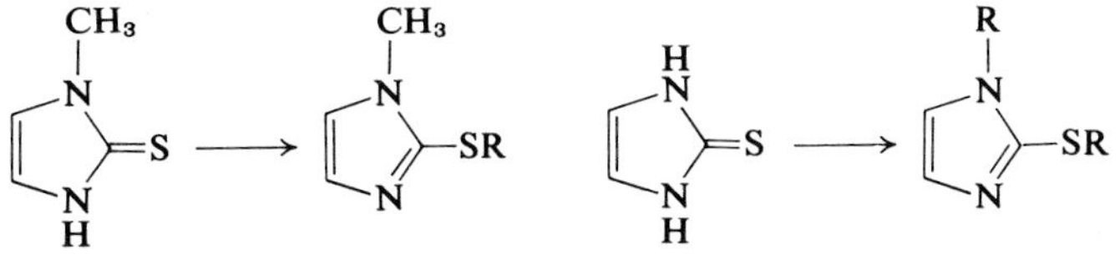

Scheme 3–77

2-Thioxo-2,3-dihydroimidazole and its 1-methyl derivative were alkylated with various reagents in a two-phase benzene/water system in the presence of only 6 mole-% of tetrabutylammonium bromide (6 h, 60 °C) [393]. With a 1-methyl substituent, only S-alkylation occurred. In the unsubstituted case only N,S-dialkylation was observed. 5,5-Dimethylisoxazolidone leads to 62–85% N- and 38–15% O-alkylation (depending on RX) under similar conditions [962]. Only S-alkylation was found with Δ4-thiazolinethiones-2 and related compounds [1086]. Furthermore, O- *vs.* N-alkylation in hydroxy pyridines in benzene/water at 60 °C is also reported [394]. 3-Hydroxypyridine gave only O-alkylation as expected. 4-Pyridone, however, led to an O/N product ratio of 1/3 O- and 2/3 N-product more or less irrespective of the alkylating agent (*n*-butyl, isobutyl and allyl bromide). 2-Pyridone gave only 15–25% O-alkylation [394].

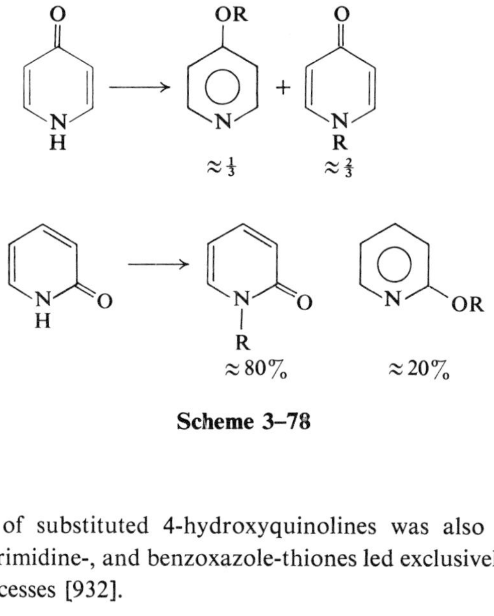

$\approx \frac{1}{3}$ $\approx \frac{2}{3}$

$\approx 80\%$ $\approx 20\%$

Scheme 3–78

PTC alkylation of substituted 4-hydroxyquinolines was also investigated [871]. 2-Pyridine-, 2-pyrimidine-, and benzoxazole-thiones led exclusively to S-alkyl derivatives in PTC processes [932].

Typical experimental procedures are presented below.

Intermediate isolation of the enolate salt. Alkylation of acetylacetone: 0.5 mole of tetrabutylammonium hydrogen sulfate was added to a cold solution of 1.1 moles of NaOH in 500 ml H_2O. After the addition of 0.5 mole acetylacetone the solution was extracted with 500 ml chloroform. After drying and evaporation the crystalline tetrabutylammonium acetylacetonate salt, m.p. 155 °C after recrystallization from acetone, was obtained in 70% yield. This salt can be alkylated in various solvents [375].

Brändström's extractive alkylation procedure (molar amount of catalyst). Alkylation of methyl acetoacetate: A solution of 0.1 mole tetrabutylammonium hydrogen sulfate and 0.2 mole NaOH in 75 ml H_2O was added to a stirred solution of 0.1 mole methyl acetoacetate and 0.2 mole alkyl iodide in 75 ml chloroform. An exothermic reaction occurred. When neutrality was reached after *ca.* 15 minutes the layers were separated, the chloroform evaporated, and the tetrabutylammonium iodide precipitated by the addition of ether. After filtration the solvent was evaporated *in vacuo* [376].

Use of small amounts of PTC catalyst. S-Alkylation of 1-methyl-2-thioxo-2,3-dihydroimidazol: 0.05 mole of the compound, 0.05 mole of alkyl halide and 0.003 mole of tetrabutylammonium bromide were dissolved in 150 ml benzene. 15 ml 40% aqueous sodium hydroxide were added, and the mixture was stirred for 6 hours at 60 °C. The organic layer was then separated, dried, and the solvent removed. The residue was distilled *in vacuo* or recrystallized from ethanol [393].

3.3.6. Isomerizations and H/D Exchange

In spite of the fact that these are the simplest possible reactions in the presence of strong aqueous alkali metal hydroxides, so far these conversions are mechanistically not well understood. Deprotonation can occur at the phase boundary, in the organic phase with extracted ammonium hydroxide, or in an inverted micellar environment. After that protonation or deuteration of the intermediate carbanion must occur either at the phase boundary or by the small amount of D_2O present in the organic medium coextracted by the anion. Further research in this area is needed before any definite conclusions can be reached.

When activated (*e.g.*, benzyl) acetylenes are stirred in dichloromethane with solid potassium hydroxide in the presence of TEBA isomerization occurs in almost quantitative yield (**A** → **B**, R = H, CH_3, C_6H_5). The mixture of benzylacetylene and phenylallene obtained from the reaction of propargyl bromide with phenylmagnesium bromide is transformed quantitatively into 1-phenyl-1-propyne [400]. All these

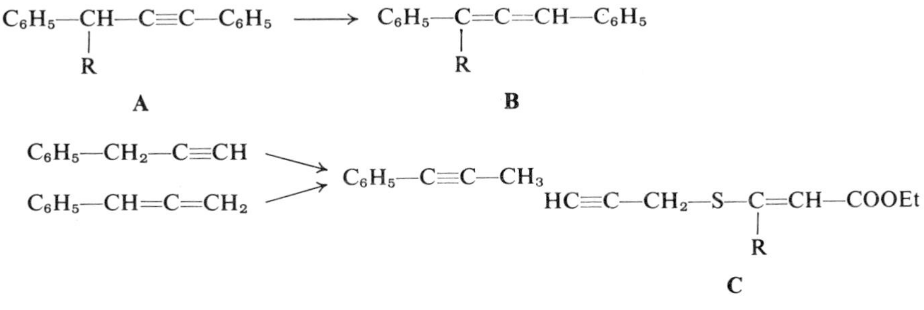

$$C_6H_5{-}\underset{\displaystyle R}{\underset{|}{CH}}{-}C{\equiv}C{-}C_6H_5 \longrightarrow C_6H_5{-}\underset{\displaystyle R}{\underset{|}{C}}{=}C{=}CH{-}C_6H_5$$

A **B**

$$\left.\begin{matrix} C_6H_5{-}CH_2{-}C{\equiv}CH \\ C_6H_5{-}CH{=}C{=}CH_2 \end{matrix}\right\} \longrightarrow C_6H_5{-}C{\equiv}C{-}CH_3$$

$$HC{\equiv}C{-}CH_2{-}S{-}\underset{\displaystyle R}{\underset{|}{C}}{=}CH{-}COOEt$$

C

Scheme 7–79A

reactions are much faster in the presence of catalyst. Less acidic compounds cannot be isomerized; methyl propargyl ether, for instance, is not converted into the more stable methoxyallene under these conditions [400].*) Thioethers **C** (R = CH_3, C_6H_5), however, could be converted into the allene compounds with 2 M NaOH/tetrabutylammonium hydrogen sulfate [209].

*) Alkyne/allene isomerizations of nonactivated compounds are possible with KO*t*-Bu/18-crown-6 in petroleum ether, however [1105].

A remarkable series of events took place when chloromethyl phenyl sulfide and the propargylic sulfide shown were reacted in the presence of NaOH and Aliquat 336 [971]:

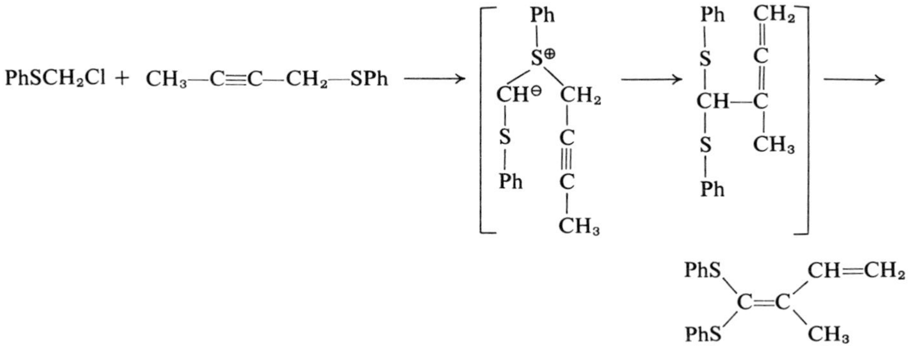

Scheme 7–79B

Using tetrabutylammonium cyanide in the absence of water, the isomerization of 3-butenenitrile to crotononitrile (which is dimerized and polymerized) and of 3-hexenedinitrile to the conjugated isomer was demonstrated [413]. Cyanide ion behaved as a base in the solvent used (acetonitrile) because no H-bonding was possible.

$$NC{-}CH_2{-}CH{=}CH_2 \longrightarrow NC{-}CH{=}CH{-}CH_3$$

$$NC{-}CH_2{-}CH{=}CH{-}CH_2{-}CN \longrightarrow NC{-}CH{=}CH{-}(CH_2)_2{-}CN$$

Starks showed that complete H/D-exchange at the C^1 and C^3 hydrogens in 2-octanone resulted when the ketone was stirred with 5% NaOD in D_2O for 0.5 hours at room temperature in the presence of 5% quaternary ammonium salt. Without the catalyst (Aliquat 336?) only 5% deuterium exchange was observed after 3 hours [4]. Threefold repeated application of this procedure gave >99% D_4 compounds with 1-acetylcyclohexane and -cyclopentane [401]. This method has also been applied to the more complex compound **D** where deuteration in the indicated positions was reported [402].

Other authors work with K_2CO_3/D_2O/Aliquat 336 [975]. H/D-exchange plus isomerization of 1-alkylindenes into the 3-alkyl compounds was effected with CH_2Cl_2/NAOD/D_2O/TEBA in 95% yield [986]. Investigation of various catalysts for this

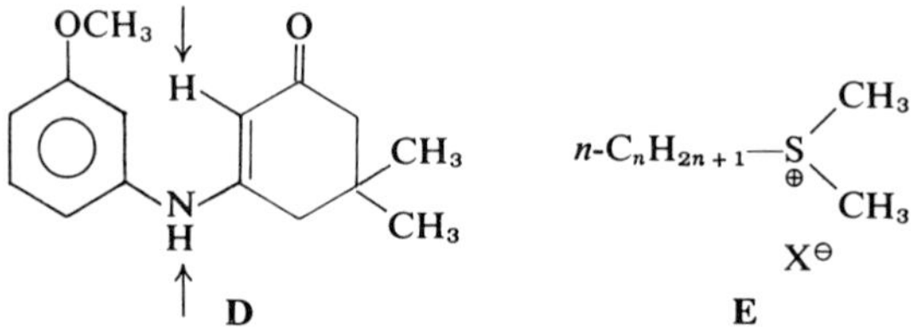

D **E**

Scheme 3–80

H/D-exchange with 5% NaOD in D_2O showed that the most lipophilic catalysts were the best: $N(C_7H_{15})_4^{\oplus}$ very efficient; $NBu_4^{\oplus}$ very slow, because this salt is distributed into the aqueous phase [853].

Base catalyzed rates of H/D-exchange of the methyl groups in sulfonium halides **E** in homogeneous D_2O systems have also been investigated [406]. The rates for longer alkyl chain compounds (n = 10, 12) were found to be much larger than those of shorter chain compounds (n = 1–8). With lauryldimethylsulfonium salts (**E**, n = 12) rates increased in the order X = I < Br < Cl, and observed effects were explained on the basis of micellar effects.

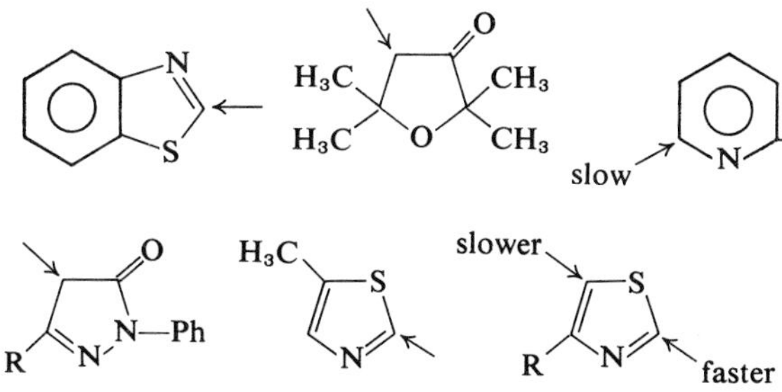

slow

slower

faster

Scheme 3–81

H/D-exchange in heterocyclic compounds with 10 M NaOD and catalyzed by tetrabutylammonium bromide as investigated [403–405]. The positions of exchange are indicated above. The best organic solvents for the reaction were benzene and cyclohexane. Tetrabutylammonium bromide was the best catalyst tested, better than cetyl-trimethylammonium bromide for example [403, 405]. Procedures have been introduced for total H/D-exchange of diazomethane [1088] and for the formation of D_2 norbornanone [1089].

3.3.7. Additions across Multiple CC-Bonds

3.3.7.1. Additions to Acetylenes

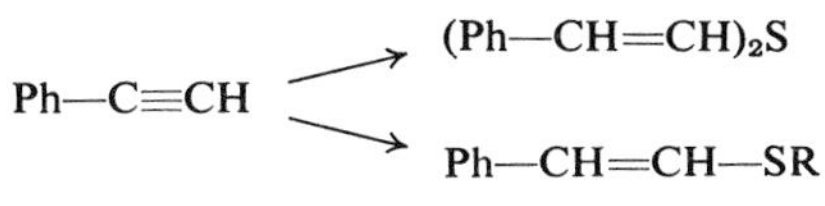

$$Ph{-}C{\equiv}CH \rightarrow (Ph{-}CH{=}CH)_2S \quad ; \quad Ph{-}C{\equiv}CH \rightarrow Ph{-}CH{=}CH{-}SR$$

Scheme 3–82

The addition of aqueous sodium sulfide or ethylthiolate to phenylacetylene has been catalyzed by dibenzo-18-crown-6 or (less efficiently) by TEBA [407]. Hydrochloric acid can be added to triple bonds with the help of $NEt_3H^{\oplus}HCl_2^{\ominus}$ [880] either without solvent or dissolved in aprotic solvents [881]. Compounds reacted include various 1-phenylalkynes, dimethyl acetylenedicarboxylate, and 3-chloropropyne. It has been shown that the same type of process can be performed using PTC with concentrated hydrochloric acid/Aliquat 336 via extraction of $NR_4^{\oplus}HCl_2^{\ominus}$ [882].

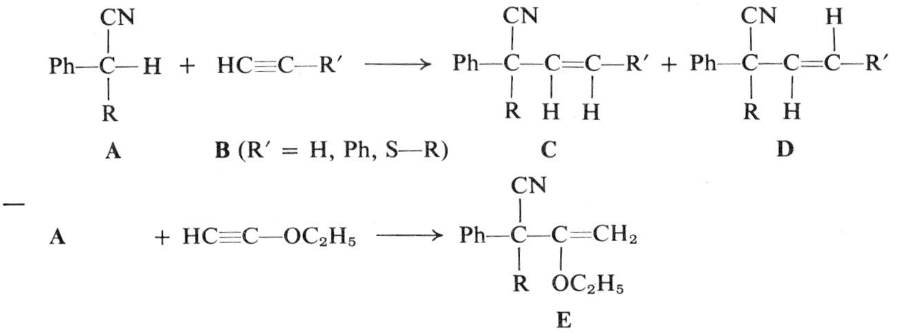

$$\underset{\mathbf{A}}{\text{Ph-C(CN)(R)-H}} + \underset{\mathbf{B}\ (R' = H,\ Ph,\ S\text{-}R)}{HC{\equiv}C\text{-}R'} \longrightarrow \underset{\mathbf{C}}{\text{Ph-C(CN)(R)-CH=CH-R'}} + \underset{\mathbf{D}}{\text{Ph-C(CN)(R)-CH=C(H)-R'}}$$

$$\mathbf{A} + HC{\equiv}C\text{-}OC_2H_5 \longrightarrow \underset{\mathbf{E}}{\text{Ph-C(CN)(R)-C(OC}_2\text{H}_5\text{)=CH}_2}$$

Scheme 3–83A

Mąkosza described the addition of 2-phenylbutyronitrile (**A**) to acetylene and phenylacetylene in an Organic Synthesis procedure [408]. Powdered solid potassium hydroxide is used in DMSO together with TEBA, and the reactions are exothermic ($\approx 60\%$ yield). Alkylphenylacetonitriles (**A**) were added to phenylacetylene [409], alkylthioacetylene [410], and ethoxyacetylene [410] under similar conditions, using NaOH sometimes instead of KOH. Yields were generally in the 80–95% range. With phenylacetylene and alkylthioacetylene the same orientation as in **C** and **D** was found. Frequently the *cis* isomers **C** predominated over the *trans* compounds. Ethoxyacetylene, however, led to **E**.

Cycloadditions of aryl phosphines and arsines to penta-1,4-diynes and its analogues with a heteroatom in the 3-position can be catalyzed by small amounts of solid

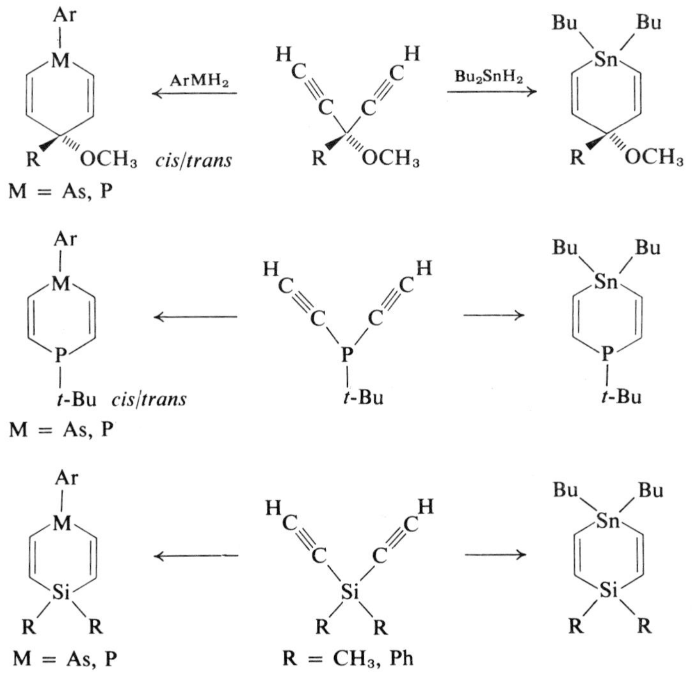

ArMH_2 ← $(HC{\equiv}C)_2C(R)(OCH_3)$ → Bu_2SnH_2: 4-R-4-OCH_3-1-Ar-1-M-cyclohexa-2,5-diene (*cis/trans*), M = As, P; 1,1-dibutyl-4-R-4-OCH_3-stannacyclohexa-2,5-diene

$(HC{\equiv}C)_2P(t\text{-Bu})$: 1-Ar-1-M-4-(*t*-Bu)-phosphacyclohexadiene (*t*-Bu *cis/trans*), M = As, P ← → 1,1-dibutyl-4-(*t*-Bu)-stannaphosphacyclohexadiene

$(HC{\equiv}C)_2SiR_2$, R = CH_3, Ph: 1-Ar-1-M-4,4-R_2-silacyclohexadiene, M = As, P ← → 1,1-dibutyl-4,4-R_2-stannasilacyclohexadiene

Scheme 3–83B

potassium hydroxide and 18-crown-6 in benzene [914]. The reactions are often exothermic and give yields of up to 90%. Contrary to expectations, even di-*n*-butyltinhydride could be added under these conditions in very satisfactory yields. These cycloadditions, however, are limited to terminal acetylenes, unlike additions in liquid ammonia with lithium amide [914].

Finally, a very unusual intramolecular PTC addition of an amide nitrogen to an allene ether formed *in situ* has been reported [1090].

3.3.7.2. Michael Additions

Triton B (benzyltrimethylammonium hydroxide) has long been used as a catalyst for addition reactions, which were performed mostly in homogeneous media. Many examples of this type of reaction can be found in the literature (for leading references, *cf.*, [411a, 412]).

Less well known are reactions catalyzed by cyanide and fluoride ions in homogeneous solution. Although some of these have been known for a long time (*cf.*, [411b]) they have only recently attracted attention. Tetrabutylammonium cyanide, for instance, brings about the addition of nitroalkanes, alcohols, and chloroform to α,β-unsaturated ketones and esters in THF or acetonitrile solution [413]. In these solvents, nitrile/quaternary ammonium ion pairs are not shielded by hydrogen bridging and thus behave as bases. Crotononitrile is dimerized, acrylonitrile polymerized [413].

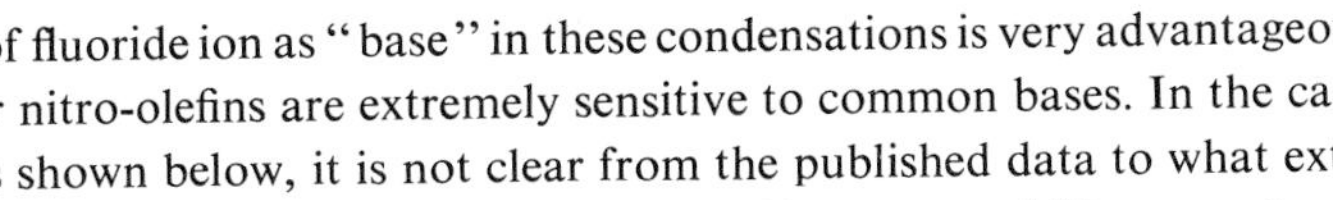

$$2H_3C{-}CH{=}CH{-}CN \xrightarrow[NBu_4CN]{} H_3C{-}CH{=}\underset{CN}{C}{-}\underset{CH_3}{CH}{-}CH_2{-}CN$$

Scheme 3–84

Addition reactions of tetraalkylammonium fluoride and reactions catalyzed by it in homogeneous solution are illustrated in Scheme 3–85.

The use of fluoride ion as "base" in these condensations is very advantageous because the lower nitro-olefins are extremely sensitive to common bases. In the case of some examples shown below, it is not clear from the published data to what extent potassium fluoride is soluble in boiling xylene. Small amounts of 18-crown-6 were used in later reactions, therefore, together with the potassium fluoride in order to make the catalyst more soluble in solvents like benzene, dichloromethane, and acetonitrile [416]. The addition of activated methylene groups bearing nitro and cyano substituents also occurred in high yield [416].

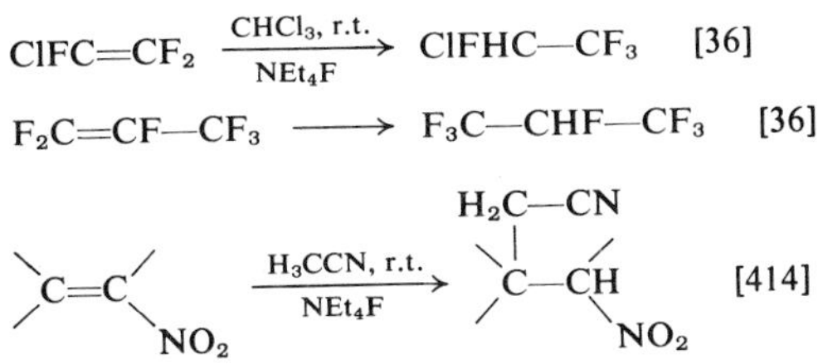

$$ClFC{=}CF_2 \xrightarrow[NEt_4F]{CHCl_3,\ r.t.} ClFHC{-}CF_3 \quad [36]$$

$$F_2C{=}CF{-}CF_3 \longrightarrow F_3C{-}CHF{-}CF_3 \quad [36]$$

$$>C{=}C<_{NO_2} \xrightarrow[NEt_4F]{H_3CCN,\ r.t.} >\overset{H_2C-CN}{C}{-}CH<_{NO_2} \quad [414]$$

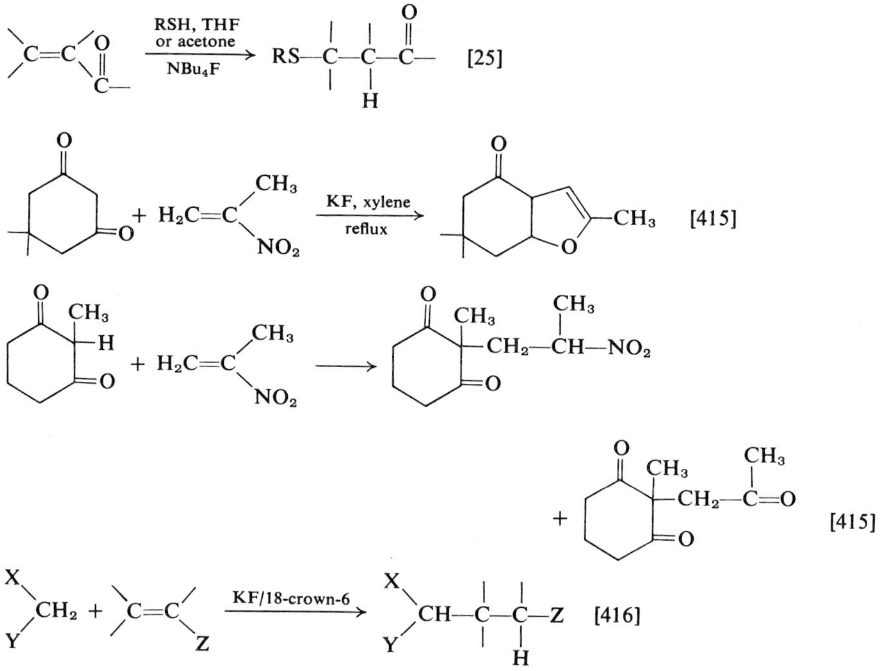

$$>C{=}C{<}C({=}O){-} \xrightarrow[NBu_4F]{RSH,\ THF \text{ or acetone}} RS{-}C{-}C(H){-}C({=}O){-} \quad [25]$$

$$\text{5,5-dimethylcyclohexane-1,3-dione} + H_2C{=}C(CH_3)NO_2 \xrightarrow[\text{reflux}]{KF,\ \text{xylene}} \text{(bicyclic furan product)}{-}CH_3 \quad [415]$$

$$\text{2-methylcyclohexane-1,3-dione} + H_2C{=}C(CH_3)NO_2 \longrightarrow \text{2-methyl-2-}(CH_2{-}CH(CH_3){-}NO_2)\text{cyclohexane-1,3-dione} + \text{2-methyl-2-}(CH_2{-}C(CH_3){=}O)\text{cyclohexane-1,3-dione} \quad [415]$$

$$XYCH_2 + >C{=}C{<}Z \xrightarrow{KF/18\text{-crown-6}} XYCH{-}C{-}C(H){-}Z \quad [416]$$

Scheme 3–85

As is shown in Scheme 3–86 (page 139), 18-crown-6/dry potassium cyanide was used for the addition of $CN^{\ominus}$ to methacrylonitrile in acetonitrile [77]. It is interesting to note that acetone cyanohydrin can serve as a carrier for hydrogen cyanide. KCN + 18-crown-6 + acetone cyanohydrin is a simple, efficient, and stereoselective hydrocyanating agent [417]. A solution of the substrate and crown ether (mole per mole) in a suitable solvent such as benzene or acetonitrile is poured over an excess of dry potassium cyanide. Acetone cyanohydrin (1.2 molar amount) is then added and the two phases are either stirred vigorously or refluxed. Yields of up to 85% are obtained. When there are two possible stereoisomers, the more thermodynamically stable product is preferred [417].

Homogeneous Michael additions of 2-nitropropane to enones were catalyzed by NEt_4F [1092].

PTC additions of malonic and acetoacetic esters to α,β-unsaturated aldehydes led to substantial resinification when carried out with concentrated NaOH/TEBA. Working in benzene with solid potassium or sodium carbonate/TEBA, however, acceptable yields of adducts were obtained [1093].

When PTC techniques are used in the Michael addition of diethyl malonate, acetylacetone, and dibenzoylmethane to 3-nitro-2-enopyranoside **G** or its precursor **F**, the thermodynamically less stable manno isomers only are formed [418]. However,

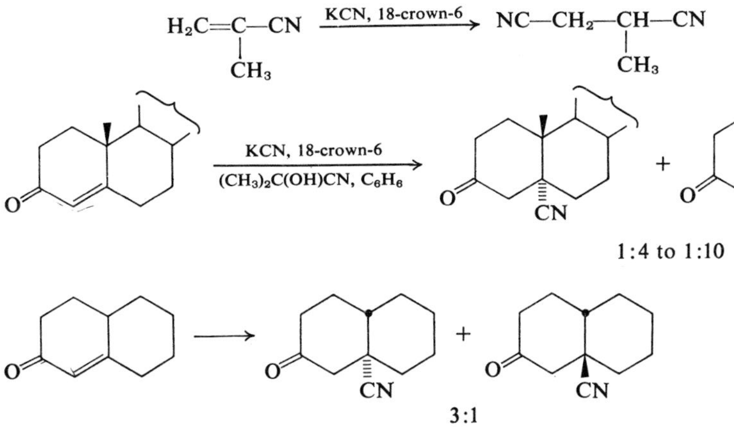

$H_2C{=}C(CH_3){-}CN \xrightarrow{KCN,\ 18\text{-crown-6}} NC{-}CH_2{-}CH(CH_3){-}CN$

KCN, 18-crown-6
$(CH_3)_2C(OH)CN$, C_6H_6

CN + CN

1:4 to 1:10

CN + CN

3:1

Scheme 3–86

malodinitrile gave a mixture of both the gluco and manno isomers. The latter were shown to isomerize to the former under the reaction conditions. These reactions were performed in benzene/0.2N NaOH at room temperature in the presence of tributylhexadecylphosphonium bromide. The same process has been applied to another series of β-glycosidic nitro sugars [419], and precursors for branched chain sugars were prepared similarly [1091]. In a very slow reaction using tributylhexadecylphosphonium (4 days at room temperature) **H** in benzene and sodium nitrate in water gave 20% **I** together with 40% **K** and 20% starting material [425]. Although the yield of **I** is low, the process proved useful since **K** can be reconverted to **H**. In another PTC Michael reaction, the acetate of alcohol **I** was transformed into **L** in 71% yield [425]. Here acetylacetone was the addend, the primary adduct being cleaved under the reaction conditions (*cf.*, Scheme 3–87, page 140).

In a one-pot saponification Michael addition-aldol condensation reaction, 5-thiacyclohexenecarboxaldehydes can be synthesized. A typical example is the reaction between **M** and crotonaldehyde to give **N** in 84% yield (CH_2Cl_2, 50% NaOH, $NBu_4^{\oplus}I^{\ominus}$) [426]. More complex product mixtures are formed from **M** and acrolein, because partial elimination of thiolacetate to form crotonaldehyde intervenes. Under similar conditions, α-mercaptocarbonyl compounds and 2-chloroacrylonitrile lead to dihydrothiophenepoxides (**O**) in excellent yields [426].

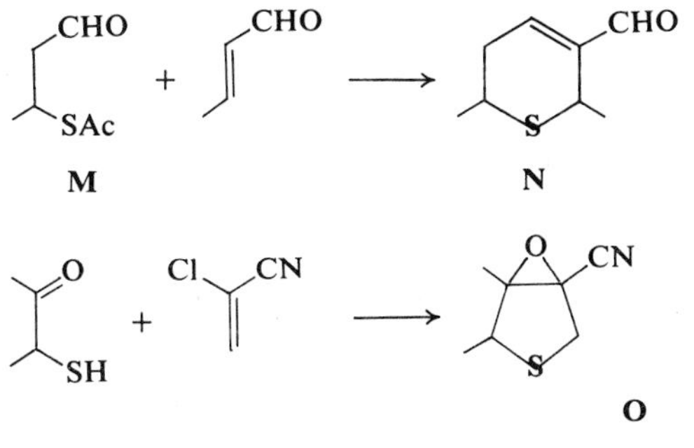

CHO SAc + CHO → CHO S

M **N**

O SH + Cl CN → O CN S

O

Scheme 3–88

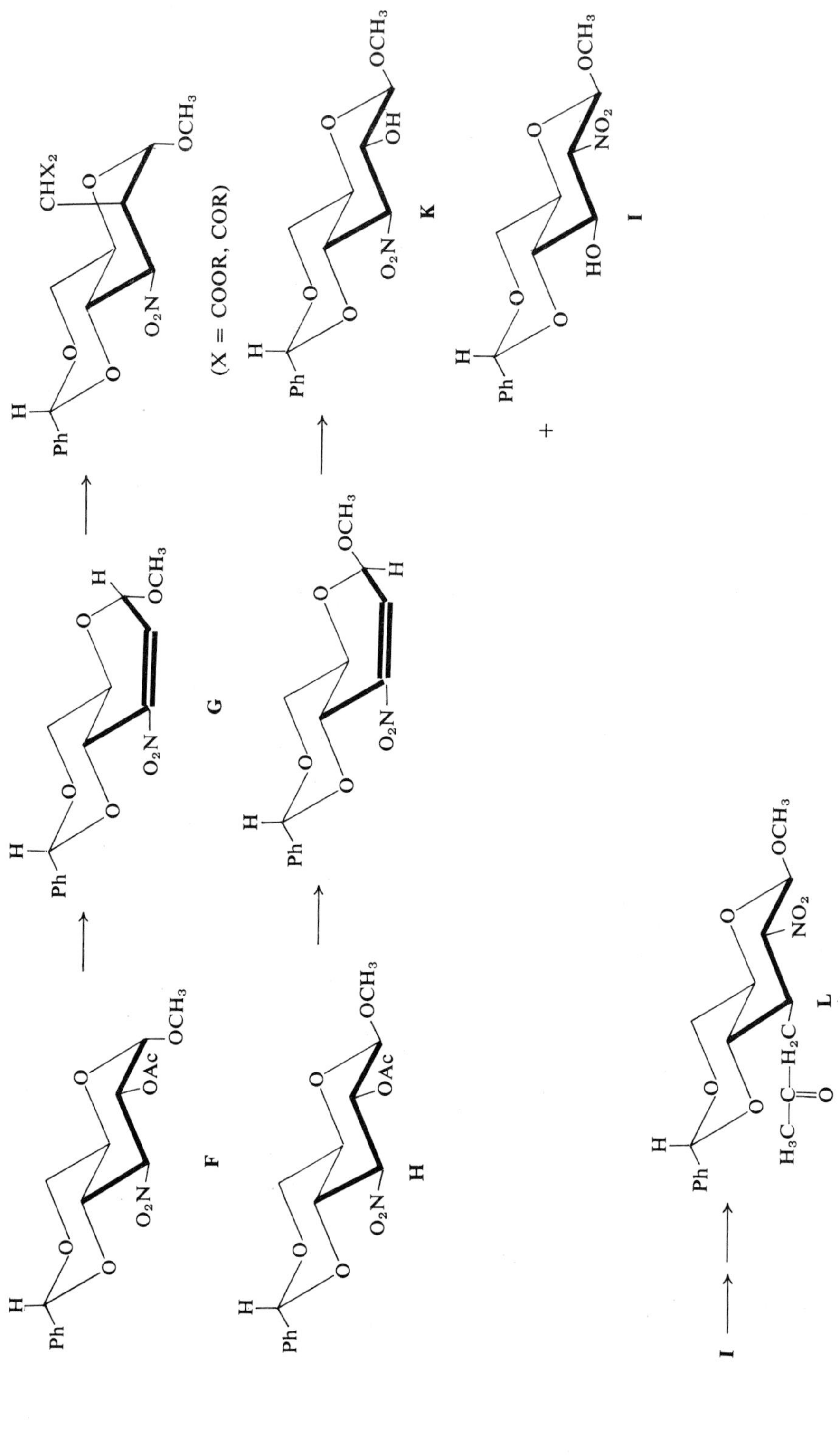

F → G → (X = COOR, COR)

H → → K + I

I → → L

Scheme 3–87

Phenyl vinyl sulfone and α-chloropropionitrile were reacted in the presence of 50% aqueous sodium hydroxide and TEBA. Compound **P** was formed along with trace amounts of stereoisomeric cyclopropanes **Q** and **R** after one hour at room temperature [421]. Under more drastic conditions (40 °C, 2 hours, a few ml DMSO as cocatalyst) only **Q** and **R** were formed. Under mild conditions very fast cyclization of the intermediate carbanions occurred with *tert*-butyl acrylate or acrylonitrile as activated olefins, so that only cyclopropane products were obtained [421].

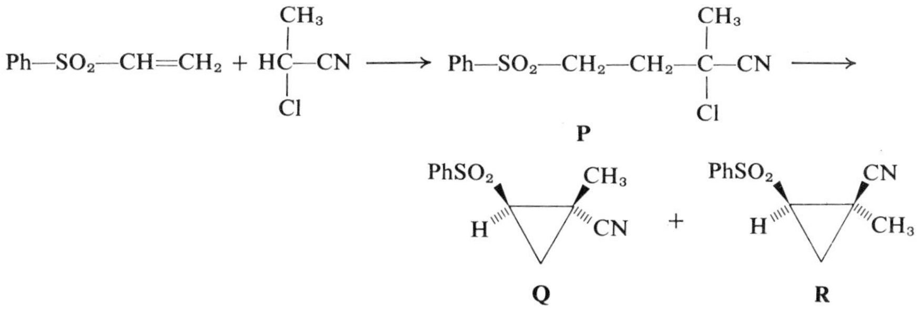

$$Ph{-}SO_2{-}CH{=}CH_2 + HC(CH_3)(Cl){-}CN \longrightarrow Ph{-}SO_2{-}CH_2{-}CH_2{-}C(CH_3)(Cl){-}CN \ (\mathbf{P}) \longrightarrow \mathbf{Q} + \mathbf{R}$$

Q: cyclopropane with $PhSO_2$, H on one carbon and CH_3, CN on the adjacent carbon; **R**: $PhSO_2$, H and CN, CH_3 (stereoisomer)

Scheme 3–89

Compounds of the type **S** (R = H, Ph; R′ = *tert*-butyl, (−)-menthyl) were reacted with *tert*-butyl acrylate, catalyzed by potassium *tert*-butoxide, in toluene, THF, or DMF, or by lithium bis(trimethylsilyl)amide in THF [422]. PTC [$NBu_4^{\oplus}Br^{\ominus}$/50% aqueous NaOH) can also be used [422].

PTC yields compared well with those of the alternative methods. Of the two possible isomers *Z* and *E*, the *E* isomer was formed preferentially in both the PTC and *tert*-butoxide reactions, whereas the *Z* isomer was preferred with the other base/solvent

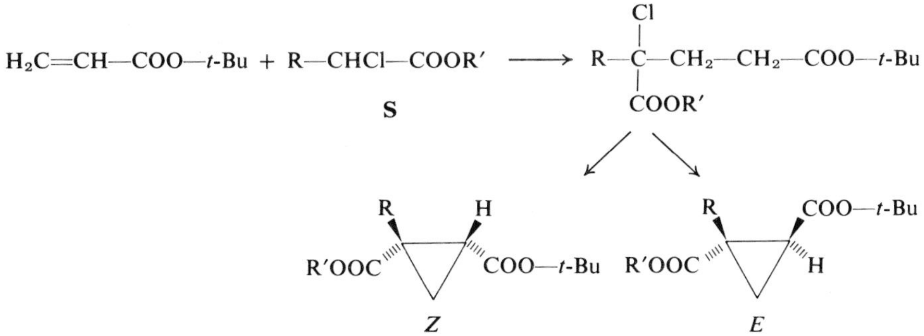

$$H_2C{=}CH{-}COO{-}t\text{-}Bu + R{-}CHCl{-}COOR' \ (\mathbf{S}) \longrightarrow R{-}C(Cl)(COOR'){-}CH_2{-}CH_2{-}COO{-}t\text{-}Bu \longrightarrow Z + E$$

Z: cyclopropane with R, R′OOC on one carbon and H, COO–*t*-Bu on the adjacent carbon; *E*: R, R′OOC and COO–*t*-Bu, H

Scheme 3–90

systems [422]. This type of Michael addition with consecutive cyclopropane ring closure has been extended to reactions between unsaturated nitriles, ketones, aldehydes, esters, phosphonates, and phosphonium salts with α-chloronitroalkanes [1094], bromomalonic esters [1095], and α-chloronitriles [1096]. Concentrated sodium hydroxide and an onium salt could be used throughout, but at least in some cases a slow conversion is possible without catalyst. Here, then, the stereoselectivity is different from the catalyzed reactions [1096] (*cf.*, similar effects in the Darzens reaction, Section 3.3.8.3).

Stereoselectivity was also reversed in ring closure reactions of **T** in chlorobenzene with catalytic amounts of potassium *tert*-butoxide either with or without 18-crown-6 at -40 °C [423]. In the former case the **U**/**V** ratio is 62:38, in the latter 38:62.

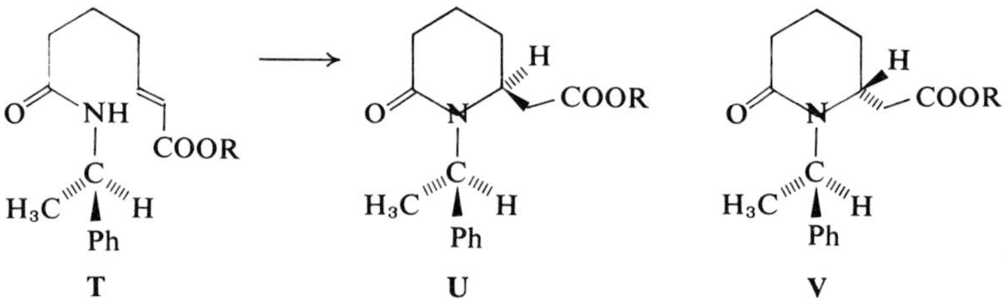

T **U** **V**

Scheme 3–91

Although not a "Michael acceptor" [1167], vinyl acetate reacts with phenylacetonitriles, Reissert compounds, and chloroform when these are stirred in benzene with concentrated sodium hydroxide and TEBA [420]. The reaction with chloroform will be considered later in connection with carbene additions.

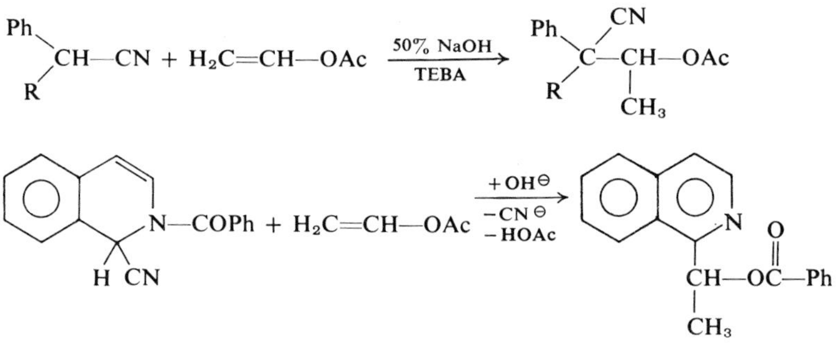

$$\text{Ph(R)CH—CN} + \text{H}_2\text{C=CH—OAc} \xrightarrow[\text{TEBA}]{50\%\ \text{NaOH}} \text{Ph(R)C(CN)—CH(CH}_3\text{)—OAc}$$

$$\text{Reissert compound (N—COPh, H, CN)} + \text{H}_2\text{C=CH—OAc} \xrightarrow[-\text{CN}^\ominus,\ -\text{HOAc}]{+\text{OH}^\ominus} \text{isoquinolyl—CH(CH}_3\text{)—OC(=O)—Ph}$$

Scheme 3–92

It would appear that stereoselective Michael additions with optically active PT catalysts are possible. Wynberg observed optical induction in homogeneous toluene solutions when substituted nitrocompounds were added to methyl vinyl ketone, cyclohexenone, or chalcone using quinine or N-benzylquininium fluoride/KF as catalysts [424, 554, 924]. The enantiomeric excess was highest (68%) in apolar solvents, it was zero in ethanol. For another system, *cf.* [1039].

Polymer-bound cinchona alkaloids have also been used as Michael addition catalysts. In homogeneous solution chemical yields were high but optical yields were disappointingly low [892].

In a number of 1,4 additions catalyzed by quinine derived compounds the enantiomeric yield of chiral products in homogeneous solution was found to be inversely proportional to the dielectric constant of the solvent [1011].

3.3.7.3. Additions across Nonactivated Double Bonds

The addition of iodine thiocyanate to alkenes is accelerated using a two-phase system plus PTC agent Adogen 464 [556]. Thus, after 2 hours at 20 °C 1.25 mmoles cyclohexene, 3 mmoles iodine, 6.25 mmoles potassium thiocyanate in 10 ml chloroform/1 ml water and 0.96 mmole Adogen 464, gave 43% of a 2:1 mixture of **W** and **X**. Other substrates include 5α-androst-2-ene, which leads to a mixture of four regioisomers. The procedure can be extended to the synthesis of vicinal iodoazides without requiring the use of either iodide chloride or sulfolane. The reaction required 48 hours at 20 °C for 88% conversion, 18-crown-6 is an alternative catalyst.

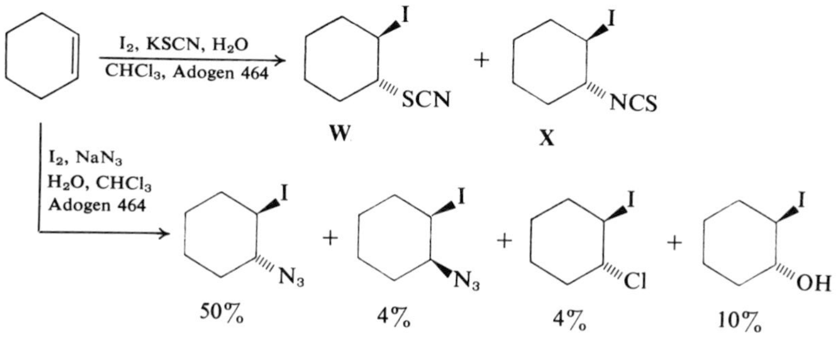

I_2, KSCN, H_2O / $CHCl_3$, Adogen 464 → **W** (I, SCN) + **X** (I, NCS)

I_2, NaN_3, H_2O, $CHCl_3$, Adogen 464 → (I, N_3) 50% + (I, N_3) 4% + (I, Cl) 4% + (I, OH) 10%

Scheme 3–93

3.3.8. Addition to C=O and C=N Bonds

3.3.8.1. Benzoin Condensation

The benzoin condensation was carried out at room temperature in 70% yield by stirring benzaldehyde with a 0.13 molar amount of tetrabutylammonium cyanide in water [435]. Only a 0.02 molar amount of quaternary ammonium cyanide was necessary for reaction at room temperature in THF or acetonitrile [413]. This is in contrast to the more conventional procedure (boiling in ethanol or methanol) where 0.2–0.4 mole alkali metal cyanide per mole benzaldehyde are used. The very hygroscopic tetraalkylammonium cyanides were prepared from the bromides by ion exchange on an IRA-400 (CN-form) column in absolute methanol [436]. Only traces of benzoin were formed if aqueous KCN was used in conjunction with Aliquat 336 [437], presumably because chloride and cyanide are similar in extractability. Benzoin condensations were also carried out with either aqueous potassium cyanide/neat aromatic aldehyde or solid KCN/aldehyde dissolved in benzene or acetonitrile at 25 to 60 °C using 18-crown-6 or dibenzo-18-crown-6 as catalysts [437].

3.3.8.2. Aldol-Type Reactions

Aromatic aldehydes and N-benzylidene-benzylamine are condensed by 50% NaOH/TEBA at room temperature to yield 2-amino-1-aryl-N-benzylidene-2-phenylethanols that can be hydrolyzed to give substituted aminoethanols [1097]:

$$ArCHO + PhCH{=}N{-}CH_2Ph \rightarrow ArCH(OH){-}CH(Ph){-}N{=}CHPh$$

Under the same conditions, arylacetonitriles and 3- or 4-substituted nitrosobenzenes give high yields of α-phenylimino-phenylacetonitriles and/or 1-anilino-2-cyano-1,2-diarylethenes, depending on the substituents present [1098]:

$$Ar^1CH_2CN + ON{-}Ar^2 \rightarrow Ar^1{-}C(CN){=}N{-}Ar^2$$

$$Ar^1{-}C(CN){=}N{-}Ar^2 + Ar^1CH_2CN \rightarrow Ar^1C(NHAr^2){=}C(CN)Ar^1$$

Knoevenagel reactions between cyanoacetic or malonic esters and α,β-unsaturated aldehydes give acceptable yields only with solid K_2CO_3/TEBA/benzene [1093]:

$$Z{-}CH_2COOR + R_2C{=}CH{-}CHO \rightarrow R_2C{=}CH{-}CH{=}C(Z){-}COOR$$

Several cases of intramolecular aldol condensations have been mentioned previously [115, 426]. 2-Methylbenzoxazole and 2-methylbenzothiazole were reacted with aromatic aldehydes, in the absence of solvent, with 50% aqueous sodium hydroxide/TEBA at room temperature for 1 to 24 hours [438]. In some cases the alcohols were isolated instead of or together with the styrene derivatives.

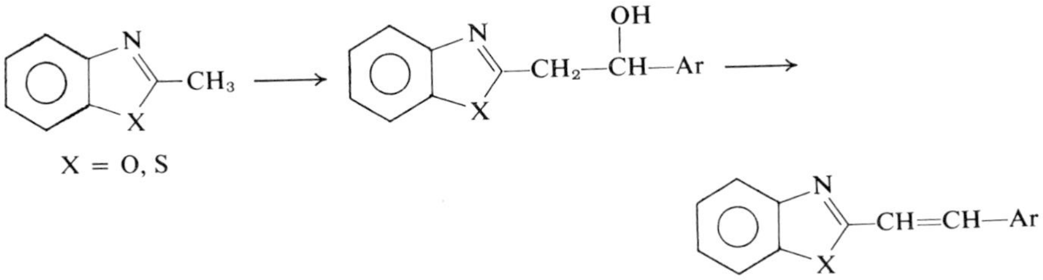

[2-methylbenzoxazole/benzothiazole ($X = O, S$) $\rightarrow$ 2-($CH_2{-}CH(OH){-}Ar$)-benzazole $\rightarrow$ 2-($CH{=}CH{-}Ar$)-benzazole]

X = O, S

Scheme 3–94

$$\underset{\mathbf{A}}{R^1{-}SO_2{-}CH_2{-}R^2} + R^3CHO \longrightarrow R^1{-}SO_2{-}C(R^2){=}CHR^3$$

Sulfones **A** (R^1 = Ph or Me_2N; R^2 = H, Ph, Me, and $Me_2C = CH$) were reacted with aromatic aldehydes in methylene chloride/50% aqueous sodium hydroxide/TEBA at room temperature for 2–6 hours [439]. Aromatic ketones proved unreactive whereas aliphatic aldehydes gave self-condensation products under these conditions.

3,5-Diphenylthiadioxane-S,S-dioxide (**B**) was synthesized from **A** ($R^1 = CH_3$, $R^2 = H$) and benzaldehyde, using the same reaction conditions [440]. Substituting 18-crown-6 for TEBA, however, the conversion is more complicated because of a competing Cannizzaro reaction. A closer investigation of the factors involved showed that the best yields in the Cannizzaro process were realized if no catalyst was present. This was interpreted in terms of the usual cyclic six-membered transition state complex in the Cannizzaro reaction. Crown ether complexation disturbs the coordinating role of the metal [440].

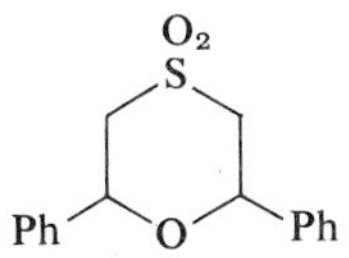

O_2 S — Ph O Ph — **B**

$$H_3C{-}CN + R{-}CO{-}R' \longrightarrow RR'C{=}CH{-}CN \quad \mathbf{C}$$

Scheme 3–95A

Solid KOH pellets in acetonitrile transform ketones directly into α,β-unsaturated ketones **C**, in some cases together with the β,γ-unsaturated compounds [215, 857]. β-Hydroxynitriles are not formed. 18-crown-6 is necessary as catalyst in mixtures containing cosolvents such as benzene; in acetonitrile no catalyst is needed. Aldehydes, methyl ketones and highly enolizable ketones give less satisfactory yields.

The following aldolcondensation-elimination reaction yielding a prostaglandin precursor was performed at 40 °C in 20 minutes in high yield [877]:

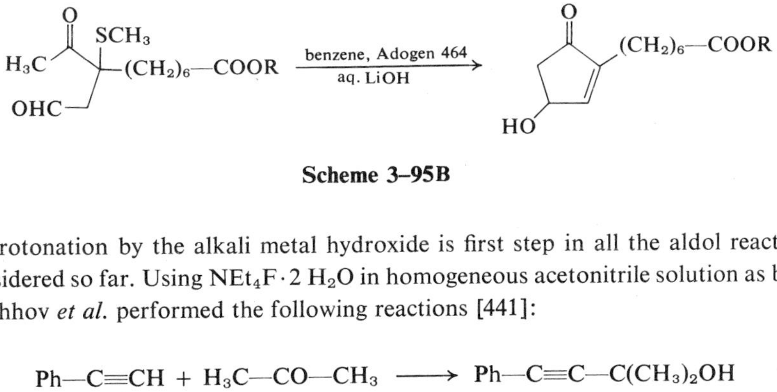

$$H_3C{-}CO{-}C(SCH_3)(CH_2CHO){-}(CH_2)_6{-}COOR \xrightarrow[\text{aq. LiOH}]{\text{benzene, Adogen 464}} \text{HO-cyclopentenone}{-}(CH_2)_6{-}COOR$$

Scheme 3–95B

Deprotonation by the alkali metal hydroxide is first step in all the aldol reactions considered so far. Using $NEt_4F \cdot 2\,H_2O$ in homogeneous acetonitrile solution as base, Rozhhov *et al.* performed the following reactions [441]:

$$Ph{-}C{\equiv}CH + H_3C{-}CO{-}CH_3 \longrightarrow Ph{-}C{\equiv}C{-}C(CH_3)_2OH$$

$$Ph{-}CO{-}R + H_3C{-}CN \longrightarrow Ph{-}CR(OH){-}CH_2CN$$

3.3.8.3. Other Types of Reactions

Related to aldol condensations and very useful preparatively, although not true PTC reactions, are reactions in which trimethylsilyl compounds are attacked by tetraalkylammonium fluorides or cyanides in catalytic amounts yielding intermediate acetylides, enolates, or alkoxylates which are immediately added across carbonyls:

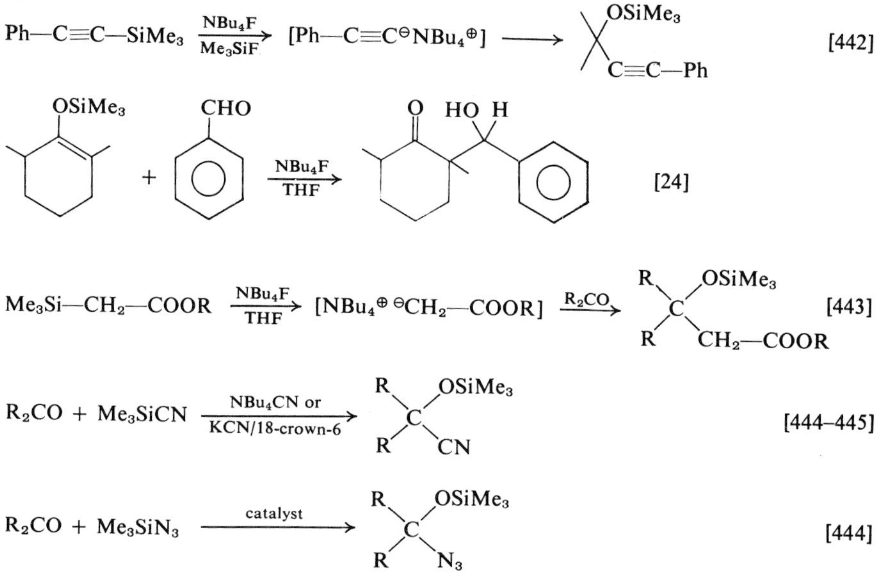

$$Ph{-}C{\equiv}C{-}SiMe_3 \xrightarrow[Me_3SiF]{NBu_4F} [Ph{-}C{\equiv}C^{\ominus}NBu_4^{\oplus}] \longrightarrow \text{(OSiMe}_3\text{)(C}{\equiv}\text{C}{-}\text{Ph) adduct} \quad [442]$$

OSiMe₃-cyclohexene + CHO-benzene $\xrightarrow[THF]{NBu_4F}$ 2-(hydroxy(phenyl)methyl)cyclohexanone (O, HO H) [24]

$$Me_3Si{-}CH_2{-}COOR \xrightarrow[THF]{NBu_4F} [NBu_4^{\oplus}\,{}^{\ominus}CH_2{-}COOR] \xrightarrow{R_2CO} R_2C(OSiMe_3){-}CH_2{-}COOR \quad [443]$$

$$R_2CO + Me_3SiCN \xrightarrow[KCN/18\text{-}crown\text{-}6]{NBu_4CN\text{ or}} R_2C(OSiMe_3)(CN) \quad [444\text{–}445]$$

$$R_2CO + Me_3SiN_3 \xrightarrow{catalyst} R_2C(OSiMe_3)(N_3) \quad [444]$$

Scheme 3–96A

The cyanohydrin, 1,2:5,6-di-O-cyclohexylidene-3-C-cyano-α-D-allofuranose, was obtained from the corresponding ketosugar derivative with aqueous KCN/benzene/TEBA in 86% yield [1099].

Esters of cyanohydrins are obtained if aromatic aldehydes, sodium cyanide, and acid chlorides are reacted in the presence of onium salts or crowns [887, 982].

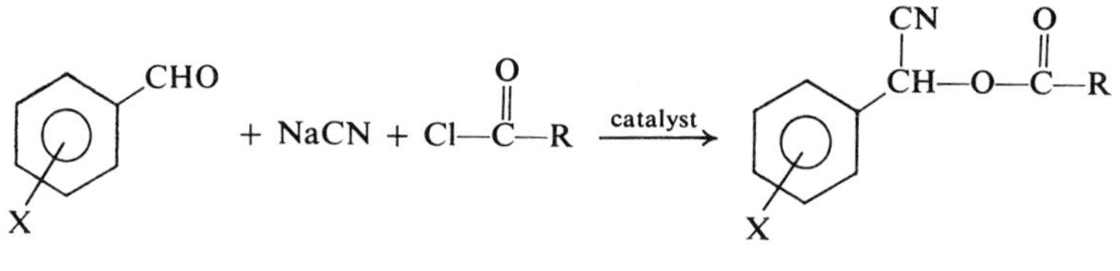

$$X{-}C_6H_4{-}CHO + NaCN + Cl{-}\overset{O}{\overset{\|}{C}}{-}R \xrightarrow{catalyst} X{-}C_6H_4{-}CH(CN){-}O{-}\overset{O}{\overset{\|}{C}}{-}R$$

Scheme 3–96B

The same type of reaction was observed with aliphatic aldehydes [963]. Furthermore, in the presence of allylic bromides, KCN, and TEBA, cyanohydrin allyl ethers were formed in 44–70% yield [963]:

$$RCHO + KCN + Br{-}CH_2{-}CH{=}CH{-}R^1 \longrightarrow R{-}CH(CN){-}O{-}CH_2{-}CH{=}CH{-}R^1$$

The concentrated sodium hydroxide/TEBA system was again used to make glycidic nitriles **D** in 55–80% yields from aromatic and aliphatic aldehydes and ketones and chloroacetonitrile at room temperature, 30 minutes stirring [448]. With asymmetrically substituted ketones a mixture of stereoisomers of **D** was obtained. An alternative to TEBA is the catalyst dibenzo-18-crown-6 [302]. The stereoselectivity of Darzens reaction between benzaldehyde and *tert*-butyl chloroacetate, *tert*-butyl 2-chloropropionate, or phenyl chloroacetonitrile under three different sets of conditions has been compared [503]. Similar ratios of stereoisomers of **E**, **F**, and **G** were obtained under PTC conditions (molar amounts of NBu_4Br and benzaldehyde, excess chloro compound and 50% NaOH) and with sodium hydride or sodium *tert*-butoxide in HMPT, indicating a similar degree of association and tightness of the primary intermediate adduct. A different ratio was observed in THF with $NaOC(CH_3)_3$ [503]. If, however, α-chlorophenylacetonitrile and benzaldehyde in benzene were reacted with 50% aqueous sodium hydroxide, the stereochemical outcome of the reaction was strongly dependent on whether TEBA was present or not [952]:

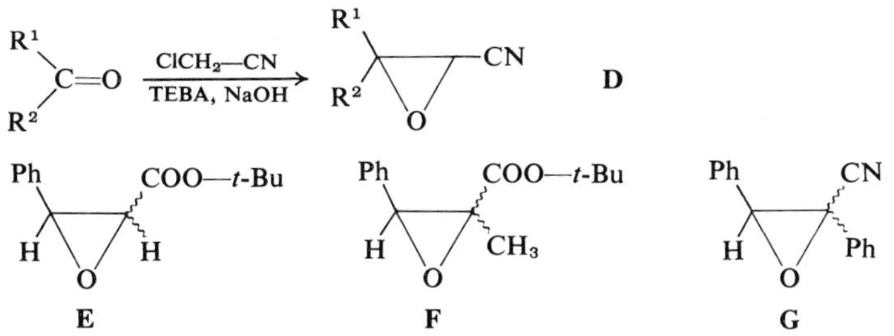

$R^1R^2C{=}O \xrightarrow[\text{TEBA, NaOH}]{ClCH_2{-}CN}$ **D**

E   **F**   **G**

Scheme 3–97A

PhCHO + Ph—CH(Cl)CN ⟶

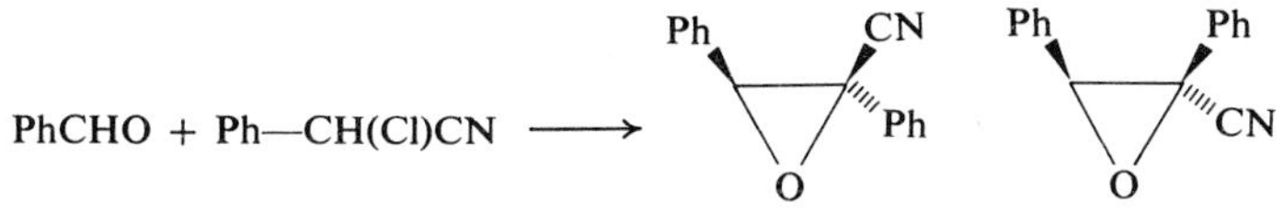

with TEBA	92	8
without TEBA	32	68

Scheme 3–97B

The cyclization step must be slow relative to the reversible primary addition, and—so it was argued—the catalyzed epoxide formation will occur in the organic medium from ion pairs, where normal steric effects favor that intermediate which leads to the *trans* product. The noncatalyzed process must occur at the interphase, and here Mąkosza and co-workers assume that ion-ion and ion-dipole interactions make the precursor and/or transition state leading to the *cis* isomer much more favorable [952].

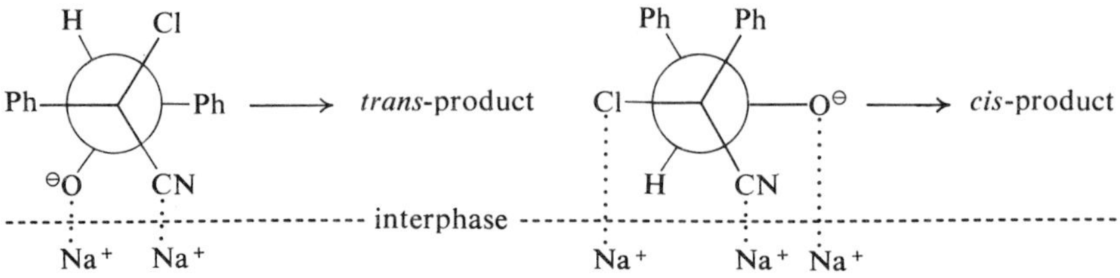

Ph, H, Cl, Ph, ⊖O, CN → *trans*-product; Ph, Ph, Cl, O⊖, H, CN → *cis*-product

interphase

Na^+ Na^+ Na^+ Na^+ Na^+

Scheme 3–97C

Further epoxides were obtained in PTC reactions of carbonyl compounds with 9-chlorofluorene [1066] and chloromethanesulfonamides [1087], and Na_2CO_3 or K_2CO_3 have been used as solid bases for PTC Darzens reactions [1048].

Low degrees of asymmetric induction (0–2.5%) were reported for similar reactions of carbonyl compounds with chloromethyl p-tolyl sulfone and α-chlorophenylacetonitrile in the presence of base and chiral, ephedrine derived catalysts. Higher (up to 23%) optical yields were sometimes observed, if these same ephedrinium salts were bound to a polymer matrix [923].

In view of the facile decomposition of ephedrinium salts to give methylphenyloxirane of very high specific rotation and/or products derived therefrom,*) these reports should be considered with caution. Higher, and possibly more reliable, optical yields for epoxides were found in condensations of *p*-chlorobenzaldehyde and phenacyl chloride after three hours of stirring in toluene/10% NaOH in the presence of quininium benzyl chloride [951]. Similarly, bromoisobutyrophenone and aqueous sodium

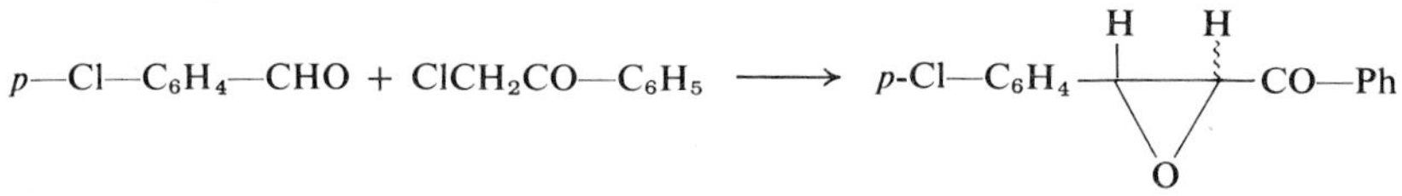

$$p\text{—Cl—}C_6H_4\text{—CHO} + ClCH_2CO\text{—}C_6H_5 \longrightarrow p\text{-Cl—}C_6H_4\text{-(epoxide, H, H)-CO—Ph}$$

Scheme 3–97D

hydroxide/toluene gave an optically active epoxide of unknown optical purity with the same catalyst [951].

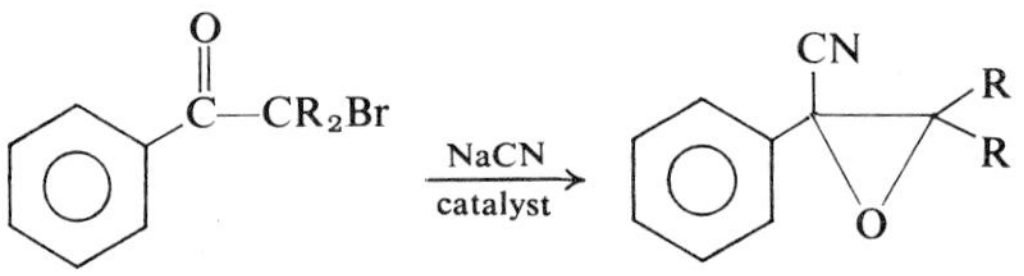

$$C_6H_5\text{—C(=O)—}CR_2Br \xrightarrow[\text{catalyst}]{\text{NaCN}} C_6H_5\text{—C(CN)(O)}CR_2$$

Scheme 3–97E

*) *Cf.*, other failures to arrive at optically active products with ephedrine derived catalysts that are detailed in various sections of the book.

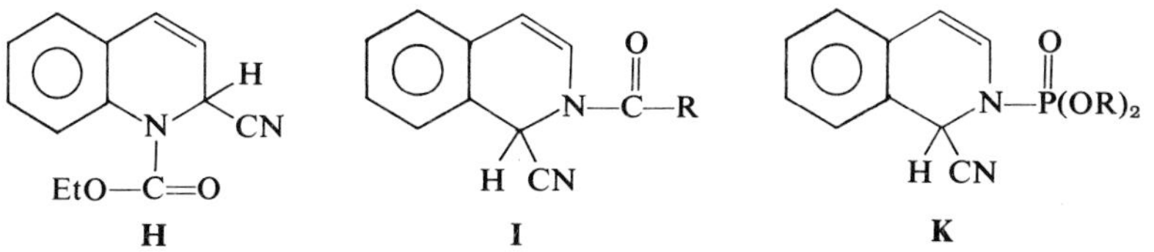

Scheme 3–97F

Reissert compounds **H–K** were obtained in much higher yields than with older procedures. A mixture of quinoline or isoquinoline, dichloromethane, potassium cyanide in water, and TEBA was stirred at room temperature while the acid chloride was added slowly [449]. Recently phthalazine Reissert compounds were also made by PTC [993]. Compounds such as **I** (R=Ph or CH_3) reacted with both aliphatic and aromatic aldehydes or ketones to give condensation products **L** and to some extent alcohols **M**, often in yields above 90% [311, 886]. These conversions were carried out in benzene or acetonitrile with 50% aqueous sodium hydroxide. Reaction occurred even in the absence of catalyst, and this result was one of the first indications that PTC reactions can in fact occur at the interphase. With ketones, the addition of TEBA increased the yields, although with aldehydes no improvement was evident [311].

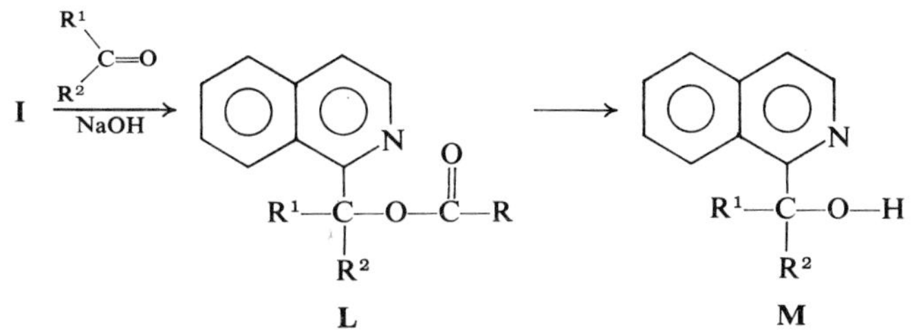

Scheme 3–98

In an addition to a triple bond, nitriles were converted into thioamides by stirring the benzene solutions at 70 °C with excess sodium sulfide in water in the presence of a quaternary ammonium salt under 760 to 1520 Torr pressure of hydrogen sulfide [1100]: $RCN + H_2S \rightarrow R{-}CS{-}NH_2$.

The addition of chloroform to CO groups will be considered in Section 3.3.13.2.

3.3.9. β-Eliminations

Elimination reactions under PTC conditions have already been encountered in conjunction with substitution reactions where they occur as undesirable side-reactions, *cf.*, Sections 3.2.1 to 3.2.3. Secondary substrates, especially cyclohexyl, give considerable amounts of elimination products, in particular with nucleophiles such as fluoride or acetate. Tertiary compounds give olefins as the only products. Detailed studies indicate that rates are faster with tetraethylammonium fluoride in acetonitrile and the proportion of Saytzeff product is much higher than with *tert*-butoxide in *tert*-butanol or DMSO, or ethoxide in ethanol [453]. For preparative purposes such eliminations have been used mainly in homogeneous solution as shown below:

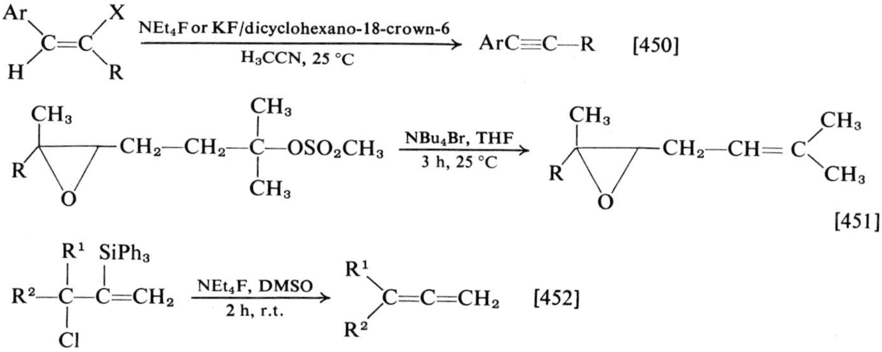

$$\mathrm{Ar(H)C{=}C(X)R} \xrightarrow[\mathrm{H_3CCN,\ 25\ ^\circ C}]{\mathrm{NEt_4F\ or\ KF/dicyclohexano\text{-}18\text{-}crown\text{-}6}} \mathrm{ArC{\equiv}C{-}R} \quad [450]$$

$$\mathrm{R(CH_3)C(O)C{-}CH_2{-}CH_2{-}C(CH_3)_2{-}OSO_2CH_3} \xrightarrow[\mathrm{3\ h,\ 25\ ^\circ C}]{\mathrm{NBu_4Br,\ THF}} \mathrm{R(CH_3)C(O)C{-}CH_2{-}CH{=}C(CH_3)_2} \quad [451]$$

$$\mathrm{R^2{-}C(R^1)(Cl){-}C(SiPh_3){=}CH_2} \xrightarrow[\mathrm{2\ h,\ r.t.}]{\mathrm{NEt_4F,\ DMSO}} \mathrm{R^1R^2C{=}C{=}CH_2} \quad [452]$$

Scheme 3–99

Using the same reagents and reaction conditions, more complex eliminations have also been successful [55, 520], *e.g.*:

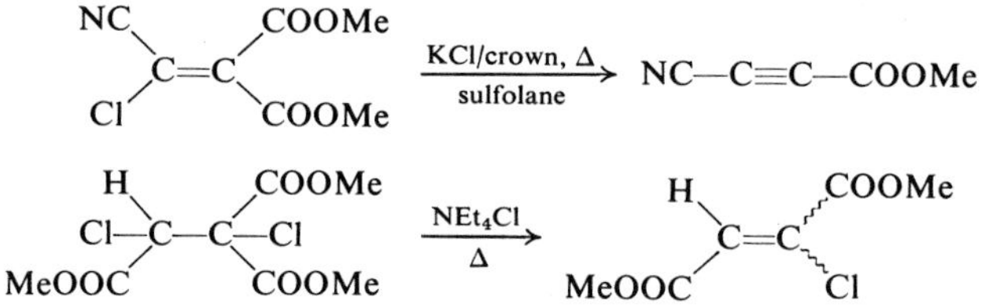

$$\mathrm{NC(Cl)C{=}C(COOMe)_2} \xrightarrow[\mathrm{sulfolane}]{\mathrm{KCl/crown,\ \Delta}} \mathrm{NC{-}C{\equiv}C{-}COOMe}$$

$$\mathrm{H(Cl)(MeOOC)C{-}C(Cl)(COOMe)_2} \xrightarrow[\Delta]{\mathrm{NEt_4Cl}} \mathrm{H(MeOOC)C{=}C(COOMe)(Cl)}$$

Scheme 3–100

Dehydrohalogenation in the presence of quaternary ammonium or phosphonium salts at temperatures up to 250 °C gives gaseous HCl and alkene. Typical examples of this type of elimination are shown below:

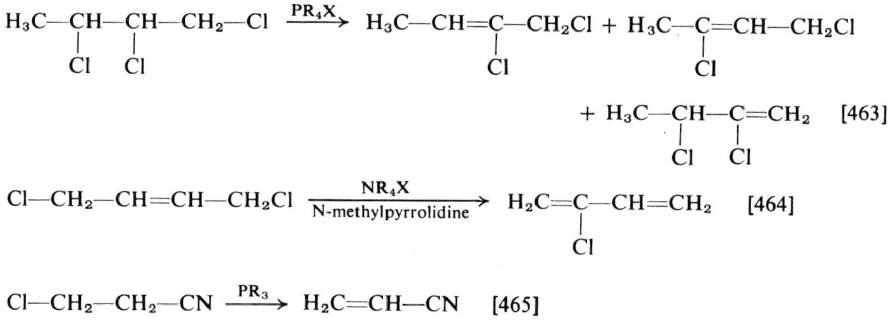

$$\mathrm{H_3C{-}CH(Cl){-}CH(Cl){-}CH_2{-}Cl} \xrightarrow{\mathrm{PR_4X}} \mathrm{H_3C{-}CH{=}C(Cl){-}CH_2Cl} + \mathrm{H_3C{-}C(Cl){=}CH{-}CH_2Cl} + \mathrm{H_3C{-}CH(Cl){-}C(Cl){=}CH_2} \quad [463]$$

$$\mathrm{Cl{-}CH_2{-}CH{=}CH{-}CH_2Cl} \xrightarrow[\mathrm{N\text{-}methylpyrrolidine}]{\mathrm{NR_4X}} \mathrm{H_2C{=}C(Cl){-}CH{=}CH_2} \quad [464]$$

$$\mathrm{Cl{-}CH_2{-}CH_2{-}CN} \xrightarrow{\mathrm{PR_3}} \mathrm{H_2C{=}CH{-}CN} \quad [465]$$

Scheme 3–101A

The sensitive compounds shown in Scheme 3–101B were generated by eliminations with KF (solid)/18-crown-6.

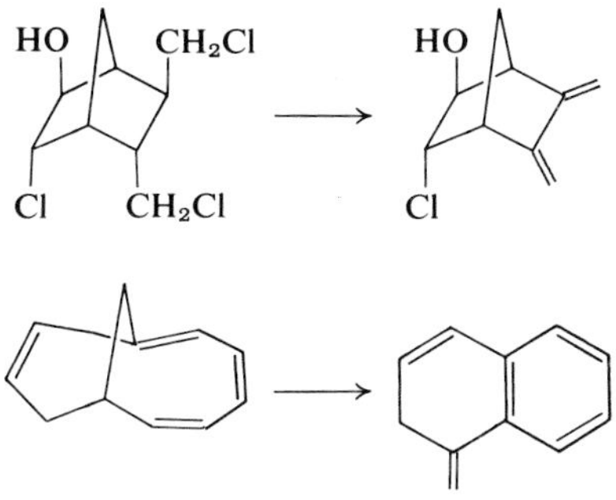

HO, CH_2Cl, Cl, CH_2Cl ⟶ HO, Cl [1102]

⟶ [1103]

Scheme 3–101B

The facile dehalogenation of *vic*-dibromoalkanes (**A**) by iodide ions is a different type of two-phase elimination reaction.

$$R^1{-}\underset{Br}{CH}{-}\underset{Br}{CH}{-}R^2 \xrightarrow[\text{catalyst}]{\text{aq. NaI, }Na_2S_2O_3} R^1{-}CH{=}CH{-}R^2$$

A

Scheme 3–102

A solution of the substrate in toluene is stirred at 90 °C together with an aqueous solution of 10 mole-% sodium iodide and excess of sodium thiosulfate in the presence of 10 mole-% tributylhexadecylphosphonium bromide [456]. Sodium thiosulfate alone can also be used but rates are much slower. *Meso*- or *erytho*-**A** react faster, giving *trans*-olefins, only, while diastereomeric *d,l*- or *threo*-compounds are converted into mixtures of *cis*- and *trans*-alkenes. The same reaction can be executed with a specially prepared cross-linked polystyrene-bound catalyst in a "triphase catalysis" process [62]—a base catalyzed mechanism was proposed for the elimination of the elements of water from benzaldoxime by KCN or KF plus catalyst in a two-phase system [929]: ArCH = NOH → ArCN.

Reactions with bases such as potassium *tert*-butoxide are often performed in polar aprotic solvents and sometimes in benzene where the solubility is not very high. In both cases the addition of crown ethers not only changes the solubility, but in addition has a profound effect on ionic association. Thus, as mentioned previously, dramatic changes in reaction rates, orientation, and stereochemistry of β-eliminations occur [454, review: 455]. Preparatively mild and simple dehydrohalogenations of chloro- and bromoalkanes with solid potassium *tert*-butoxide and 1 mole-% 18-crown-6 can be effected by heating in petroleum ether of a boiling range far from that of the alkene formed. In 6 hours at 120 °C, bornyl chloride, for example, gave 92% bornene free from camphene and tricyclene [1104]. 1-Alkynes can be prepared from 1,2-dihalides and 1,1-dihalides under similar conditions. Internal geminal dichlorides (from ketones and PCl_5) give internal alkynes in excellent yields. Isomerizations of the alkynes to allenes or isomeric alkynes are much slower than the elimination processes normally. (E)-Haloalkenes and the reagent system KO*t*Bu/crown also yield alkynes in a *syn* elimination [1105].

Turning now to the most frequently used PTC bases, alkali metal hydroxides, dehydrohalogenation occurs only in those cases where the proton is activated by electron withdrawing substitutents or by a benzylic or vinylogous benzylic position. These PTC processes are very similar to those under homogeneous conditions mentioned above where HCl gas was formed and where the same catalysts were used. In general, however, much lower temperatures are necessary in the presence of aqueous sodium hydroxide.

$$Cl{-}CH_2{-}CH(Cl){-}CH{=}CH_2 \xrightarrow[\text{aq.NaOH}]{\text{onium salt}} H_2C{=}CCl{-}CH{=}CH_2 \quad [457\text{–}461]$$

$$Cl_2CH{-}CH_2Cl \xrightarrow[\text{aq. NaOH}]{\text{Aliquat 336}} Cl_2C{=}CH_2 \quad [462]$$

$$PhO{-}CF_2{-}CHCl_2 \xrightarrow[\text{aq. KOH}]{\text{Triton B}} PhO{-}CF{=}CCl_2 \quad [472]$$

The elimination of HBr from phenethyl bromide with 50% NaOH was complete after 2 hours at 90 °C with *n*-butyltriethylammonium chloride as catalyst. Without catalyst only 1% styrene was isolated [466, 467]. Compound **B** gave 60% of the substituted butadiene after 4 hours at room temperature [466, 467], and **C** needed only 3 hours reflux for 83% conversion [468].

$$Ph{-}CH_2{-}CH_2Br \longrightarrow Ph{-}CH{=}CH_2$$

$$\underset{\textbf{B}}{(p\text{-}F{-}C_6H_4)_2C{=}CH{-}CH_2{-}CH_2Cl} \xrightarrow{\text{NaOH/TEBA}} (p\text{-}F{-}C_6H_4)_2C{=}CH{-}CH{=}CH_2$$

$$\underset{\textbf{C}}{Ph{-}CH(Br){-}CH_2Br} \xrightarrow[\text{aq. KOH}]{\text{benzene, Aliquat 336}} Ph{-}(Br)C{=}CH_2$$

Under standard conditions (60 °C, conc. NaOH) the yields of styrene were examined for a series of catalysts of varying chain length [467]:

Catalyst $R\text{-}NEt_3^{\oplus}Br^{\ominus}$

R	C_2H_5	$n\text{-}C_3H_7$	$n\text{-}C_4H_9$	$n\text{-}C_5H_{11}$	$n\text{-}C_6H_{13}$
yield %	3	7	12	50	53
R	$n\text{-}C_7H_{15}$	$n\text{-}C_8H_{17}$	$n\text{-}C_{10}H_{21}$	$n\text{-}C_{12}H_{25}$	
yield %	45.5	44	42	38	

The increase at the beginning of the series is not unexpected for solubility reasons. The subsequent slight decrease, however, is harder to explain. Similar effects were noted with dihalocarbene reactions (*vide infra*).

The preparation of acetylenes by dehydrohalogenation with Triton B (benzyltrimethylammonium hydroxide) has been published [469, 470]. The commercially available solution of base in methanol was transformed into a toluene, benzene, or pyridine solution by pouring it into an excess of solvent and distilling off most of the methanol. The Triton B was not always fully dissolved in benzene or toluene which in many cases did not disturb. For sensitive ketones or esters (which were not saponified) working in pyridine proved better. Temperatures and reaction times ranged between 70 °C/30 minutes and −10 °C to −30 °C/1–5 minutes. Some typical processes giving yields of 40–85% are shown below:

$$p\text{-X—C}_6\text{H}_4\text{—CCl}_2\text{—CH}_3 \searrow$$
$$\qquad\qquad\qquad\qquad p\text{-X—C}_6\text{H}_4\text{—C}\equiv\text{CH}$$
$$p\text{-X—C}_6\text{H}_4\text{—CCl=CH}_2 \nearrow$$

$$\text{R—CH=CBr}_2 \longrightarrow \text{R—C}\equiv\text{C—Br (R = Ph, CH(OEt)}_2\text{, CCl=CCl}_2\text{)}$$

$$p\text{-X—C}_6\text{H}_4\text{—CH=CBr—COOMe} \longrightarrow p\text{-X—C}_6\text{H}_4\text{—C}\equiv\text{C—COOMe}$$

$$\text{Cl}_2\text{C=CCl—CH=CX—COMe} \longrightarrow \text{Cl}_2\text{C=CCl—C}\equiv\text{C—COMe}$$

Recently, these same reactions were performed under true PTC conditions in higher yields (70 to 87%) [471]. Substrates dissolved in pentane were stirred with 50% aqueous sodium hydroxide and molar amounts of tetrabutylammonium hydrogen sulfate. Some of the eliminations were exothermic. It is mechanistically important that no reaction occurred under these mild conditions when tetrabutylammonium hydrogen sulfate was replaced by the bromide, iodide, TEBA, or hexadecyltrimethylammonium bromide [471].

Let us now consider the mechanistic details of PTC elimination processes with NaOH. It seems quite clear that at least two reaction pathways exist. Path (a). The extraction of $OH^{\ominus}$ into the organic phase and subsequent elimination requires a molar amount of catalyst together with an original counterion more hydrophilic than hydroxyl. These reactions proceed readily at low temperature. A molar amount of Triton B, not fully dissolved in toluene or a similar solvent, is less appropriate since the halogenide formed is more soluble than the hydroxide. Path (b) involves elimination *via* the ion pair $NR_4^{\oplus}X^{\ominus}$:

$$\text{>C(X)—CH<} + NR_4^{\oplus}X^{\ominus} \longrightarrow \text{>C=C<} + NR_4^{\oplus}HX_2^{\ominus}$$

Scheme 3–103

Additional proof that this process is possible is given by the fact that elimination also occurs in the absence of additional base with the evolution of HCl gas. The existence of $NR_4^{\oplus}HX_2^{\ominus}$ in organic media has been discussed in Section 3.2.1. In the presence of base, $NR_4^{\oplus}HX_2^{\ominus}$ is neutralized at the interphase, thereby displacing the elimination-addition equilibrium and increasing the overall reaction rate.

3.3.10. Hydrolysis Reactions

Hydrolysis reactions are mechanistically more sensitive than most other PTC reactions in the presence of aqueous base/onium salt. Whereas in most of the other conversions deprotonation is the first step, here initial attack is by hydroxide ion for which the possibilities are somewhat more limited. Many hydrolysis reactions have been characterized under micellar [473] or inverted micellar [473, 474] conditions. It is not possible, however, to directly compare these rate studies and physicochemical measurements in dilute "homogeneous" solutions with preparative saponifications and other PTC two-phase hydrolysis reactions. So far there has been very little systematic study of PTC hydrolysis reactions.

The hydrolysis of alkyl halides under PTC conditions to form alcohols is not a synthetically useful process because ethers are the main products, sometimes in very satisfactory yields (*cf.*, Section 3.3.2). Only in a few special cases are these hydrolysis reactions successful. For example, ammonium salts accelerate the hydrolysis of $H_3C{-}C(Cl){=}CH{-}CH_2Cl$ by aqueous alkali metal hydroxide [225]. **A** (X = Br or Cl) gave **B** in 50 or 64% yield, respectively, after boiling in benzene with aqueous sodium hydroxide/Aliquat 336 for 48 hours [246]. Rather surprisingly, no normal Favorskii products were found. 1,2-Dichloropropane was hydrolyzed to the diol in aqueous $NaHCO_3$ with hexadecyltributylphosphonium bromide at 100 °C under pressure of carbon dioxide [1106].

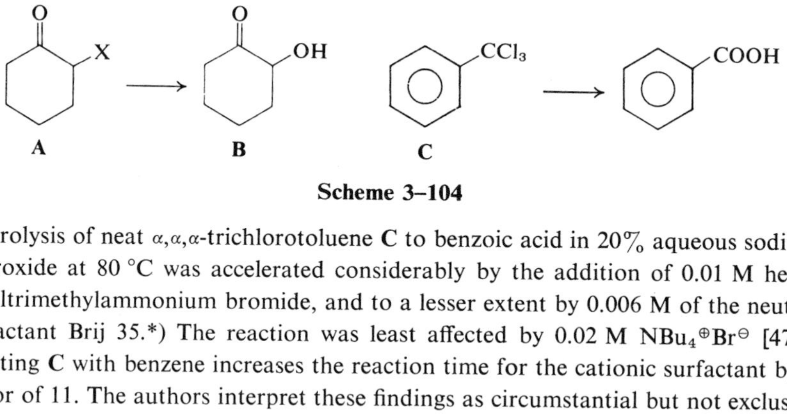

Scheme 3–104

Hydrolysis of neat α,α,α-trichlorotoluene **C** to benzoic acid in 20% aqueous sodium hydroxide at 80 °C was accelerated considerably by the addition of 0.01 M hexadecyltrimethylammonium bromide, and to a lesser extent by 0.006 M of the neutral surfactant Brij 35.*) The reaction was least affected by 0.02 M $NBu_4^{\oplus}Br^{\ominus}$ [475]. Diluting **C** with benzene increases the reaction time for the cationic surfactant by a factor of 11. The authors interpret these findings as circumstantial but not exclusive evidence for emulsion or micellar catalysis rather than true PTC.

Prolonged refluxing of 2-nitropropane with aqueous tetraethylammonium fluoride hydrolyzes off the nitro group *via* the *aci* tautomer; acetone is the product [1092].

---

*) $C_{12}H_{25}(OCH_2{-}CH_2)_{23}OH$

3.3.10.1. Hydrolysis of Sulfuryl Chlorides

The hydrolysis of sulfuryl chlorides in a two-phase system is markedly accelerated by quaternary ammonium salts. Early kinetic studies on SO_2Cl_2 in carbon tetrachloride with detergent-type catalysts were interpreted in terms of an interphase reaction where the detergent helps to lower the surface tension [476]. Starks showed that the hydrolysis of pure or dissolved long chain alkanesulfonyl chlorides with aqueous sodium hydroxide is very slow. However, addition of a little quaternary ammonium catalyst made the exothermic reaction proceed rapidly [4, 38, 39].

3.3.10.2. Saponification of Esters

Crown ethers and cryptates have proved valuable in the saponification of sterically hindered esters not easily hydrolysable under standard conditions. Pedersen and co-workers [504, 505] found that dicyclohexano-18-crown-6 dissolves potassium and sodium hydroxide in benzene, thus making it possible to prepare solutions as concentrated as 1 N. In most cases the complex is first formed in methanol, which is subsequently exchanged for benzene, however, about 1% CH_3OH is retained. Thus, although the resultant solution contains both $OH^{\ominus}$ and $^{\ominus}OCH_3$ ions [43, 504], it is still capable of hydrolyzing esters of 2,4,6-trimethylbenzoic acid. A benzene solution of methyl tetradecanoate was hydrolyzed by stirring at room temperature with powdered potassium hydroxide/catalytic amounts of 18-crown-6 for 12 hours [969].

A one-pot reaction for the hydrolysis and decarboxylation of esters containing α-activating groups has also been developed [35]:

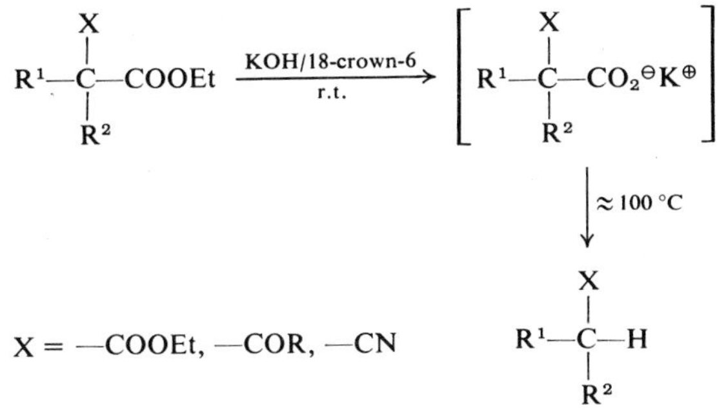

$$R^1-\underset{R^2}{\overset{X}{C}}-COOEt \xrightarrow[\text{r.t.}]{\text{KOH/18-crown-6}} \left[R^1-\underset{R^2}{\overset{X}{C}}-CO_2^{\ominus}K^{\oplus}\right] \xrightarrow{\approx 100\ ^\circ C} R^1-\underset{R^2}{\overset{X}{C}}-H$$

X = —COOEt, —COR, —CN

Scheme 3–105

In this case the saponification step is carried out with potassium hydroxide/equimolar amount of 18-crown-6 in ≈6 vol-% ethanol/benzene at room temperature. With nonenolizable substrates the hydrolysis is very rapid, but up to 20 hours at room temperature are needed with acidic substrates like diethyl phenylmalonate. The decarboxylation step involves refluxing for 2–20 hours. If the carboxylate is difficult to decarboxylate (*e.g.*, monoethyl acetamidomalonate), dioxane is used as the solvent. This one-pot procedure was not successful for the decarboxylation of very acidic substrates (pK_a < 10), where the unreactive enolate form hindered the reaction. Other reactions where crown ethers have a pronounced effect on the rate of decarboxylation are also known [506, 507, 1101].

Whereas the above experiments were performed in homogeneous solution, Dietrich and Lehn [359] compared the effectiveness of potassium hydroxide/dicyclohexano-18-crown-6 in homogeneous toluene/1% methanol with that of a mixture of solid powdered potassium hydroxide Kryptofix[222] (**5**). The results for the saponification of methyl mesitoate are shown in Table 3–7a.

Table 3–7a. Saponification of Methyl Mesitoate.

Catalyst	Solvent	Time	Temperature	Yield (%)
none	1-propanol (homogeneous)	5 h	75 °C	0
dicyclohexano-18-crown-6	toluene (homogeneous)	31 h	74 °C	58
Kryptofix[222]	toluene (heterogeneous)	12 h	25 °C	80
Kryptofix[222]	DMSO (homogeneous)	2 min	25 °C	50

It is clear from these results that the mild PTC process with Kryptofix[222] as catalyst is very useful for sterically hindered compounds, although the cost of the reagent may deter from its widespread use. It should be noted that equivalent amounts of crown ether or cryptate relative to ester are used.

Cheap onium salt catalysts have proven useful for simple saponifications. Thus, the hydrolysis of dimethyl adipate by 50% aqueous sodium hydroxide is slow in the absence of catalyst at room temperature but becomes fast and exothermic with the addition of Aliquat 336 [4, 38, 39]. It appears, however, that if the carboxylic part of the ester has a long chain, the carboxylate formed associates with the catalyst cation, and the catalytic effect is destroyed [4].

PTC saponifications have been studied extensively [477, 1109], *cf.*, Table 3–7b. The influence of the solvent on the saponification of diethyl adipate with 50% aqueous sodium hydroxide/Aliquat 336 is as follows: petroleum ether, better than benzene, better than ether, better than methylene chloride. This result is surprising on first sight as in "normal" PTC reactions the opposite order of solvents is observed, the most unpolar solvent being the most useful. This can be interpreted as further evidence that $OH^{\ominus}$ is not transferred to the organic phase, contrary to reactions with crown ethers and cryptates as mentioned above. Under standard conditions (0.01 mole ester, 0.05 mole concentrated sodium hydroxide, 10^{-4} mole catalyst, 1 hour, room temperature) esters of hydrophilic hydroxycarboxylic and dicarboxylic acids were hydrolyzed much faster than simple long chain analogues. With the latter, the catalytic effect of ammonium salts was very slight or negligible. The saponification of sterically hindered methyl mesitoate was not greatly accelerated by tetrabutylam-

Table 3–7b. Phase-Transfer-Catalysed Saponification of Esters Using Aliquat 336 as Catalyst under Standard Conditions (10^{-2} mol ester, 5 ml light petroleum, 3 ml 50% NaOH, 10^{-4} mol Aliquat, 1 h, room temperature) [1109].

		Yields (%)		
No.	Ester	(a) without catalyst	(b) with catalyst	Acceleration (b/a)
1	Diethyl adipate	11	50	4.5
2	Diethyl phthalate	14	46	3.2
3	Ethyl decanoate	7	7	1
4	Ethyl stearate	13	20	1.5
5	*n*-Octyl stearate	5	13	2.6
6	Ethyl 2-phenylbutanoate	3	8	2.7
7	Ethyl 3-phenylpropionate	20	32	1.6
8	Ethyl p-nitrobenzoate	10	47	4.7
9	*n*-Butyl benzoate	27	33	1.2
10	Isopropyl benzoate	15	19	1.3
11	Benzyl benzoate	40	56	1.4
12	Ethyl benzoate	5	24	4.8
13	Methyl benzoate	12	34	2.8
14	Ethyl 2,4,6-trimethylbenzoate	8	9	1.1
15	Methyl salicylate	95	100	1.05
16	Ethyl phenylacetate	20	98	4.9
17	*n*-Octyl acetate	6	42	7.0
18	Cyclohexyl acetate	13	44	3.4
19	Ethyl propionate	37	85	2.3
20	Ethyl 2,2-dimethylpropionate	27	41	1.5
21	Isopropyl 2,2-dimethylpropionate	54	72	1.3
22	Ethyl 3,3-dimethylbutanoate	19	36	1.9
23	Ethyl 3-methylbutanoate	24	27	1.1
24	Cyclohexyl 3-methylbutanoate	20	22	1.1
25	Ethyl 3-methylbut-2-enoate	20	26	1.3
26	Ethyl toluene-*p*-sulphonate	9	12	1.3
27	*n*-Octyl toluene-*p*-sulphonate	5	6	1.2

monium catalyst at 100 °C. When the catalyst anion in the saponification of diethyl adipate in petroleum ether was varied, effectiveness of the tetrabutylammonium salts increased in this order: iodide < bromide < chloride < hydrogen sulfate < hydrogen adipate.

When the symmetrical cations were changed under standard conditions (0.01 mole diethyl adipate, 5 ml petroleum ether, 0.05 mole concentrated sodium hydroxide, 10^{-4} mole catalyst, 1 hour, room temperature), the following results were obtained:

Catalyst	$N(C_4H_9)_4Br$	$N(C_5H_{11})_4Br$	$N(C_6H_{13})_4Br$	$N(C_7H_{15})_4Br$	$N(C_8H_{17})_4Br$
Yield	18	20	45	46	39

Tetrabutylammonium hydrogen sulfate was the best commercially available ammonium salt catalyst. Just as effective, however, was the nonionic surfactant (**D**). Compounds (**E**) (R = H, CH_3) (*cf.*, [477]) proved reasonably good also.

$$H_{33}C_{16}{-}O(CH_2{-}CH_2{-}O)_{29}{-}O{-}CO{-}CH_3 \quad \textbf{D}$$

$$H_{33}C_{16}{-}O(CH_2{-}CH_2{-}O)_{30}{-}OR \quad \textbf{E}$$

Although anionic surface active sodium stearate proved a rather good catalyst, the crown ethers were not very useful under the reaction conditions. For a more detailed discussion of the various factors involved see Section 2.2.

In summary, it can be concluded that PTC saponifications are synthetically useful for sterically hindered esters with crown ethers or cryptates as catalysts using solid potassium hydroxide/toluene. Furthermore, the hydrolysis of simple esters by concentrated aqueous sodium hydroxide is accelerated markedly for hydrophilic carboxylates. Quaternary ammonium salts, in particular NBu_4HSO_4, and some anionic and nonionic surfactants are good catalysts. There are indications, therefore, that any one of the three possible mechanisms: surface reactions, micellar catalysis, or "true" PTC reactions can be operative, depending on the conditions.

Although rarely executed, acid PTC saponifications are possible too [57, 1202].

3.3.11. Generation and Conversion of Phosphonium and Sulfonium Ylides

3.3.11.1. Wittig Reactions

The first step in a normal Wittig reaction is the deprotonation of a phosphonium salt; in most cases of a triphenylphosphonium salt [478, 480]:

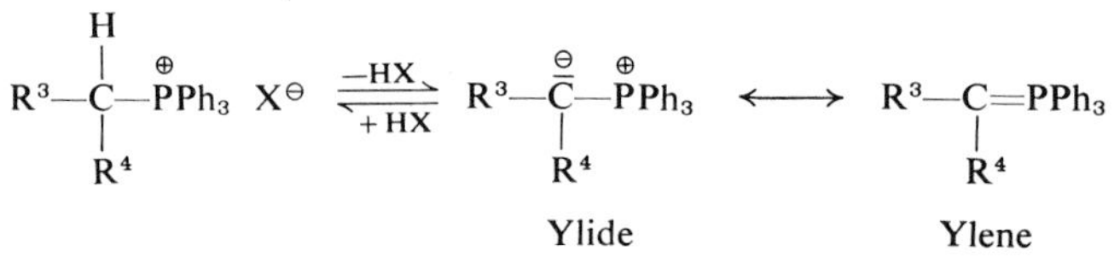

$$R^3{-}\overset{\displaystyle H}{\underset{\displaystyle R^4}{C}}{-}\overset{\oplus}{P}Ph_3\ X^{\ominus} \underset{+HX}{\overset{-HX}{\rightleftharpoons}} R^3{-}\underset{\displaystyle R^4}{\overset{\ominus}{C}}{-}\overset{\oplus}{P}Ph_3 \longleftrightarrow R^3{-}\underset{\displaystyle R^4}{C}{=}PPh_3$$

Ylide Ylene

Scheme 3–106

The acid/base character of the system is governed by the substituents R^3 and R^4: electron attracting groups enhance the acidity of the salt, or the basicity of the corresponding ylide. In such cases, weak bases, *e.g.*, potassium carbonate, suffice in removing the α-hydrogen. In the more general case, where little or no extra acidification is present, stronger bases are commonly employed: organolithium compounds, sodamide in liquid ammonia, alkali metal alkoxides in a hydroxylic solvent or in dimethyl sulfoxide, or the dimsyl anion in DMSO. Stabilized ylides (R^3 = COOR, CN, etc.) are isolable. On the other hand, it is well known that normal phosphonium ylides are sensitive both to water and oxygen, so that standard procedures call for scrupulously dry solvents and an inert atmosphere. With water, irreversible decomposition occurs to give alkyl diphenyl phosphine oxide and benzene. In air the following reactions are known to occur:

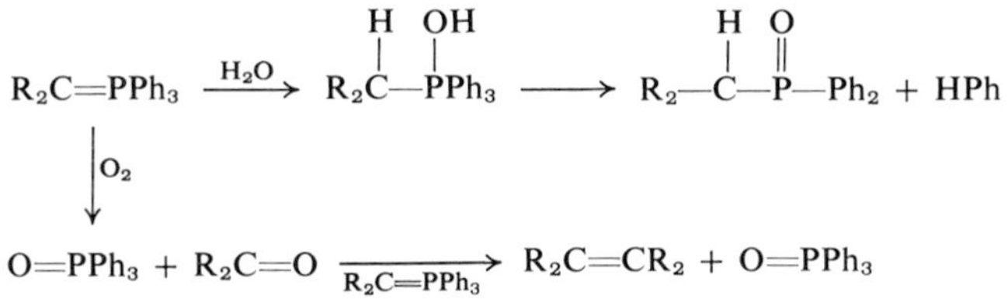

$$R_2C{=}PPh_3 \xrightarrow{H_2O} R_2\overset{H}{C}{-}\overset{OH}{P}Ph_3 \longrightarrow R_2{-}\overset{H}{C}{-}\overset{O}{\overset{\|}{P}}{-}Ph_2 + HPh$$

$$R_2C{=}PPh_3 \xrightarrow{O_2} O{=}PPh_3 + R_2C{=}O \xrightarrow[R_2C=PPh_3]{} R_2C{=}CR_2 + O{=}PPh_3$$

Scheme 3–107

In the second step, the ylide and the carbonyl compound combine to give a betaine in either the *threo* or *erythro* stereoisomeric form. These are subsequently converted to either the *E* or *Z* olefins, respectively. Interconversion of the betaines occurs only through reconversion to the educts. Since this process is often very slow or impossible, the Z/E ratio is difficult to influence. Resonance-stabilized ylides (e.g., R^3 = COOEt) yield mainly *E* products, whereas nonstabilized ylides (*e.g.*, R^3 = alkyl) yield mainly *Z* olefins. The Z/E ratio can be changed to a limited extent by the choice of solvent and the addition of lithium salts [480, 481, 482].

In view of what is known about standard Wittig reactions, as summarized above, prospects for such conversions under PTC conditions did not seem promising at first. In 1973, however, Märkl and Merz showed that concentrated sodium hydroxide/organic solvent could be used in Wittig reactions even with nonactivated phosphonium salts. Since then this technique has been widely exploited because of its preparative simplicity [483]. Mechanistically the situation is still rather obscure. Some authors use catalysts (ammonium salts, crown ethers), and others do not, since phosphonium salts themselves are known to be PTC agents. This, of course, is valid only for the transfer of anions more lipophilic than the halide brought in with the starting salt. Thus, in many cases with 5N to concentrated aqueous sodium hydroxide, the primary deprotonation step seems to occur at the interphase. The ylene formed must then diffuse deeply enough into the organic layer so that reaction with water cannot effectively compete. In other cases solid potassium *tert*-butoxide or potassium carbonate are used as bases [484].

In a variety of solvents (THF, benzene, methylene chloride) salts such as $[Ph_3P{-}CH_2{-}R^3]^{\oplus}X^{\ominus}$ (where R^3 = aryl) together with various aldehydes gave yields of 60–80% olefin, with aqueous or solid bases both with or without a catalyst [483–486]. When R^3 = CH_3, however, yields of only 20–30% were reported in the absence of catalyst in concentrated sodium hydroxide/methylene chloride [483]. Working with solid potassium *tert*-butoxide or potassium carbonate and 18-crown-6 in THF or methylene chloride, some substrates give yields of over 90% [484]. These results are understandable provided that in the latter cases a genuine PTC mechanism, involving transfer of complexed base, is operating.

Double Wittig conversions between glyoxal and arylmethylphosphonium salts in the presence of aqueous sodium hydroxide gave about 20% diarylbutadienes [487].

With benzene as the organic phase it was found that olefin yields depended on the sodium hydroxide concentration, first rising to a maximum, then decreasing with increasing concentration. Optimum conditions depend not only on the anion of the phosphonium salt but also on R^3 [485]:

Optimum NaOH concentration/maximum yields for the reaction C_6H_5—CHO + $R^3CH_2PPh_3{}^{\oplus}X^{\ominus}$.

R^3	X	NaOH	%
C_6H_5	Cl	6N	80
H	I	6N	80
H	Cl	1N	70
CH_2=CH	Cl	1N	60

The *Z*/*E* ratio in PTC Wittig reactions is similar to that observed with the alkoxide method. $Ph_3P^{\oplus}$—CH_2—R^3 and R^1CHO give $Z:E \approx 1:1$ with R^3 = phenyl or *m*-, *p*-substituted phenyl. If R^3 = C_6H_5—CH=CH or 9-anthracenyl, the percentage of *Z* decreases to 36 and 0%, respectively [483].

The effect of the solvent on the stereochemistry of the olefins was noted when 18-crown-6/potassium carbonate or potassium *tert*-butoxide were used. In the reaction R^1CHO + R^3CH_2—$PPh_3{}^{\oplus}X^{\ominus}$ with R^1 = C_6H_5 or C_2H_5, R^3 = C_6H_5, both in THF and CH_2Cl_2, the *trans* isomer predominated, being somewhat higher in the latter solvent. In contrast, for R^3 = CH_3, R^1 = C_6H_5, the $Z:E$ proportion was 85:15 in THF, and 22:78 in CH_2Cl_2 [484].

Even NH-bonds adjacent to a phosphonium center are acidic enough to be deprotonated by solid potassium hydroxide. Separation of the resulting phosphine alkylimines from the alkali and subsequent reaction with diphenylketene gives keteneimines. These can be prepared in optically active form without any measurable racemization at the asymmetric center directly attached to the nitrogen [488]:

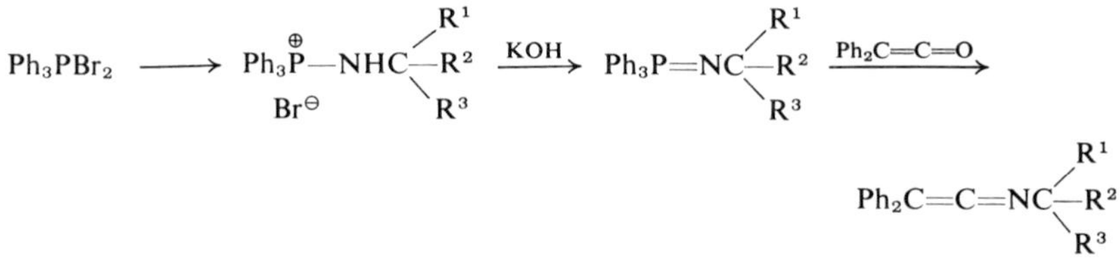

$$Ph_3PBr_2 \longrightarrow Ph_3\overset{\oplus}{P}\text{—NHC}(R^1)(R^2)(R^3)\ Br^{\ominus} \xrightarrow{KOH} Ph_3P\text{=NC}(R^1)(R^2)(R^3) \xrightarrow{Ph_2C=C=O} Ph_2C\text{=C=NC}(R^1)(R^2)(R^3)$$

Scheme 3–108

Table 3–8 gives a survey of published work.

Because of the many factors involved such as structure of either the carbonyl compound or the phosphonium salt, the halide, base, catalyst, solvent, reaction time and temperature, there remains much work to be done before the mechanisms involved are clearly understood. Tentatively it could be suggested that reactions in the absence of water (solid alkali hydroxides, *tert*-butoxide, potassium carbonate) might raise yields considerably if vigorous mixing is ensured. Methylene chloride may not be the best choice of solvent as to some extent it is attacked by strong bases like potassium *tert*-butoxide/crown ether.

Reaction variables were investigated in more detail recently [1113]. In general, the presence of an additional catalyst was not necessary. The PTC Wittig reaction could be extended to reactions of ketones, although yields were low in most cases. Best results were obtained with aromatic aldehydes and benzyltriphenylphosphonium salts as reagents using such bases as solid potassium *tert*-butoxide or fluoride. Aliphatic aldehydes and/or alkyltriphenylphosphonium salts gave lower yields in most cases, and the very high literature values (determined by gas chromatography [484]) could not be reproduced preparatively.

3.3.11.2. Horner (PO Activated) Olefination

In this reaction that is related to the Wittig reaction, phosphonate carbanions are used instead of phosphorus ylides [489]. The advantages of this process are: firstly, the phosphonate carbanions are more nucleophilic and react with a wider variety of aldehydes and ketones under milder conditions; secondly, the water soluble phosphates allow a better separation of reaction mixtures in the workup; thirdly, the phosphonates available from the Arbusov reaction are cheaper and easier to obtain. Normally phosphonates which are successfully employed in the Horner reaction have R^3 stabilizing the carbanion through resonance. When R^3 = H or alkyl, lower olefin yields are obtained. Stereochemically, the formation of *trans*-

$$(EtO)_3P + R^3CH_2X \longrightarrow R^3CH_2{-}P(OEt)_3^{\oplus}X^{\ominus} \xrightarrow{-EtX}$$

$$R^3CH_2{-}PO(OEt)_2 \xrightarrow{-H^{\oplus}} R^3CH^{\ominus}{-}PO(OEt)_2$$

R^3 = COR, COOR, CN, aryl, *etc.*

olefins is favored if the substituent on the α-carbon of the phosphonate is small. Steric hindrance in both the phosphonate and the carbonyl reactant favor the formation of a betaine intermediate which eventually leads to *cis*-olefin [490, 491].

Recently the Horner reaction has been performed under PTC conditions. As in the Wittig process some authors do not use additional catalysts, and others employ ammonium salts or crown ethers. Preliminary results would seem to indicate a preference for *trans*-olefin formation. In the system CH_2Cl_2/50% NaOH, 100%

pure *E*-isomers are formed from aldehydes and $(EtO)_2POCH_2X$ (X = COOR, SO_2CH_3, $PO(OEt)_2$) without added catalyst [492]. The reaction

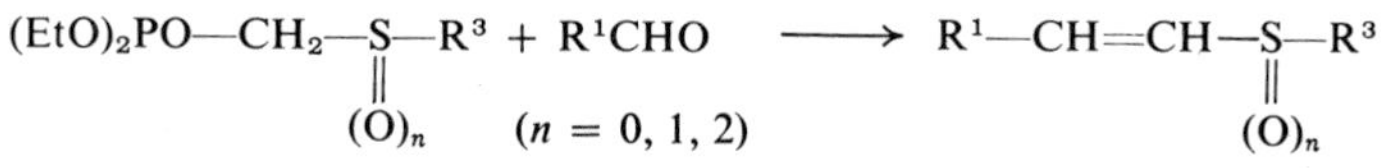

$$(EtO)_2PO{-}CH_2{-}\underset{\underset{(O)_n}{\parallel}}{S}{-}R^3 + R^1CHO \longrightarrow R^1{-}CH{=}CH{-}\underset{\underset{(O)_n}{\parallel}}{S}{-}R^3 \quad (n = 0, 1, 2)$$

Scheme 3–109

in CH_2Cl_2/aqueous NaOH with TEBA was found to be specific for aromatic aldehydes. Ketones and aliphatic aldehydes were unreactive. With $n = 0$ or 1 mixtures of *cis/trans*-isomers were obtained while $n = 2$ led to pure *trans*-compounds. In the case $n = 0$ a strange catalytic effect on the stereochemistry of the olefin was noted by Mikolajczyk *et al.* [493]. Whereas quaternary ammonium chloride yielded almost equal amounts of *Z*- and *E*-alkene, with the corresponding bromide, the iodide, and with crown ether catalysis $Z < E$. It would appear that the salts influence the relative stability of the intermediate betaines.

The stereochemistry of the reaction

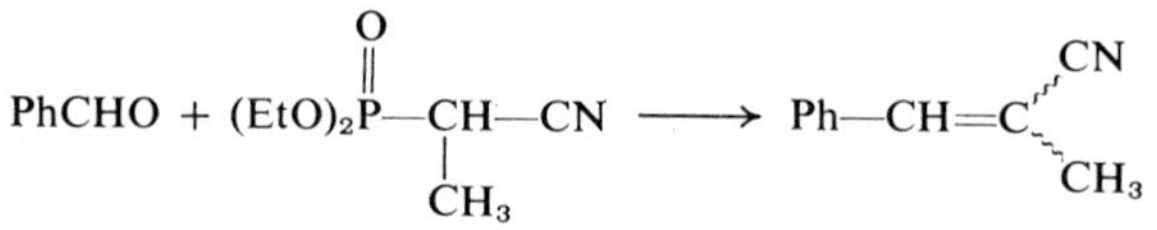

$$PhCHO + (EtO)_2\overset{\overset{O}{\parallel}}{P}{-}\underset{\underset{CH_3}{|}}{CH}{-}CN \longrightarrow Ph{-}CH{=}C\begin{smallmatrix}CN\\CH_3\end{smallmatrix}$$

Scheme 3–110A

was also studied [496]. Slight but distinct changes in the *E/Z* ratio were effected by varying the solvent and catalyst. The *E/Z* ratio was, however, very different from that observed in HMPT.

Horner PTC reactions proved useful also for the synthesis of conjugated dienes and polyenes. In aqueous sodium hydroxide/benzene with NBu_4I, such compounds were synthesized by two routes [494]:

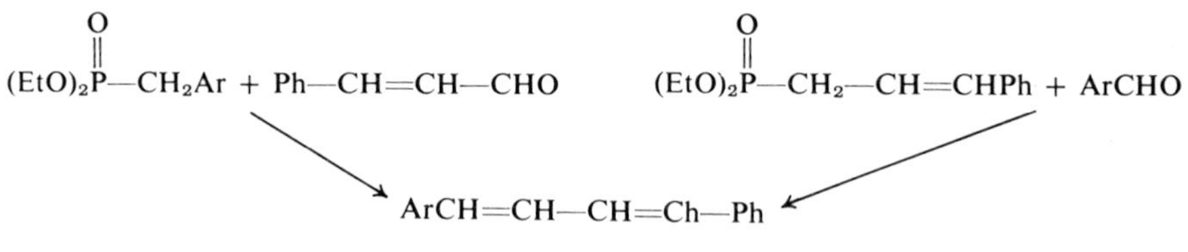

$$(EtO)_2\overset{\overset{O}{\parallel}}{P}{-}CH_2Ar + Ph{-}CH{=}CH{-}CHO \longrightarrow ArCH{=}CH{-}CH{=}Ch{-}Ph \longleftarrow (EtO)_2\overset{\overset{O}{\parallel}}{P}{-}CH_2{-}CH{=}CHPh + ArCHO$$

Scheme 3–110B

Here, as in the Wittig-process, further systematic work is needed to get a clearer understanding of all factors involved. Table 3–8 gives the pertinent literature references.

Table 3–8. Wittig and Horner Reactions under PTC-Conditions.

Reaction	Base	Solvent	Catalyst	Yield (%)	Reference
$R^1CHO + Ph_3P^{\oplus}CHR^3R^4X^{\ominus}$ (R^1 = Aryl, X = Cl, Br, I R^3, R^4 = H, CH_3, C_6H_5)	conc. NaOH	H_2O/CH_2Cl_2	none $NBu_4^{\oplus}I^{\ominus}$	20–90	[483, 1110]
$R^1CHO + R^2\text{—}CH_2PPh_3^{\oplus}X^{\ominus}$	solid KO-*t*-Bu or K_2CO_3	THF or CH_2Cl_2	18-crown-6	82–97	[484]
(R^1, R^2 = Alkyl, Phenyl)	5 N NaOH	H_2O/C_6H_6	none	60–99	[485]
$Ar\text{—}CH_2PPh_3^{\oplus}X^{\ominus} + CH_2O$	conc. NaOH	—		80–90	[486]
$2Ar\text{—}CH_2PPh_3^{\oplus}X^{\ominus} + OCH\text{—}CHO$	conc. NaOH	H_2O/CH_2Cl_2	none	13–23	[487]
$Ph_3P{=}NCR^1R^2R^3 + Ph_2C{=}C{=}O$ (R^1, R^2, R^3 = H, CH_3, C_6H_5)	solid KOH	ether	none	65–85	[488]
$Ar\text{—}COR + PPh_3^{\oplus}\text{—}CH{=}CH\text{—}NHAr\ Cl^{\ominus}$	NaOH	H_2O/CH_2Cl_2	none	low	[1111]
$RCHO + R^1S\text{—}CH(R^2)\text{—}PPh_3^{\oplus}\ Cl^{\ominus}$	NaOH	H_2O/CH_2Cl_2	none	~90	[1112]
$Ar\text{—}CHO + (EtO)_2PO\text{—}CH_2\text{—}X$ (X = CN, COOR, COPh, S—Ph, SO—R, SO_2R)	conc. NaOH	H_2O/CH_2Cl_2	none	55–95	[492]
$Ar\text{—}CHO + (EtO)_2POCH_2\text{—}X$ (X = —SR, —SO_2R, —SOR)	conc. NaOH	H_2O/CH_2Cl_2	various $NR_4^{\oplus}X^{\ominus}$ crown ethers	55–87	[493]
$Ar\text{—}CH{=}CH\text{—}CHO + (EtO)_2POCH_2Ar$ $Ar\text{—}CHO + (EtO)_2POCH_2\text{—}CH{=}CHAr$	conc. NaOH	H_2O/C_6H_6	$NBu_4^{\oplus}I^{\ominus}$	72–81	[494]
$R^1R^2CO + (EtO)_2POCH_2\text{—}X$ (R^1, R^2 = H, Alkyl, Aryl, X = CN, COOR, heteroaromatics)	conc. NaOH	H_2O/CH_2Cl_2	$NBu_4^{\oplus}I^{\ominus}$	30–77	[495, 1114]
$Ar\text{—}CHO + (EtO)_2PO\text{—}CH(CH_3)\text{—}CN$	dilute NaOH	H_2O/CH_2Cl_2	$NBu_4^{\oplus}Br^{\ominus}$	80	[323]
	dilute NaOH	H_2O/various solvents	various onium salts	high	[496]

Optically active 3-substituted dihydroisocarbostyrils (**A**) react with diethyl cyanomethylphosphonate and 50% NaOH at room temperature, yielding the diastereomers **B** and **C**. The diastereomeric excess of **B** depends on the solvent. Without catalyst, the reaction is much slower, but a little more selective [985].

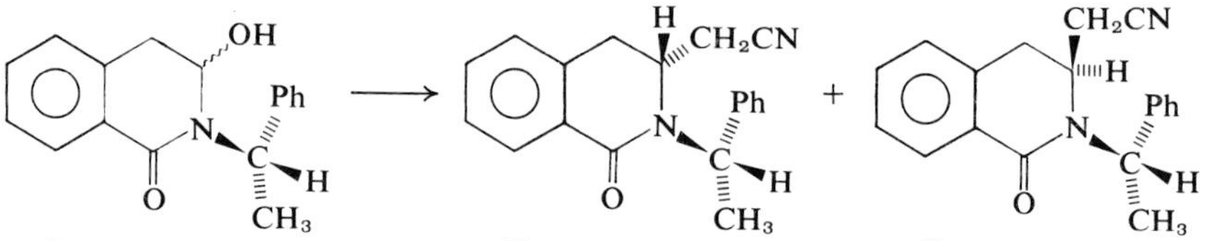

A → **B** + **C**

	B	C
NBu_4Br, CH_2Cl_2	70.5	29.5
NBu_4Br, dioxane	75.5	24.5
no catalyst, CH_2Cl_2	77	23

Scheme 3–110C

3.3.11.3. Sulfonium Ylide and Oxosulfonium Ylide Reactions

Similar to P-ylides, S-ylides can also be reacted with carbonyl compounds. It is known, however, that in the latter case dimethylsulfonium methylide gives oxiranes with all carbonyl compounds. In contrast, dimethyloxosulfonium methylide yields oxiranes from normal carbonyls and cyclopropyl substituted compounds from conjugated starting materials [498]:

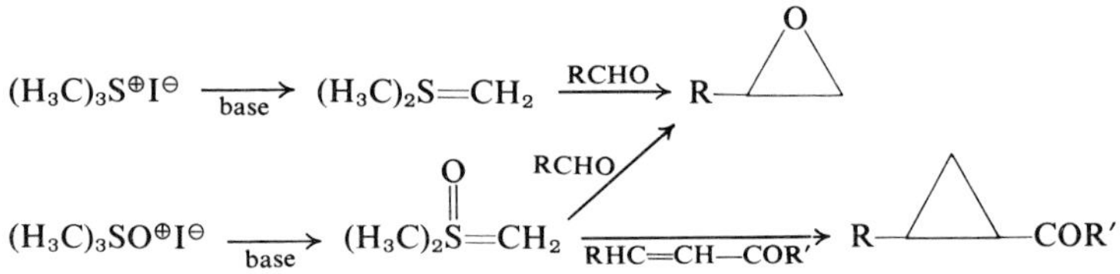

$$(H_3C)_3S^{\oplus}I^{\ominus} \xrightarrow{\text{base}} (H_3C)_2S{=}CH_2 \xrightarrow{\text{RCHO}} \text{R-oxirane}$$

$$(H_3C)_3SO^{\oplus}I^{\ominus} \xrightarrow{\text{base}} (H_3C)_2S(=O){=}CH_2 \xrightarrow[\text{RHC=CH—COR}']{} \text{R-cyclopropane-COR}'$$

($(H_3C)_2S(=O){=}CH_2$ + RCHO → R-oxirane)

Scheme 3–111

Sulfonium salts are prepared by reacting organic sulfides with alkyl halides. These in turn are deprotonated in nonaqueous base according to older preparative procedures. PTC methods are again applicable in some cases, either with or without additional catalyst. Merz and Märkl [498] showed that trialkylsulfonium salts with small alkyl groups cannot be transformed in methylene chloride/50% aqueous sodium hydroxide without the addition of a catalyst. For example, benzaldehyde and cinnamyl aldehyde (50 °C, 48 h) give yields of over 90% with 1–5 mole-% of NBu_4I. With ketones, the yields are lower: only 36% oxiranes from acetophenone and 18% from benzophenone after 72 h. The catalytic effect of the added ammonium salts on the reaction mechanism is of an obscure nature. Of all halides, iodide is the least likely to transfer hydroxide. Merz and Märkl assume that trimethylsulfonium iodide itself shows no catalytic activity because of its small alkyl groups.

It was found in PTC displacement reactions [28] that onium salts with small substituents are distributed into the aqueous phase and therefore show little catalytic effect. This cannot be the case here, however, because with concentrated sodium hydroxide even small onium salts remain in the organic phase. The answer probably lies in as yet unexplained solute-solvent interactions.

Yano *et al.* [497] investigated the reaction between benzaldehyde and trimethylsulfonium iodide (**A**) and lauryldimethylsulfonium iodide (**B**) in the absence of catalyst in a two-phase, benzene/aqueous sodium hydroxide system of varying concentration. In 5N NaOH, **A** did not give any oxirane; presumably **A** was dissolved in the water phase. With 15N NaOH, 83% phenyloxirane was obtained. The only difference to previous unsuccessful attempts [498] appears to be the use of benzene

$$\underset{\mathbf{A}}{(CH_3)_3S^{\oplus}I^{\ominus}} \qquad \underset{\mathbf{B}}{C_{12}H_{25}S^{\oplus}(CH_3)_2I^{\ominus}}$$

instead of methylene chloride. With compound **B**—which can function as a typical detergent and micelle-forming agent—even in 5N NaOH, 23% phenyloxirane was obtained, and the yield rose to 81% in 15N NaOH. High conversions were realized between trimethylsulfonium chloride or iodide and ketones after 10 hours of stirring with 15N NaOH. Apparently chloride gives higher yields than iodide [497] (Table 3–9).

Table 3–9. Oxirane Yields for the Reaction

$$R^1R^2C{=}O + (CH_3)_3S^{\oplus}X^{\ominus}.$$

Carbonyl Compound	$X^{\ominus}$	Yield (%)
C_6H_5—CHO	Cl	86
cyclohexanone	Cl	88
	I	27
acetophenone	Cl	85
camphor	Cl	40

Oxiranes [499] were prepared from benzaldehyde, formaldehyde and acetaldehyde using a number of sulfonium ylides (Table 3–10). The following experimental conditions were used: an excess of 50% aqueous sodium hydroxide was added to a warm mixture of the aqueous sulfonium salt solution, the carbonyl compound and (usually) an immiscible solvent. Various solvents such as benzene, toluene, *n*-hexane, methanol, isopropanol, as well as mixtures of these were used. The *trans*-oxirane isomers

Table 3–10.

Yields of Oxiranes $R{-}HC\overset{O}{\frown}CH{-}R'$ [499].

Sulfonium Salt	$C_6H_5{-}CHO$ (%)	CH_2O (%)	CH_3CHO (%)
$Me_3S^{\oplus}Cl^{\ominus}$	68	—	—
$Et_3S^{\oplus}Cl^{\ominus}$	80	—	—
$CH_3{-}CH{=}CH{-}SMe_2^{\oplus}Cl^{\ominus}$	70	8	—
$C_6H_5{-}CH_2SMe_2^{\oplus}Cl^{\ominus}$	85	87	48

predominated. In the case of aliphatic aldehydes, Cannizzaro and aldol-type reactions intervened unfavorably. An undesirable side reaction also occurred with allyldimethylsulfonium chloride, giving 2-methyloxirane:

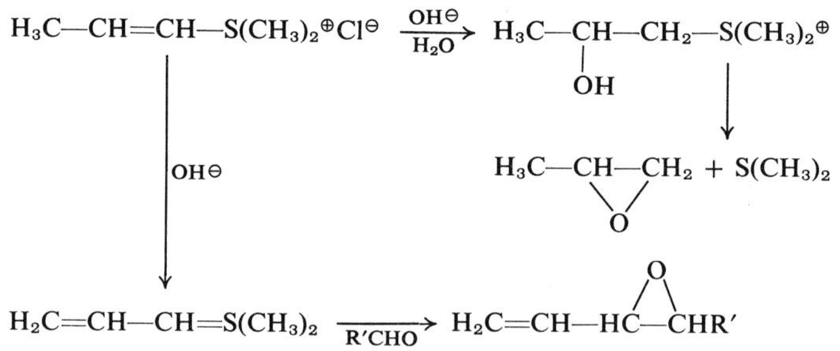

$$H_3C{-}CH{=}CH{-}S(CH_3)_2^{\oplus}Cl^{\ominus} \xrightarrow[H_2O]{OH^{\ominus}} H_3C{-}\underset{OH}{CH}{-}CH_2{-}S(CH_3)_2^{\oplus} \longrightarrow H_3C{-}CH\overset{O}{\frown}CH_2 + S(CH_3)_2$$

$$H_3C{-}CH{=}CH{-}S(CH_3)_2^{\oplus}Cl^{\ominus} \xrightarrow{OH^{\ominus}} H_2C{=}CH{-}CH{=}S(CH_3)_2 \xrightarrow[R'CHO]{} H_2C{=}CH{-}HC\overset{O}{\frown}CHR'$$

Scheme 3–112

An enantioselective phenyloxirane synthesis from trimethylsulfonium iodide and benzaldehyde/50% aqueous sodium hydroxide using the chiral catalyst **C**, which is derived from ephedrine, has also been reported [500, 555].

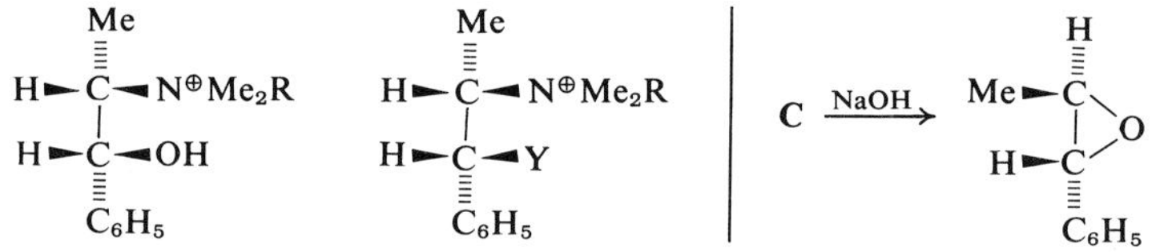

C: Me / H–C–$N^{\oplus}Me_2R$ / H–C–OH / C_6H_5

D: Me / H–C–$N^{\oplus}Me_2R$ / H–C–Y / C_6H_5

$$C \xrightarrow{NaOH} E$$

E: H / Me–C–O–C–H / C_6H_5 (oxirane)

R = Me, Et, *n*-Pr, *n*-Bu   Y = H, OCH_3

Scheme 3–113

Catalysts without a hydroxyl group, *i.e.*, **D**, were said to give no enantioselectivity. The claim of optical induction was later disproved [551]. Optical activity of the phenyloxirane is in fact due to small amounts of an impurity, methylphenyloxirane **E**, formed by elimination of amine from the catalyst **C**, which is a well-known reaction [501].

Trimethyloxosulfonium halides are also precursors of dimethyloxosulfonium methylides which can in turn react with carbonyl compounds to yield oxiranes. Brändström points out that the chloride of this system has the advantage of being soluble in THF. The normal synthetic procedure leads to the iodide. Until recently the conversion of the iodide to the chloride had to be carried out with elemental chlorine. Using Brändström's ion pair extraction process, the transformation becomes very easy. The iodide and an equimolar amount of benzyltributylammonium chloride are distributed between methylene chloride and water. The water contains the trimethyloxosulfonium chloride while the organic phase consists of a $C_6H_5—CH_2NBu_3^{\oplus}I^{\ominus}$ solution [1].

Merz and Märkl [498] observed that the formation of phenyloxirane under the conditions indicated below was accompanied by an aldol-type reaction:

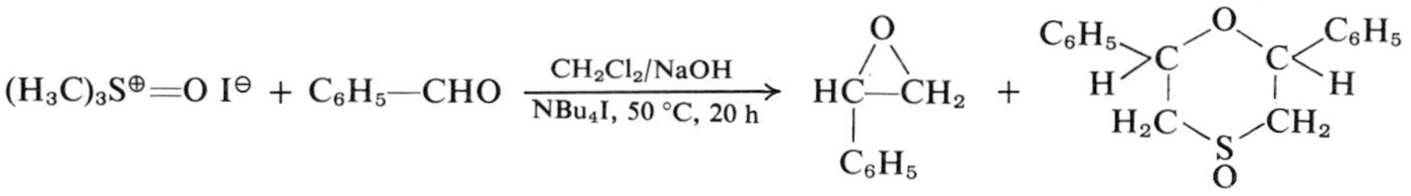

$$(H_3C)_3S^{\oplus}{=}O\ I^{\ominus} + C_6H_5—CHO \xrightarrow[NBu_4I,\ 50\ °C,\ 20\ h]{CH_2Cl_2/NaOH} \text{HC(O)CH}_2\text{ (oxirane), } C_6H_5 + \text{cyclic sulfoxide: } O\text{-bridged } C_6H_5(H)C\text{—}C(H)C_6H_5,\ H_2C\text{—}S(O)\text{—}CH_2$$

Scheme 3–114

α,β-Unsaturated aromatic ketones gave Z/E mixtures of cyclopropyl ketones in yields between 70 and 80%.

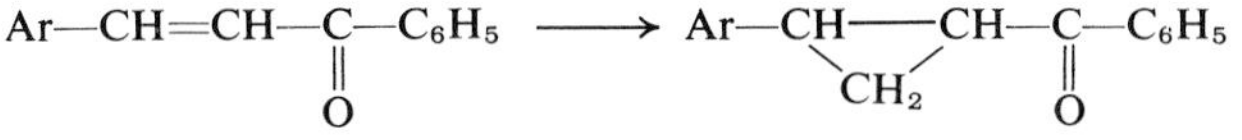

$$Ar—CH{=}CH—C({=}O)—C_6H_5 \longrightarrow Ar—\underset{\diagdown CH_2 \diagup}{CH——CH}—C({=}O)—C_6H_5$$

Scheme 3–115

1,5-Diphenylpenta-1,4-dien-3-one gave a mixture of mono- and biscyclopropanation products [498]. Table 3–11 summarizes sulfonium ylide reactions.

Typical experimental procedures are presented below.

Olefins (Wittig reaction): A heterogeneous mixture of 3 mmols aldehyde, *e.g.*, benzaldehyde or cinnamyl aldehyde, 5 mmoles alkyltriphenylphosphonium halide, 5 ml benzene and 10 ml aqueous sodium hydroxide is stirred a few minutes to several days at room temperature, then worked up in the usual manner.

cis/trans-stilbene: 10 ml 50% sodium hydroxide is added to a well stirred mixture of 20 mmoles benzaldehyde and 20 mmoles benzyltriphenylphosphonium chloride in 20 ml dichloromethane. The reaction is finished after 10 minutes. Workup gives about equal amounts of *cis*- and *trans*-compound [483].

Olefins (Horner reaction): 20 mmoles phosphonate and 20 mmoles aldehyde dissolved in 5 ml benzene are added, at room temperature, to the two-phase system 20 ml benzene/20 ml 50% aqueous sodium hydroxide containing 1 mmole NBu_4I as catalyst. The mixture is stirred vigorously and refluxed for about 30 minutes and then worked up [494].

2-phenyloxirane: 100 ml 50% aqueous sodium hydroxide is added to a stirred mixture of 0.1 mole benzaldehyde, 0.1 mole trimethylsulfonium iodide and 1 mmole

Table 3–11. Reactions of Sulfonium and Oxosulfonium Ylides and Carbonyl Compounds under PTC-Conditions.

Reaction	Base	Solvent	Catalyst	Yield (%)	Reference
$R^1R^2CO + (O)_nSMe_3{}^{\oplus}I^{\ominus}$ ($n = 0, 1$; R^1, R^2 = H, Ph, Ph—CH=CH, $R-C_6H_4-$ ($R = CH_3, OCH_3$))	conc. NaOH	H_2O/CH_2Cl_2	$NBu_4{}^{\oplus}I^{\ominus}$ and $SMe_3{}^{\oplus}I^{\ominus}$	20–90	[498]
$R^1R^2CO + R^3SMe_2{}^{\oplus}X^{\ominus}$ (R^3 = Me, lauryl; X = Cl, I; R^1, R^2 = H, Me, Ph, cyclohexanone, camphor)	5, 10, 15 N NaOH	H_2O/C_6H_6	none	30–90	[497]
$R^1HCO + R^3R^4R^5S^{\oplus}Cl^{\ominus}$ (R^1 = H, Me, Ph; $R^3 = R^5$ = Me, Et R^4 = Me, Et, allyl, benzyl)	conc. NaOH	H_2O/benzene H_2O/toluene H_2O/methanol H_2O/ethanol H_2O/isopropanol H_2O/*n*-hexane	none	0–90	[499]
$PhHCO + SMe_3{}^{\oplus}I^{\ominus}$	conc. NaOH	H_2O/CH_2Cl_2	(−)N,N-dimethyl-ephedrinium bromide	67	[500, 555]
ArCOR + polystyrene-$SMe_2{}^{\oplus}X^{\ominus}$ ($-SEt_2{}^{\oplus}X^{\ominus}$)	conc. NaOH	H_2O/CH_2Cl_2	$NBu_4{}^{\oplus}I^{\ominus}$	86–99	[1115]

NBu_4I in 100 ml methylene chloride. The reaction mixture is heated with vigorous stirring at 50 °C for 48 hours. Workup gives 92% 2-phenyloxirane [498].

3.3.12. Miscellaneous Reactions in the Presence of Strong Base

In this section various PTC reactions are collected.

3.3.12.1. Diazo Group Transfer

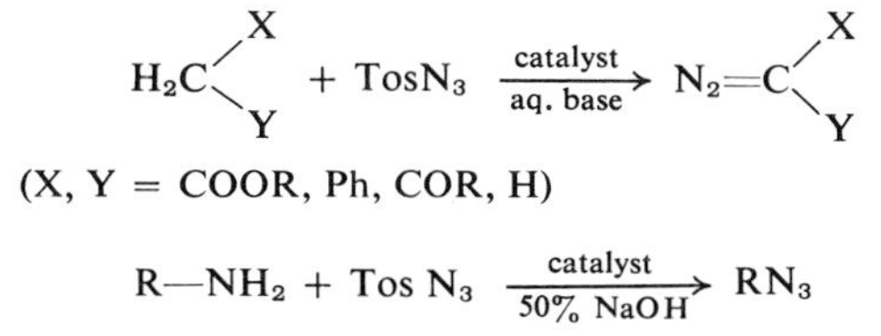

$$H_2C(X)(Y) + TosN_3 \xrightarrow[\text{aq. base}]{\text{catalyst}} N_2{=}C(X)(Y)$$

(X, Y = COOR, Ph, COR, H)

$$R{-}NH_2 + Tos\,N_3 \xrightarrow[50\%\ NaOH]{\text{catalyst}} RN_3$$

Scheme 3–116

α-Diazocarbonyl compounds were obtained after shorter reaction times and uncontaminated by unreacted tosyl azide using the PTC procedure with benzene, pentane, or dichloromethane/aqueous sodium carbonate or 10 N NaOH with Aliquat 336 or tetrabutylammonium bromide [508]. The mixture was usually stirred for 15 hours at room temperature, and yields ranged between 77 and 100%. Aryl azides were prepared in yields of up to 94% from a 10 to 50 molar excess of amine to tosyl azide with 50% NaOH/TEBA. Tetrahexylammonium bromide as catalyst gave somewhat better yields. With aliphatic amines yields were generally lower [509].

Monosubstituted hydrazino compounds can be reduced to hydrocarbons by refluxing with tosyl azide in benzene/concentrated aqueous NaOH in the presence of tetraethylammonium bromide for 2 to 10 hours [1116]:

$$RNHNH_2 + TosN_3 \rightarrow RH + 2N_2 + TosNH_2$$

3.3.12.2. γ-Elimination

Several γ-eliminations following Michael additions were described in Section 3.3.7. The process

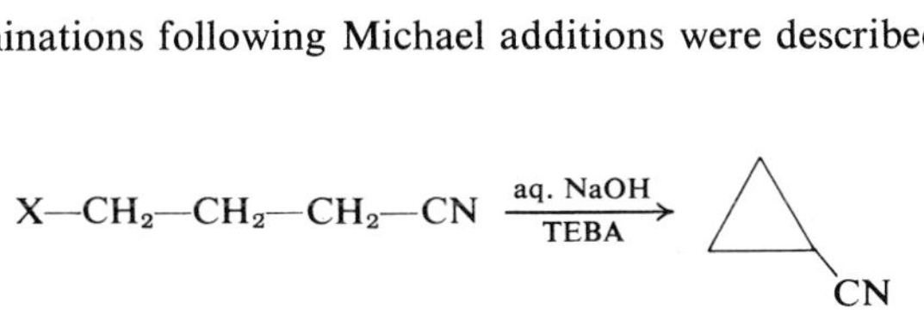

$$X{-}CH_2{-}CH_2{-}CH_2{-}CN \xrightarrow[\text{TEBA}]{\text{aq. NaOH}} \triangle{-}CN$$

Scheme 3–117

is described in a recent patent [513], and the formation of an epoxide from a chlorohydrin was achieved by heating with 20% aqueous sodium hydroxide in the presence of tetrabutylammonium hydroxide [1117].

3.3.12.3. Preparation of Acid Fluorides

A PTC process for the preparation of aliphatic and aromatic acyl fluorides from the corresponding acids, potassium carbonate and the dimer of hexafluoropropene, perfluoro-2-methyl-2-pentene (**A**), or its trimeric mixture in dichloromethane has been developed [514]. At the beginning of the reaction, the mixture consists of three

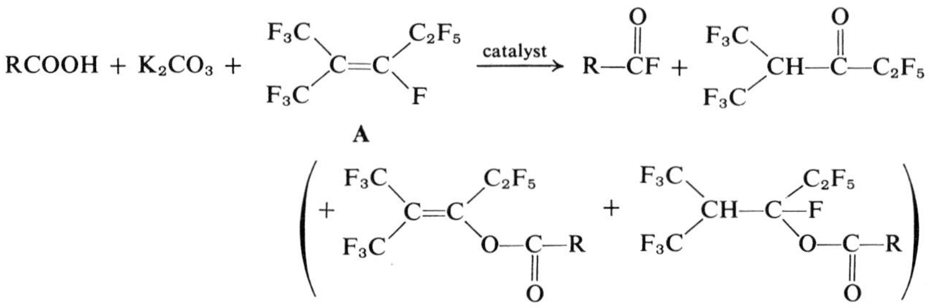

$$RCOOH + K_2CO_3 + (F_3C)_2C{=}C(C_2F_5)F \ (\mathbf{A}) \xrightarrow{\text{catalyst}} R{-}CF{=}O + (F_3C)_2CH{-}C({=}O){-}C_2F_5$$

$$\left(+ (F_3C)_2C{=}C(C_2F_5){-}O{-}C({=}O){-}R + (F_3C)_2CH{-}CF(C_2F_5){-}O{-}C({=}O){-}R \right)$$

Scheme 3–118

phases, **A**, the potassium salt, and dichloromethane. Addition of 18-crown-6, Aliquat 336, or a glyme with more than 4 oxyethylene units, catalyzes the conversion, so that reaction times of 2 to 24 hours at room temperature are possible.

3.3.12.4. Rearrangements

A Smiles rearrangement of compound **B** to **C** is a result of stirring the sulfonamide in dichloromethane with 18-crown-6 and solid potassium hydroxide. In aqueous alkali hydroxide (without crown ether) **C** is not isolable because of a second rearrangement yielding **D** [515].

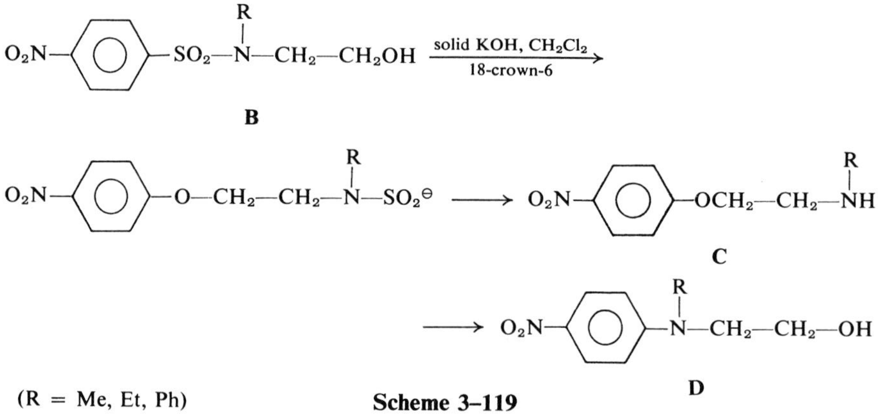

$$O_2N{-}C_6H_4{-}SO_2{-}N(R){-}CH_2{-}CH_2OH \ (\mathbf{B}) \xrightarrow[\text{18-crown-6}]{\text{solid KOH, } CH_2Cl_2}$$

$$O_2N{-}C_6H_4{-}O{-}CH_2{-}CH_2{-}N(R){-}SO_2^{\ominus} \longrightarrow O_2N{-}C_6H_4{-}OCH_2{-}CH_2{-}NHR \ (\mathbf{C})$$

$$\longrightarrow O_2N{-}C_6H_4{-}N(R){-}CH_2{-}CH_2{-}OH \ (\mathbf{D})$$

(R = Me, Et, Ph)

Scheme 3–119

$$Ar_2C{=}CHX \longrightarrow Ar{-}C{\equiv}C{-}Ar$$

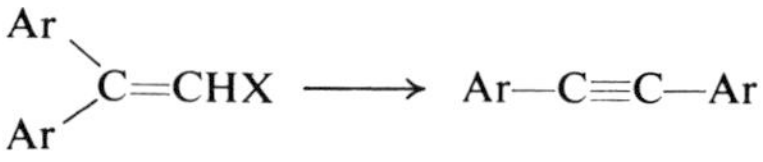

Scheme 3–120

In a borderline PTC reaction, Fritsch-Buttenberg-Wiechell rearrangements occur at room temperature in toluene/20% DMSO with 3 equivalents of potassium *tert*-butoxide and 1 equivalent of cryptate [222] (**5**) [552]. Finally, a PTC Wittig re-arrangement was described recently [1204].

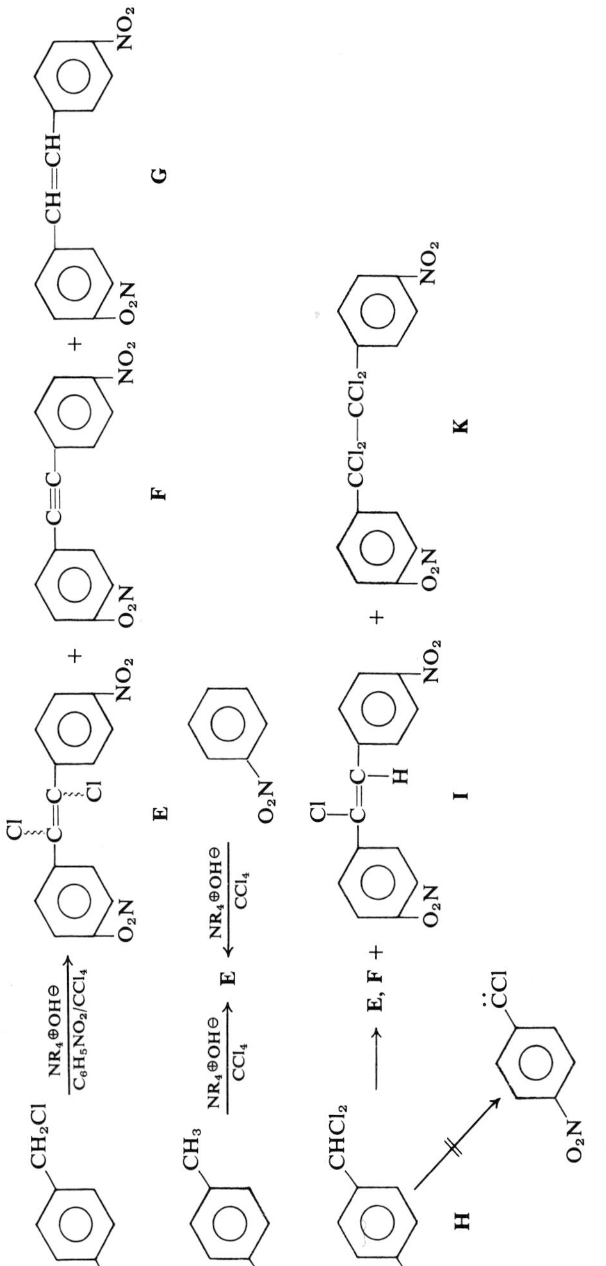

O_2N–C_6H_4–CH_2Cl $\xrightarrow[C_6H_5NO_2/CCl_4]{NR_4^{\oplus}OH^{\ominus}}$ O_2N–C_6H_4–C(Cl)=C(Cl)–C_6H_4–NO_2 (**E**) + O_2N–C_6H_4–C≡C–C_6H_4–NO_2 (**F**) + O_2N–C_6H_4–CH=CH–C_6H_4–NO_2 (**G**)

O_2N–C_6H_4–CH_3 $\xrightarrow[CCl_4]{NR_4^{\oplus}OH^{\ominus}}$ **E** $\xleftarrow[CCl_4]{NR_4^{\oplus}OH^{\ominus}}$ O_2N–C_6H_5

O_2N–C_6H_4–$CHCl_2$ (**H**) ⟶ **E, F** + O_2N–C_6H_4–C(Cl)=C(H)–C_6H_4–NO_2 (**I**) + O_2N–C_6H_4–CCl_2–CCl_2–C_6H_4–NO_2 (**K**)

H ↛ O_2N–C_6H_4–$\ddot{C}Cl$

Scheme 3–121

3.3.12.5. Radical Reactions (*cf.*, Scheme 3–121)

p-Nitrobenzyl chloride and benzyltriethylammonium hydroxide in a mixture of carbon tetrachloride and nitrobenzene (1:2) gave up to 98% **E**. **F**, and **G** were formed as by-products [516]. In the absence of nitrobenzene, **E** was formed in only 25% yield, whereas in the absence of carbon tetrachloride no **E** was observed. **E**, together with chloroform and hexachloroethane, were also generated from *p*-nitrotoluene or nitrobenzene and carbon tetrachloride under similar conditions.

Attempts to trap a possible carbene intermediate in the PTC reaction of **H** were unsuccessful. Instead the stereoisomers of **E**, **F**, **I**, and **K** were formed [517]. Reactions between nitrobenzyl chlorides and substituted phenylacetonitriles in the presence of concentrated sodium hydroxide/catalyst [308, 319] yield mainly substitution, together with some radical coupling products (for details *cf.*, Section 3.3.4).

3.3.12.6. Reactions with Carbon Tetrahalides

Mąkosza and co-workers found the following high yield reactions between aryl-acetonitriles, carbon tetrachloride, and concentrated sodium hydroxide/TEBA [522]:

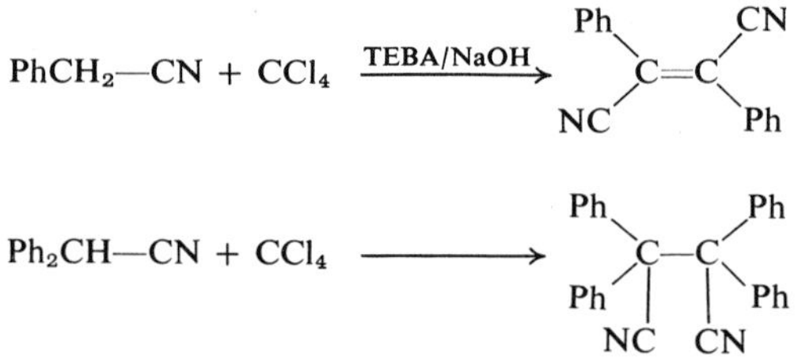

$$PhCH_2{-}CN + CCl_4 \xrightarrow{TEBA/NaOH} (Ph)(NC)C{=}C(CN)(Ph)$$

$$Ph_2CH{-}CN + CCl_4 \longrightarrow Ph_2C(CN){-}C(CN)Ph_2$$

Scheme 3–122

Apparently a $Cl^{\oplus}$ is transferred from CCl_4 to a carbanion intermediate in a complex series of consecutive steps. Chloro-phenylacetonitrile anions could be trapped by aldehydes and acrylonitrile [1120]. A similar transfer of positive chlorine must occur in the synthesis of di-*tert*-butyl phosphorohalidates which gives ≈90% yields [523]. The diester-monohalides of phosphoric acid were often prepared previously in an impure state by N-halosuccinimide reactions.

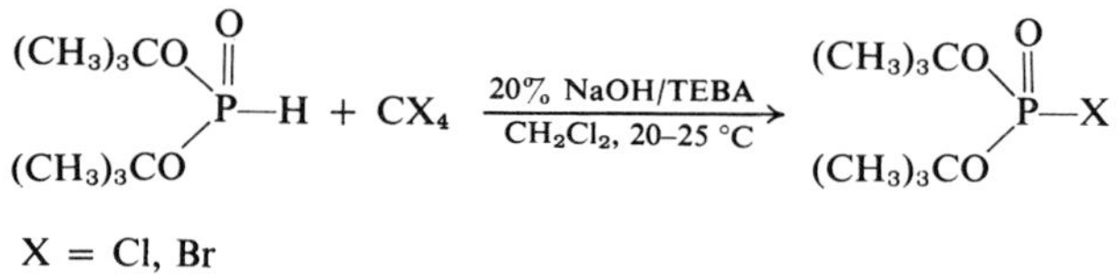

$$((CH_3)_3CO)_2P(O){-}H + CX_4 \xrightarrow[CH_2Cl_2,\ 20\text{–}25\ °C]{20\%\ NaOH/TEBA} ((CH_3)_3CO)_2P(O){-}X$$

X = Cl, Br

Scheme 3–123

Without isolating the dialkoxy phosphorohalidate, this process has been used for the preparation of mixed esters [129], diester-monoamides [126, 272], diester-hydrazides [128], and the diester-alkoxy-amides [127] of phosphoric acid (*cf.*, Section 3.2.3.).

Finally, phenylacetylene was transformed into phenylchloroacetylene by adding

50% sodium hydroxide to a vigorously stirred mixture of the alkyne, CCl_4, and TEBA at 35 °C [769], and a number of other compounds could be chlorinated in this way also [1120].

$$Ph—C{\equiv}CH \longrightarrow Ph—C{\equiv}C—Cl$$

3.3.12.7. Generation of Benzonitrile-N-Sulfide

Reacting N-benzyliminosulfurdifluoride (**L**) and dimethyl acetylenedicarboxylate (12 h, 132 °C) in chlorobenzene with 2 equivalents of NaF and 0.1 equivalent of 18-crown-6 gave a 65% yield of adduct (**M**) [518].

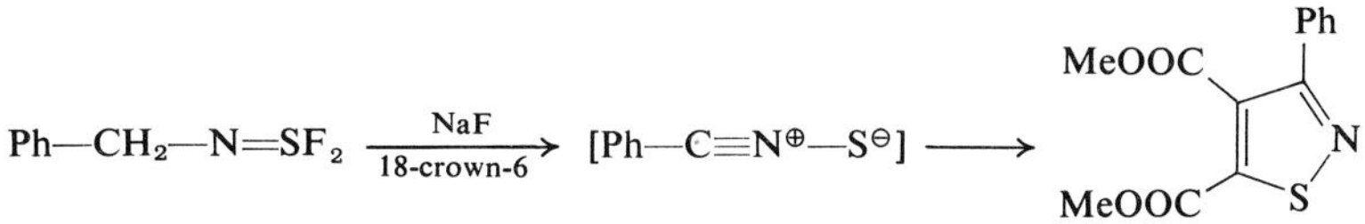

$$\underset{\mathbf{L}}{Ph—CH_2—N{=}SF_2} \xrightarrow[\text{18-crown-6}]{\text{NaF}} [Ph—C{\equiv}N^{\oplus}—S^{\ominus}] \longrightarrow \mathbf{M}$$

(**M**: 3-phenyl-4,5-bis(MeOOC)-isothiazole; labels: Ph, MeOOC, MeOOC, N, S)

Scheme 3–124

3.3.12.8. Generation of Nitrenes

The use of fluoride in α-limeinations yielding transient nitrenes was first described in heterogeneous or homogeneous solution

$$Ph—CO—NHCl \xrightarrow[\text{benzene}]{\text{KF, boiling}} Ph—N{=}C{=}O \quad \text{secondary products [519]}$$

$$p\text{-}O_2N—C_6H_4—SO_2—NH—COOEt \xrightarrow{NEt_4F} \overline{\underline{N}}—COOEt \longrightarrow \text{insertion products}$$

N — **O** — [520]

Using cetyltrimethylammonium bromide as catalyst, the nitrene **O** was generated from **N** in the presence of aqueous sodium hydroxide [521]. Addition-rearrangement or insertion products **P**, **Q**, and **R** were obtained in 13, 18, and 12% yield, respectively.

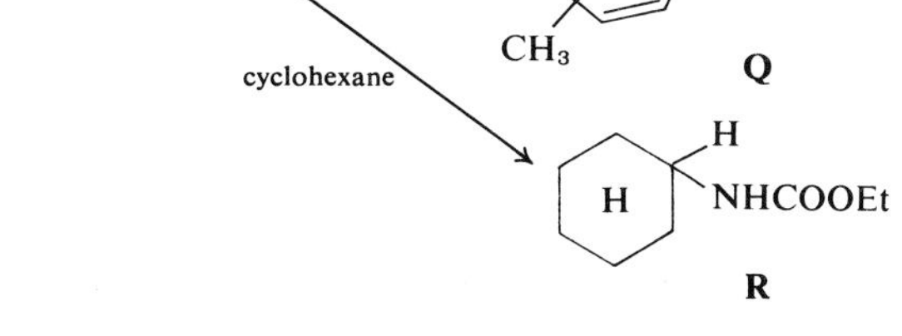

$$N \xrightarrow[\text{NaOH}]{\text{catalyst}}$$

benzene → **P** (azepine N—COOEt)
toluene → **Q** (methylazepine N—COOEt, CH_3)
cyclohexane → **R** (cyclohexyl-NHCOOEt)

Scheme 3–125

In reactions with olefins, the ratio of stereospecific (singlet nitrene) addition and insertion to nonstereospecific (triplet) addition depended on reaction variables including the PT catalyst used. Ammonium iodides gave significantly more of the singlet nitrene derived products [1118].

3.3.12.9. PTC in Diazo Couplings

Arenediazonium tetrafluoroborates are normally insoluble in chloroform, but the presence of dicyclohexano-18-crown-6 (*cf.*, Sections 3.2.1.3 and 3.2.1.4) allows diazo couplings in this solvent with pyrroles that cannot be executed easily otherwise [1119]. Azo couplings *via* gegenion exchange with PTC ammonium salts are known too [81].

3.3.12.10. Organometallic PTC Applications

Sulfur-donor ligand *ortho*-metalated complexes **S** were obtained readily using this reaction at room temperature [510]:

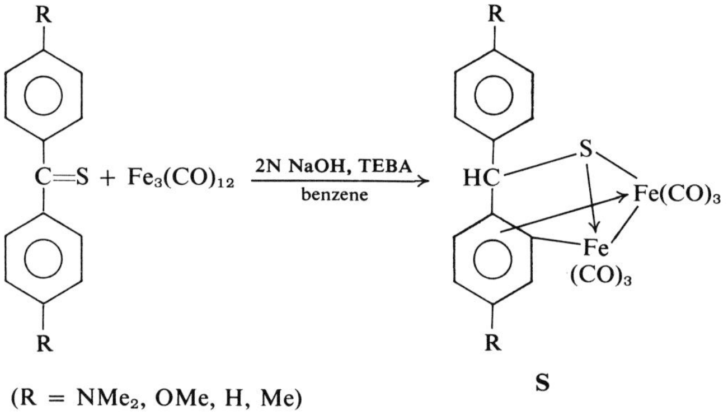

$(R\text{-}C_6H_4)_2C{=}S + Fe_3(CO)_{12} \xrightarrow[\text{benzene}]{\text{2N NaOH, TEBA}}$ **S**

(R = NMe_2, OMe, H, Me)

Scheme 3–126

It is assumed that $HFe_3(CO)_{11}^{\ominus}$ is the important species which is transferred into the organic phase. The benzylic hydrogen in the product comes from the O-position of the starting material. Treatment of the same thiobenzophenones with cyclopentadienyliron dicarbonyl dimer, 50% sodium hydroxide, and a PT catalyst in benzene at room temperature affords fulvenes in superior yields. The ion pair $Q^{\oplus}$ $[Fe(C_5H_5)(CO)_2]^{\ominus}$ seems to be the intermediate [1122].

π-Allylcobalt tricarbonyl complexes **T** were prepared in 70–80% yield by reacting a benzene solution of an allyl bromide with dicobalt octacarbonyl, 5N NaOH, and TEBA at room temperature [511]. Again the PTC process is superior to the older synthetic methods in terms of yield, reaction rates, simplicity, and mildness. $Co(CO)_4^{\ominus}$ appears to be the extracted anion.

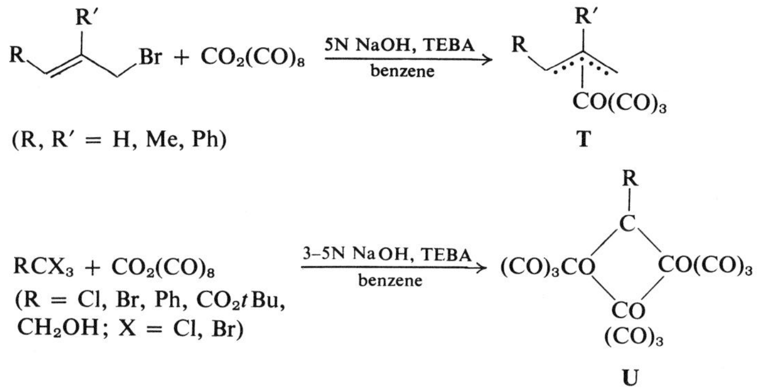

$$\text{R}\diagdown\!\!=\!\!\underset{}{\overset{\text{R}'}{|}}\diagdown\text{Br} + CO_2(CO)_8 \xrightarrow[\text{benzene}]{\text{5N NaOH, TEBA}} \text{R-}\eta^3\text{-allyl(R')}\text{-}CO(CO)_3 \quad \textbf{T}$$

(R, R′ = H, Me, Ph)

$$RCX_3 + CO_2(CO)_8 \xrightarrow[\text{benzene}]{\text{3–5N NaOH, TEBA}} (CO)_3CO\text{–}C(R)\text{–}CO(CO)_3,\ CO(CO)_3 \quad \textbf{U}$$

(R = Cl, Br, Ph, CO_2tBu, CH_2OH; X = Cl, Br)

Scheme 3–127

Under similar conditions, the clusters **U** were prepared in good yields from *gem*-trihalocompounds [511]. The palladium-catalyzed carbonylation of halides (benzyl, aryl, vinyl, or heterocyclic) is also accelerated by PTC [512]. Working with 30% sodium hydroxide/xylene and small amounts of triphenylphosphine (to avoid precipitation of metallic Pd), $Pd[P(C_6H_5)_3]_2Cl_2$, and tetrabutylammonium iodide, the organic halide is dropped slowly into the reaction mixture at 95 °C under 5 bar pressure of carbon monoxide. The action of the catalyst is to liberate the carboxylate anion from an intermediate Pd-complex and bring it into the aqueous layer.

$$RX + CO + 2NaOH \xrightarrow[\text{Pd-complex, xylene}]{\text{30\% NaOH, }NBu_4I} RCO_2^{\ominus}Na^{\oplus} + NaX + H_2O$$

PTC substitutions of group VI metal carbonyls were observed too [551]. An example is given in the following equation:

$$Mo(CO)_6 + Ph_2P\text{—}(CH_2)_2\text{—}PPh_2 \xrightarrow[\text{50\% aq. NaOH, }NBu_4I]{\text{benzene, 80 °C, 2 h}} Mo(CO)_4(Ph_2P\text{—}(CH_2)\text{—}PPh_2)$$

Similar reactions were performed with chromium, molybdenum, and wolfram carbonyls, and other ligands introduced were $AsPh_3$, PPh_3, and dipyridine. The PT reactions were much faster than either the older procedures or a newer homogeneous catalytic reaction with sodium hydroxide in aqueous methanol. It was suggested that the hydroxide functions in such a way as to give a substitution labile species $[M(CO)_5COOH]^{\ominus}$, which after substitution by the new ligand loses hydroxide ion again [551].

Chloro platinum(II) complexes of tertiary phosphines are soluble in benzene or dichloromethane, but not in water. Using 18-crown-6 or dicyclohexano-18-crown-6 in the presence of aqueous potassium hydroxide or solid potassium cyanide, substitutions of hydroxide or cyanide, respectively, were executed [944]:

$$\textit{sym-trans}\text{-}[Pt_2Cl_4L_2] \longrightarrow [Pt_2Cl_2(OH)_2L_2] + [K(18\text{-crown-6})][PtCl_3PEt_3]$$

$L = PEt_3, PMe_2Ph, PEt_2Ph$

$$[PtR_2(\text{cyclooctadiene})] \longrightarrow [K(18\text{-crown-6})]_2[Pt(CN)_2R_2]$$

R = 2-thienyl, benzofuran-2-yl, 4-methoxyphenyl, methyl

$$\textit{cis}\text{-}[PtMe_2(PEt_3)_2] \longrightarrow [K(18\text{-crown-6})][Pt(CN)Me_2(PEt_3)]$$

Similarly, long chain ammonium ions were PT catalysts for halide ligand exchanges of the following type [1123]:

$$[OsCl_6]^{2\ominus} \rightarrow [OsFCl_5]^{2\ominus} \rightarrow [OsF_2Cl_4]^{2\ominus}$$

Complex anions $[W(CO)_5OH]^\ominus$, $[Cr(CO)_5OH]^\ominus$, $[W(CO)_5F]^\ominus$, and $[Cr(CO)_5F]^\ominus$ with $[(\text{dibenzo-18-crown-6})K]^\oplus$ counterions were prepared by irradiation of methylene chloride/THF solutions of the metal hexacarbonyls in the presence of crown ether and potassium hydroxide or potassium fluoride for 2 hours with a Hanovia high pressure quartz mercury vapour lamp [922].

A phase transfer catalyzed ^{18}O exchange process in CO ligands bound to neutral group 6b and iron metals has been described too. An 11.3 M $Na^{18}OH$ solution was added to the mixture of the metal carbonyl and NBu_4I in benzene and heated to 75 °C with vigorous stirring [928]. Complexes thus exchanged included: $M(CO)_5L$ (M = Cr, Mo, W; L = CO, PPh_3, PBu_3) and $Fe(CO)_4PPh_3$. The exchange was thought to involve an equilibrium of the following type in the organic phase.

$$[M(CO)] + NBu_4^\oplus OH^\ominus \rightleftharpoons [M\ COOH]^\ominus NBu_4^\oplus$$

It is now known, however, that hardly any hydroxide can be extracted in the presence of iodide. Thus, the addition of $OH^\ominus$ might in fact occur at the interphase. Aided by the presence of the cation which detaches the anion from the interphase, the scrambling will take place faster with than without catalyst.

Benzyl bromides can be converted into aryl acetic acids by a PTC carbonylation in the presence of 2 mole-% dicobalt octacarbonyl, 5 N NaOH, and TEBA (room temperature, 10–12 hours, atmospheric pressure of carbon monoxide) [920, 996]. It is believed that the intermediate ion pair $[NR_4^\oplus[Co(CO)_4]^\ominus]$ is benzylated

$$ArCH_2Br + [NR_4^\oplus[CO(CO)_4]^\ominus] \longrightarrow ArCH_2\text{—}Co(CO)_4 + [NR_4^\oplus Br^\ominus]$$

This in turn inserts CO and subsequently hydrolyzes

$$ArCH_2\text{—}Co(CO)_4 + CO \longrightarrow Ar\text{—}CH_2\text{—}CO\text{—}Co(CO)_4$$

$$ArCH_2\text{—}CO\text{—}Co(CO)_4 + 2\,OH^\ominus \longrightarrow ArCH_2\text{—}CO_2^\ominus + [Co(CO)_4]^\ominus + H_2O$$

Likewise, *o*-dihalomethylaryl compounds, dicobalt octacarbonyl, and crown complexed KOH yield cyclic ketones [1124]. Finally, 4-hydroxy-4-methyl-2-phenylbut-2-enolide is formed in 44% yield, when phenylacetylene is reacted with methyl iodide and carbon monoxide (1 bar) in 5 M NaOH/benzene in the presence of small amounts of dicobalt octacarbonyl and cetyltrimethylammonium bromide [1125], and benzyl halides give arylpyruvic acids with TEBA/$Co_2(CO)_8$ [446].

Other applications of organometallic compounds in PTC include: alkylation of aryl acetic esters complexed with $Cr(CO)_3$ as a convenient preparative method (Section 3.3.4.4), reduction of nitro compounds with $Fe_3(CO)_{12}$/catalyst (Section 3.4.2), and dehalogenation of α-bromoketones with $Co_2(CO)_8$/NaOH/TEBA (Section 3.4.2).

Concluding this section on miscellaneous types of reactions, it should be mentioned that quaternary ammonium salts may find applications in electrochemistry beyond the well established use as supporting electrolytes. Thus, it was found that a direct current applied across the liquid-liquid interphase of a heterogeneous, unreactive $Cu^{2+}/[V(CO)_6]^-$ redox system, causes deposition of a copper layer at the interphase [524]. Also, a system consisting of 3 M aqueous NaCN, naphthalene or anisol in methylene chloride, and a PT catalyst was oxidized at a platinum anode [79]. Monocyano products were formed in isolated yields up to 70%. Acyloxylations were also possible using this technique (*cf.*, p. 68).

3.3.13. α-Eliminations

Concentrated inorganic bases can frequently effect α-eliminations. As a matter of fact the base-catalyzed formation of dihalocarbenes is one of the most common PTC reactions. The first step, deprotonation, is normally much faster than the subsequent loss of halide. Thus, provided a suitable acceptor is present, reactions of the anions initially formed can compete with, or even dominate over, the carbene reactions. In this subchapter we shall consider these two related reactions together. Although not strictly belonging here, a modern PTC technique to generate dihalocarbenes without the presence of base will be covered. Finally, reactions of carbenes other than dihalocarbenes will be treated.

3.3.13.1. Generation and Addition Reactions of Dichlorocarbene

The generation of dichlorocarbene from chloroform by base was long believed to be rather moisture sensitive since the hydrolysis of chloroform, eventually yielding carbon monoxide and formate, is well known (for a review of dihalocarbene chemistry up to ≈1970, *cf.*, [605]). Although the classical Reimer-Tiemann reaction is performed in aqueous sodium hydroxide, Doering and Hoffmann [606] observed only 0.5% dichloronorcarane when reacting cyclohexene with chloroform and aqueous potassium hydroxide in their pioneering study on CCl_2 in 1954. After that there were vague observations that CX_2 might not be all that sensitive to water. For example, olefins were reacted in tetraglyme at 95 °C with sodium hydroxide and chloroform [607] or with $CHCl_3$/fused 85% KOH at 110 °C [840]. However, it was not until the now famous paper of Mąkosza and Wawrzyniewiez appeared in 1969 [608] that the generation of dihalocarbenes in the presence of strong aqueous bases and a catalyst (usually TEBA) became an accepted method.*) Since then this method has found the widest possible application because of its simplicity and efficiency. The "Mąkosza method" is now the reaction of choice for almost all dichlorocarbene additions. It gives acceptable to very high yields even in cases where conventional dichlorocarbene reagents fail, *e.g.*, with sterically or electronically deactivated alkenes. Only a small number of cases are known where Seyferth's reagent ($PhHgCBrCl_2$) or the decarboxylation of sodium trichloroacetate are superior to the Mąkosza method. These cases are usually base-sensitive substrates. It should be added immediately, though, that Seyferth's reagents themselves can be prepared in a PTC process and that the sodium trichloroacetate decarboxylation can also be executed advantageously in a PTC reaction, for example, at lower temperatures. It has been shown by careful selectivity comparisons that dichlorocarbene generated from various precursors is always the same free dichlorocarbene, and not a carbenoid [609–611]. This applies to the PTC-formed CCl_2 too [4]. More recently, Moss and co-workers compared the selectivities of several carbenes/carbenoids both in the presence and absence of crown ethers. Unlike other systems, CCl_2 always had the same selectivity, thus reinforcing the idea that it is really free [612]. Why then does the system PTC—CCl_2 frequently give higher yields or is it the last resort even with hopelessly deactivated substrates?

Several factors contribute to this difference: Firstly, it is cheap and easy to use a huge excess of chloroform and sodium hydroxide. Secondly, PTC—CCl_2 is not irreversibly formed and momentarily consumed as is the case with the other CCl_2 methods. Thus, unwanted side reactions (attack on either the solvent, or the CCl_2 precursor, dimerization, polymerization) do not disturb to as great an extent. Finally, chloroform as a solvent is rather inert against attack by CCl_2. The advantage of this method over the others is clearly seen if the second factor above is more clearly examined:

(a) $LiCCl_3$ decomposes above −73 °C [610].
(b) The reaction between chloroform and potassium *tert*-butoxide is instantaneous at −20 °C or higher.
(c) Sodium trichloroacetate loses CO_2 in glyme at 80 °C. In the case of the PTC reaction, in nonpolar solvents, dry ammonium trichloroacetate decomposes slowly even at room temperature. No stable intermediate $NR_4^{\oplus}CCl_3^{\ominus}$ is demonstrable [31].
(d) Seyferth's reagent, phenylmercuric bromodichloromethylide, decomposes at around 80 °C in PhHgBr and CCl_2. In the presence of iodide the decomposition is partly reversible at lower temperatures [611].

In all cases the dichlorocarbene formed has a very short lifetime. In the absence of a reactive olefin it "fizzles out." In the case of sodium trichloroacetate, for example, the

*) It is interesting to note that very recently powdered $NaOH/TEBA/CHCl_3$ has been suggested as a very effective CCl_2 precursor [868]. A comparison with the original Mąkosza method showed similar yields, but the process with solid base is a little faster. With slowly reacting olefins this could be a disadvantage, as the CCl_2 may be hydrolyzed instead of kept in a "waiting position" until needed.

multistep, complicated reactions with the precursor are well understood [614]. With the Mąkosza reaction, however, the dichlorocarbene is not all formed at once. Hydrolysis and side reactions are slow, and the system remains reactive for long periods of time even in the absence of a good carbene acceptor. Thus, the CCl_2, trapped in an equilibrium, can "wait" for its consumer, and in this way conversions with very poor substrates become possible. In practice 50% (concentrated) aqueous sodium hydroxide is used together with TEBA as catalyst and an excess chloroform as solvent. The general trend of CCl_2 formation, addition, and hydrolysis is clarified in Table 3–12. The slow hydrolysis of chloroform is accelerated about sixfold by TEBA in the absence of olefin. Addition of olefin brings about a further increase in chloroform consumption, and this rate increase depends on the nature of the olefin. More importantly, however, the ratio of carbene addition *vs.* chloroform hydrolysis is dependent on the nucleophilicity of the olefin and can vary greatly [384].

Table 3–12. Reaction between 0.4 mole $HCCl_3$, 0.2 mole 50% Aqueous NaOH, and 0.1 mole Olefin in the Presence of 1 mole TEBA at 23 °C after 4 hours [384].

Olefin	$HCCl_3$ Consumed (mmoles)	$HCCl_3$ Hydrolyzed		CCl_2 Adduct (mmoles)
		(mmoles)	(%)	
cyclohexene (no TEBA)	3.4	3.4	100	0.0
none (TEBA present)	20.8	20.8		—
3,3-dimethyl-1-butene	25.0	14.5	58	12.0
1-hexene	53.6	8.4	15.7	45.2
cyclohexene	67.3	11.85	17.6	55.45
2-methyl-2-butene	80.4	3.3	4.1	77.1

Sluggish substrates, therefore, should be stirred for longer periods with a large excess of both base and chloroform. From the data in Table 3–12 it is apparent that there is no optimal set of reaction conditions for all olefins. Nevertheless, extensive work [32, 613] has succeeded in establishing good reaction conditions for alkenes of mediocre reactivity. The following main points should be noted:

Rate of Stirring. Stirring has to be vigorous. 800 rpm or more give constant conditions. It does not matter whether a mechanical or magnetic stirrer is used. Vibro-mixing gives a slightly faster rate which is due to heat transfer, not to better mixing [32].

Temperature, Reaction Time. An increase in temperature accelerates the reaction, and often a higher yield of adduct is found after a shorter reaction time at 50 °C, for example. It must be remembered though that hydrolysis is competing with carbene addition, and the thermal response of both may be different. Mąkosza reactions sometimes proceed erratically and violently. There have been cases where one of two seemingly identical reactions became unexpectedly hot and went out of control, the other remaining normal. Thus, two precautions are generally taken. In the first, the reagents are mixed at ≈5 °C and stirred with normal substrates, first in the ice bath, then at room temperature. After a few hours, the mixture is heated for 1 hour at 50 °C [613]. Alternatively, the substrate and catalyst in chloroform solution are stirred at elevated temperatures, and the concentrated sodium hydroxide is added dropwise over a period of several hours. The rationale for this is the observation that the yields in a standard reaction (0.1 mole cyclohexene, 0.4 mole $HCCl_3$, 0.2 mole sodium hydroxide, 1 mmole TEBA, 23 °C) increased with time: 1 h 20%, 4 h 46%, 24 h 74%, 72 h 86%. Using 0.4 mole sodium hydroxide at 50 °C, the same reaction yielded more than 90% after only 2 hours. Similar effects were observed with 1-hexene, 3-methyl-2-butene, and 3,3-dimethyl-1-butene [613]. With sluggish, thermally unstable, or hydrolysis sensitive substrates, the normal procedure is to stir for prolonged reaction times (up to 100 hours) at room temperature.

Amount of Chloroform, Cosolvents. For small-scale reactions (1 to 20 mmoles) an excess of chloroform is used as both solvent and reagent. Working with 0.1 to 1.0 mole olefin, the total volume becomes more critical. It was found that a three- to fourfold excess of $CHCl_3$ over alkene gives the best results under the conditions described in Table 3–12. In Mąkosza reactions deposits of brownish insoluble by-products (polymeric CCl_2 ?)—together with consistent emulsions—may make stirring difficult, especially in slow reactions. In such cases cosolvents may prove useful. Some authors use benzene, but dichloromethane is preferable as it seems to have a slight accelerating effect on the main reaction. This desirable action is even more pronounced with other dihalocarbenes.

Amount of Base. Most workers use concentrated (50%) sodium hydroxide. However, concentrated potassium hydroxide and $Ba(OH)_2$ have been used occasionally. Yields under standard conditions (4 hours, 4 moles $HCCl_3$ per mol alkene) rise sharply up to a 4:1 ratio of NaOH to olefin. After that further increase is small [613].

Amount of Catalyst. One mole-% catalyst relative to olefin is generally used. The yield of 7,7-dichloronorcarane under standard conditions was virtually unaffected by the concentration of TEBA used between 0.5 and 20 mole-%. It decreased rapidly with lower catalyst concentrations, possibly because smaller quantities of catalyst are situated mainly in the aqueous phase [613].

Type of Catalyst. For an extensive discussion of the effects of catalyst cation and anion on rates, see Chapter 2. In practice many quaternary ammonium salts have been used successfully, and small rate differences level off with normal reaction times of a few hours. The catalyst salts used should preferably be chlorides, and certainly not iodides. TEBA is the most widely used ammonium salt, but Aliquat 336, cetyl-trimethylammonium bromide, and many others, including some phosphonium and arsonium compounds, perform similarly [32, 613, 616].

An *additive* sometimes employed (but giving only very slight improvements with CCl_2) is a small amount of ethanol (1 m per mmole TEBA) and, as mentioned before dichloromethane as cosolvent.

Other types of catalysts sometimes employed are specially prepared polystyrene-bound ammonium salts [62, 68, 902], polymer supported polyoxyethylenes [1121], and commercial Amberlite IRA-400 ion exchange resin [616]. An amine oxide is used in a patent [1000], but in our experience such catalysts do not compare well with onium salts. In another patent a sulfonium salt plus a nonionic surfactant cocatalyst are recommended [620]. Tertiary but not primary or secondary amines were found to be very effective catalysts also [617]. Mąkosza and co-workers explained this rather surprising result by demonstrating that a series of labil intermediates is involved [433] (*cf.*, Section 3.3.13.2). Tri-*n*-propylamine and tri-*n*-butylamine belong to the most active of all catalysts [613] (*cf.*, Table 2–8). It is surprising, therefore, that certain tertiary amines have been recommended as selective catalysts for monoadditions of CCl_2 to polyenes [866]. Recently, Reeves *et al.* [432] showed that even primary and secondary amines can be used if an alkylating agent (*e.g.*, ethyl bromide) is present to generate the actual catalyst.

Crown ethers have been used as catalysts by a number of groups but seem to offer no real advantages [45, 302, 618]. Recently, however, they have been found to have a substantial effect on dichlorocarbene selectivity [619]. Selective catalysis has also been found with trialkyl-β-hydroxyethyl-ammonium ions [621, 622]. Catalyst **A** was reported to give monoadducts **B**, **C**, **D**, and **E** from the corresponding olefins, while cetyltrimethylammonium bromide gave bis- or tris-adducts. However, attempts to confirm this selectivity failed [619].

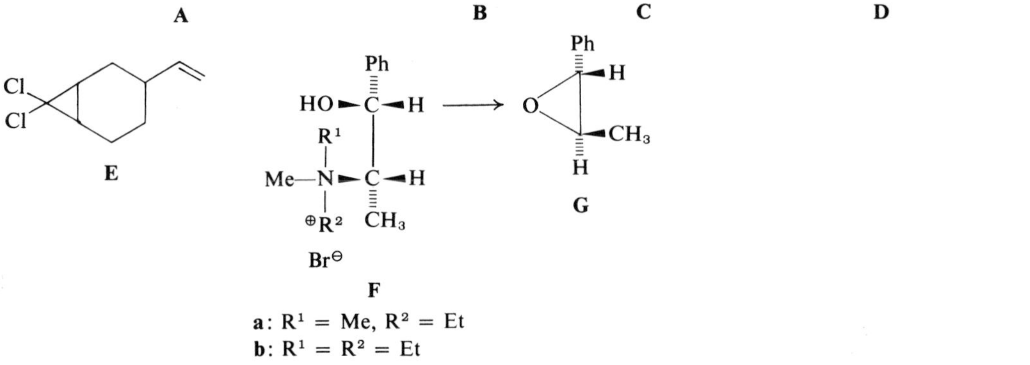
Ph—CH₂—N(Me)₂—CH₂—CH₂OH
⊕
OH⊖
A
Cl
Cl
B
Cl
Cl
C
Cl Cl
OH
D
Cl
Cl
E
Ph
HO—C—H
R¹
Me—N—C—H
⊕R² CH₃
Br⊖
F
a: R¹ = Me, R² = Et
b: R¹ = R² = Et
Ph
H
O
CH₃
H
G

Scheme 3–128

Moreover, it was claimed in the original paper that the use of (+)-N-trimethyl-α-phenethylammonium bromide [(−)-N-ethyl-N-methylephedrinium bromide, **Fa**] brought about a small asymmetric induction in dichlorocarbene adducts [622].

Careful workup of similar reactions with the optically active catalyst **Fb** [384] showed that the optical activity of the adduct was present only in the raw material which was contaminated with epoxide **G** formed from the catalyst, which has a very high specific rotation. The formation of optically active dichlorocarbene adducts seemed unlikely for mechanistic reasons. Free dichlorocarbene has been demonstrated time and time again as an intermediate. Therefore, another report on small asymmetric inductions in CCl_2 additions with optically active amines **H** (R = Et, Ph; R′ = Me, Et) [623] should be considered with some reservation. In spite of the fact that intermediates of the type (**I**), in which the carbene is transferred by a nitrogen ylide, have been discussed and dismissed as possible by Mąkosza [433], this work has been ignored in some later work, and claims for such intermediates are still made [623].

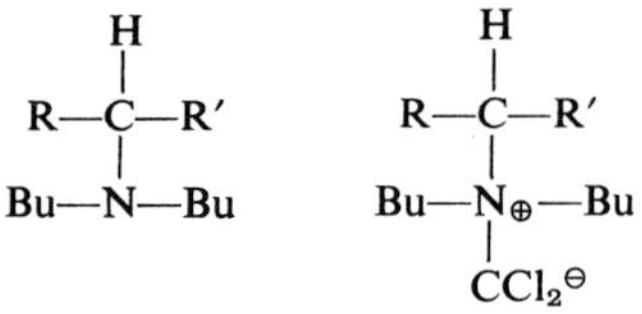

```
      H                 H
      |                 |
  R—C—R'            R—C—R'
      |                 |
  Bu—N—Bu          Bu—N⊕—Bu
                        |
                       CCl2⊖

      H                 I
```

Scheme 3–129

In an effort to verify a possible optical induction of this type, styrene was reacted with concentrated sodium hydroxide and chloroform in the presence of (S)-(+)-N,N-dimethyl-phenylethylamine. The distilled reaction product actually had a small optical rotation which vanished, however, on careful purification [843].

Typical Experimental Procedure for the Mąkosza method [624, 625]. 0.1 mole substrate and 1 mmole TEBA (or tri-*n*-propylamine) are dissolved in 0.4 mole chloroform and 1 ml ethanol. 0.4 mole ice-cold, freshly prepared 50% NaOH are added, and the mixture is stirred, first in an ice bath, then at room temperature for 1–2 hours. The reaction mixture is then heated and stirred for a further 3–5 hours at 50 °C. The reaction mixture is then poured into a large amount of water. Disregarding good phase separation and brownish polymer deposits, the organic phase is repeatedly poured into fresh water (or dilute HCl) to destroy any persistent emulsion. After drying the solvent is removed. The products are isolated by distillation if possible. Nonpolar solid or high boiling products are separated from darkish side-products by passing a petroleum ether/ether or a toluene solution through a short silica gel column.

Generation of Dichlorocarbene from sodium tri-chloroacetate.

$$NaO_2CCCl_3 \longrightarrow NaCl + CO_2 + CCl_2$$

Sometimes the use of strong bases in dichlorocarbene generations is undesirable. Here the well-known sodium trichloroacetate decarboxylation may be an alternative.

In the classical procedure this method requires expensive anhydrous dimethoxyethane as solvent and temperatures around 80 °C. An improved PTC version [675, 31] has been developed which does away with the expensive solvent and allows lower temperatures; with thermolabile products such as **K** it is even possible to work at room temperature. In the latter case, however, the reaction requires a few days. Yields of 65–88% are obtained in boiling chloroform solution with 1–2 mole-% TEBA or

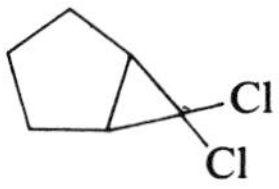

Cl
Cl

K

Scheme 3–130

Aliquat 336, and 2 moles of solid sodium trichloroacetate per mole double bond. No decarboxylation occurs if a concentrated aqueous solution of sodium trichloroacetate is used under PTC conditions, although a (hydrated) anion is extracted. The amount of catalyst is rather critical here: With too much catalyst a relatively large amount of the trichloroacetate is in solution and is decomposed per unit of time so that the well-known side reactions [614] (attack of CCl_2 or $CCl_3^{\ominus}$ on trichloroacetate and tarring) become prominent; the CCl_2 is "fizzling out." In contrast to the Mąkosza method the olefin has no influence on the rate of consumption of the carbene precursor [675]. The rate effect of cations ($K^{\oplus}$ > $Na^{\oplus}$ > $Li^{\oplus}$) both in the presence and absence of crown ethers on the decarboxylation and dichlorocarbene addition has also been studied [676]. The same conclusions were reached: a fast decarboxylation gives a relatively poor yield of adduct.

Typical Experimental Procedure for the Trichloroacetate Method [675]. One mmole catalyst and 0.05 mole olefin are dissolved in 50 ml $CHCl_3$. 0.1 mole solid, finely ground sodium trichloroacetate (commercial weed-killer, 96.7% without further purification) is added, and the mixture is heated and stirred vigorously at 80 °C until the CO_2 evolution ceases (5–6 hours). During the process the mixture turns brown or black. After adding 100 ml water the phases are separated. An insoluble flaky precipitate is disregarded. The organic phase is dried and worked up either by distillation or crystallization.

Of the very many Mąkosza–CCl_2 reactions found in the literature, general reactions are mentioned in the tables and only special cases will be discussed in detail. Table 3–13 contains the CCl_2 reactions of compounds with one double bond. With simple olefins, yields in the 75–95% range are reached easily, with severely sterically hindered alkenes (*e.g.*, 3,3-dimethyl-1-butene) yields may be as low as 40%, even under favorable conditions. Triphenylethene but not tetraphenylethene reacts.

Limits of applicability are found with certain olefins carrying electron-withdrawing substituents. No conversion of the Mąkosza reagent is observed with di-, tri-, or tetrachloroethenes, but the trichloroacetate method works with the first two sub-

Table 3-13. Addition Reactions of PTC—CCl_2 to Monoolefins.

Substrate	Yield	Remarks	Reference
Various aliphatic and aromatic alkenes	generally very high	(If sterically hindered, lower yields)	[4, 31, 32, 433, 608, 617, 636, 641–643, 675, 736]
$Ar(R^1)C{=}C(R^2)R^3$ (R^1, R^2, R^3 = H, alkyl, aryl)	generally very high	tetraphenylethene does not react	[626–633, 732]
S	42%		[735]
ferrocene substituted alkenes	generally very high		[627, 634]
R, R′	generally very high	adducts of limited stability	[31, 635]
$(CH_2)_n$, R (n = 3, 4, 5, 6, 10)	generally very high		[801]
steroids		side chain nucleus	[639] [640, 644]
	82%		[646]
R, Ph (R = H, Me, Ph)	generally very high		[647]

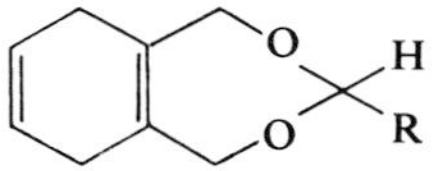 O, O, H, R	—	4 stereoisomeric mono adducts	[945]
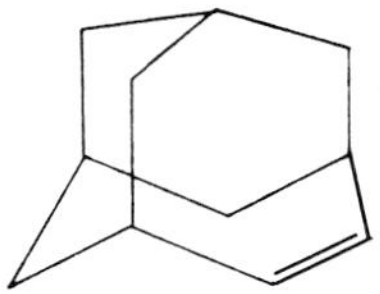		competing C—H insertion	[649]
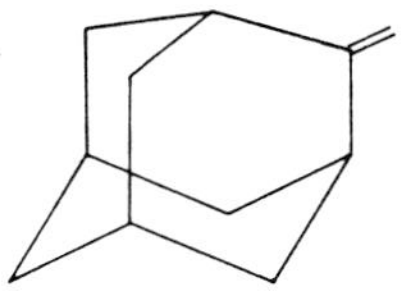	40%		[648]
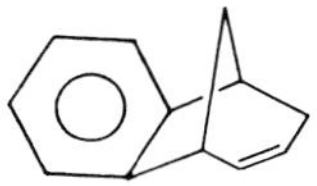	92%		[947]
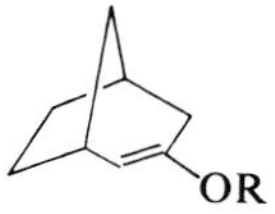 OR (R = Me, Ac)	85; 75%		[955]
 OAc	81.5%		[956]

(continued)

Table 3-13. Addition Reactions of PTC—CCl_2 to Monoolefins.

Substrate	Yield	Remarks	Reference
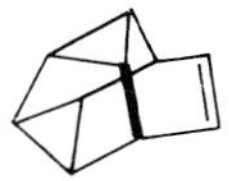	45%		[693]
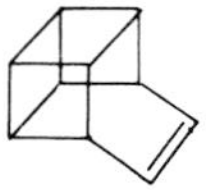	60%		[693]
$R^1HC{=}C(SiMe_3)R^2$	55–72%		[650]
$ClH_2C{-}CH{=}CH{-}CH_2Cl$	45%		[652, 1131]
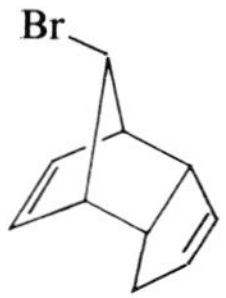 Br	71%	add. to nonhindered double bond	[695, 968]
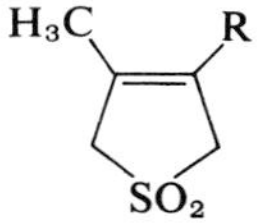 H_3C, R, SO_2			[653, 1126]
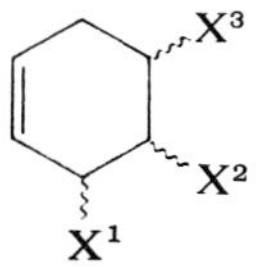 X^3, X^2, X^1 (X^1, X^2, X^3 = H, CN, or COOR)	up to 85%		[654, 655]

$R_2C{=}C(R^1){-}COOR^2$ ($R^1 \neq H$)	up to 92%		[616,671–673, 677, 872]
$R_2C{=}C(R^1){-}CONH_2$			[616]
$H_2C{=}C(CH_3){-}CN$	42%	by-prod.: 4,4,4-trichloro-2-methylbutyronitrile	[671]
$R{-}C({=}O){-}C(R^1){=}CR^2R^3$ ($R^1 \neq H$)	up to 90%	simple ketones tend to tar	[674]
	75–83%		[694]
$R_2C{=}CH{-}CH(R'){-}OH$	60–92%		[616, 621]
$R_2C{=}C(R){-}CR_2OH$	70–90%		[502]
$R_2C{=}C(R){-}CH_2{-}OCH(OR')CH_3$			[663]
$R^1HC{=}CH{-}C(R)(OR^2)_2$	38–82%		[660–662, 674]

(continued)

Table 3-13. Addition Reactions of PTC—CCl_2 to Monoolefins.

Substrate	Yield	Remarks	Reference
Cyclohexenone ketal	71%		[884]
	78%	attack only from β-side	[651]
enol ethers of cyclic ketones	70–80%		[656, 722, 1127]
(X = H, OR)	up to 95%		[657–659]
enole acetates of ketones	up to 70%		[420, 686]
enamines	up to 86%	some adducts rearrange easily under react. cond.	[664–666]
imines Ar—CH=N—Ar′ (R)	up to 90%	some adducts rearrange easily under react. cond.	[667, 669, 670, 964, 1128]
imines Ar—CH=NR	low yields	unstable adducts	[667]
imines R—CH=NR	low yields	unstable adducts	[667]
dihydropyranes	75%		[987, 1130]
unsaturated sugar derivatives	high yields		[1132]

strates. As stated before, the dichlorocarbene of the Mąkosza process is in equilibrium with the trichloromethylide anion:

$$CHCl_3 \xrightarrow[\text{fast}]{-H^{\oplus}} CCl_3^{\ominus} \rightleftharpoons CCl_2 + Cl^{\ominus}$$

Depending upon whether electron-rich or electron-poor olefins are present, either the electrophile CCl_2, or the nucleophile $CCl_3^{\ominus}$, or in some cases both, are scavenged. Thus, small structural changes in the olefin will determine the path of reaction towards dichlorocyclopropane (a) or addition of chloroform across the double bond (b):

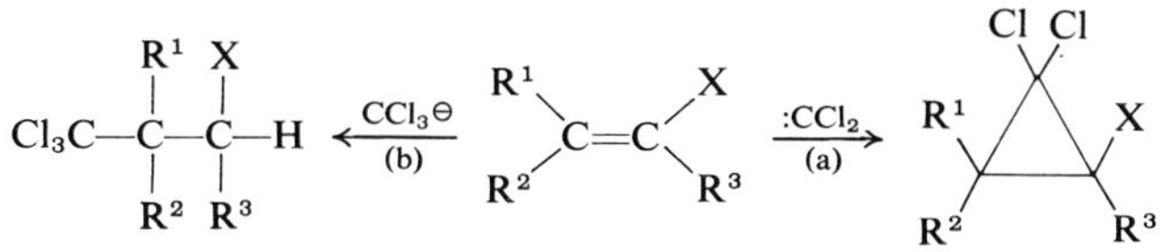

$$Cl_3C{-}C(R^1)(R^2){-}C(X)(R^3){-}H \xleftarrow[(b)]{CCl_3^{\ominus}} R^1R^2C{=}C(X)R^3 \xrightarrow[(a)]{:CCl_2} \text{1,1-dichloro-2-}R^1\text{-2-}R^2\text{-3-}X\text{-3-}R^3\text{-cyclopropane}$$

Scheme 3–131

As a general rule when $R^1 = R^2 = R^3 = H$, only path (b) is observed. But R^3 = alkyl, or phenyl; R^1, R^2 = H or alkyl, leads to path (a) [420, 671, 674, 677, 686, 768]. The situation becomes more complex if one or both of R^1, R^2 are alkyl or phenyl and R^3 is H: Mixtures of adducts and secondary reactions appear which are considered in Section 3.3.13.2. Allylic hydroxyls somewhat deactivate double bonds towards CCl_2 attack [502]. An extensive study was made on the influence of a neighboring cyclopropane ring on the reactivity of a double bond towards PTC CCl_2 [990]. It turned out that cyclopropyl alkenes react about 10 times faster than isopropyl alkenes and about as fast as the corresponding methyl alkenes.

Acetylenes are poor nucleophiles too, and their reactions with dichlorocarbene are usually sluggish. The Mąkosza reagent is no exception. In addition, the primary reaction products, dichlorocyclopropenes, are partly hydrolyzed during the reaction to give α,β-acetylenic ketones and cyclopropenones, which in turn can also ring-open:

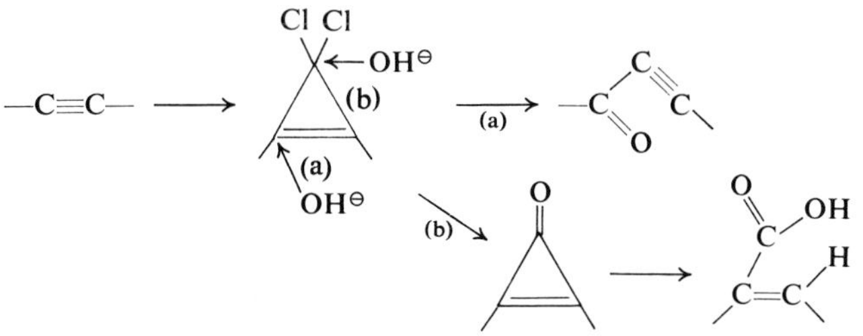

$$-C{\equiv}C- \longrightarrow \text{3,3-dichlorocyclopropene} \;(\text{attack of } OH^{\ominus} \text{ at (a) or (b)}) \xrightarrow{(a)} -C(=O){-}C{\equiv}C- ;\quad \xrightarrow{(b)} \text{cyclopropenone} \longrightarrow HO{-}C(=O){-}C{=}C(H)$$

Scheme 3–132

Thus, the conversion of acetylenes with the Mąkosza reagent does not give very satisfactory yields, but because of its simplicity it has been used in the synthesis of phenyl, *tert*-butyl, *tert*-butoxy, and cyclopropyl-substituted cyclopropenones [626, 678, 891]. Reaction of alkynes with sodium trichloroacetate leads to dichlorocyclobutenone as the main product [679].

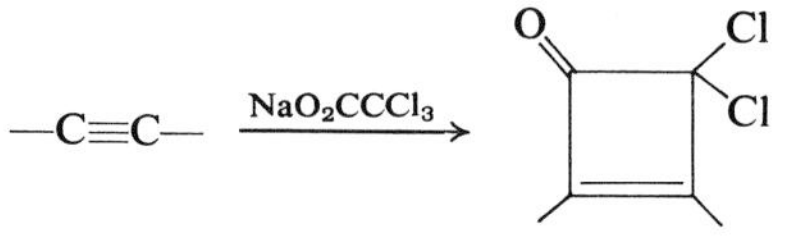

$$-C{\equiv}C- \xrightarrow{NaO_2CCCl_3}$$

Scheme 3–133

The addition products of polycyclic bridged alkenes exhibit facile rearrangements, if sterically demanding situations are provided. In such cases, rearrangements may occur under the reaction conditions, and the products may take up another carbene so that complex mixtures of products occur. Mechanistically, the *syn* chlorine is temporarily expelled as chloride, and disrotatory opening of the three membered ring leads to the allyl cation, which picks up the chloride again on the same face of the molecule:

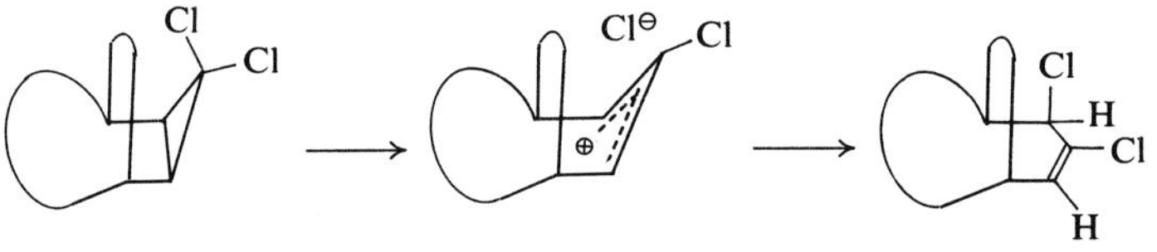

Scheme 3–134

Examples are given below:

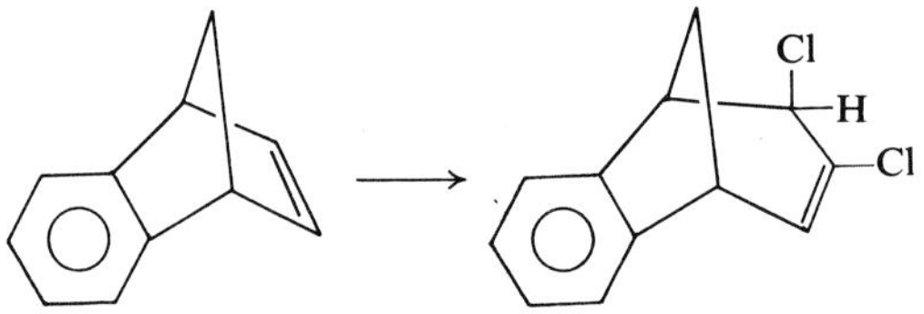

[696, 697, 1129]

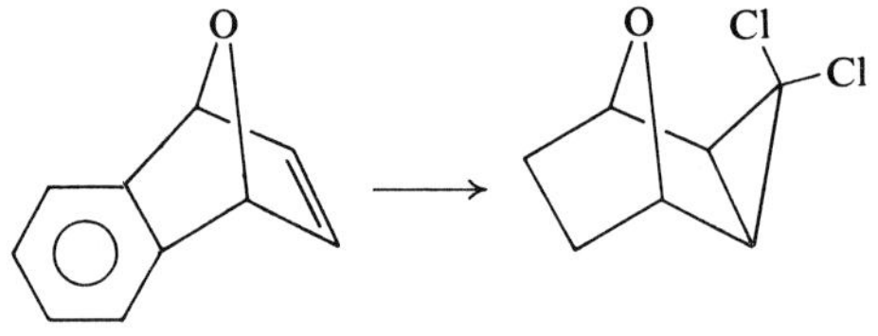

[698]

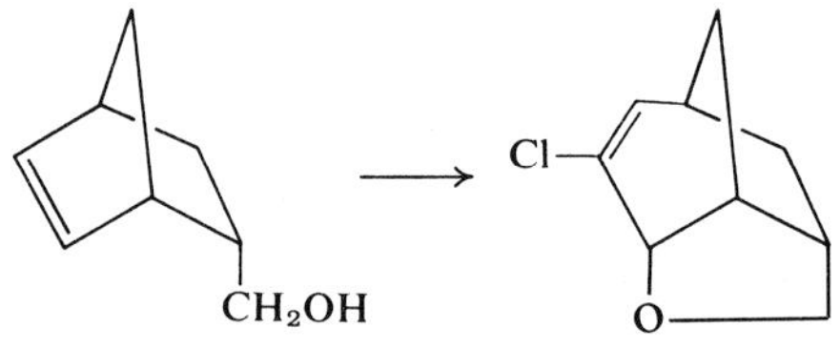

(main product) [699]

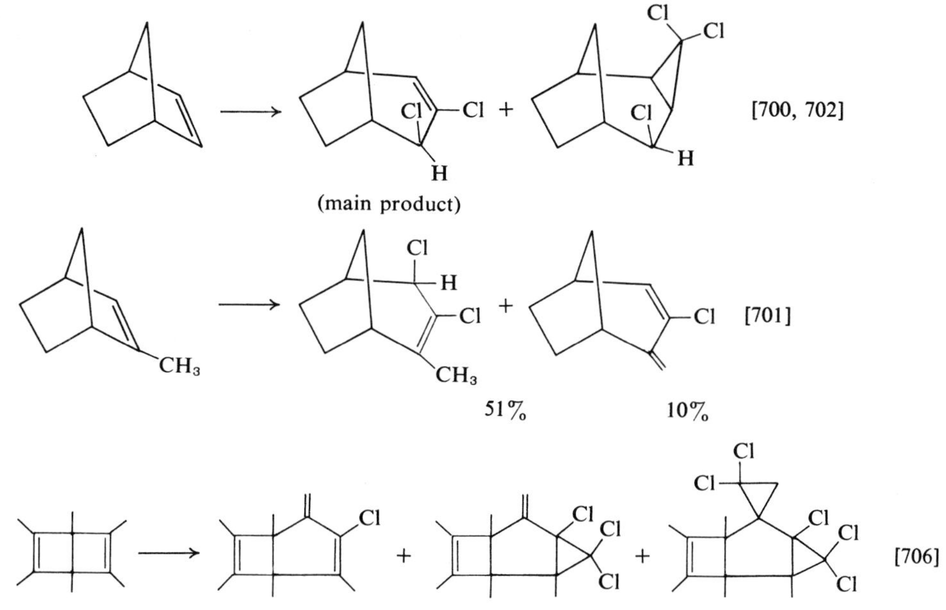

(main product)

51% 10%

[700, 702]

[701]

[706]

Scheme 3–135

Less strained cyclopropanes remain unchanged:

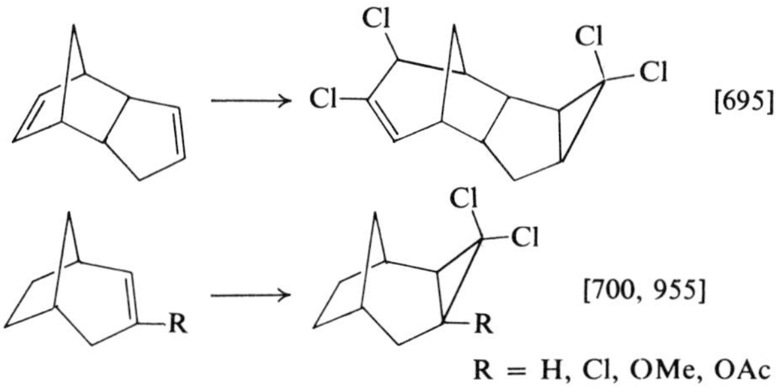

[695]

[700, 955]

R = H, Cl, OMe, OAc

Scheme 3–136

Most interesting of all is the bicyclo[2.2.1]heptadiene ring system where three competing primary additions—1.2-*exo*- and *endo*-, and the novel homo-1,4 addition —lead to a large number [*cf.*, 687] of products [703–705]; especially with substituted compounds.

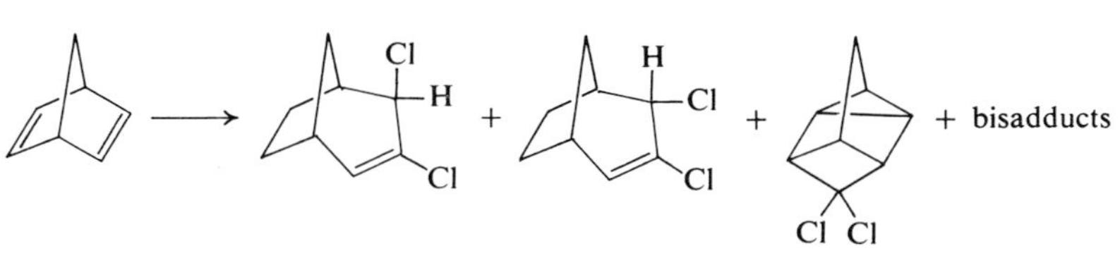

+ bisadducts

Scheme 3–137

Systems with more than one double bond react only once with normal dichlorocarbene reagents as the electron-withdrawing effect of the chlorines deactivates the system towards further attack. With the PTC process, however, multiple additions are common, and it is often difficult to stop a process at the mono addition stage. If there are isolated multiple bonds, the more electron-rich one will be attacked first. With conjugated systems the more activated part might in fact be the only one to react as exemplified below:

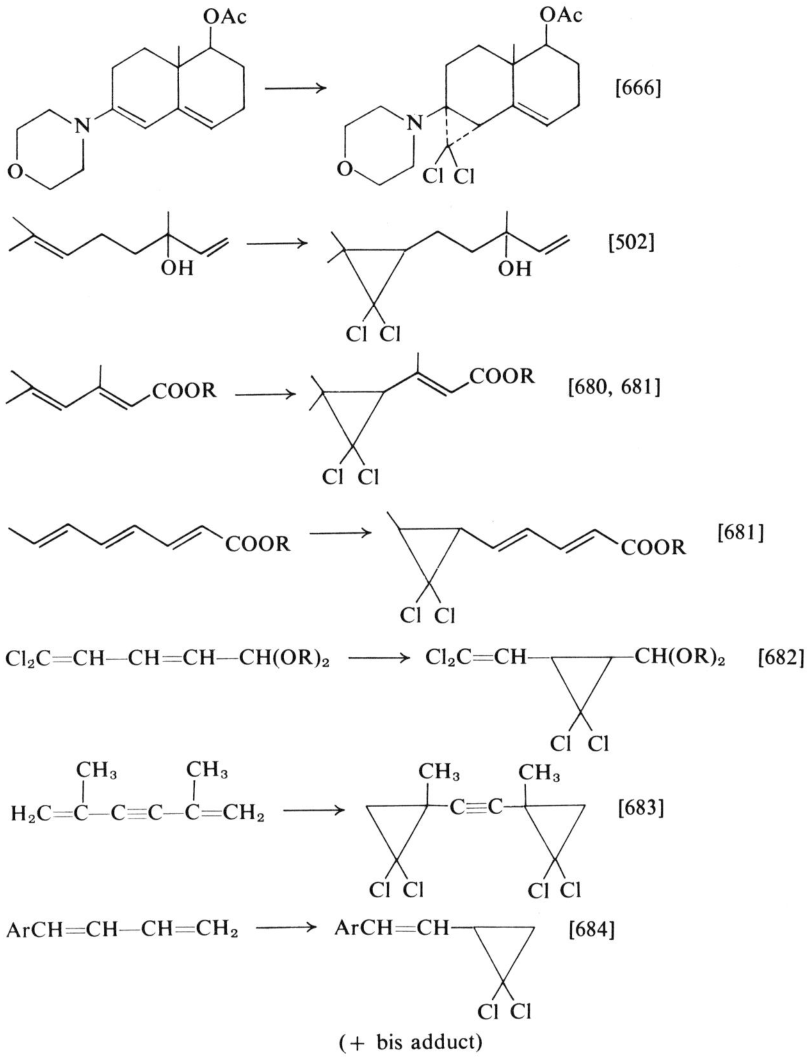

OAc → OAc, Cl Cl [666]

OH → OH, Cl Cl [502]

COOR → COOR, Cl Cl [680, 681]

COOR → COOR, Cl Cl [681]

$Cl_2C{=}CH{-}CH{=}CH{-}CH(OR)_2 \longrightarrow Cl_2C{=}CH{-}$(dichlorocyclopropane)$-CH(OR)_2$ [682]

$H_2C{=}C(CH_3){-}C{\equiv}C{-}C(CH_3){=}CH_2 \longrightarrow$ (bis-dichlorocyclopropane) $-C{\equiv}C-$ [683]

$ArCH{=}CH{-}CH{=}CH_2 \longrightarrow ArCH{=}CH{-}$(dichlorocyclopropane) [684]

(+ bis adduct)

Scheme 3–138

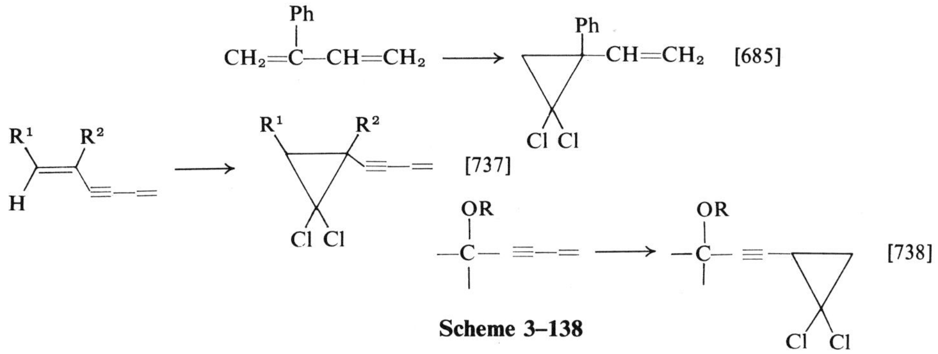

$$CH_2{=}C(Ph){-}CH{=}CH_2 \longrightarrow$$ [685]

[737]

[738]

Scheme 3–138

Normally, however, multiple additions abound, although the yields of the higher adducts tend to be lower. Known examples of PTC—CCl_2 multiple additions are collected in Table 3–14.

Sometimes unexpected rearrangements of primary adducts occur. A rather unique reaction of phenylallene is also known [707–708].

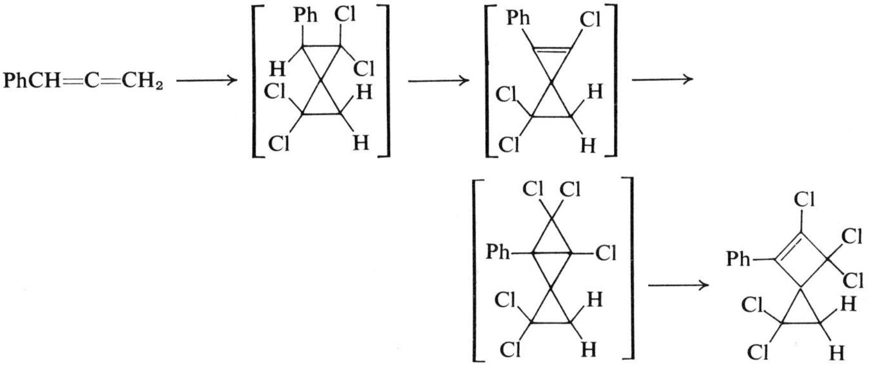

$$PhCH{=}C{=}CH_2 \longrightarrow$$

Scheme 3–139

Mechanisms have also been suggested for these conversions:

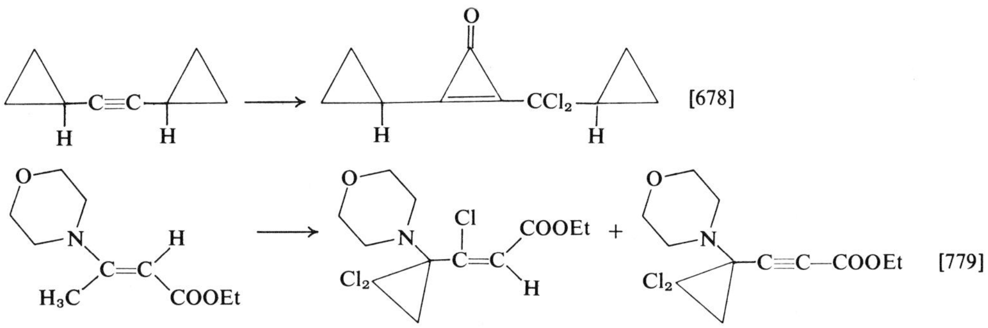

[678]

[779]

Scheme 3–140

Table 3-14. Multiple Additions of PTC—CCl_2.

Reaction	Remarks	Reference
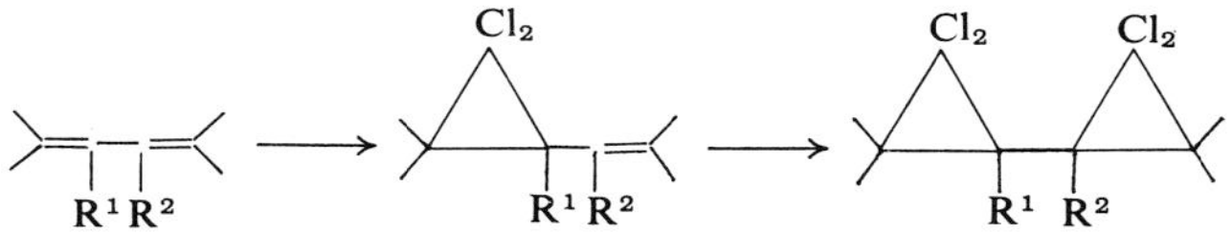 $(CH_3)_2C{=}C(R^1){-}C(R^2){=}C(CH_3)_2$ ⟶ (mono-Cl_2-cyclopropane) ⟶ (bis-Cl_2-cyclopropane) (R^1, R^2 = H, alkyl, aryl)	mixtures of stereoisomers	[684, 685, 687, 688]
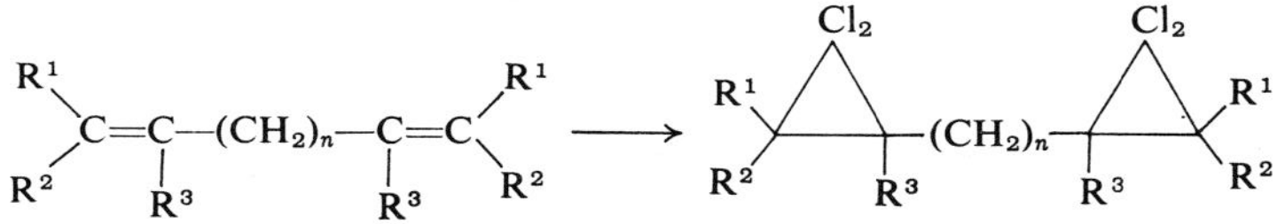 $R^1R^2C{=}C(R^3){-}(CH_2)_n{-}C(R^3){=}CR^1R^2$ ⟶ (Cl_2-cyclopropane with R^1, R^2, R^3)$-(CH_2)_n-$(Cl_2-cyclopropane with R^3, R^1, R^2) (R^1, R^2, R^3 = H, alkyl, aryl; n = 0, 1, 2)	mixtures of stereoisomers	[626, 689, 1146, 1147]
$(CH_3)_2C{=}CH{-}(CH_2)_2{-}C(CH_3){=}CH{-}CH(OEt)_2$ ⟶ (Cl_2-cyclopropane)$-(CH_2)_2-C(CH_3){=}CH-CH(OEt)_2$ ⟶		[758]
(Cl_2-cyclopropane)$-(CH_2)_2-$(Cl_2-cyclopropane with CH_3)$-CH(OEt)_2$	77%	
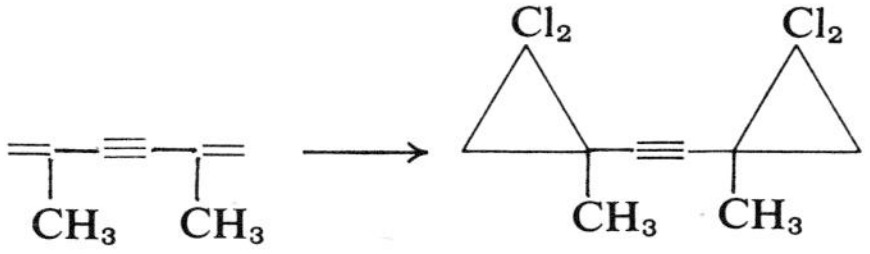 $CH_2{=}C(CH_3){-}C{\equiv}C{-}C(CH_3){=}CH_2$ ⟶ (Cl_2-cyclopropane with CH_3)$-C{\equiv}C-$(Cl_2-cyclopropane with CH_3)	62%, 2 isomers	[683]

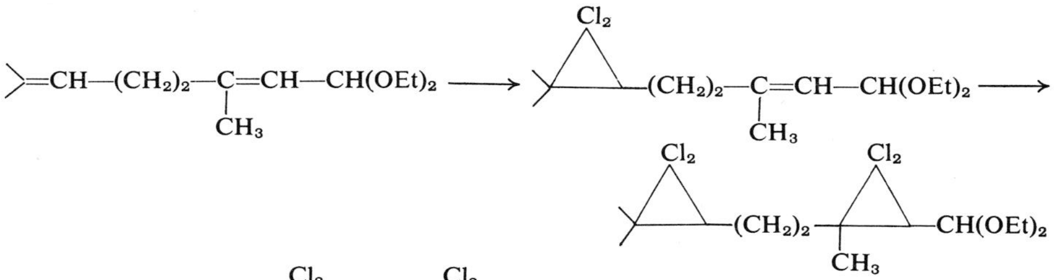

Reaction	Yield / remarks	Ref.
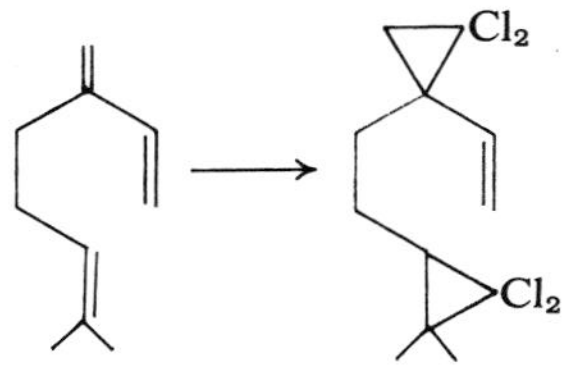 Cl_2 Cl_2	42%, 2 isomers	[692]
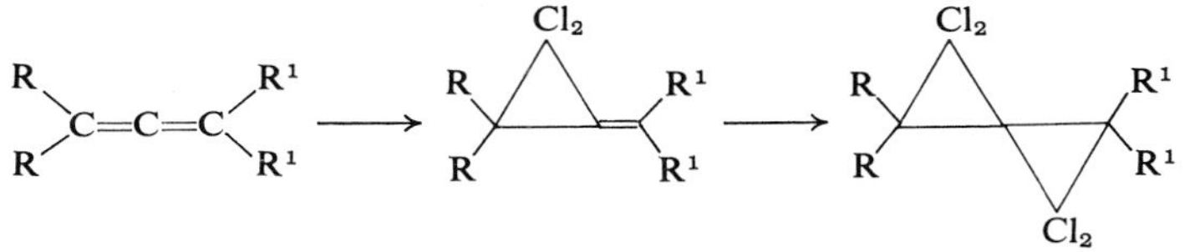 Cl_2 Cl_2 Cl_2 (R = H, alkyl, aryl; R′ = H, alkyl)	mixtures of stereoisomers	[707, 708, 1144, 1145]
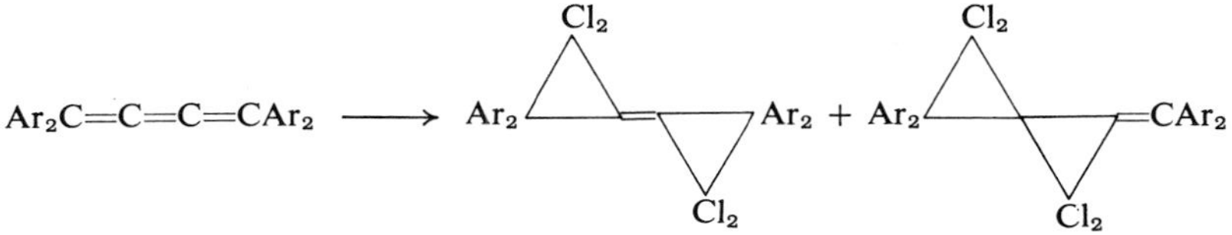 $Ar_2C{=}C{=}C{=}CAr_2$ ⟶ Ar_2 Cl_2 Cl_2 Ar_2 + Ar_2 Cl_2 Cl_2 $=CAr_2$		[626]
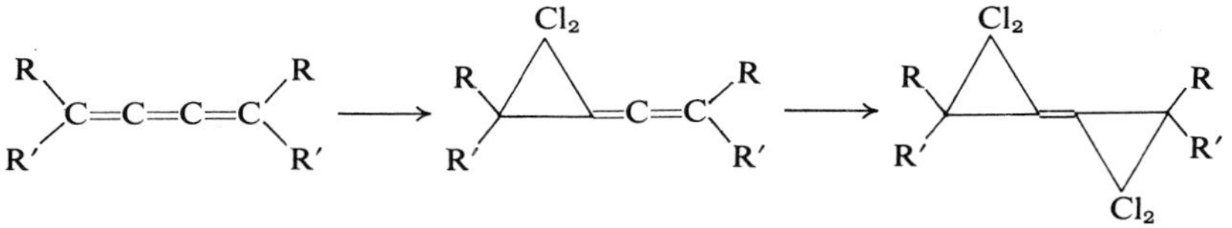 Cl_2 $=C{=}C$ Cl_2 Cl_2 (R = Ph; R′ = alkyl or R = R′ = adamantyl)		[709, 710]
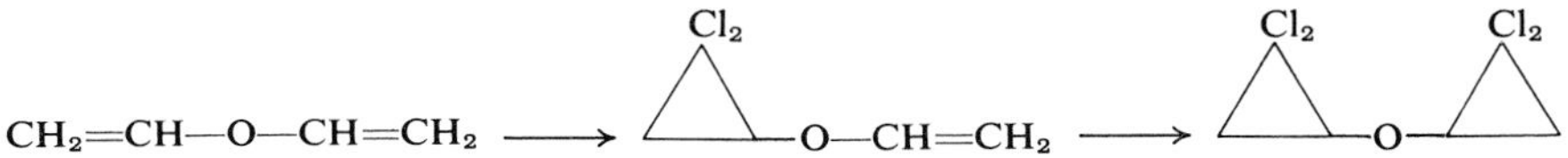 $CH_2{=}CH{-}O{-}CH{=}CH_2$ ⟶ Cl_2 $-O-CH{=}CH_2$ ⟶ Cl_2 $-O-$ Cl_2		[690]

(*continued*)

Table 3-14. Multiple Additions of PTC—CCl_2.

Reaction	Remarks	Reference
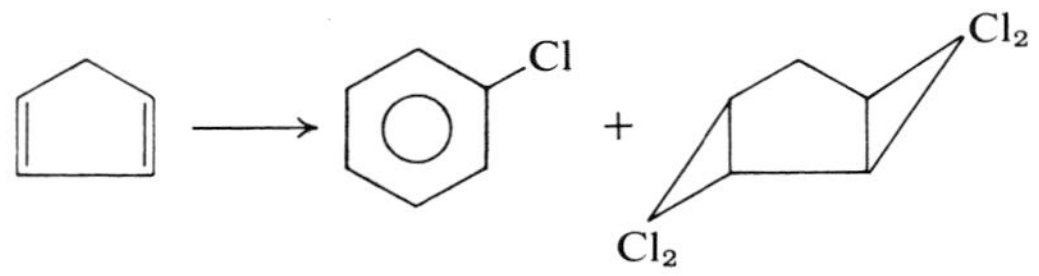 → Cl + Cl_2 Cl_2	only one isomer	[687]
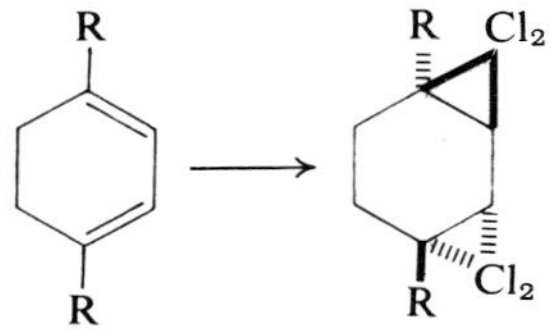 R R → R Cl_2 R Cl_2 (R = H, Ph)	only one isomer	[685, 689]
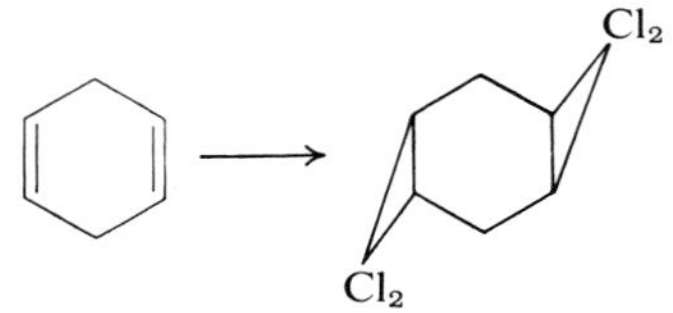 → Cl_2 Cl_2	only one isomer	[689]
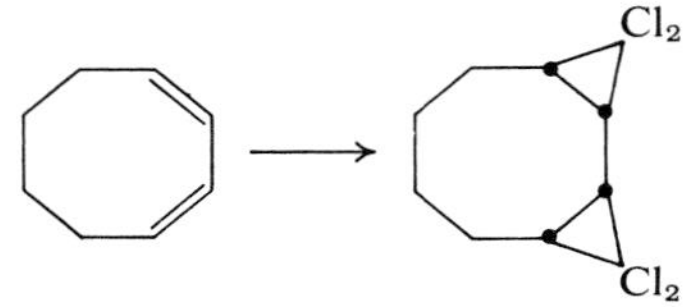 → Cl_2 Cl_2	only one isomer	[689]
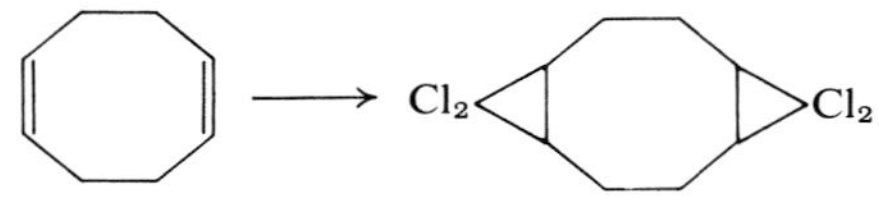 → Cl_2 Cl_2	2 stereoisomers	[31]

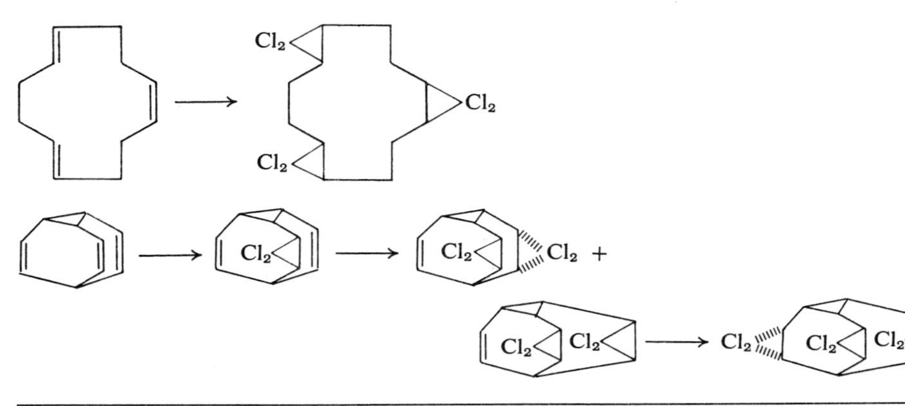

[711, 713]

[626, 712, 713]

[622]

[730, 693]

(continued)

Table 3-14. Multiple Additions of PTC—CCl_2.

Reaction	Remarks	Reference
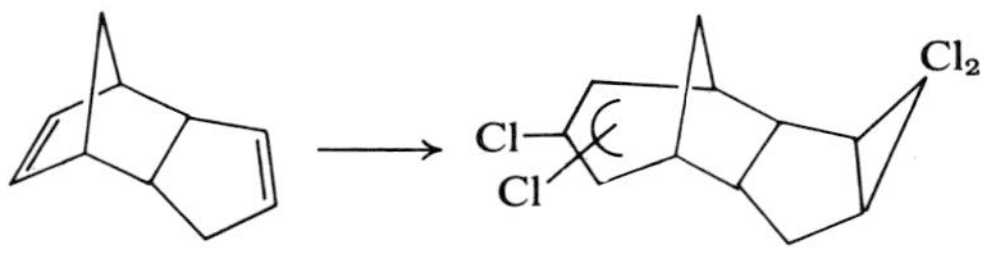		[968]
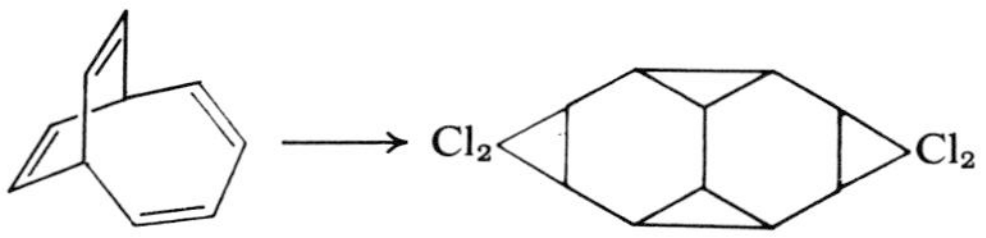		[693]
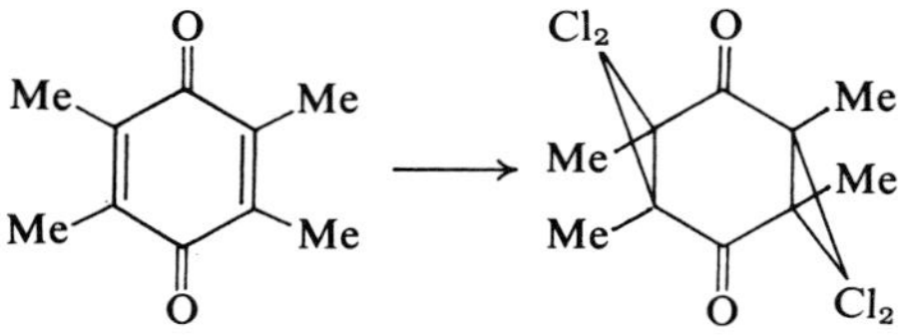		[691]
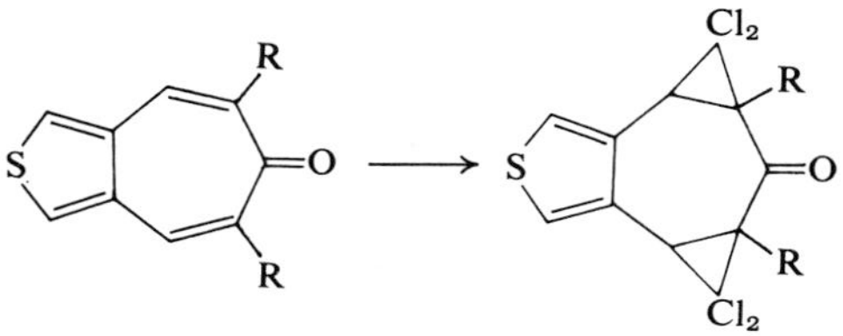		[948]
Further polycyclic compounds		[1148–1150]

Reactions of PTC generated dichlorocarbene with aromatic and heterocyclic compounds are collected in Table 3–15.

Generally, cycloaddition of CCl_2 to a double bond forming part of an aromatic system is not favored, and such reactions occur only where the double bond has some olefinic character (*e.g.*, phenanthrene), or a high electron density and partial localization (*e.g.*, 2-methoxynaphthalene). Yields are low to very low in most cases. Unusual transformations of alkylated aromatic hydrocarbons lead to mixtures of spirononatrienes [718–720] as exemplified for 2-methyl-naphthalene. Many other compounds have been investigated, often complex mixtures and low yields are observed.

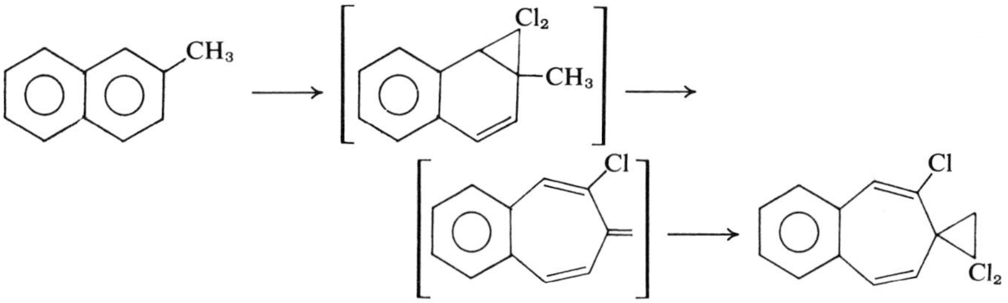

Scheme 3–141

Side Reactions with Chloroform. As stated before, the $NaOH/HCCl_3$/catalyst system stays ready for reaction for long periods of time. Nevertheless, slow side processes of CCl_2 with chloroform occur if no good carbene acceptors are present. Small amounts of CH_2Cl_2, CCl_4, C_2Cl_4, C_2HCl_5, and a polymer are formed. Mechanisms for these conversions have been advanced [1143].

3.3.13.2. Other Reactions of Dichlorocarbene and Trichloromethylide Anion

Besides addition to C=C double bonds, the $CCl_2/CCl_3^{\ominus}$ equilibrium system exhibits a wealth of other synthetic possibilities to be covered in this subsection.

Insertion Reactions. C—H insertions of conventionally generated dichlorocarbene are rare and give poor yields if *tert*-butoxide and chloroform or sodium methoxide and trichloroacetic ester are used. Somewhat better results are observed with sodium trichloroacetate in glyme. Even better is the bromodichloromethyl(phenyl)mercury reagent when applied at somewhat elevated temperatures. It was assumed therefore that the carbene needs excess thermal energy in order to insert. Experiments with PTC—CCl_2 around 0 °C disproved this assumption [739]. Nevertheless, even with this method yields of preparative interest occur only with insertion into bridgehead positions or α to activating groups. The less sterically hindered the bridgehead, the higher the yield. A striking difference is found with the isomeric decalins: only the *cis* compound, with a relatively "free" tertiary hydrogen, reacts. The procedures

Table 3-15. PTC-Dichlorocarbene Additions to Aromatic and Heterocyclic Systems.

Reaction	Remarks	Reference
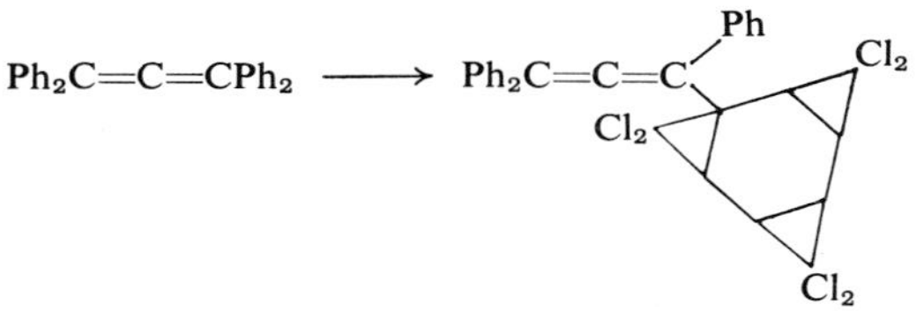 $Ph_2C{=}C{=}CPh_2 \longrightarrow Ph_2C{=}C{=}C$ (Ph, Cl_2, Cl_2, Cl_2)	low yield	[708]
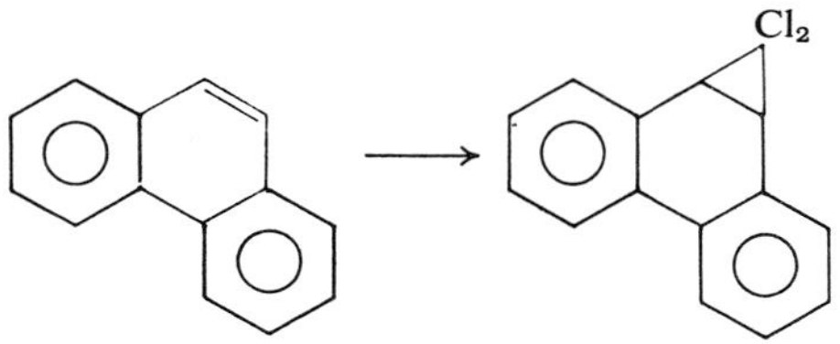 $\longrightarrow$ (Cl_2)	79% yield	[636, 637, 638, 733]
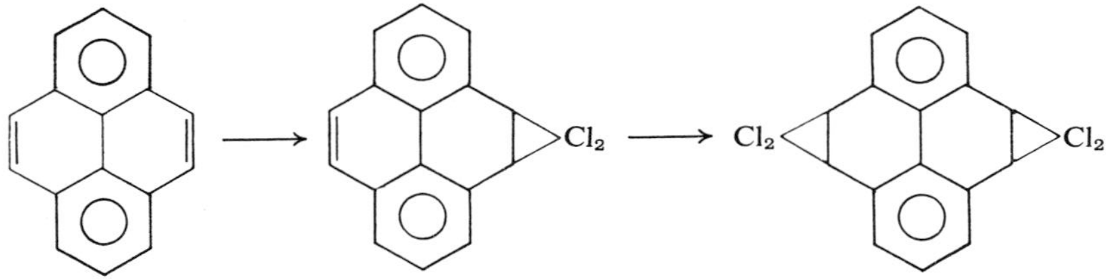 $\longrightarrow$ Cl_2 $\longrightarrow$ Cl_2 Cl_2	mostly monoprod.	[730]
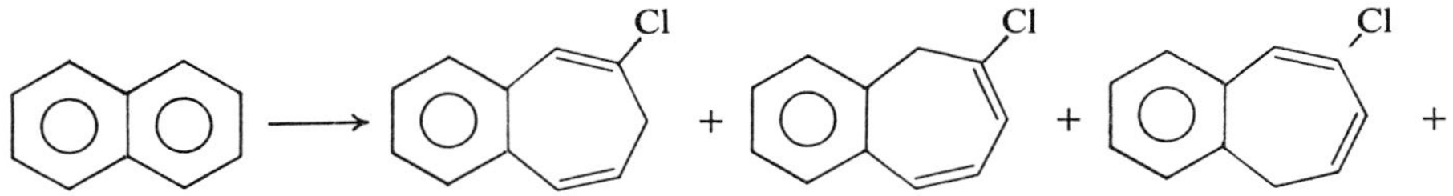 $\longrightarrow$ Cl + Cl + Cl +		

Reaction	Yield	Ref.
Cl, Cl (+ isomers)	low yield	[719]
→ H, R + (H)$_2$ R = H, Cl, OC_2H_5	low yield	[719]
alkylated aromatics ⟶ spiro[2,6]nonatrienes		[718, 719, 720]
OMe, OMe → O, Cl, OMe + OMe, Cl, =O + O, OMe, Cl	low yields	[723]

(continued)

Table 3-15. PTC-Dichlorocarbene Additions to Aromatic and Heterocyclic Systems.

Reaction	Remarks	Reference
further examples		[723, 734]
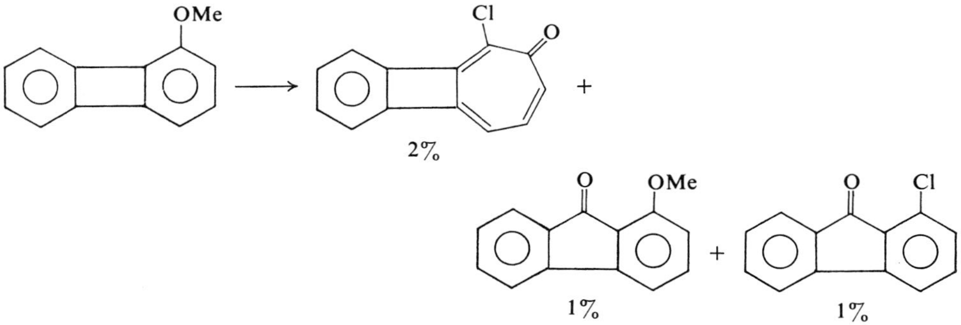 OMe → Cl, O; 2% + O, OMe; 1% + O, Cl; 1%		[721]
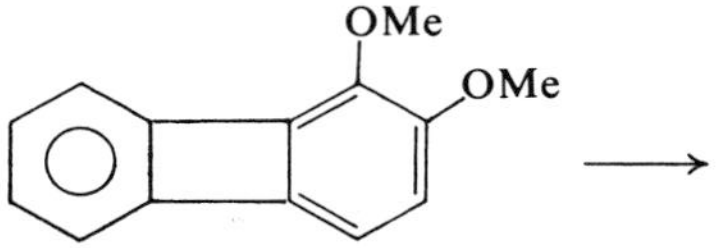 OMe, OMe → 7 products in very low yield		[724]
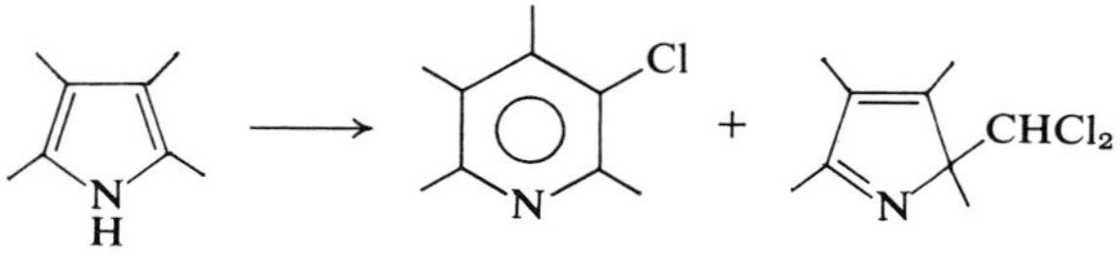 N H → Cl, N + N, $CHCl_2$	37 and 44% lower if less subst.	[728] [731]
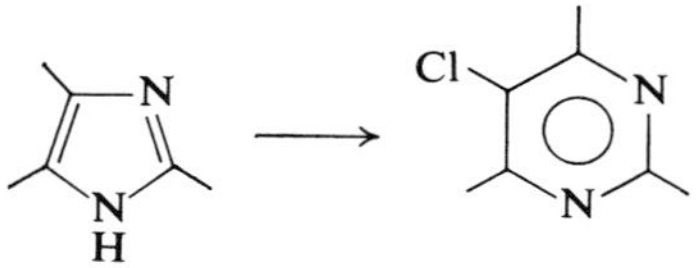 N, N H → Cl, N, N	39%	[728]

Reaction	Yield	Ref.
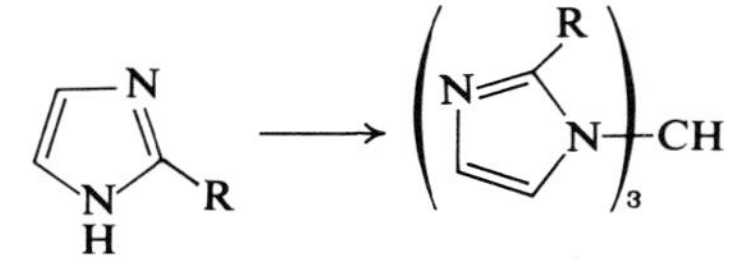	6%	[731, 987]
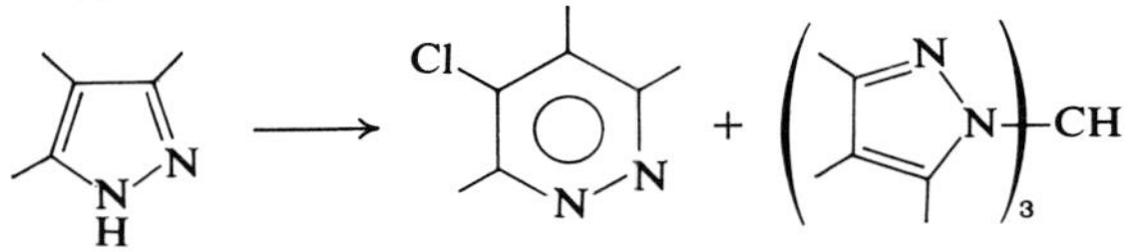	3 and 63%	[728]
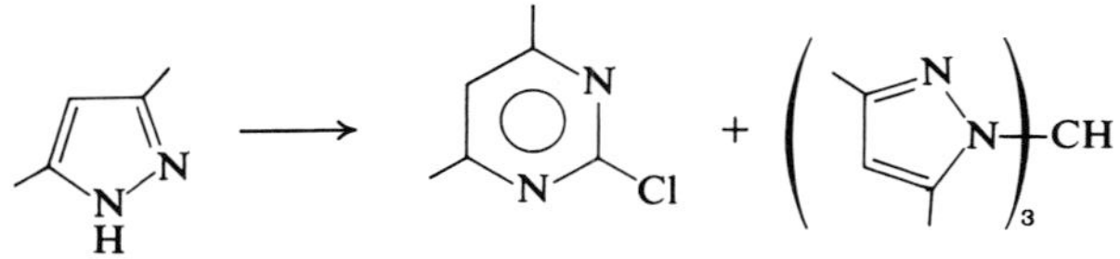	1 and 33%	[987]
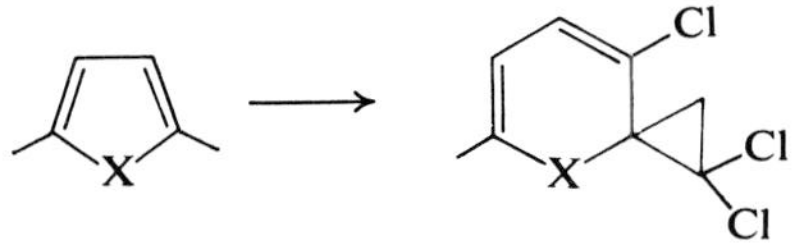 X = O, S	—	[718, 731]
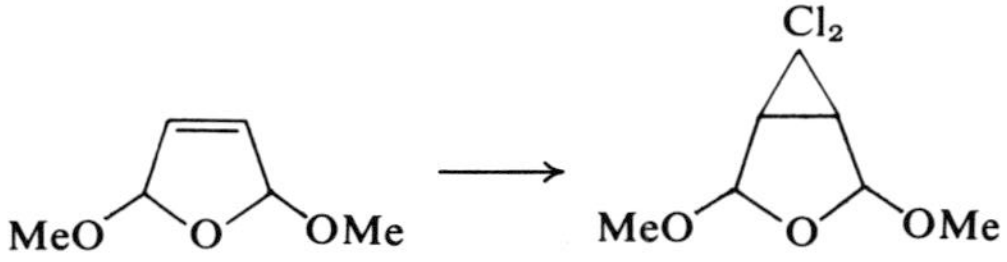	15%	[987]

(continued)

Table 3-15. PTC-Dichlorocarbene Additions to Aromatic and Heterocyclic Systems.

Reaction	Remarks	Reference
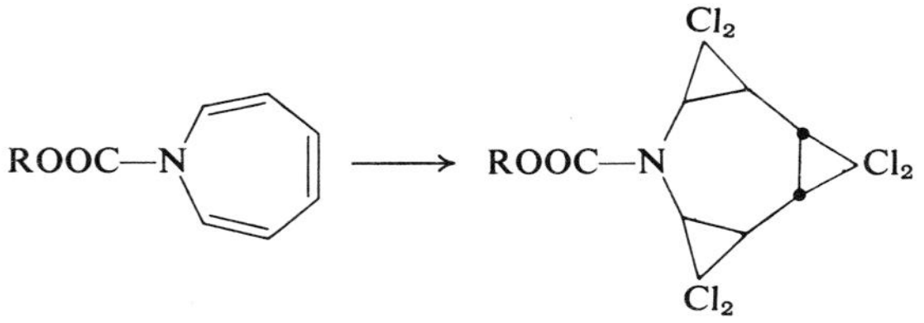 ROOC—N → ROOC—N (Cl_2, Cl_2, Cl_2)		[713, 714]
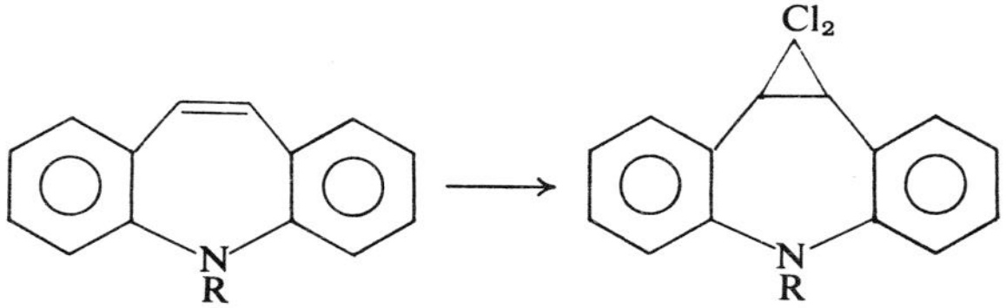 N R → (Cl_2) N R	61–78%	[645, 1136]
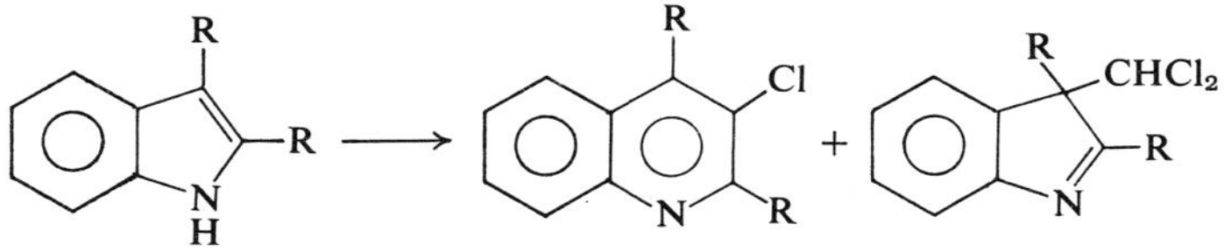 R, R, N H → R, Cl, N, R + R, $CHCl_2$, R, N	30–70%	[728, 729, 987]
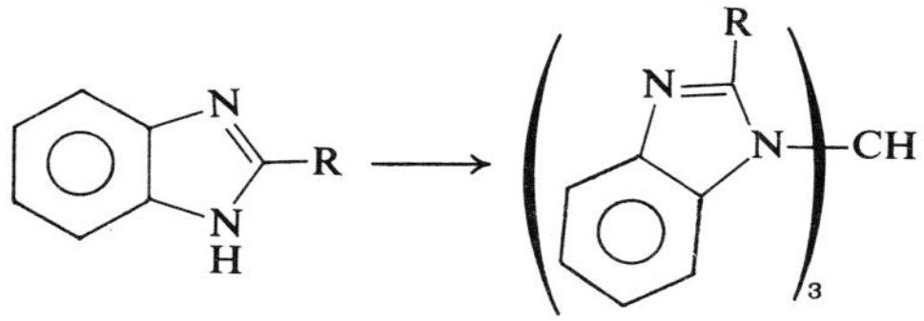 N, R, N H → (N, R, N)$_3$ CH	14%	[731, 979, 987]

Reaction	Yield	Ref.
O, CH_3 → O, Cl, Cl_2		[718]
O, O → OHC, O, O	3%	[987]
N → N, Cl	low yield	[735]
O, N, CH_3 → Cl_2, O, N, CH_3	82%	[1137]
O → O, Cl_2	72%	[987]
O → O, $CHCl_2$, H + Cl_2, O	83%	[987]

Table 3–16. C—H Insertions of PTC Dichlorocarbene.

Compound	Position of Insertion	Yield (%)	Reference
cumene	*tert*-H	24	[739, 741, 742]
ethylbenzene	CH_2	2	[741]
tetralin	1	21	[741]
isopentane	*tert*-H	12	[742]
methylcyclohexane	*tert*-H	5	[739, 742]
cis-decalin	9	29	[739, 742]
trans-decalin	9	< 3	[739]
adamantane	1	54	[740]
diamantane	1 4	63 37	[747]
trishomo-barrelane	1	64	[744, 745]
trishomo-bullvalene	1	59	[745]
benzhydryl methyl ether	*tert*-H	17	[742]
p-chlorobenzyl methyl ether	CH_2	18	[742]
dibenzyl ether	CH_2	26	[742]
diethyl ether	CH_2	8	[742]
methoxycyclohexane	*tert*-H	5	[739]
THF	2	18	[741, 742]
diisopropyl ether	*tert*-H	43	[739, 742, 743]
1-methyl-1,2-dihydroquinoline	2- and 4-	30	[746]
2-alkyl-1,3-dioxanes	CH	5–18	[980, 1133]
2-alkyl-1,3-dioxolanes	CH	30–90	[980, 1133–1135]

adopted are similar to those in CCl_2 additions. Table 3–16 gives a survey of known yields.

Nitriles by Dehydration with PTC-Dichlorocarbene. Saraie *et al.* [748] and Höfle [749] independently discovered that aliphatic and aromatic amides, thioamides, and aldoximes, as well as certain ureas ($R_2N—CONH_2 \rightarrow R_2N—CN$) can be converted into nitriles by the Mąkosza reagent. The substrate is dissolved or suspended in chloroform, TEBA and warm 40% sodium hydroxide is added, and the mixture is refluxed for 30 minutes. For lipophilic compounds, yields range up to 95%, lower aliphatic amides give only 10 to 20%. In the latter case, and with some bisamides, hydrolysis predominates. One possible mechanism is shown:

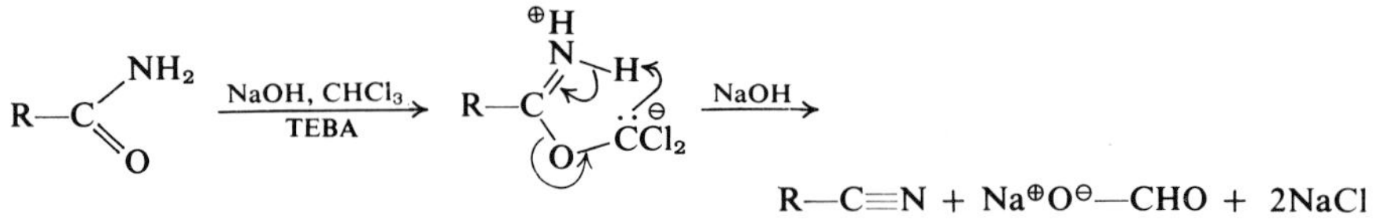

$$R—C(=O)NH_2 \xrightarrow[\text{TEBA}]{\text{NaOH, CHCl}_3} R—C(=N^{\oplus}H—H)—O—\overset{\ominus}{C}Cl_2 \xrightarrow{\text{NaOH}} R—C{\equiv}N + Na^{\oplus}O^{\ominus}—CHO + 2NaCl$$

Scheme 3–142

As dehydrations of this type are normally effected under acidic conditions, this basic process might be of occasional merit. Graefe [750] reacted amidines too (yields: 61–92%):

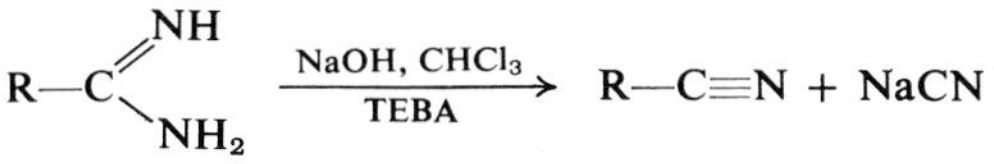

$$R{-}C({=}NH){-}NH_2 \xrightarrow[\text{TEBA}]{\text{NaOH, CHCl}_3} R{-}C{\equiv}N + NaCN$$

Scheme 3–143

Finally, α-hydroxyketoximes can be fragmented into nitriles and ketones by PTC-CCl_2 [1139]:

$$R_2C(OH){-}C(R'){=}NOH \rightarrow R_2CO + R'CN$$

Reactions with Amines. The well known long-established carbylamine reaction

$$RNH_2 \xrightarrow[\text{NaOH}]{\text{HCCl}_3} R{-}N{=}C{:}$$

usually gives modest but poorly reproducible yields. These shortcomings are overcome in the PTC version where yields between 20 and 76% are realized with TEBA as catalyst [751–754]. An Organic Syntheses procedure for this conversion is available [754], in which a methylene chloride solution of amine (1.93 moles), chloroform (0.98 mole), and TEBA (2g) is added dropwise to concentrated sodium hydroxide at 45 °C.

$$R{-}NH_2 \xrightarrow{\text{SOCl}_2} R{-}N{=}SO \xrightarrow[\text{crown ether}]{\text{CHCl}_3\text{, solid KOH}} R{-}N{=}C{:}$$

An alternative synthesis of isocyanides is believed to give better yields: the amines are first converted with thionyl chloride into N-sulfinylamines, which in turn are reacted with chloroform, solid potassium hydroxide, and dibenzo-18-crown-6—or even better dicyclohexano-18-crown-6—in cyclohexane or benzene [755]. In this case the PTC process with TEBA/aqueous sodium hydroxide is not more advantageous and even yields less pure products.

$$N_2H_4 \xrightarrow[\text{catalyst}]{\text{HCCl}_3\text{/KOH}} CH_2N_2$$

The formation of diazomethane from hydrazine, chloroform, and sodium hydroxide was first noted by Staudinger in 1912. In the PTC version, this reaction gives 48% yield with potassium hydroxide/18-crown-6 or approximately 35% with sodium hydroxide/quaternary ammonium catalyst [756]. Either etheral or dichloromethane solutions of diazomethane are obtained which contain some hydrazine carried over with the diazomethane in the azeotropic distillation. From methylene chloride solutions the hydrazine can be removed by extraction with water [756]. At a time when N-nitroso compounds are unpopular because of the cancerogenic dangers, this method may find application.

$$R{-}CO{-}NH{-}NH_2 \xrightarrow[\text{NaOH}]{\text{CHCl}_3\text{/TEBA}} R{-}CO{-}NH{-}NH{-}OC{-}R + CH_2N_2$$

Acid hydrazides (R = aryl, benzyl) are transformed into diacyl hydrazines (40–50% yield) and diazomethane (27%) by the PTC/CCl_2 system [757]. The mechanism of this reaction is still uncertain.

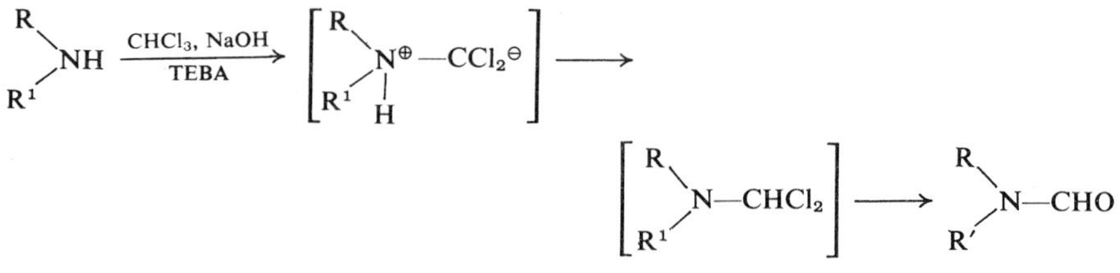

$$\begin{matrix} R \\ \quad \diagdown NH \\ R^1 \diagup \end{matrix} \xrightarrow[\text{TEBA}]{\text{CHCl}_3,\ \text{NaOH}} \left[\begin{matrix} R \\ \quad \diagdown N^{\oplus}{-}CCl_2^{\ominus} \\ R^1 \diagup \ | \\ \quad H \end{matrix} \right] \longrightarrow \left[\begin{matrix} R \\ \quad \diagdown N{-}CHCl_2 \\ R^1 \diagup \end{matrix} \right] \longrightarrow \begin{matrix} R \\ \quad \diagdown N{-}CHO \\ R' \diagup \end{matrix}$$

Scheme 3–144

With PTC—CCl_2, secondary amines lead to formamides [759, 760]. Yields are in the range 39–91%. Sometimes methylene chloride is used as cosolvent. Attack at a double bond is much slower than attack at the nitrogen lone pair of electrons. Thus, diallylamine gives N,N-diallylformamide.

Reaction of dichlorocarbene with tertiary amines must give an ammonium ylide (**A**) as a first step. This has been mentioned in the preceding subsection in connection with the ability of tertiary amines to catalyze the PTC dichlorocarbene generation [617, 433] and the supposed generation of optically active carbene adducts [623]:

$$\underset{}{R_3N + CCl_2} \longrightarrow \underset{\mathbf{A}}{R_3N^{\oplus}{-}CCl_2^{\ominus}}$$

It was shown that no dichlorocarbene adducts are formed from tertiary amine, olefin, and chloroform in the absence of sodium hydroxide after 2 hours at 50 °C [617]. Furthermore, no reaction was observed after prolonged boiling of tri-*n*-propylamine and chloroform [447]. Kimura *et al.* [623] believe that **A** transfers dichlorocarbene to olefins. Mąkosza showed, however, that the formation of **A** is not reversible: If a mixture of equimolar amounts of tertiary amine and chloroform is reacted with styrene and NaOH it does not give a styrene dichlorocarbene adduct [433]. Mąkosza believes that under normal conditions (with an excess of chloroform) **A** deprotonates chloroform to give an ion pair **B** which can then transfer CCl_2 to give **C**. **C** then decomposes to chloroform and amine, and the cycle can begin again

$$\underset{\mathbf{A}}{R_3N^{\oplus}{-}CCl_2^{\ominus}} + HCCl_3 \longrightarrow \underset{\mathbf{B}}{R_3N^{\oplus}{-}CHCl_2\ CCl_3^{\ominus}} \longrightarrow$$

$$\underset{\mathbf{C}}{R_3N^{\oplus}{-}CHCl_2\ Cl^{\ominus}} + CCl_2 \longrightarrow \text{addition products}$$

[433]. It was shown recently, however, that NR_3, chloroform, and sodium hydroxide give a somewhat darkish syrupy material, containing a quaternary ammonium

chloride, which is soluble in water and is capable of functioning as a PTC catalyst in a new CCl_2 addition reaction. If this material is dissolved in dichloromethane and stirred with cyclohexene/sodium hydroxide no dichloronorcarane is formed [447]. It is probable, therefore, that this material contains **C** and that the catalysis of tertiary amines in dichlorocarbene reactions must involve the generation of **C**, the true **PT** catalyst. This explanation varies slightly from Mąkosza's original proposal as **C** does not decompose again but rather acts as a normal catalyst. Reaction of **D** with the PTC system involves the attack of two molecules of CCl_2 and ring fission to give **E** [716].

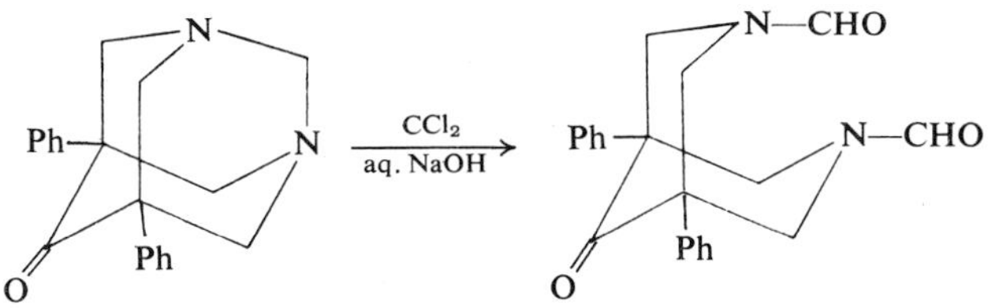

Ph N Ph O **D** $\xrightarrow[\text{aq. NaOH}]{CCl_2}$ N—CHO N—CHO Ph Ph O **E**

Scheme 3–145

Other Compounds. Attack of compound **F** is at sulfur rather than at the double bond. Andrews and Evans observed a [2, 3] sigmatropic rearrangement [715]:

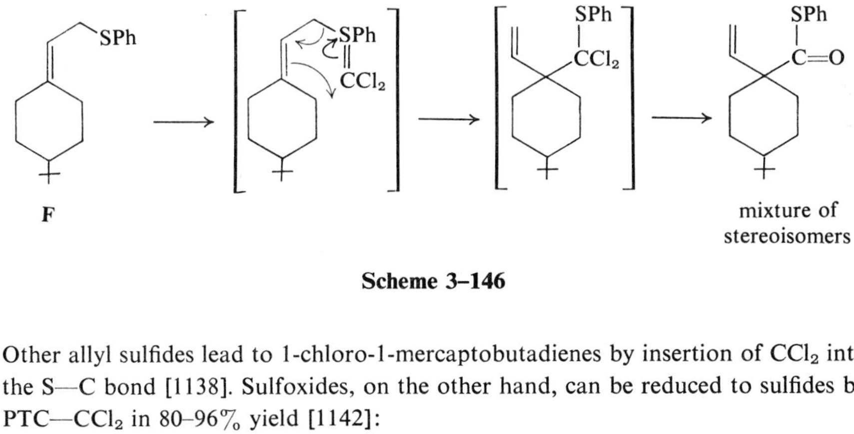

SPh **F** → [SPh CCl_2] → [SPh CCl_2] → SPh C=O mixture of stereoisomers

Scheme 3–146

Other allyl sulfides lead to 1-chloro-1-mercaptobutadienes by insertion of CCl_2 into the S—C bond [1138]. Sulfoxides, on the other hand, can be reduced to sulfides by PTC—CCl_2 in 80–96% yield [1142]:

$$R_2SO + CCl_2 \rightarrow R_2S + COCl_2$$

The Reimer-Tiemann reaction usually occurs without catalyst. For highly lipophilic substrates, however, it was found useful to work in the presence of cetyltrimethylammonium catalyst [761]. In fact, an "abnormal" PTC Reimer-Tiemann reaction occurs at room temperature with 10% NaOH solution, whereas without catalyst higher temperatures and more concentrated base are required [841].

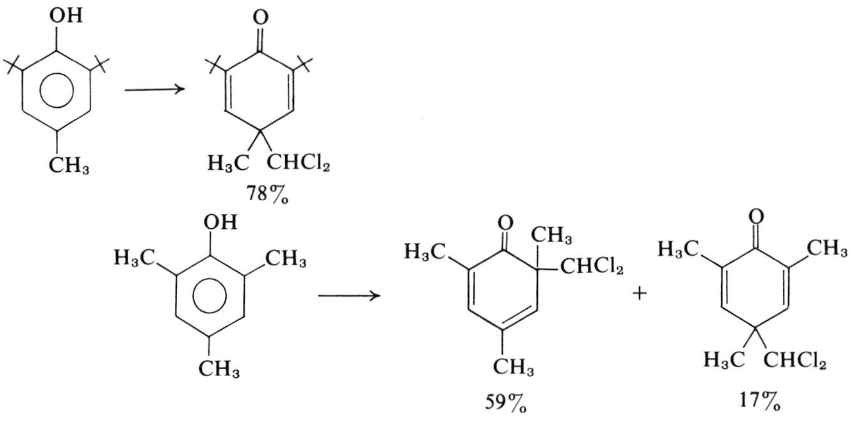

78%

59% 17%

Scheme 3–147

Reactions with Alcohols.

$$RCH_2\text{—}OH \xrightarrow[NaOH/TEBA]{HCCl_3} (R\text{—}CH_2\text{—}O)_3CH$$

The reaction between the lower alcohols, excess chloroform, and 50% aqueous sodium hydroxide in the presence of TEBA leads to complex mixtures. Only ethanol and 2,2,2-trifluoromethanol are preparatively interesting, giving orthoformates in yields of 36% and 33% [762]. In the case of ethanol, ethyl chloride, carbonate, and oxalate were by-products.

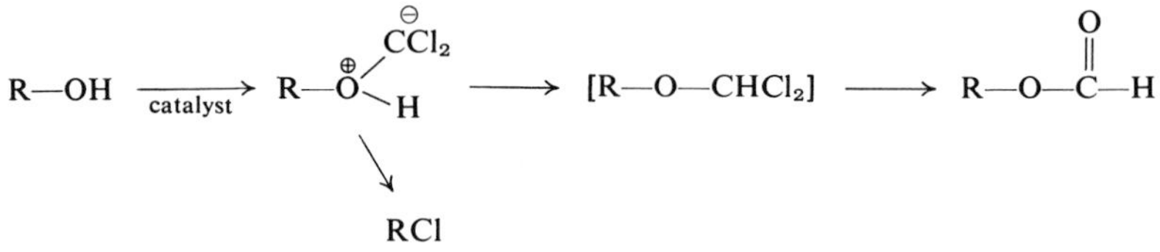

$$R\text{—}OH \xrightarrow[catalyst]{} R\text{—}\overset{\oplus}{O}(H)\text{—}\overset{\ominus}{C}Cl_2 \longrightarrow [R\text{—}O\text{—}CHCl_2] \longrightarrow R\text{—}O\text{—}C(=O)\text{—}H$$

$$\searrow RCl$$

Scheme 3–148

Analogous to the secondary amines, formates might be expected as major products with the higher alcohols. These, however, are formed to a minor extent only, Tabushi *et al.* [763] discovered that chlorides are formed instead in high yields. Except when the reaction is exothermic stirring is continued for about 5 hours at room temperature. This method has been applied to a number of steroids and terpenes [3, 644], and retention of configuration has been observed in support of an S_Ni mechanism. In certain cases, however, rearrangements and inversions have been found [763], and the details of the mechanism await further investigation. Nevertheless, the method is preparatively interesting because it allows the alcohol → chloride conversion to proceed under basic conditions at room temperature. Compounds with double bonds are cyclopropanated usually with retention of the hydroxy group.

Phenols, including steroids, can also be converted into the corresponding chlorides [3]. This reaction can be extended to the preparation of alkyl bromides [764].

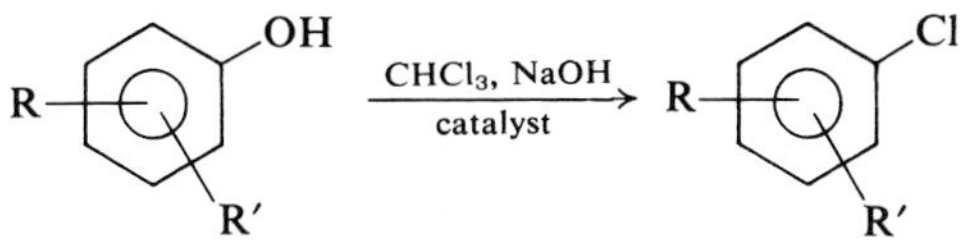

OH; R; R′ $\xrightarrow[\text{catalyst}]{CHCl_3,\ NaOH}$ Cl; R; R′

Scheme 3–149

Complex reactions occur if glycols are reacted with the two-phase dichlorocarbene system. Among the products formed are ketones, epoxides, olefins, and dichlorocyclopropanes, and mechanisms to explain these have been advanced [765]. The dichlorocarbene system is capable of deoxygenating epoxides. The olefins formed

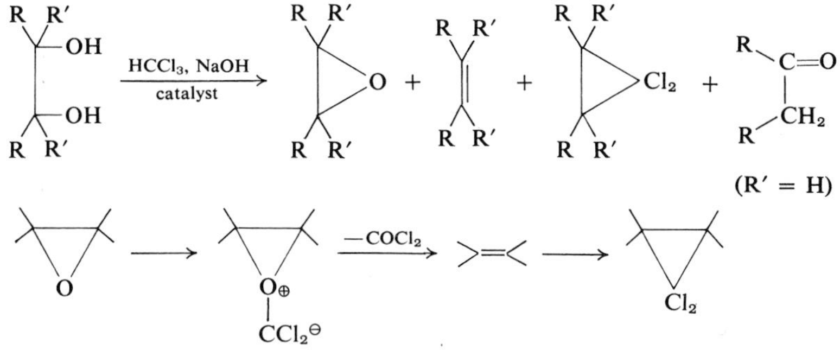

R R′; OH; OH; R R′ $\xrightarrow[\text{catalyst}]{HCCl_3,\ NaOH}$ R R′; O; R R′ + R R′; R R′ + R R′; Cl_2; R R′ + R; C=O; CH_2; R

(R′ = H)

O ⟶ $O^{\oplus}$; $CCl_2^{\ominus}$ $\xrightarrow{-COCl_2}$ >=< ⟶ Cl_2

Scheme 3–150

are mainly transformed into dichlorocyclopropanes, but overall yields are not very impressive [766]. Both deoxygenation and addition of CCl_2 are stereospecific.

Reactions of Trichloromethylide Anions. In the preceding subsection (3.3.13.1) it was mentioned that the two-phase $NaOH/CHCl_3$/catalyst system provides both $CCl_3^{\ominus}$ and CCl_2 and that depending on the substrate, either of these can be used. Very electron deficient olefins such as vinyl acetate, acrylonitrile, or acrylic esters, give addition of chloroform across the double bond only, but an α substituent may suffice to shift the reaction over to cyclopropane formation. With methacrylonitrile both possibilities occur simultaneously:

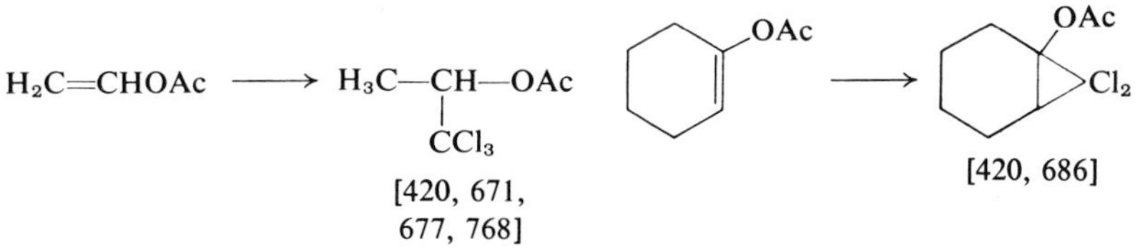

$H_2C{=}CHOAc \longrightarrow H_3C{-}CH(CCl_3){-}OAc$ [420, 671, 677, 768]

OAc ⟶ OAc; Cl_2 [420, 686]

$H_2C{=}CH{-}COOR \longrightarrow Cl_3C{-}CH_2{-}CH_2{-}COOR$ [671]

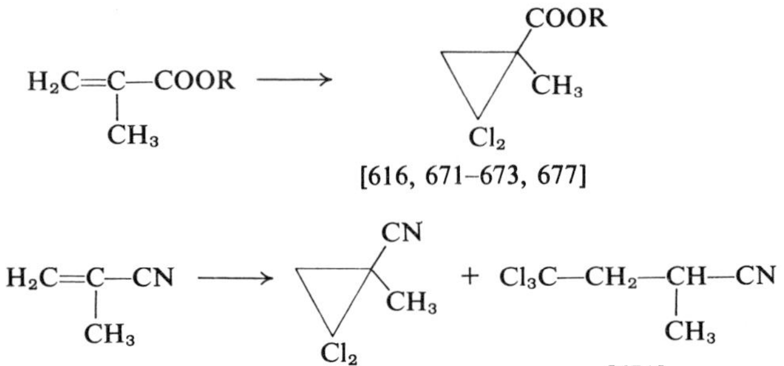

$$H_2C{=}C(CH_3){-}COOR \longrightarrow \text{1,1-dichloro-2-methyl-2-(COOR)cyclopropane}$$

[616, 671–673, 677]

$$H_2C{=}C(CH_3){-}CN \longrightarrow \text{1,1-dichloro-2-methyl-2-cyanocyclopropane} + Cl_3C{-}CH_2{-}CH(CH_3){-}CN$$

[671]

Scheme 3–151

The addition to vinyl acetate (Scheme 3–151) proceeds *via* a small concentration of free acetaldehyde as shown [1167]:

$$CH_2{=}CH{-}OAc \longrightarrow CH_2{=}CH{-}O^{\ominus} \longrightarrow CH_3{-}CHO$$
$$\longrightarrow CH_3{-}CH(CCl_3){-}O^{\ominus}$$
$$CH_3{-}CH(CCl_3){-}O^{\ominus} + CH_2{=}CH{-}OAc$$
$$\longrightarrow CH_3{-}CH(CCl_3)OAc + CH_2{=}CH{-}O^{\ominus}$$

It has been found that α,β-unsaturated esters of the type **G** were transformed under these conditions into spiropentanes **H** if at least one of the substituents R^1 and R^2 was alkyl or phenyl:

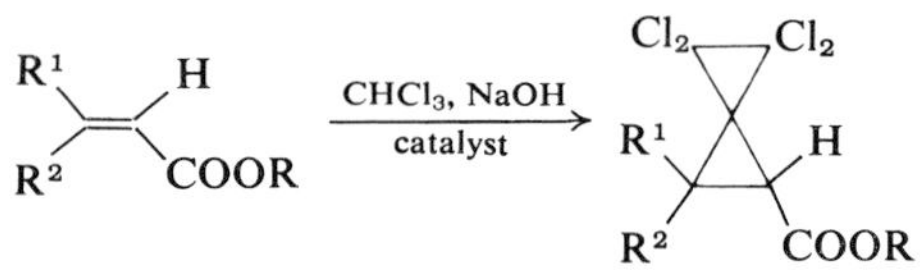

$$\underset{\mathbf{G}}{R^1R^2C{=}CH{-}COOR} \xrightarrow[\text{catalyst}]{CHCl_3,\ NaOH} \underset{\mathbf{H}}{\text{Cl}_2\text{,Cl}_2\text{-spiropentane}(R^1,R^2,H,COOR)}$$

[671, 680]

Scheme 3–152

Tetrachloro- and tetrabromo-spiro compounds are the products from some α,β-unsaturated steroidal ketones [644], but simpler unsaturated carbonyl compounds give mainly resinification. The mechanism of the formation of **H** [680] initially involves normal dichlorocarbene addition. These adducts (**I**) are then dehydrochlorinated in the strongly basic medium to give **K**. Cyclopropene carboxylic esters like **K** are known to add any available nucleophile instantaneously. Thus, the addition of hydroxide eventually leads to the substituted succinic half ester **L**, formed as a by-product. Alternatively, addition of trichloromethylide gives **M**, and a second elimination-addition sequence eventually leads to **N**. This in turn is attacked by another $CCl_3^{\ominus}$ which pulls off a positive chlorine giving carbon tetrachloride (by-

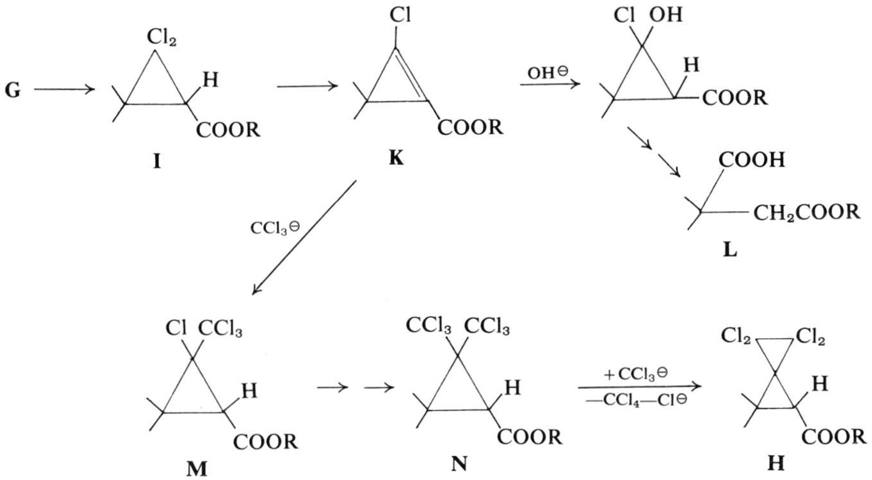
G
Cl2
H
COOR
I
Cl
COOR
K
OH⊖
Cl OH
H
COOR
COOH
CH2COOR
L
CCl3⊖
Cl CCl3
H
COOR
M
CCl3 CCl3
H
COOR
N
+CCl3⊖
—CCl4—Cl⊖
Cl2 Cl2
H
COOR
H

Scheme 3–153

product) and the spiropentane **H**. This ability of the PTC system to dehalogenate is also found in the following reaction sequence [680]:

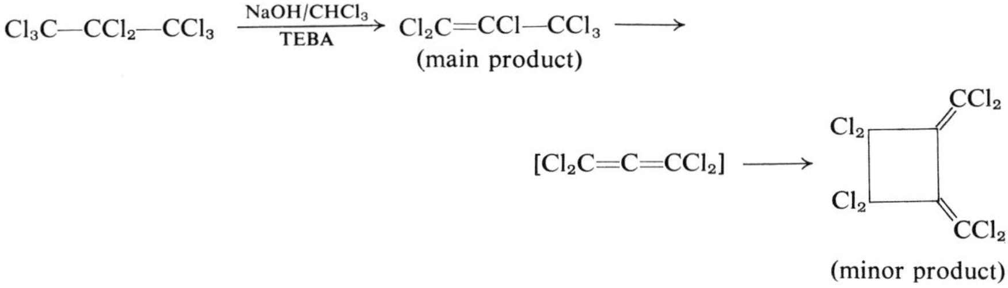
Cl3C—CCl2—CCl3
NaOH/CHCl3
TEBA
Cl2C=CCl—CCl3
(main product)
[Cl2C=C=CCl2]
CCl2
Cl2
Cl2
CCl2
(minor product)

Scheme 3–154

$$RS{-}CN + CHCl_3 \xrightarrow[\text{TEBA}]{\text{NaOH}} RS{-}CCl_3 + NaCN$$

The reaction between alkyl thiocyanates and chloroform in the presence of NaOH/TEBA is moderately exothermic and (by the displacement of cyanide) gives trichloromethyl sulfides in excellent yields [300].

$$PhHgCl \longrightarrow PhHgCX_3$$

Trihalomethyl(phenyl)mercurials (Seyferth's reagents) are used as alternative dihalocarbene precursors for difficult substrates in neutral medium. The synthesis of these reagents is usually rather delicate. Mąkosza, however, found an easy alternative [767]: A solution of 20 g sodium hydroxide and 37 g potassium fluoride per 100 g of water is dropped slowly into a suspension of 0.05 mole phenylmercuric chloride and 0.4 g TEBA in 70 ml chloroform at room temperature and stirred for 2 hours. A

72% yield is obtained. Other trihalomethyl(phenyl)mercurials (X_3 = $BrCl_2$, Br_3) can be prepared similarly. Fluoride apparently functions as additional strong base.

Quaternary nicotinamides **O** are transformed by chloroform and solid $Ba(OH)_2 \cdot 8H_2O$ into trichloromethyl compounds, some of which can in turn rearrange to cyclopropanes [725]. Further transformations of these systems under PTC conditions have been investigated [726, 727].

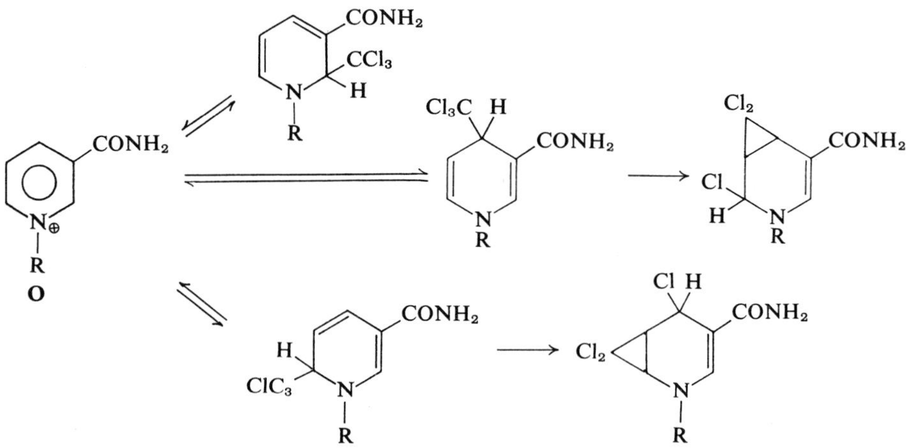

CONH2 CCl3 H N R
CONH2 N⊕ R **O**
Cl3C H CONH2 N R
Cl2 CONH2 Cl H N R
H ClC3 CONH2 N R
Cl H CONH2 Cl2 N R

Scheme 3–155

The PTC dichlorocarbene reaction of dehydronuciferine (**P**) gave the aldehyde shown instead of the expected cyclopropane [778].

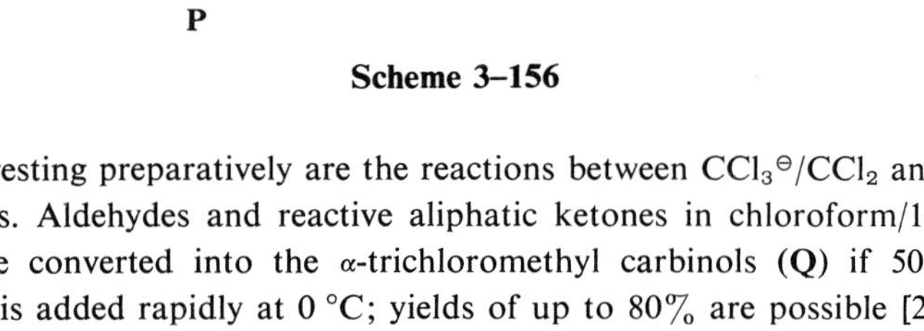

MeO MeO N—Me **P**
MeO MeO N—Me CHO

Scheme 3–156

More interesting preparatively are the reactions between $CCl_3^{\ominus}/CCl_2$ and carbonyl compounds. Aldehydes and reactive aliphatic ketones in chloroform/1–5 mole-% TEBA, are converted into the α-trichloromethyl carbinols (**Q**) if 50% sodium hydroxide is added rapidly at 0 °C; yields of up to 80% are possible [235]. In the presence of dimethyl sulfate the corresponding ethers are formed.

If the temperature is raised, the yield of products **Q** decreases rapidly because of retrocleavage in the aqueous phase. A competing reaction with aromatic aldehydes

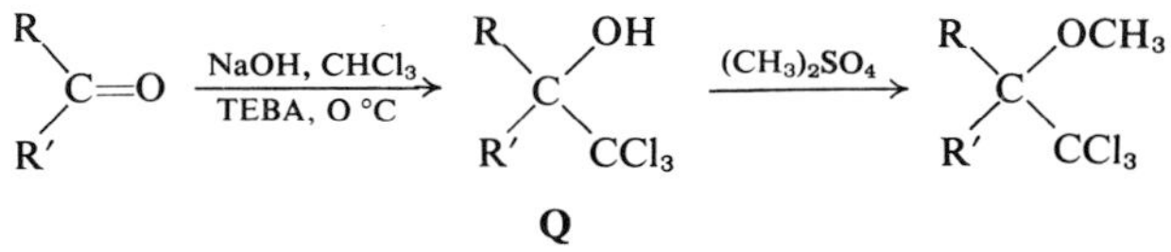

Scheme 3–157

is the Cannizzaro reaction. In the narrow temperature range of 56 ± 2 °C, however, aromatic aldehydes give 75–83% yields of mandelic acids [235, 770, 772]. Acetone and alicylic ketones give mixtures of α-hydroxy-(**R**), α-chloro-(**S**), and α,β-unsaturated acids (**T**) [235, 771]. Here the temperature appears to be not too critical [771]. At 20 °C the chloroacids **S** are almost the only products; at 56 °C **R** predominates

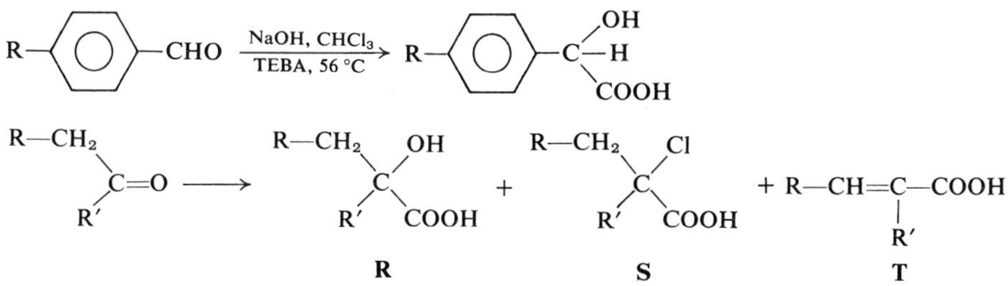

Scheme 3–158

(13–55% yield), Furthermore, **Q** (R—R′ = $(CH_2)_5$) can be transformed into **S** (R—R′ = $(CH_2)_4$) at 20 °C [771], but at higher temperatures the conversion of 2,2,2-trichloro-1-phenylethanol into mandelic acid was not possible [235, 770]. Mechanistically, the intermediacy of a dichloro oxirane has been discussed:

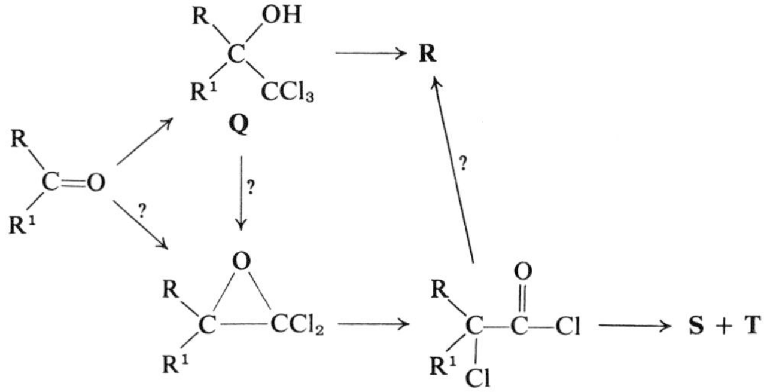

Scheme 3–159

Aromatic aldehydes can be transformed into α-aminoarylacetic acids (29–81% yield) with $KOH/LiCl/NH_3$ in the aqueous phase, $CHCl_3/CH_2Cl_2$ as organic phase, and TEBA as catalyst [1141].

3.3.13.3. Dibromocarbene

Dibromocarbene adducts are often preferred to their chlorine analogues because they are much easier to modify in subsequent reactions. The PTC chemistry of dibromocarbene is generally very similar to that of CCl_2. From selectivity data on bromoform/potassium *tert*-butoxide reactions both in the presence and absence of crown ethers, Moss concluded that there is "minimal kinetic-effective carbenoid involvement" [780]. Dibromocarbene, however, should be more reactive than dichlorocarbene. This prediction is borne out by competition experiments: equal amounts of chloroform and bromoform, styrene, TEBA, and concentrated sodium hydroxide give 1,1-dibromo-2-phenyl-cyclopropane as the major product [384]. An earlier report giving the dichloro compound as the main product [781] was subsequently withdrawn (*cf.*, Footnote 16 in ref. [384]). Since CBr_2 is more reactive, one suspects that it will also have a greater sensitivity towards hydrolysis. Indeed, Hine found that the relative rate of hydrolysis of $CHBr_3$ compared to $CHCl_3$ is six times larger in homogeneous aqueous medium [782]. Under PTC conditions, the rates of hydrolysis—in the absence of olefin—do differ as well at room temperature [1140]. In the presence of carbene acceptors other than water, however, there is a marked difference in the temperature sensitivity of the CCl_2 and the CBr_2 addition reactions. Rates and absolute yields of dichlorocarbene additions rise in most cases when the temperature is raised from room temperature to 40–60 °C. With analogous dibromocarbene reactions, a rise in temperature may accelerate the rate of hydrolysis more than the rate of addition, depending on the nucleophilicity of the olefin. It was found that in practice yields sometimes rise, sometimes stay constant, and sometimes fall if the temperature is raised with various olefins [783]. It is clear therefore that experiments to find optimum conditions for CBr_2 addition are more complicated than in the case of CCl_2, and no single set of optimum conditions exists. From the early PTC dibromocarbene publications, it appeared that the CBr_2 adduct yields were always lower than the corresponding CCl_2 yields. There are several reports in the literature on improved reactions conditions, but as there are many variables involved, it is not always easy to compare results. Skattebøl *et al.* [784] found that long reaction times (up to 96 hours) at room temperature are best. Efficient stirring was necessary, and less reactive olefins required up to a fourfold excess of $CHBr_3$. Styrene turned out to be more reactive than 2-methyl-2-butene needing less bromoform and a shorter reaction time. Mąkosza and Fedoryński [785] observed that the addition of small amounts of ethanol (0.8 ml for a 0.2 molar run) to the reaction mixture improved the yield by between 10 and 30%. They recommended a 100% excess of $CHBr_3$ at 40–45 °C for 3 hours. The influence of ethanol is not well understood. It was assumed that specific solvation of the $NR_4^{\oplus}CBr_3^{\ominus}$ ion pair may occur, stabilizing it and allowing it to travel deeper into the organic phase before decomposition, and this slows down hydrolysis. Another possible explanation is that ethoxide ions are more readily extracted into the organic medium than hydroxyl ions, thereby changing the reaction kinetics to some extent. Even better conditions for sluggish olefins have been found [786]. Conditions are as before but now tri-*n*-butylamine is used as catalyst and methylene chloride (volume about equal to bromoform volume) as cosolvent. Most workers use at least 2 moles $HCBr_3$ per mole olefin. With some α-substituted α,β-unsaturated ketones, however, excellent yields have been found with only 1.25 moles haloform per mole substrate [674].

The influence of the many variables involved has been investigated further [783, 787, 1140], but two important precautions should be mentioned:

(1) Since bromoform is rather high boiling, it is often difficult to purify reaction products with similar boiling points from the last traces of bromoform, so some literature yields may not be exact.
(2) Vigorous stirring is extremely important. At least 600–800 rpm is recommended when using a magnetic stirrer.

Other variables investigated were:

—*Amount of sodium hydroxide:* Best results are obtained if 5–10 equivalents of concentrated sodium hydroxide per mole $CHBr_3$ are present.
—Amount of $CHBr_3$/cosolvent: Two moles $CHBr_3$ per mole double bond were sufficient in most cases. To improve stirring an equal volume of methylene chloride can be added although the effect on yields is not very large.
—*Temperature, time:* Room temperature is recommended. Normal reaction times are (styrene, cyclohexene) 4–6 hours. In more sterically hindered cases (3,3-dimethylbutene) or in electron-poor electronically deactivated systems (poly-addition to cyclooctatetraene) much longer times (2 to 6 days) are necessary. In such cases, although the actual yield increases day by day, it may save time if the mixture is worked up after 1–2 days and the crude product freshly reacted.
—*Amount of catalyst:* 1 mole-% (referred to olefin) suffices.
—*Type of catalyst, cocatalyst:* Good results are observed with many catalysts including symmetrical tetraalkylammonium halides, Aliquat 336, TEBA, tri-*n*-butylamine, and tri-*n*-propylamine, as well as crown ethers. Addition of a little ethanol has a very pronounced beneficial effect with the quaternary ammonium salts.

Typical experimental procedures for CBr_2 additions are presented below.

—*Mąkosza and Fedorynski method* [786]: 0.1 mole alkene, 0.2 mole bromoform, 25 ml methylene chloride, 0.4 ml tri-*n*-butylamine, and 0.5 mole (40 ml) 50% aqueous sodium hydroxide are stirred vigorously at 40–45 °C for 4 hours. The mixture is diluted with water and methylene chloride and separated. The aqueous layer is extracted twice with methylene chloride, and the combined organic layers are washed with dilute HCl, and water, dried, and worked up.
—*Dehmlow procedure* [783, 787]: 0.1 mole alkene, 0.2 mole bromoform, 20 ml methylene chloride, 1 ml ethanol, 250 mg TEBA, and 1–2 mole 50% aqueous sodium hydroxide are stirred vigorously at room temperature for 4–6 hours to 2–6 days, workup as before.

The first part of Table 3–17 collects reactions under partially optimized conditions, other dibromocarbene reactions are mentioned subsequently.

Several points are evident from the results in Table 3–17. A free hydroxyl group in the olefin seems to lead to preferential or exclusive formation of the isomer with the dibromocyclopropane ring *trans* to OH [502, 788, 789, 790]. Allylic hydroxyls

Table 3–17. PTC Dibromocarbene Additions.

	Olefine	Temper-ature	Moles $CHBr_3$/Moles Alkene	Catalyst/Spec. Cond.	Time (h)	Yield (%)	Reference
(a) Optimized Reactions							
C_4	1-chloro-2-methyl-1-propene	45 °C	2	NBu_3/CH_2Cl_2	4	65	[786]
C_5	2-methyl-2-butene	r.t.	4	TEBA	72	73	[784]
		45 °C	2	TEBA/trace EtOH	3	81	[785]
	1-chloro-3-methyl-2-butene	45 °C	2	TEBA/trace EtOH	3	89	[785]
C_6	1-hexene	45 °C	2	NBu_3/CH_2Cl_2	4	61	[786, 797]
	2-hexene	45 °C	2	NBu_3/CH_2Cl_2	4	67	[786]
	cyclohexene	r.t.	3.5	TEBA	96	72	[784]
		r.t.	2	TEBA/trace EtOH	20	86	[787]
		r.t.	2	TEBA/trace EtOH	4	73.5	[787]
		45 °C	2	TEBA/trace EtOH	3	92	[797]
		45 °C	2	NBu_3/CH_2Cl_2	4	76	[786]
		45 °C	4	PMe_4I/CH_2Cl_2	2	~100	[616]
		40 °C	2.5	dibenzo-18-crown-6		74	[618]
	n-butyl vinyl ether	45 °C	2	TEBA/trace EtOH	3	50	[785]
	3,3-dimethyl-1-butene	r.t.	4	TEBA/trace EtOH 10 moles NaOH	24	30	[792, *cf.*, 797]
C_7	1-heptene	45 °C	2	NBu_3/CH_2Cl_2	4	67	[786]
	tetramethylallene	r.t.	6?	TEBA	72?	10	[784]
	ethyl 3-methyl-2-butenoate	45 °C	2	TEBA/trace EtOH	3	89	[785]

C_8	styrene	r.t.	2	TEBA	24	66	[784]
		45 °C	2	TEBA/trace EtOH	3	78	[785, 797]
		45 °C	2	NBu_3/CH_2Cl_2	4	88	[786]
		45 °C	2	TEBA/trace EtOH + CH_2Cl_2 + 10-fold mol. amount NaOH	4	95	[783]
	1-octene	45 °C	2	NBu_3/CH_2Cl_2	4	75	[786]
	2-octene	45 °C	2	NBu_3/CH_2Cl_2	4	75	[786]
	cyclooctene	r.t.	3	TEBA	48	70	[784]
	1,5-octadiene (bisadduct)	45 °C	4	NBu_3/CH_2Cl_2	4	60	[786, 626]
	n-butyl methacrylate	45 °C	2	TEBA/trace EtOH	3	76	[785]
C_9	α-methylstyrene	45 °C	2	TEBA/trace EtOH	3	80	[785]
	isopropenylbenzene	45 °C	2	NBu_3/CH_2Cl_2	4	89	[786]
C_{10}	1-decene	45 °C	2	NBu_3/CH_2Cl_2	4	77	[786]
	o-divinylbenzene (bisadduct)	r.t.	6?	TEBA	< 24	50	[784]
C_{12}	1-dodecene	45 °C	2	TEBA/trace EtOH	3	78	[798]
(b) Nonoptimized Reactions							
	$Ph_2C{=}CH_2$	r.t.	7	TEBA	15	34	[626]
	stilbene	r.t.	7	TEBA	18	32	[626]
		40 °C	2.5	dibenzo-18-crown-6	9	35	[618]
	1-phenyl-1-propene	r.t.	2.5	TEBA	16	23	[626]
	cyclooctatetraene	r.t.	4	TEBA/trace EtOH CH_2Cl_2	240	$+1CBr_2$ 38 $+2CBr_2$ 25 $+3CBr_2$ 1 $+4CBr_2$ 0.3	[626, 800, 730]

(*continued*)

Table 3–17. PTC Dibromocarbene Additions.

Olefine	Temperature	Moles $CHBr_3$/ Moles Alkene	Catalyst/ Spec. Cond.	Time (h)	Yield (%)	Reference
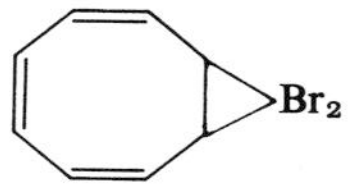	r.t.	4	TEBA/trace EtOH/ CH_2Cl_2	240	$+1CBr_2$ 42 $+2CBr_2$ 10 $+3CBr_2$ 1.2	[730]
			TEBA/benzene		?	[711]
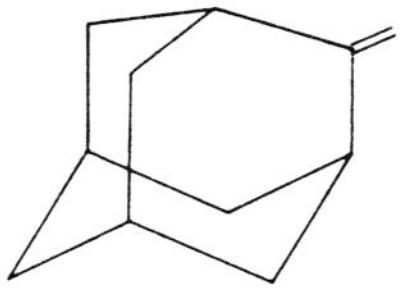	5–10 °C		TEBA/benzene		54	[648]
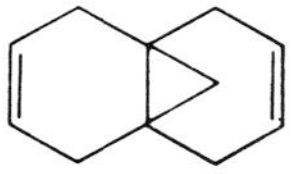 (bisadduct)	r.t.		TEBA		59	[794]
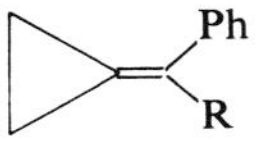 (R = H, Me, Ph)	r.t.	2	TEBA/trace EtOH	3.5	22–60	[647]
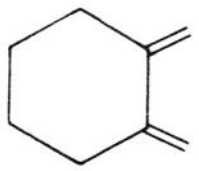 (bisadduct)	45 °C	4	TEBA/trace EtOH		?	[799]

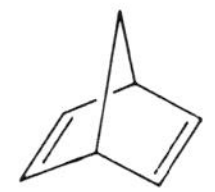	50 °C	?	cetyltrimethylammonium chloride	2.5	80–90	[704]
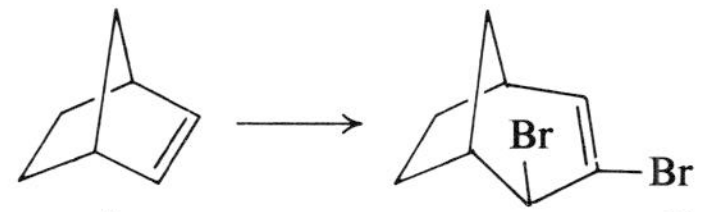	r.t.	11	TEBA	94	94	[700]
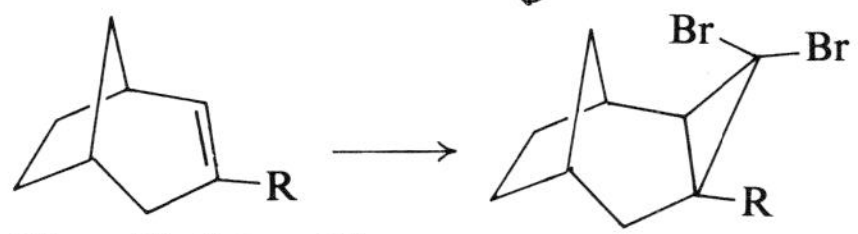 (R = H, OAc, Cl)	r.t.	7–10	TEBA	91; 72	48–94	[700, 955, 1153]
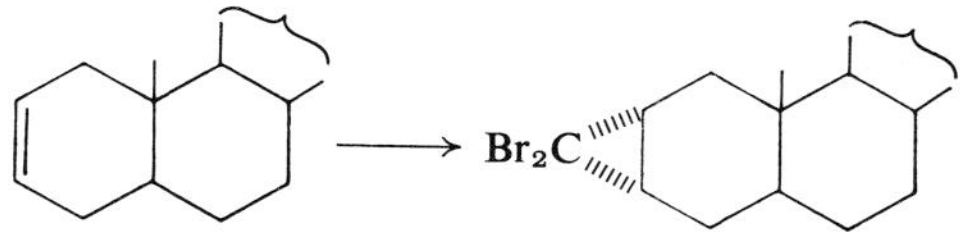	r.t.	≈10	TEBA/trace EtOH	18	64	[802]
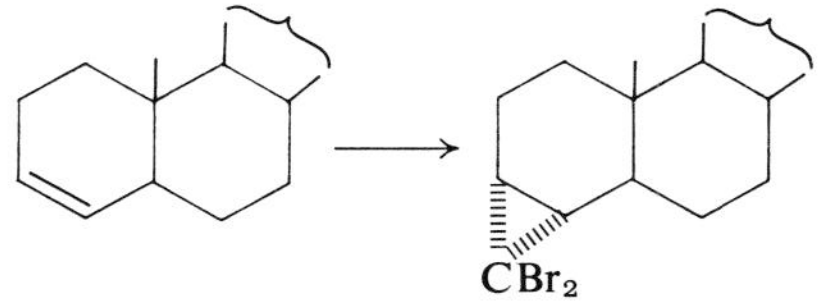	r.t.	≈10	TEBA/trace EtOH	18	57	[802]
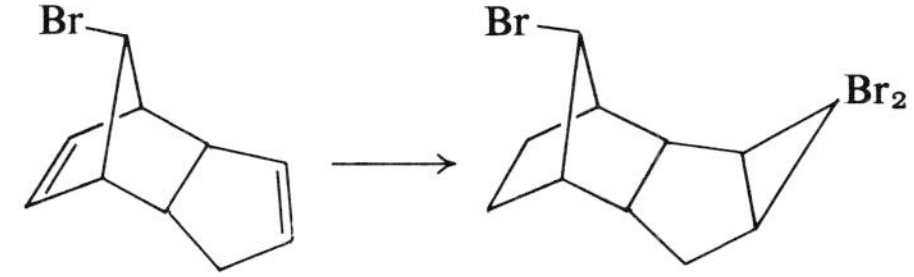	r.t.	>10	TEBA	48	55	[695, 968]

(continued)

Table 3–17. PTC Dibromocarbene Additions.

Olefine	Temper-ature	Moles $CHBr_3$/Moles Alkene	Catalyst/Spec. Cond.	Time (h)	Yield (%)	Reference
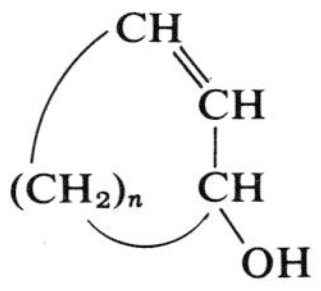 CH=CH–CH(OH)–$(CH_2)_n$ (ring)	r.t.	5	cetyltrimethylamm. bromide	≈12	25–63	[789, 790]
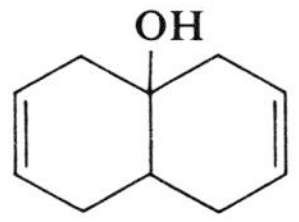 OH (bisadduct)	r.t.	5	TEBA		53	[788]
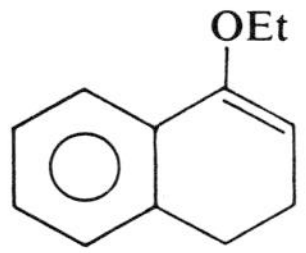 OEt	40 °C	2	TEBA	3	75	[722]
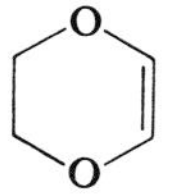 O, O	r.t.	2.5	TEBA	24	39	[658]
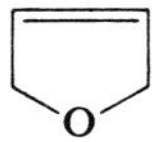 O	O °C	1	$C_{14}H_{29}NMe_3Cl$	7	17	[791]
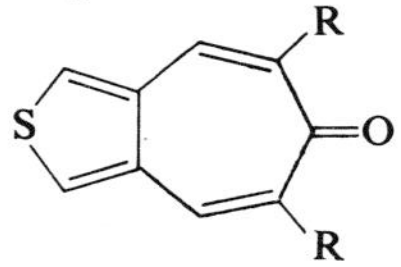 S, R, O, R	45 °C	4	TEBA/trace EtOH	30	40	[948]

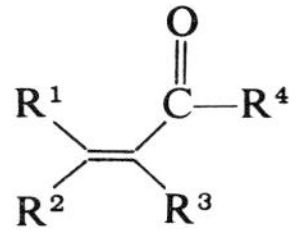 $R^1R^2C{=}CR^3{-}C({=}O){-}R^4$	r.t.	1.25	TEBA	2–24	20–90	[674, 796, 1151]
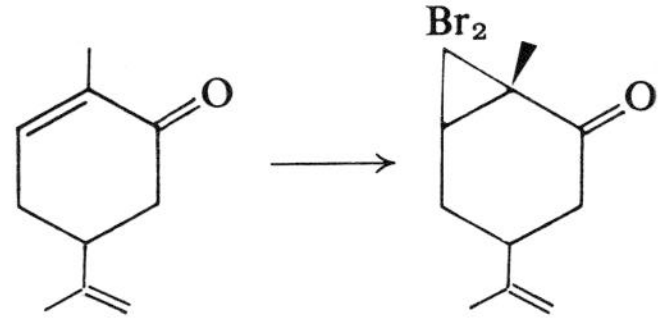 (carvone) → (Br_2 cyclopropane adduct)	r.t.	1.25	TEBA	2?	87	[796]
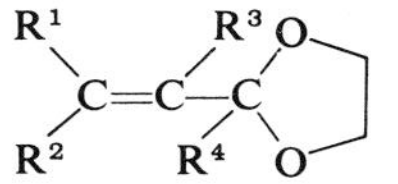 $R^1R^2C{=}C{-}CR^3R^4$(dioxolane)	r.t.	1.25	TEBA	2	30–60	[674, 1152]
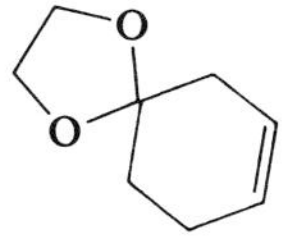 (spiro dioxolane cyclohexene)	no details given					[793]
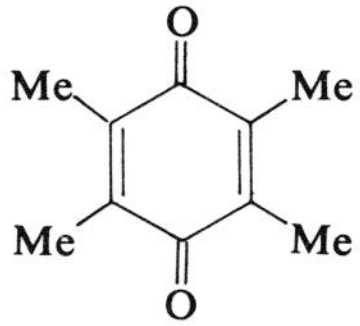 (tetramethyl-p-benzoquinone) (*anti* bisadduct)	45 °C	5.5	TEBA/trace EtOH/ CH_2Cl_2	4	57	[691]
$H_2C{=}C(R^1){-}COOR$			TEBA		up to 90	[795, 672]

(*continued*)

Table 3-17. PTC Dibromocarbene Additions.

Olefine	Temperature	Moles $CHBr_3$/ Moles Alkene	Catalyst/ Spec. Cond.	Time (h)	Yield (%)	Reference
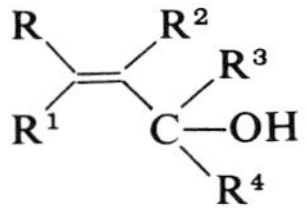 $RR^1C=CR^2-CR^3R^4-OH$	r.t.	2	TEBA	16	34–88	[502]
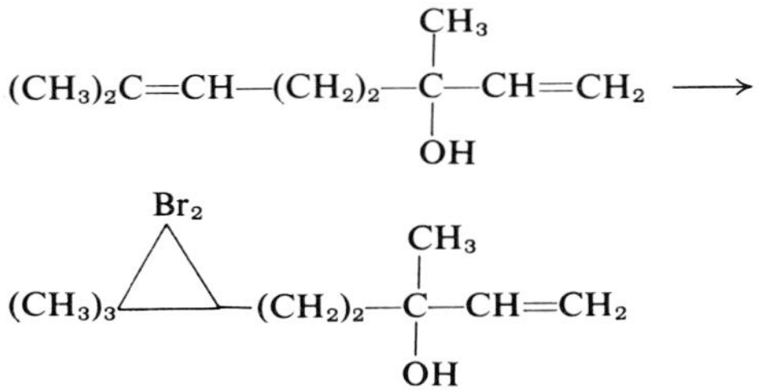 $(CH_3)_2C=CH-(CH_2)_2-C(CH_3)(OH)-CH=CH_2$ ⟶ $(CH_3)_2C(CBr_2)CH-(CH_2)_2-C(CH_3)(OH)-CH=CH_2$ (dibromocyclopropane)	r.t.	2	TEBA	16	93	[502]
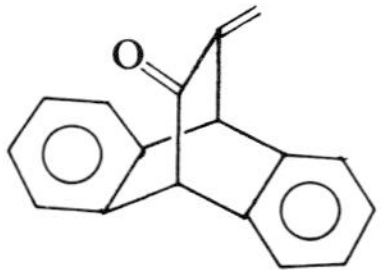 [structure: dibenzobicyclic ketone with exocyclic methylene]	7 °C	10.7	NBu_4Br	4	90	[867]
(c) Recent Additions						
allylic halides						[1131, 1154]
allylic ethers						[1155, 1156]
butadienes						[1157]
4-vinylcyclohexene						[1159]
bullvalene						[730]
various heterocycles						[987]

deactivate double bonds compared to related hydrocarbons [502].—On the other hand, saturated alcohols can be transformed into bromides with CBr_2, even in cases where conventional reagents ($SOBr_2$ and Ph_3PBr_2) fail [764]. This is analogous to the situation for CCl_2.

Insertion of CBr_2 into CH-bonds has not been investigated extensively, and the yields are often poorer than with CCl_2 [742]. With 2,5-dihydrofurane, the ratio of addition to insertion is 65:35 at 0 °C, and 40:60 at room temperature [791, 987]. With 2,3-dihydro-1,4-dioxane an insertion product is apparently formed thermally from the carbene adduct [658]. 2-Alkyl-1,3-dioxolanes exhibit insertion into the activated CH-bond (30% yield) [980, 1134].

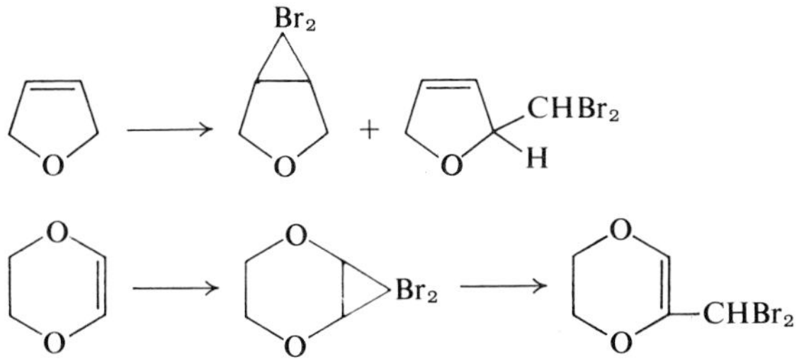

Br2
CHBr2
H
O O O
O O O
Br2
CHBr2
O O O

Scheme 3–160

With very unreactive substrates an unusual side reaction may occur: tribromomethylide anions can attack bromoform yielding carbon tetrabromide.

$$CBr_3^{\ominus} + HCBr_3 \longrightarrow CBr_4 + HCBr_2^{\ominus} \longrightarrow H_2CBr_2$$

In the presence of light and air, CBr_4 and $CHBr_3$ can add to the olefin in a radical reaction. Thus, with 3,3-dimethylbutene these three products are formed, the ratio depending on the reaction conditions [1143]:

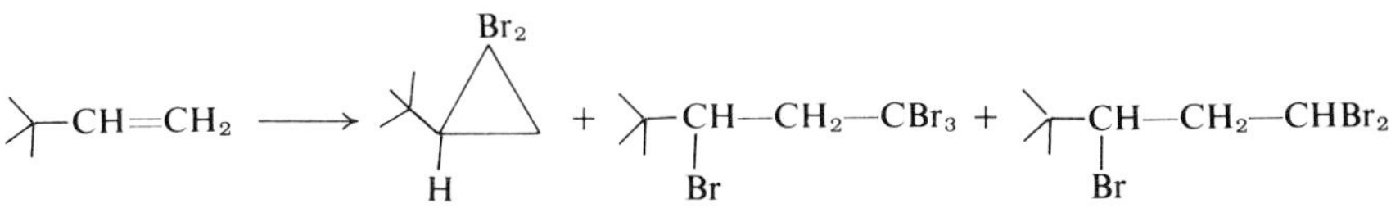

Br2
—CH=CH2 → + —CH—CH2—CBr3 + —CH—CH2—CHBr2
H Br Br

Scheme 3–161

This radical process can be avoided by working in the dark under nitrogen. Another unusual reaction seems to be dependent on the catalyst used: when allylic bromides are reacted in the presence of cetyltrimethylammonium bromide the normal CBr_2 adducts are formed in medium yields, whereas a substitution of bromide by tribromomethyl was the main reaction course with TEBA or tetrabutylammonium bromide as catalysts [1154, *cf.*, 1131]:

$$R_2C{=}C(R')\text{—}CH_2Br \longrightarrow R_2C{=}C(R')\text{—}CH_2\text{—}CBr_3$$

This reaction was extended to benzyl bromides recently [1205].

In considering new unexpected CBr_2 reactions, note should be taken of the possibility that triplet dibromocarbene might be in equilibrium with the singlet, as the energy gap between these species seems to be small [1158].

Contrary to CF_2, CFCl, and CCl_2, dibromocarbene and norbornadiene give four rearranged products, none of which arises from a linear cheletropic reaction. Apparently the steric requirements for a homo-1,4-addition are too severe [704]:

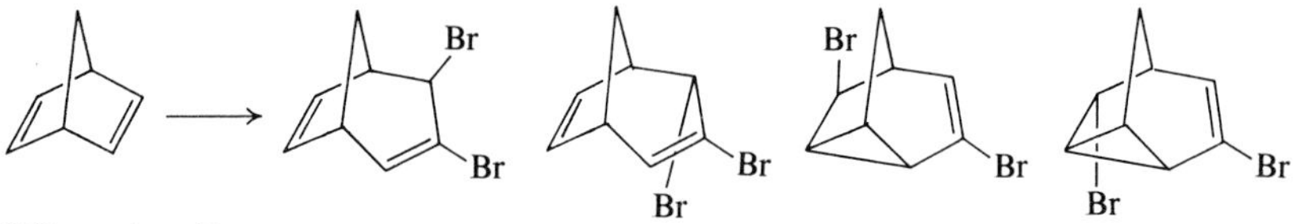

KO-*tert*-butyl/ $CHBr_3$	36%	26%	25%	13%
PTC-method	51%	22%	19.5%	7.5%

Scheme 3–162

Multiple addition of CBr_2 to compounds with several double bonds is possible, as in the case of CCl_2. Electronic and sterical effects disfavor these processes somewhat, however. Thus, cyclooctatetraene gives extremely low yields of tris- and tetra adducts in the direct process [730]. To get appreciable amounts of the desired products, the isolated mono- or bisadducts have to be rerun. After removing two of the bisadduct bromines the yields of fresh CBr_2 additions are dramatically improved [792] (*cf.*, Scheme 3–163, p. 227).

As in the case of chloroform/base, electron-poor, acceptor-substituted olefins add $CBr_3^{\ominus}$ instead of CBr_2. Thus, vinyl acetate, acrylonitrile and acrylic esters give the products shown [384, 768]:

$$R{-}CH{=}CH{-}O{-}CO{-}R' \longrightarrow R{-}CH_2{-}CH(CBr_3){-}O{-}CO{-}R'$$

$$H_2C{=}CH{-}X \longrightarrow Br_3C{-}CH_2{-}CH_2{-}X \quad (X = CN, COOR)$$

Compounds with α-methyl groups give cyclopropanes [616, 672, 674, 691, 795, 796]. Methacrylonitrile leads also to the dibromocarbene adduct only [384]:

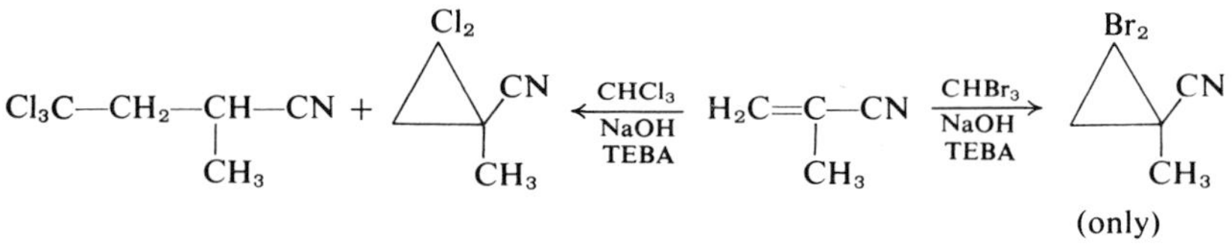

$$Cl_3C{-}CH_2{-}CH(CH_3){-}CN + \text{1,1-dichloro-2-methyl-2-cyanocyclopropane} \xleftarrow[\text{NaOH, TEBA}]{CHCl_3} H_2C{=}C(CH_3){-}CN \xrightarrow[\text{NaOH, TEBA}]{CHBr_3} \text{1,1-dibromo-2-methyl-2-cyanocyclopropane (only)}$$

Scheme 3–164

This is in contrast to the similar chloroform reaction, where both open chain and cyclic adducts were observed. Suitable precursors may give rise to tetrabromospiropentanes, a reaction discussed previously in connection with chloroform [644].

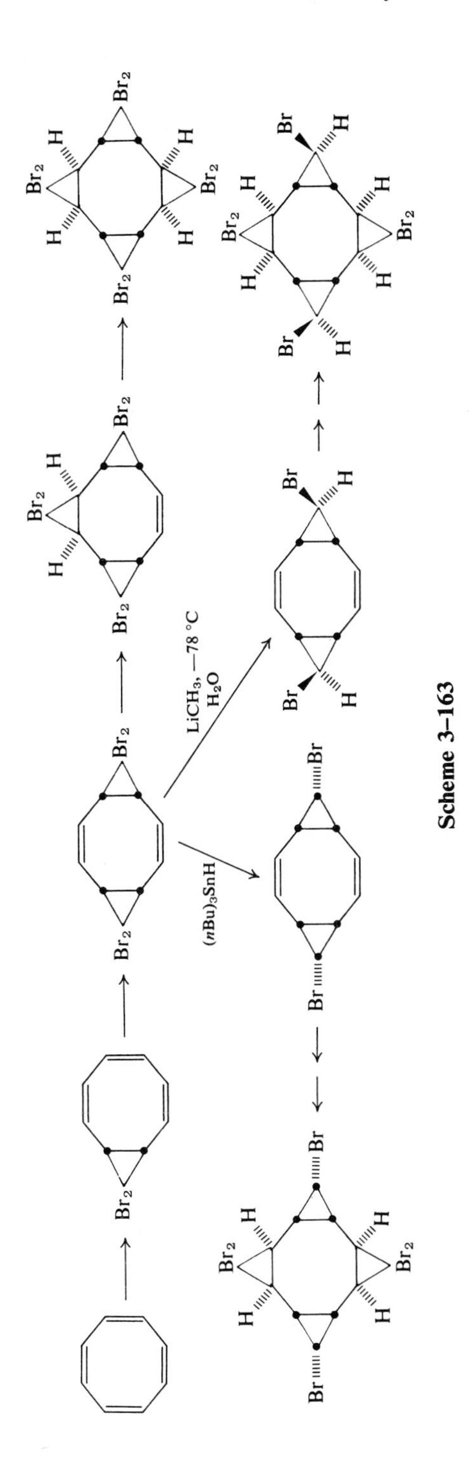

Scheme 3–163

An entirely different PTC method for preparing dibromocarbene adducts consists of the decomposition of sodium tribromoacetate in the presence of a quaternary ammonium salt in a nonpolar solvent. As detailed in the section on CCl_2, this procedure avoids the basic conditions of the Mąkosza process. Cyclohexene gave 78% dibromononorcarane [31].

3.3.13.4. Other Dihalocarbenes

All possible dihalo combinations of CX_2 can be prepared by PTC. For good results, however, the preparative details of the CCl_2 reactions cannot always be applied without modification. Firstly, some of the haloforms are rather expensive and cannot be used in a large excess, and secondly the reactivity of other carbenes differs from that of CCl_2, as already mentioned for CBr_2. The order of reactivities is well established (notwithstanding recent qualifications and improvements on the reactivity–selectivity principle [815, 1160]):

$$\underleftarrow{CH_2 \quad CBr_2 \quad CCl_2 \quad CFCl \quad CF_2} \text{ increasing electrophilicity}$$
$$\xrightarrow{\hspace{6cm}} \text{ increasing selectivity}$$

From the now classical pioneering work of Hine on base-induced α-eliminations, a few facts are brought to attention here (surveys: *cf.*, [782, 803]):

(i) Rates of carbanion formation ($HCXYZ \rightarrow CXYZ^{\ominus}$) are $I \approx Br > Cl > F$. This step is much faster than the subsequent ones. pK_a values are: $CHBr_3$ 9, $CHCl_3$ 15, CHF_3 26.5 [1161].
(ii) The relative ease with which a halogen anion separates from the intermediate trihalocarbanion ($CXYZ^{\ominus} \rightarrow CXY + Z^{\ominus}$) is $Br \geq I > Cl \gg F$.
(iii) The ability of the halogen to stabilize CXY is $F \gg Cl > Br > I$. The most stable carbene therefore is the most selective one.
(iv) Halogen exchange with external halogenide is also possible:

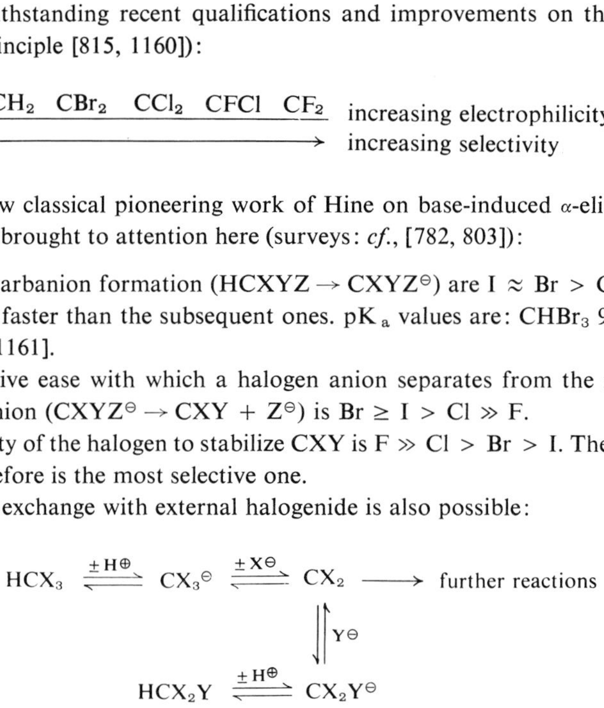

$$HCX_3 \underset{}{\overset{\pm H^{\oplus}}{\rightleftharpoons}} CX_3^{\ominus} \overset{\pm X^{\ominus}}{\rightleftharpoons} CX_2 \longrightarrow \text{further reactions}$$
$$CX_2 \overset{Y^{\ominus}}{\rightleftharpoons} CX_2Y^{\ominus}$$
$$HCX_2Y \overset{\pm H^{\oplus}}{\rightleftharpoons} CX_2Y^{\ominus}$$

Scheme 3–165

Bromochlorocarbene and Halogenide Exchange Processes. Halogen exchange was mentioned as a possibility above. For mixed haloforms in theory, the processes of Scheme 3–166 are conceivable.

If these processes are fast, then even a foreign inorganic halide in the medium (for instance one brought in with the catalyst) can exchange. Experimentation partly confirmed these expectations: When using aqueous sodium hydroxide saturated with sodium chloride and TEBA for a dibromocarbene addition to cyclohexene, the

$HCX_2Y \rightleftharpoons CX_2Y^{\ominus} \rightleftharpoons Y^{\ominus} + :CX_2 \longrightarrow X_2$-cyclopropanes

$CX_2Y^{\ominus} \rightleftharpoons X^{\ominus} + :CXY \longrightarrow XY$-cyclopropanes

$CX_2Y^{\ominus} \longrightarrow$ open chain final products $\longleftarrow CXY_2^{\ominus}$

$CXY_2^{\ominus} \rightleftharpoons Y^{\ominus} + :CXY \longrightarrow XY$-cyclopropanes

$HCXY_2 \rightleftharpoons CXY_2^{\ominus} \rightleftharpoons X^{\ominus} + :CY_2 \longrightarrow Y_2$-cyclopropanes

Scheme 3–166

product contained 3% bromochloronorcarane **(B)** formed in an exchange reaction. The reverse reaction with sodium bromide saturated sodium hydroxide in a dichlorocarbene reaction did not show any exchange. As mentioned before, bromide is eliminated more easily than chloride, and the small amount of the anion $CBrCl_2^{\ominus}$ probably formed leads to CCl_2 again. Local concentration of foreign halide can only be very small because TEBA concentration is low. A control experiment showed that no nucleophilic halide exchange occurs in the absence of sodium hydroxide with a PT catalyst [384]. Higher local concentrations of various halides and possibly better chances for exchange processes can be expected from mixed haloforms. Indeed, in spite of bromide being the better leaving group, both $CHBrCl_2$ and $CHBr_2Cl$ give all three possible combinations **A**–**C** with cyclohexene and TEBA [384]. Dichlorofluoromethane, on the other hand, leads only to chlorofluoro carbene adducts. With

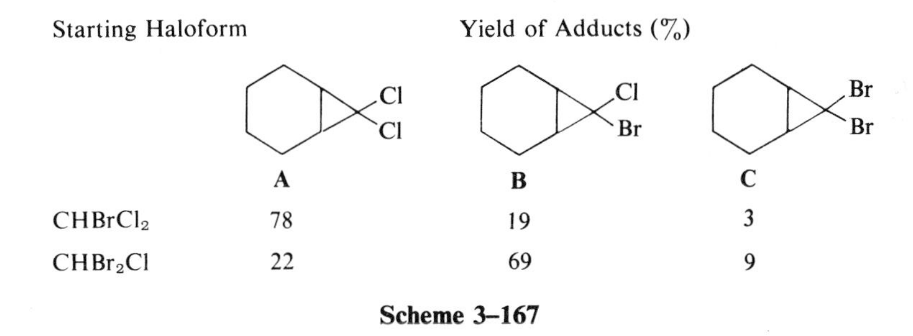

Starting Haloform	Yield of Adducts (%)		
	A (Cl, Cl)	**B** (Cl, Br)	**C** (Br, Br)
$CHBrCl_2$	78	19	3
$CHBr_2Cl$	22	69	9

Scheme 3–167

$CHBr_2Cl$, the extent of exchange and combined yields are strongly dependent on the nucleophilicity and the steric accessibility of the double bond (see Table 3–18) [989].

Insertion into C—H bonds by the species derived from $CHBr_2Cl$ in the presence of TEBA gave these results [989]:

	Yield of Insertion Products (%)			
	Combined	CCl_2	CBrCl	CBr_2
cumene	7.0	2.0	1.3	1.3
adamantane	15.6	3.7	7.2	4.7

Table 3–18. Reaction Products of $CHBr_2Cl$ with Various Alkenes under Standardized Conditions [989] (Catalyst: TEBA).

	Yields of Adducts			
Alkene	Combined	CCl_2	CBrCl	CBr_2
3,3-dimethyl-1-butene	8.1	4.9	2.8	0.4
1-hexene	10.5	3.1	5.7	1.7
cis-2-butene	27.1	8.9	15.7	2.4
isobutene	54.3	17.9	33.7	2.7
cyclohexene	54.6	13.1	32.2	9.2
2-methyl-2-butene	62.6	25.7	33.2	3.8

Preparatively then, the production of bromochlorocyclopropanes is not a straightforward process under normal PTC conditions, because the resulting mixtures of adducts are often difficult to separate. Interestingly, however, Fedoryński found that the outcome of an alkene/$CHBr_2Cl$/NaOH reaction with crown ether as catalyst depended on the type of crown. With 18-crown-6 or 15-crown-5 a mixture of the three dihalocyclopropanes was always obtained, although the selectivity was usually higher than that in reactions with ammonium salt catalysts. Dibenzo-18-crown-6 had the singular property of leading only to bromochlorocyclopropanes [835]: Molar amounts of alkene and $CHBr_2Cl$, 2 mole-% dibenzo-18-crown-6, and 50% NaOH were stirred vigorously at 45 °C for 4 hours; 43–76% yields [835]. Control experiments, however, revealed that: (a) dibenzo-18-crown-6 catalyzes halide exchange among haloforms in the absence of alkenes; (b) even with dibenzo-18-crown-6 pure CBrCl adducts are obtained only within a relatively narrow range of experimental conditions; and (c) the relative rates of exchange and addition in the presence of olefins depend on the nature of the alkene. Sometimes pure CBrCl adducts are formed, and in other cases extensive scrambling of halides is found [989, 1012].

Reasons for the specific effect of dibenzo-18-crown-6 are not yet clear. More understandable is the fact that potassium *tert*-butoxide/CHBrCl gives only CBrCl adducts (*cf.*, for instance [842]), since under these conditions carbene formation is virtually irreversible. The same is true for the bromochlorocyclopropane synthesis from Seyferth's phenylmercuric dibromochloromethylide, $PhHgCBr_2Cl$ [804].

PTC exchange processes occur even in the competing generation of carbenes from equal molar amounts of bromoform and chloroform: reaction for instance with styrene in the presence of sodium hydroxide/TEBA under specific conditions gave 78% CBr_2-adduct, 15% CCl_2-adduct and 7% CBrCl-adduct [384]. In the absence of a carbene acceptor this exchange process can be utilized to prepare mixed haloforms from chloroform and bromoform. Under suitable conditions hydrolysis did not exceed 16%. Reacting 2 moles each of $CHCl_3$ and $CHBr_3$, 3 g TEBA, and 200 ml 50% NaOH at 35 °C for 1–2 hours (cooling is necessary in the beginning) and distilling through an efficient column, 0.66 mole $CHBrCl_2$ and 1.08 moles $CHBr_2Cl$ are obtained along with starting material and residue [836]. Working with bromoform/

NaOH saturated with NaCl/TEBA or with chloroform/NaOH saturated with NaBr/TEBA at 150 °C for up to 24 hours, only about 10% yield each of $CHBr_2Cl$ and $CHBrCl_2$ were found [843].

Conventionally these exchanges are performed using aluminium trichloride catalysis. Recently ethylene oxide/tetraalkylammonium halide was employed for this purpose [844] in a non-PTC reaction.

Fluorine Containing Carbenes: CF_2, CClF, CBrF, CFI. All attempts to generate CF_2 either from $CHClF_2$/aqueous NaOH/catalyst [808, 845] or from NaO_2CCF_3/catalyst [31] by PTC methods proved futile. But with $HCClF_2/CH_2Cl_2$/solid NaOH/catalyst low yields of CF_2 adducts ($\approx 10\%$) could be obtained recently [988]. CClF, however, was easy to work with. The adducts are valuable intermediates for further transformations into fluoroalkenes, fluorodienes, and fluoroalkylalkohols. Schlosser *et al.* [805–807, 834] used concentrated potassium hydroxide as base and either TEBA,

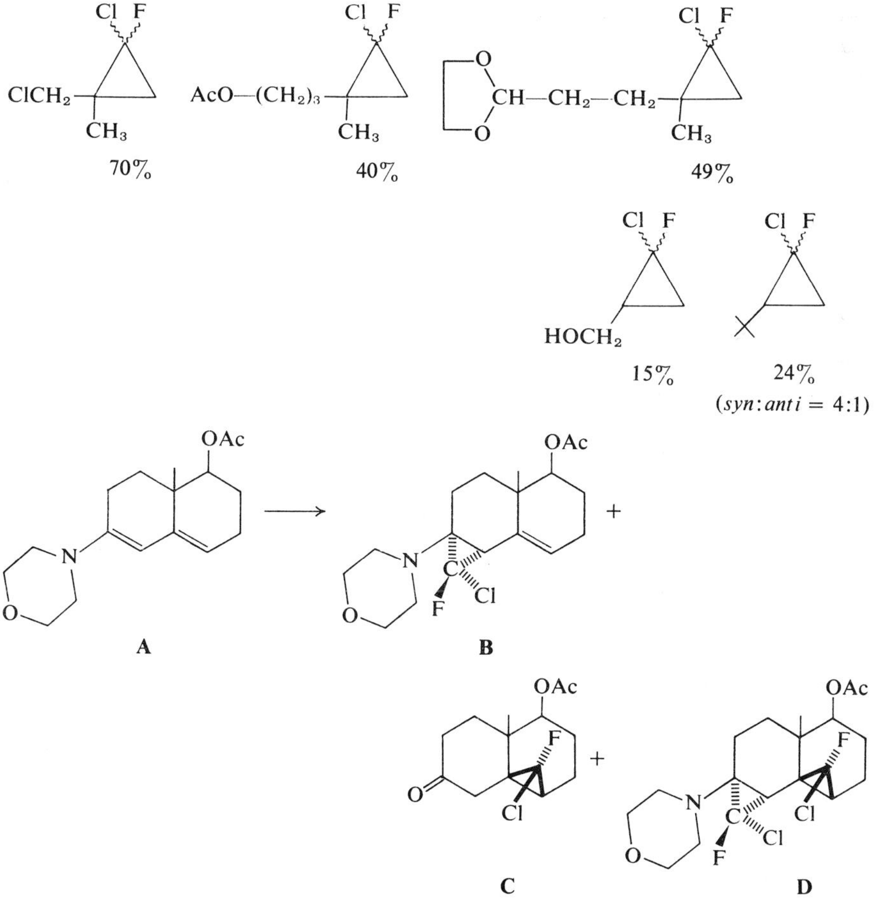

Scheme 3–168 A

dicyclohexano-18-crown-6 or 18-crown-6 as catalyst. The solvent was an excess of $HCCl_2F$, and the reactions proceeded at 0 °C [condenser cooled to −30 °C) with vibromixing in 0.5–2 hours. Many different olefins were reacted (*cf.*, [1162, 1163]). Even products containing critical side groups (allylic halogen, ester, acetal, hydroxyl, see below) were converted, and yields were between 30 and 70%. Weyerstahl *et al.* recommend stirring at 0–20 °C for 2–3 days [808]. Their preferred catalyst was TEBA-bromide. With this catalyst in sodium hydroxide, compound **A** gives (depending on the excess of $HCCl_2F$) only monoadducts **B** and **C** or bisadduct **D** after 2 days [666].

While stable adducts of CClF were formed from 2,5-dihydrofurane and dihydropyrane, one of the 2,3-dihydrofurane adducts and both of the 2,3-dimethylindole adducts underwent further conversions [987]:

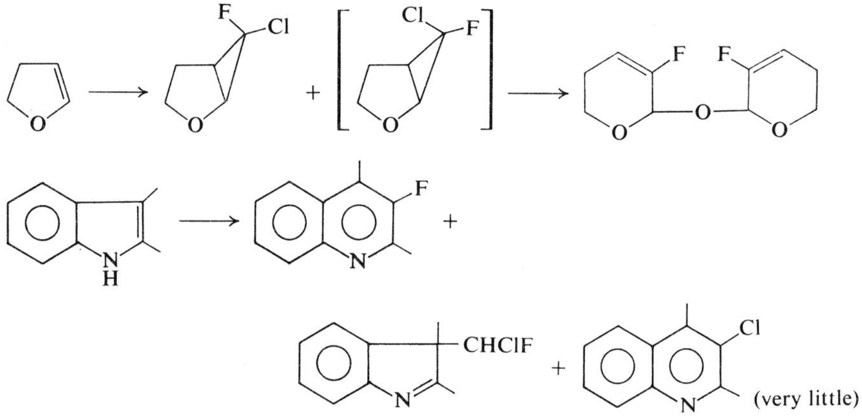

F Cl Cl F F F O O O O O O N H N F N CHClF N Cl (very little)

Scheme 3–168 B

1-Bromo-1-fluorocyclopropanes obtained with CBrF are interesting intermediates for further selective conversions. In Friedel-Crafts reactions the fluorine atom is lost [811, 801], nucleophilic attack yields open chain fluoro compounds, and exchange of bromide gives fluoro-substituted cyclopropanes [812]. The preparation is carried out with an excess of $HCBr_2F$ or, even better, with methylene chloride as cosolvent and TEBA as catalyst. Excess concentrated sodium hydroxide is dropped in at such a rate that the mixture slowly comes to the boil. It is further refluxed for 3 hours and then stirred 1.5 hours. Yields are between 51% (3,3-dimethyl-1-butene) and 88% (α-methylstyrene) [808, 811, 812, 1164]. Dibenzo-18-crown-6 was used as catalyst also [959]. syn/anti-Ratios of adducts to arylethenes varied widely with different substituents on the aromatic ring [1164].

CFI adducts have been formed from $HCFI_2$ in methylene chloride with TEBA/concentrated sodium hydroxide by stirring for 4 hours at 20 °C. The haloform was made from $HCBr_2F$ and sodium iodide in acetone (7 days, 100 °C). As this is the less easily accessible reaction partner, the olefins were used in excess. With styrene,

60% conversion (based on $HCFI_2$) was achieved but yields were not quite so good for the adducts of α-methylstyrene, cyclohexene, 2-methyl-1-butene, 3,3-dimethyl-1-butene [809] and diphenylethene [810].

Diiodocarbene and Chloroiodocarbene. The cyclopropane derivatives of these carbenes can be made from iodoform or chlorodiiodomethane, TEBA, and concentrated aqueous sodium hydroxide with methylene chloride as cosolvent at 60 °C (stirring for 3–5 hours) [810, 813]. The crude products are rather unstable, and some have caused explosions on attempted distillation. The instability is probably due to very sensitive by-products formed in the alkaline medium by oxidative processes which possibly involve iodates. These are believed to initiate decomposition. Even the "purified" alkyl diiodocyclopropanes made by the PTC method decompose rather fast in the pure state as well as in solution. The aryl compounds and the chloro-iodo cyclopropanes are somewhat more stable [810]. It must be mentioned that the same chemicals are much more stable if made by an alternative method without using aqueous alkali hydroxide [814]. The yields of diiodocyclopropanes were generally low (2–20%) except for the *p*-chlorophenyl derivative where 59% was obtained [813]. Homogeneous chloro-iodo cyclopropanes or mixtures of *syn/anti* isomers respectively were prepared in yields between 50–65%. Even 3,3-dimethyl-1-butene gave a *syn/anti* mixture [810]. It was with some surprise that Weyerstahl observed the PTC effect of TEBA in the formation of CFI, CClI, and CI_2 [809, 810, 813], because "catalyst poisoning" by iodide ion was known from some PTC substitution and alkylation reactions. The authors believe that the cosolvent methylene chloride circumvents this difficulty. An alternative explanation is found in the relative lipophilicities exhibited by the trihalomethylide/iodide pair. It was shown that the generation of CCl_2 is slowed down by iodide [32], *i.e.*, anion exchange of cation $NR_4^{\oplus}$ at the interphase ($I^{\ominus} \rightleftharpoons$ anchored $CCl_3^{\ominus}$) is not as favorable as ($Cl^{\ominus} \rightleftharpoons$ anchored $CCl_3^{\ominus}$). Iodine-carrying methylide anions, however, are much more lipophilic than $CCl_3^{\ominus}$ and should exchange easily, even with iodide.

3.3.13.5. Other Mixed Carbenes

Monohalocarbenes cannot be synthesized by phase transfer methods even at elevated temperatures. Concentrated sodium hydroxide/catalyst only causes a slow hydrolysis of the methylene dihalides. Similarly, attempts at trapping arylhalocarbenes and styrylhalocarbenes from precursors **A** and **B** were futile [816]. In all these cases it appears that rates of deprotonation cannot compete with rates of hydrolysis.

$Ar{-}CHCl_2$	$C_6H_5{-}CH{=}CH{-}CHCl_2$	$C_6H_5SCHCl_2$
A	**B**	**C**
$C_6H_5{-}S{-}CH_2Cl$	$CH_3S{-}CHCl_2$	$RS(Cl)C{=}C(Cl)SR$
D	**E**	**F**

Chloromercaptocarbenes. Experiments on chloro(phenylthio)carbene, generated and trapped in an exothermic reaction of olefin/**C**-mixtures with 50% aqueous sodium hydroxide and catalytic amounts of TEBA were more successful. Yields of adducts

Table 3–19. Preparation of Phenylthiocyclopropanes from C_6H_5—S—CH_2Cl and Olefins in Methylene Chloride with TEBA/NaOH.

Substrate	Yield of **G** (%)	*exo/endo* or *cis/trans* Ratio
trans-stilbene	78	—
styrene	70	9.0
cyclohexene	67	1.8
cis-2-butene	65	5.3
trans-2-butene	79	—
ethyl vinyl ether	60	5.5

ranged from 20% (1-hexene) to 63% (styrene). In addition *ca.* 10% of the dimerization product **F** (R = C_6H_5) was formed [817]. Phenylthiocyclopropanes (**G**) were likewise prepared from **D** and olefin in methylene chloride with TEBA/sodium hydroxide by vibro-mixing for 4 hours (Table 3–19) [818]:

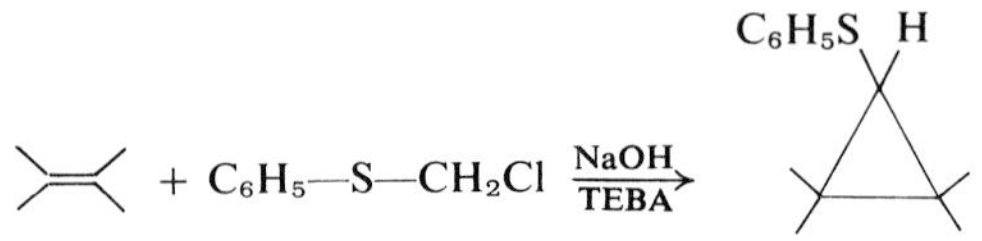

$$\text{olefin} + C_6H_5\text{—S—}CH_2Cl \xrightarrow[\text{TEBA}]{\text{NaOH}} \text{cyclopropane}(C_6H_5S, H)$$

D **G**

Scheme 3–169

One would expect that chloro(methylthio)carbene is less electrophilic than chloro(phenylthio)carbene. Moss and Pilkiewicz [819] trapped this species with a number of olefins in low to moderate yield when working with **E** and potassium *tert*-butoxide as base. The unwanted side product, formed in higher yield, was the formal carbene dimer **F** (R = CH_3) (*cf.*, Scheme 3–169). It was shown in a single experiment with tetramethylethylene, that a higher adduct yield and a lower conversion to **F** was possible with the PTC technique (conc. NaOH/TEBA, 50 °C, 3 hours) [819]. Chloro(methylthio)carbene was shown to be a free carbene, not a carbenoid [846].

Alkylidene Carbenes. M. S. Newman *et al.* have extensively researched the alkaline decomposition of N-nitrosooxazolidones **H**. In the presence of olefins, the products **I** were obtained. These probably arise from alkylidenecarbenes [820, 821]. Besides these, however, in the presence of methanol compounds of the type **J** and **K** are formed. They later found that adducts of unsaturated carbenes were obtained in better yields from open chain N-[2-hydroxyalkyl]-N-nitrosoacetamides **L** using Aliquat 336 together with aqueous sodium hydroxide as base [821, 822]. This reaction favors the formation of compounds **I** relative to the competing rearrangement giving **J**. An exception is **L** (R = C_6H_5, R′ = cyclo-C_3H_5), which yields only **J** (R = C_6H_5, R′ = cyclo-C_3H_7) [821].

In a more recent study, isopropylidenecarbene was prepared from dimethylvinyl trifluoromethanesulfonate and potassium *tert*-butoxide both with and without

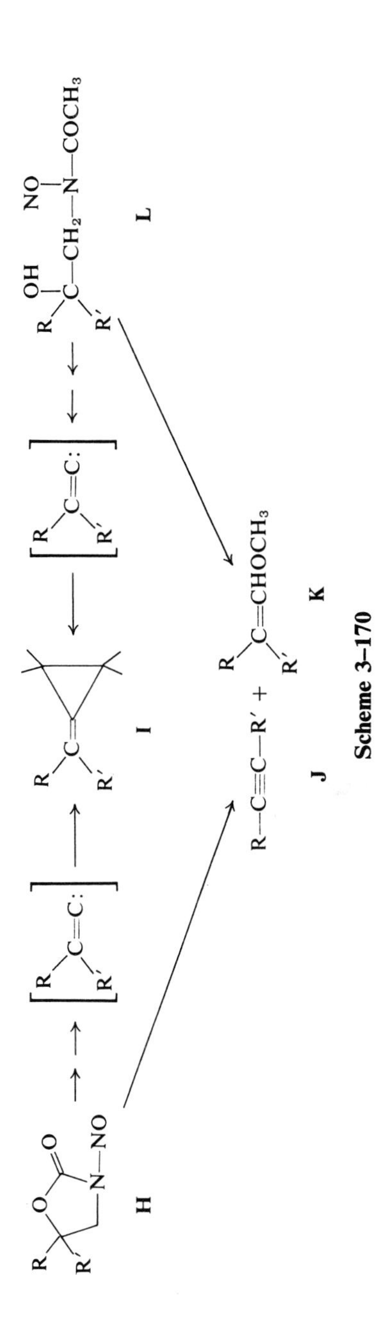

Scheme 3–170

18-crown-6. Its electrophilicity was established by means of competition reactions with substituted styrenes. A Hammett correlation gave $\rho = -0.75$, whether or not the crown was present. This was taken as evidence for the intermediacy of free carbene [823].

Similar experiments with halophenylcarbenes showed that carbene and carbenoid have different selectivities and that these can be distinguished by using crown ethers [612]. The isopropylidene-carbene generated from oxazolidone **H** has a ρ value of -3.4 [820], indicating that this species is not a free carbene. The species generated by PTC from **L** has not been compared directly, but from the ρ values it can be assumed that it is "free." Using the PTC technique the compounds shown below were formed in yields of between 35 and 80% [824].

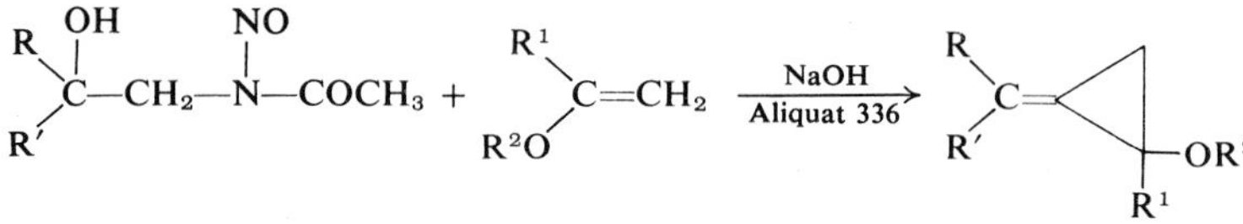

$$RR'C(OH)-CH_2-N(NO)-COCH_3 + R^1(R^2O)C=CH_2 \xrightarrow[\text{Aliquat 336}]{\text{NaOH}} RR'C=C(cyclopropane)(R^1)(OR^2)$$

L

$R/R' = (CH_2)_5;\ (CH_2)_4;\ CH_3, CH_3$
$R^1 = CH_3, Ph, OC_2H_5, H$
$R^2 = CH_3, C_2H_5$

Scheme 3–171

Ring opening with mercuric acetate and treatment with hydrogen sulfide or KI_3, transforms these products into β,γ-unsatured ethers or their β-iodine derivatives in high yield. Cyclohexylidenecarbene is generated at -10 to -5 °C from **L** [R—R′ = $(CH_2)_5$] (*cf.*, Schemes 3–171 or 3–173) in olefinic solution with Aliquat 336 by the slow addition of aqueous sodium hydroxide. It adds to 1,4-cyclohexadiene and norbornadiene to give **M** and **N**. With cyclopentadiene, however, instead of the

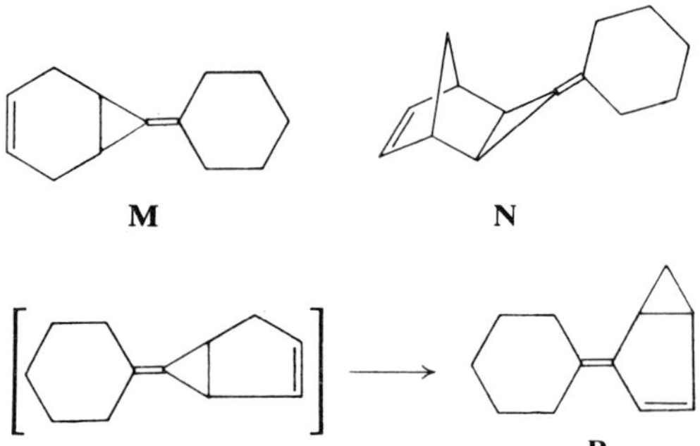

M **N**

[**O**] → **P**

Scheme 3–172

expected **O**, **P** is obtained in 76% yield [825]. Rearrangement occurs even at 30 °C. The same PTC reaction with **L** [R—R′ = $(CH_2)_5$] in the presence of nucleophiles gave compounds **Q** in 40–60%. Similarly, **L** and aldehydes RCHO or ketones R—CO—CH_2—R′ gave compounds **S** or **T**, respectively, in rather low yields. It has not been established whether or not products **Q**, **S**, and **T** were formed *via* a carbene [826].

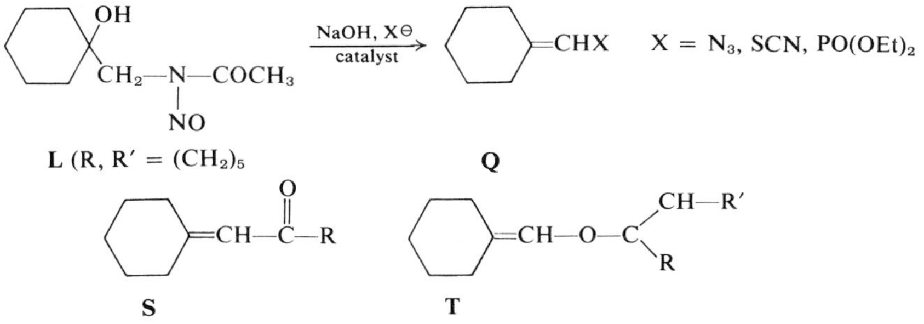

$$\text{L (R, R}' = (CH_2)_5\text{)}: \; \text{cyclohexane(OH)}(CH_2\text{—N(NO)—}COCH_3) \xrightarrow[\text{catalyst}]{\text{NaOH, X}^{\ominus}} \text{cyclohexylidene}{=}CHX \;\; (\textbf{Q}) \qquad X = N_3, SCN, PO(OEt)_2$$

$$\text{cyclohexylidene}{=}CH\text{—}C(=O)\text{—}R \;\; (\textbf{S}) \qquad \text{cyclohexylidene}{=}CH\text{—}O\text{—}C(=CH\text{—}R')\text{—}R \;\; (\textbf{T})$$

Scheme 3–173

The most convenient and mildest alkylidenecarbene generation published to date consists of the reaction between (trimethylsilyl)vinyl trifluoromethanesulfonates and fluoride ion at −20 °C to 0 °C in 1–2 hours. Quantitative yields of cyclohexene or ethyl vinyl ether adducts with potassium fluoride/18-crown-6, anhydrous $NR_4^{\oplus}F^{\ominus}$, or Aliquat 336/potassium fluoride were obtained by the following reaction sequence [901]:

$$Me_2C{=}C(SiMe_3)OSO_2CF_3 + F^{\ominus} \longrightarrow Me_2C{=}C\!: + FSiMe_3 + {}^{\ominus}OSO_2CF_3$$

Isopropylidenecarbene so generated could be trapped with isonitriles also, and the final products of this PTC process were substituted acrylamides in 35–52% yield [1165]:

$$Me_2C{=}C\!: + \,:C{=}N\text{—}R \longrightarrow [Me_2C{=}C{=}C{=}N\text{—}R] \longrightarrow Me_2C{=}CH\text{—}CO\text{—}NHR$$

α-Elimination from 1-bromovinyl compounds results in rearrangement normally, but ω-bromomethyleneadamantane can be converted into intermediate adamantylidene carbene with potassium *tert*-butoxide and 18-crown-6. Olefin adducts of this carbene were obtained in yields up to 97% [1166].

Alkenylidenecarbenes. Adducts **U** of alkenylidenecarbenes have been prepared by the PTC method from either propargyl halides **V** or allyl halides **W** (Sheme 3–174).

A great number of olefins were transformed into vinylidenecyclopropanes. Table 3–20 gives a selection of yields obtained from **V** and **W**. Yields were better than with older methods mostly, and the PTC generated species seems to be a free carbene [970].

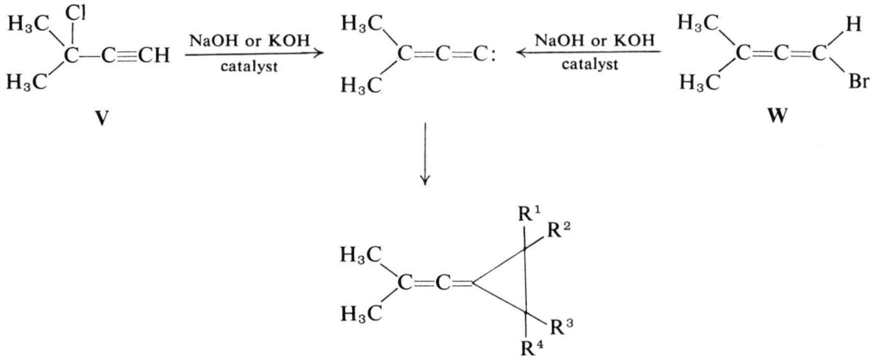

Scheme 3–174

Reaction variables were investigated to some extent. Thus, starting from **V** with TEBA as catalyst: room temperature is better than 45 °C, concentrated potassium hydroxide gave better yields than more dilute solutions, benzene proved to be a better organic phase than hexane, and *ca.* 10 hours seemed to be the optimum reaction time required [828]. Under certain conditions some crown ethers gave as good or much better yields [829]. It was found, however, that the yields in 18-crown-6-catalyzed reactions decreased more or less sharply after having reached a maximum value with increasing reaction time [829]. This effect was most pronounced at 45 °C where 85% maximum yield was observed after 2 hours, and only 53% after 10 hours. Low boiling (*e.g.*, butadiene) or hydrophilic substrates can be reacted with solid potassium hydroxide/crown ether/at temperatures as low as −78 °C [829].

Starting from **W**, Aliquat 336 [830] and tetrabutylammonium iodide [831] were used as catalysts with NaOH. For the norbornadiene reaction with the latter catalyst, yields and workup were comparable with other ammonium salts [831]. Here 6 days at 60 °C are required for maximum yields.

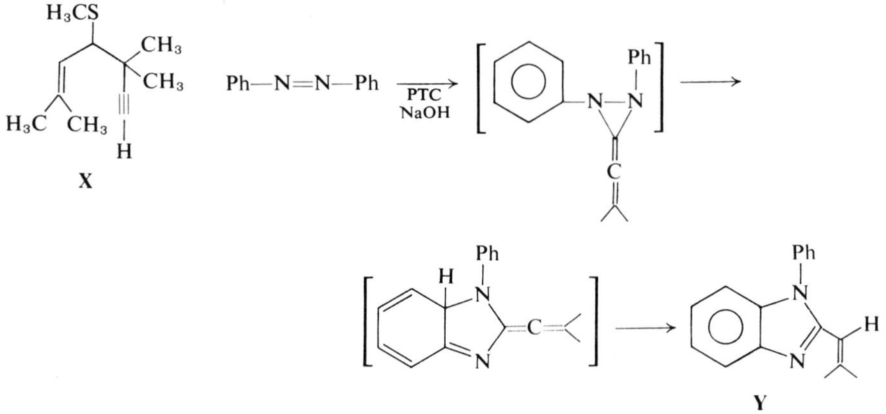

Scheme 3–175

Table 3–20. Selected Typical Olefin Transformations into (Dimethylvinylidene) cyclopropanes.

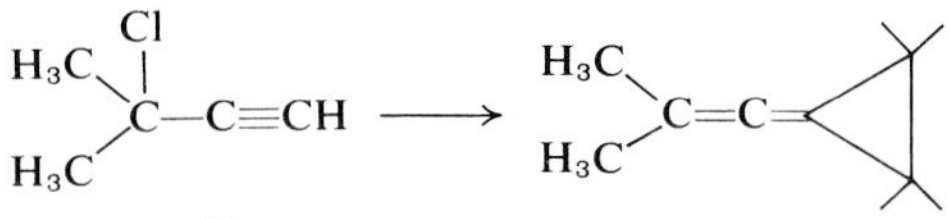

$$(H_3C)_2C(Cl)-C\equiv CH \longrightarrow (H_3C)_2C{=}C{=}\text{(cyclopropylidene)}$$

V

Olefin	Yield (%)	Catalyst	Remarks	Reference
styrene	81	Aliquat 336	NaOH/pentane	[827]
	60	TEBA	KOH/benzene	[828]
	69	dicyclohexano-18-crown-6	KOH/benzene	[829]
isobutoxyvinyl ether	48	Aliquat 336	NaOH/pentane	[827]
cyclopentene	27	Aliquat 336	NaOH/pentane	[827]
cyclohexene	37	Aliquat 336	NaOH/pentane	[827]
3-methylbut-2-en-1-ol	16	Aliquat 336	NaOH/pentane	[827]
2,3-dimethyl-2-butene	87	Aliquat 336	NaOH/pentane	[827]
α-pinene	20	TEBA	KOH/benzene	[828]
β-pinene	31	TEBA	KOH/benzene	[828]
1,5-cyclooctadiene	19	TEBA	KOH/benzene	[828]
cinnamyl alcohol	21	TEBA	KOH/benzene	[828]
isoprene	26	TEBA	KOH/benzene	[828]
	38	dicyclohexano-18-crown-6	KOH/benzene	[829]
indene	16	TEBA	KOH/benzene	[828]
β-methylstyrene	36	TEBA	KOH/benzene	[828]
dihydropyran	38	dicyclohexano-18-crown-6	KOH/benzene	[829]
norbornadiene	18	dicyclohexano-18-crown-6	KOH/benzene	[829]

$$(H_3C)_2C{=}C{=}C(H)Br \longrightarrow (H_3C)_2C{=}C{=}\text{(cyclopropylidene)}$$

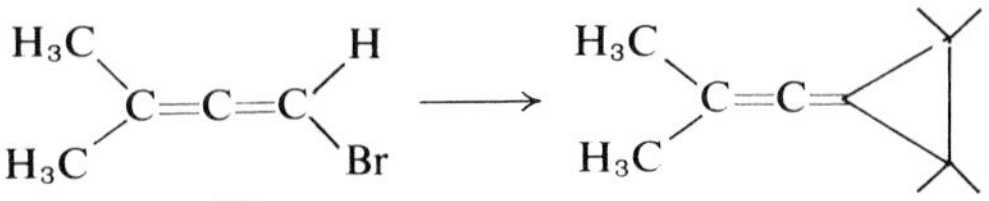

W

Olefin	Yield (%)	Catalyst	Remarks	Reference
2,3-dimethyl-2-butene	68	Aliquat 336	16 hours	[830]
cyclohexene	61	Aliquat 336	2 days	[830]
1-hexene	25	Aliquat 336	3 days	[830]
norbornadiene	21	NBu_4J	6 days	[831]

Methyl(dimethyl)allyl sulfide and **V** gave the alkylation product **X** instead of the carbene adduct [827]. Azobenzene gave the benzimidazole derivative **Y**. A probable mechanism for this conversion is presented [832] (Scheme 3–175, page 238).

Even more highly unsaturated carbenes **Z** were trapped in rather low yields by cyclohexene [833].

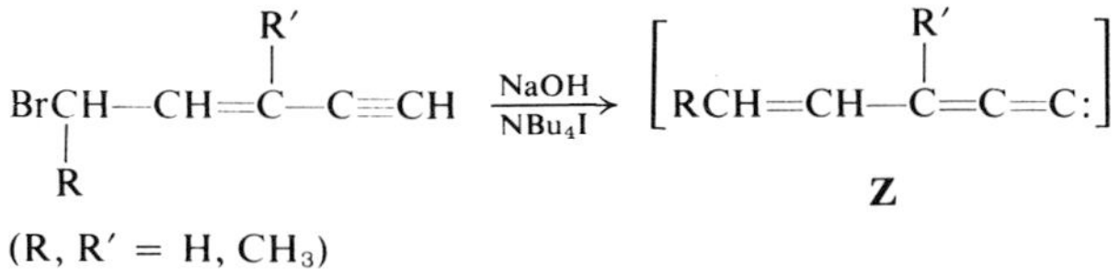

$$BrCH(R)-CH{=}C(R')-C\equiv CH \xrightarrow[NBu_4I]{NaOH} \left[RCH{=}CH-C(R'){=}C{=}C{:}\right]$$

Z

(R, R′ = H, CH_3)

Scheme 3–176

Typical experimental procedures are presented below.

(a) *Dimethylvinylidenecyclopropane derivatives from* 3-*chloro*-3-*methyl*-1-*butyne* (**V**) [828]: 30 mmoles olefin and 4 ml benzene are stirred vigorously with 150 mg TEBA and 50 ml 50% potassium hydroxide at room temperature under N_2. Ten mmoles **V** in 6 ml benzene are added slowly over a period of 2.5 hours. After the addition is completed, stirring is continued for a further 10–13 hours. Dilution with water and extraction with ether is followed by normal workup procedures.
(b) *Dimethylvinylidenecyclopropane derivatives from* 1-*bromo*-3-*methyl*-1,2-*butadiene* (**W**) [830]: 20 ml olefin, 7 ml concentrated aqueous sodium hydroxide, 1 g Aliquat 336 and 3 g **W** are stirred vigorously at 0–60 °C for up to 3 days. Workup as before.

3.4. Reduction Reactions

3.4.1. Complex Hydrides and Diborane

Of the more common complex hydrides, lithium aluminum hydride appears a likely candidate for improvements with solid-liquid PTC. It decomposes with hydroxylic solvents and has a low solubility in many nonhydroxylic ones: (g/100 g at 25 °C) THF 13, dioxane 0,1,di-*n*-butyl ether 2, diethyl ether 35–40 (only if more dilute solutions are concentrated). At first sight, replacement of the metal by an ammonium group or masking with a crown ether seems fine. Tetra-*n*-butyl ammonium aluminum hydride can be prepared from NBu_4Br and $LiAlH_4$ in tetrahydrofurane by precipitation with ether. NBu_4GaH_4 was made similarly. Other synthetic processes start from sodium or potassium alanates [525]. Among the known compounds are $NMe_4^{\oplus}AlH_4^{\ominus}$, $NEt_4^{\oplus}AlH_4^{\ominus}$, $N(n\text{-}C_3H_7)(n\text{-}C_8H_{17})_3^{\oplus}AlH_4^{\ominus}$, and $NMe_4^{\oplus}GaH_4^{\ominus}$. There have been no reports about these compounds being used for reductions, so it would appear that quaternary ammonium alanates cannot be used for this purpose. On the contrary, it has been shown that the presence of an alkali metal cation is absolutely necessary for the reduction. Pierre and co-workers [526, 527, 528] reported on the role of the cation in the reduction of ketones by $LiAlH_4$ and $NaBH_4$. They compared the reactions between cyclohexanone and the two complex hydrides in diglyme in both the presence and absence of [2.1.1]-cryptate (**A**) (complexing $Li^{\oplus}$) or [2.2.1]-cryptate (**B**) (masking $Na^{\oplus}$). In the case of $LiAlH_4$, equimolar amounts of **A** suppress reduction completely. Addition of lithium or sodium iodide restored the

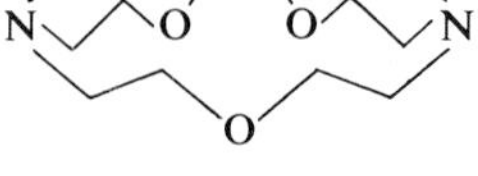

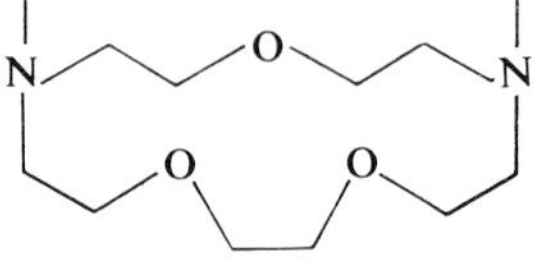

Scheme 3–177

reducing power of the system. From this it can be concluded that the metal ion has a vital function. Catalytic amounts of it, however, are sufficient. The following mechanism was proposed:

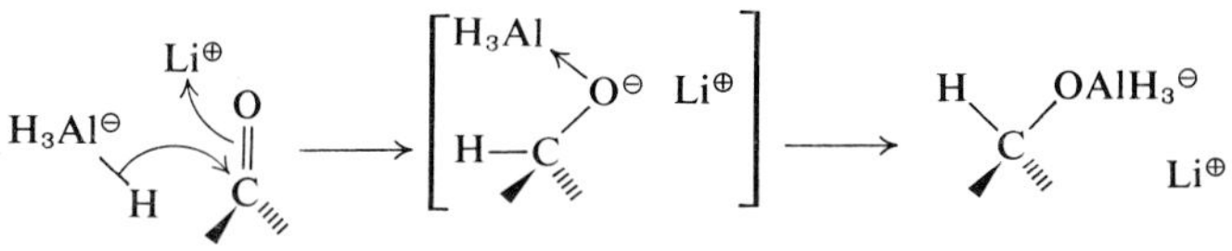

$$H_3Al^{\ominus} \; (Li^{\oplus}) \; H \; O{=}C \longrightarrow \left[\begin{array}{c} H_3Al \leftarrow O^{\ominus} \; Li^{\oplus} \\ H{-}C \end{array} \right] \longrightarrow H{-}C{-}OAlH_3^{\ominus} \; Li^{\oplus}$$

Scheme 3–178

Loupy and co-workers found a similar retardation of aldehyde and ketone reductions in ether and THF on the addition of **A** [529].

From these investigations it is clear that contrary to first expectations, $MAlH_4$ ($M = Na, Li$) cannot be used under PTC conditions since the metal cation itself is necessary for the reduction process, and in nonpolar solvents this is always complexed. The only possible solution could be the presence of an additional lithium salt. Turning now to borohydrides, the case for PTC is more promising. On the other hand, the need for improvement over conventional processes is not as urgent since $NaBH_4$ decomposes in hydroxylic solvents only very slowly.

The reduction of 2-octanone with aqueous $NaBH_4$ in 2N NaOH either with or without Aliquat 336 has been investigated [4, 38, 39]. Although the catalyzed reaction was much faster, it was still not as effective as the normal homogeneous process.

Other results showed that $NaBH_4$ masked with one equivalent of **B** in absolute diglyme does not reduce cyclohexanone directly [526]. Here the addition of small amounts of water or—what is more usual—the addition of water in the workup catalyze the conversion:

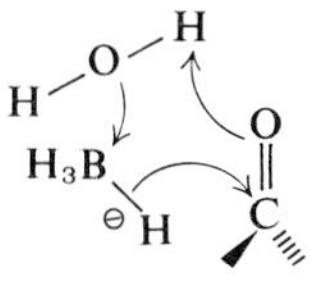

$$H{-}O{-}H \;\; H_3B{-}H^{\ominus} \;\; O{=}C$$

Scheme 3–179

$NaClO_4$ or $LiClO_4$ catalyze also.

Tetraalkylammonium borohydrides have been advocated for several years as reducing agents in non-hydroxylic solvents [530]. The solubility and reducing power depends somewhat on the size of the cation. Thus, $NMe_4^{\oplus}BH_4^{\ominus}$ is a white stable solid whose reducing power equals $NaBH_4$ but with the advantage that it is soluble even in benzene. It also decomposes less easily in solvents like ethanol and ethylene glycol. Higher members such as cetyltrimethylammonium boranate (a white granular solid) and tricaprylmethylammonium boranate (a wax) are even more soluble in hexane or mineral oil [530]. The reducing power of these reagents, however, is less than that of $NaBH_4$, as demonstrated by the fact that when working fast enough $NBu_4^{\oplus}BH_4^{\ominus}$ can be recrystallized from ethyl acetate or even acetone [531]. They can reduce aldehydes at room temperature, ketones only slowly, and esters slowly even at elevated temperatures. Nitriles are not reduced, but peroxides and acid halides react readily. Aromatic nitro groups are resistant to reduction at low temperatures and are only partly reduced at 65 °C [530]. The real reason for this selectivity of ammonium boranates may be related to the fact that the reagents contain different amounts of residual water. The lower members contain more than those with higher alkyl groups. It has not been investigated whether only carbonyl reductions or all reductions require catalytic amounts of water or alkali metal cations. In addition, some functional groups appear to form loose complexes with the reagent which readily decomposed reductively in the workup. Some ammonium boranates are commercially available although at a relatively high price (1976), for example:

$NMe_4^{\oplus}BH_4^{\ominus}$ 25 g *ca.* $75.00

$NEt_4^{\oplus}BH_4^{\ominus}$ 25 g *ca.* $65.00

Traditionally they were prepared by precipitation after a double transformation:

$$NR_4^{\oplus}X^{\ominus} + NaBH_4 \longrightarrow NR_4^{\oplus}BH_4^{\ominus} + NaX$$

Brändström's ion pair extraction technique greatly simplifies the preparation.

Tetra-n-butylammonium boranate [531]: A mixture of 1 mole NBu_4HSO_4 dissolved in 200 ml water and 250 ml 5N NaOH is combined with a solution of 1.1 moles $NaBH_4$ in 100 ml water at room temperature. Extraction with CH_2Cl_2 and evaporation after drying gives a quantitative yield of the ammonium boranate. This solution can be used directly in the reduction. The exothermic process shown below for example, was performed in 90% yield [532].

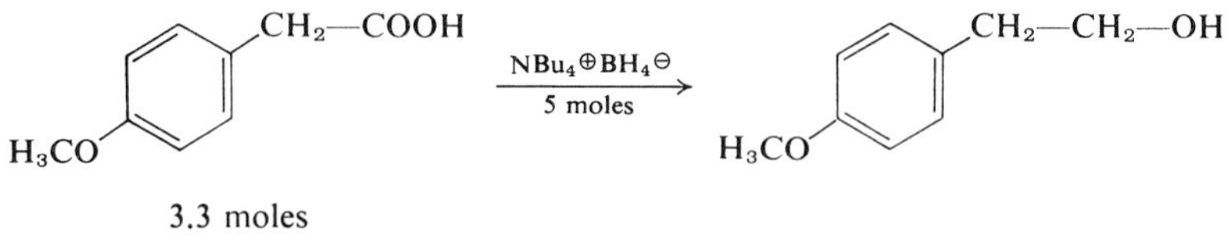

$$H_3CO\text{-}C_6H_4\text{-}CH_2\text{—}COOH \xrightarrow[\text{5 moles}]{NBu_4^{\oplus}BH_4^{\ominus}} H_3CO\text{-}C_6H_4\text{-}CH_2\text{—}CH_2\text{—}OH$$

3.3 moles

Scheme 3–180

In a recent alkaloid transformation, the reagent was used to cleave a C—Hg bond hydrogenolytically in CCl_4/CH_3OH [1169]. While Sullivan and Hinckley [530] worked in homogeneous benzene, *n*-hexane, or mineral oil, Raber and Guida [533] prefer methylene chloride. In this solvent ketones are still reduced at room temperature conveniently quickly whereas esters react quite slowly.

Two alternative mechanisms were proposed: direct hydride reduction of the carbonyl compound (1) and initial reaction of the borohydride ion with CH_2Cl_2 to give the actual reducing agent diborane (2). It was assumed that the reduction of aldehydes proceeds largely by pathway (1) whereas pathway (2) may be significant for the ketone reduction.

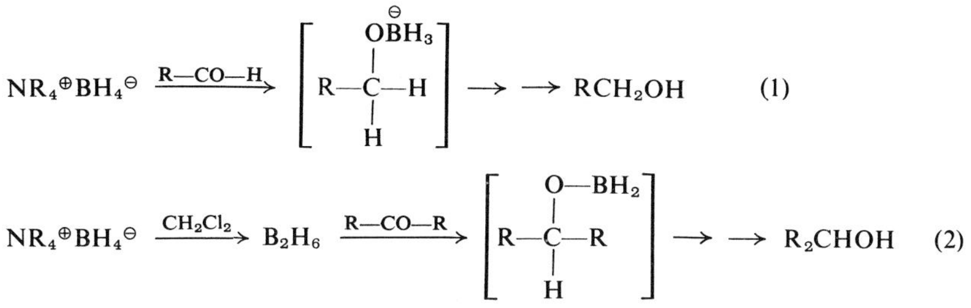

$$NR_4^{\oplus}BH_4^{\ominus} \xrightarrow{R-CO-H} \left[R-\overset{O^{\ominus}BH_3}{\underset{H}{C}}-H \right] \rightarrow \rightarrow RCH_2OH \quad (1)$$

$$NR_4^{\oplus}BH_4^{\ominus} \xrightarrow{CH_2Cl_2} B_2H_6 \xrightarrow{R-CO-R} \left[R-\overset{O-BH_2}{\underset{H}{C}}-R \right] \rightarrow \rightarrow R_2CHOH \quad (2)$$

Scheme 3–181

The conversion of esters to aldehydes was also carried out in the following way, and overall yields were between 75 and 85% [534]:

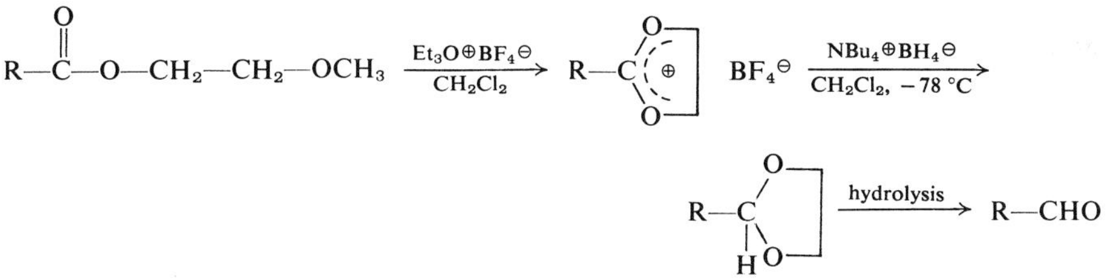

$$R-\overset{O}{\overset{\|}{C}}-O-CH_2-CH_2-OCH_3 \xrightarrow[CH_2Cl_2]{Et_3O^{\oplus}BF_4^{\ominus}} R-C^{\oplus}(O-CH_2CH_2-O) \; BF_4^{\ominus} \xrightarrow[CH_2Cl_2,\ -78\ °C]{NBu_4^{\oplus}BH_4^{\ominus}} R-\underset{H}{C}(O-CH_2CH_2-O) \xrightarrow{\text{hydrolysis}} R-CHO$$

Scheme 3–182

Again in homogeneous methylene chloride solution trithiocarbonate-S,S-dioxides were reduced at $-30°$ C [535]:

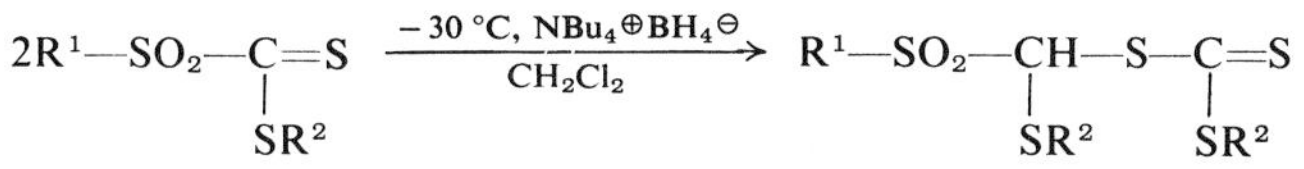

$$2R^1-SO_2-\underset{SR^2}{\underset{|}{C}}{=}S \xrightarrow[CH_2Cl_2]{-30\ °C,\ NBu_4^{\oplus}BH_4^{\ominus}} R^1-SO_2-\underset{SR^2}{\underset{|}{CH}}-S-\underset{SR^2}{\underset{|}{C}}{=}S$$

Scheme 3–183

The cyanotrihydridoboranate is a less vigorous reducing agent. The strong electron withdrawing group CN makes the anion more reluctant to deliver a hydride. The reagent can be prepared easily by ion pair extraction.

Tetrabutylammonium cyanoboranate [536]: A suspension of 0.1 mole $NBu_4^{\oplus}HSO_4^{\ominus}$ in 50 ml water is treated with 35 ml 5N NaOH. At room temperature, a solution of 0.1 mole $NaBH_3CN$ in 50 ml water is added, and after some minutes the mixture is extracted with methylene chloride.

In HMPA this mild reagent reduces primary iodides to the corresponding hydrocarbons without elimination to olefins (*e.g.*, 1-iododecane → decane, r.t., 21 h, 81%). Bromo, chloro, and tosyl compounds react more slowly. With secondary halides hydride displacement occurs predominantly with inversion of configuration; rates are slower than with primary compounds [536, 537]. Primary and secondary benzyl halides are smoothly reduced. In the latter case no evidence of double bond formation is found. Although allylic halides are converted to the alkenes, vinylic halides are resistant towards reduction. Whereas geminal dihalides give monohalides easily with NBu_4BH_3CN, 1,2-dihalides react less well [537]. Aldehydes, ketones, esters, amides, cyanides, and aromatic nitro compounds are inert at room temperature [536]. However, if a catalytic amount of 0.1N HCl is added, aldehydes may be reduced in the presence of ketones [536]. Increasing the concentration of the acid also increases the reduction rate of ketones. Other substituents like cyano, ester, amide, and nitro are inert under these conditions. In a more recent publication [538] the reduction of unsaturated carbonyl compounds in acidified methanol (pH *ca.* 3) with $NaBH_3CN$ or NBu_4BH_3CN at room temperature has been reported. The ammonium compound is not superior in this case. Within 1–2 hours most unsaturated aldehydes and ketones yield allylic alcohols with allylic methyl ethers formed as by-products. Changing the solvent to HMPT improves the reaction rate. Here, then, no products are found that result from solvent intervention, but the amount of allylic hydrocarbon is substantial. 1,4-addition giving ultimately the saturated alcohol becomes prominent when the α,β-unsaturated carbonyl is further conjugated with an aromatic ring, an additional cyano, carboxyl, or ester group, or when it is in a five- or six-membered ring [538].

Another powerful reducing reagent is B_2H_6, which was mentioned above as a possible intermediate. It can be obtained easily in solution by adding alkyl halides (*e.g.*, methyl iodide) to an anhydrous solution of tetra-alkylammonium boranate in methylene chloride. The by-product, NR_4-halide, can readily be removed [531].

$$2BH_4^{\ominus} + 2CH_3I \longrightarrow B_2H_6 + 2CH_4 + 2I^{\ominus}$$

Because of the ease and safety of this method it compares favorably with older procedures.

Reactions with diborane formed in situ: A dried methylene chloride solution of $NBu_4^{\oplus}BH_4^{\ominus}$ (prepared as described above) and the substrate are stirred under argon, while cooled in an ice bath. Two equivalents of CH_3I are added slowly so that methane evolution is not excessive. After stirring for 30 minutes at room temperature, the excess hydride is destroyed by the addition of ethanol, and the mixture is hydrolyzed with HCl.

Examples of reduction and hydroboration reactions performed similarly are [531, 540]: 3-nitrobenzoic acid → 3-nitrobenzyl alcohol (90%), ethyl(4-benzyloxyphenyl)-acetate → 2-(4′-benzyloxyphenyl)ethanol (87%), benzonitrile → benzylamine (95%), *p*-chlorobenzaldehyde → *p*-chlorobenzyl alcohol (94%), cyclohexene → cyclohexanol (98%), and the reduction of an N-formyl to an N-methyl group in a heterocyclic compound [912]. A more complicated example of hydroboration/oxidation is the transformation of 3-cyclododecen-1-one into a mixture of 1,3- and 1,4-cyclododec-andiols [541].

The investigations mentioned so far in this section usually involved working with reducing agents in homogeneous media. Matsuda and Koida [539] worked on ketone reductions in boiling xylene or toluene in the presence of equimolar amounts of catalysts to aid in solubilization. Catalysts used were diglyme, dimethoxyethane, and dibenzo-18-crown-6. The last mentioned gave the best results but yields were not overwhelmingly good because of condensation side reactions.

Starting with aqueous sodium borohydride, the PT catalyst cation can be expected to coextract some water of solvation molecules along with the $BH_4^{\ominus}$ and this suffices to improve the reduction. These expectations are fully borne out by experiment.

Landini and co-workers tested the catalytic activity of the following crown ethers in the reduction of 2-octanone in benzene/water with KBH_4 or $NaBH_4$ at 80 °C: 18-crown-6, dibenzo-18-crown-6, benzo-15-crown-5, and dicyclohexano-18-crown-6. The latter was the most active. Apparently the increase in lipophilic character is the decisive factor [44, 45].

Schick [120, 999] reported the reduction of the ketones shown below which gave different stereoselectivities under PTC conditions and in homogeneous methanol medium:

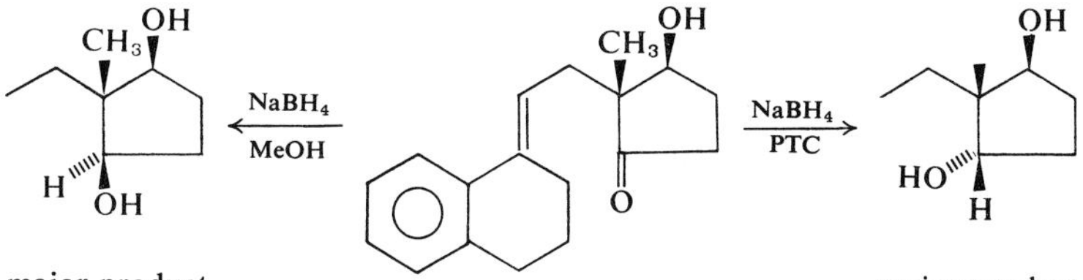

OH CH$_3$ H OH — $NaBH_4$ / MeOH ← OH CH$_3$ O → $NaBH_4$ / PTC — OH HO H

major product major product

Scheme 3–184

Balcells *et al.* [542] investigated possible asymmetric induction in the $BH_4^{\ominus}$ reduction of carbonyl compounds by means of chiral ephedrine derived PT catalysts. Agents used were **Ca**–**Cc** and the experiments were carried out at 0 °C or room temperature in benzene/water. The authors were unable to promote asymmetric induction in the reaction of simple ketones like 2-octanone, acetophenone, or propiophenone using

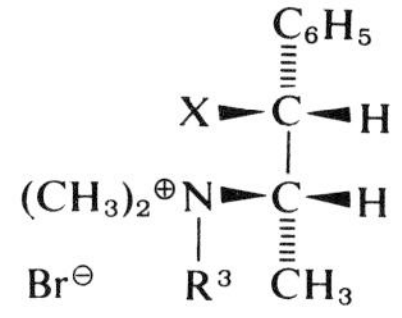

C_6H_5, X–C–H, $(CH_3)_2{}^{\oplus}N$–C–H, $Br^{\ominus}$ R^3 CH_3

C

	R^3	X
a	*n*-$C_{12}H_{25}$	OH
b	C_2H_5	OH
c	*n*-$C_{12}H_{25}$	H
d	*n*-$C_{16}H_{33}$	OH

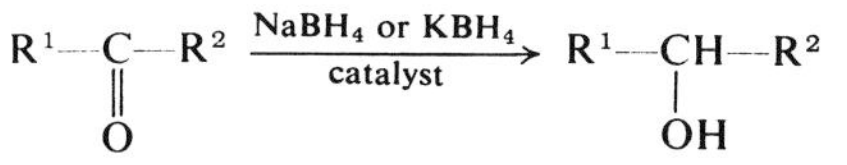

$$R^1–C(=O)–R^2 \xrightarrow[\text{catalyst}]{NaBH_4 \text{ or } KBH_4} R^1–CH(OH)–R^2$$

D **E**

	R^1	R^2
a	C_6H_5	*iso*-C_3H_7
b	C_6H_5	*tert*-C_4H_9
c	mesityl	CH_3
d	α-phenylethyl	CH_3
e	α-phenylpropyl	CH_3
f	C_6H_5	CH_3

Scheme 3–185

any of these catalysts [773]. With sterically more demanding ketones **Da–e**, catalyst **Ca** is believed to induce small rotations. Optical purities of 2–3%, and in one example 13% have been claimed. The authors consider it significant that slower reaction rates and practically racemic alcohols were observed with **Cc** which does not contain an OH group. It is assumed that interaction of the OH group of the catalyst with the carbonyl favors one of the two possible diastereomeric transition states.

In partial contradiction to these observations, Massé *et al.* [543] contend that even acetophenone and isobutyl methyl ketone give optically active alcohols (0.35–8.95% optically pure) in 1,2-dichloromethane/2N $NaOH/NaBH_4$ with catalysts **Ca** and **Cd**, but not with **Cb**. The somewhat higher optical yields were obtained with a catalyst to substrate ratio of 1:10. It is possible that an ammonium salt needs relatively long groups to make it more bulky and soluble in the organic phase. In subsequent work, Massé's highest rotations could not be reproduced, but some optical induction was still found [1014].

It should be remembered, though, that the elimination of amine from the catalysts **Ca**, **Cc**, and **Cd** gives an optically active oxirane (**F**), in this case of very high specific rotation ($[\alpha]_D$ + 117.6°, neat) [544]. A small amount of impurity introduced by this oxirane, its reduction or ring opening products, or optically active impurities in the catalyst **C** could in fact easily simulate a small optical yield of the main reaction product. It should be remembered that a similar situation exists for the alleged "optically active" dichlorocarbene adducts which were shown to be in error (*cf.*, Section 3.3.13). Control experiments with catalyst **Cd** and preformed inactive **Ef**, under conditions similar to the reduction conditions but in the absence of $NaBH_4$, lead to small positive rotations in **Ef** [545]. It may very well be that the low rotations found are in fact due to some impurity which is difficult to separate [545].

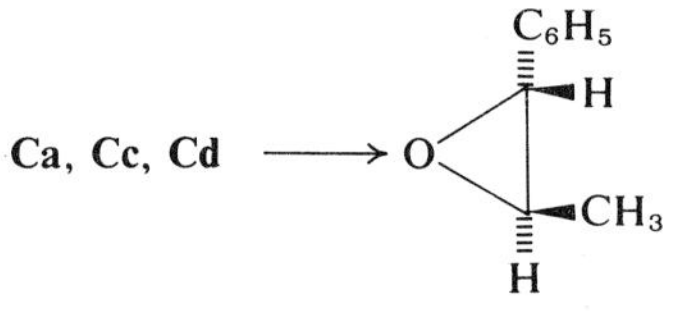

Ca, **Cc**, **Cd** ⟶ O< (epoxide with C_6H_5, H, CH_3, H)

F

Scheme 3–186

Keeping these concerns in mind, it must be stated that a number of additional papers on supposed and/or real asymmetric reductions have been published. Thus, extremely low optical inductions were reported in the reduction of acetophenone. A 10% induction was observed with pivalophenone. In both cases the presence of hydroxyl groups in the catalyst was necessary [905]. Further ketone reductions with (1R,2S)-ephedrinium derived salts and (1S,2S)-ψ-ephedrinium derivatives as catalysts [1170] are known. In contrast to the predictions by Prelog's rule, borohydride reactions of (3R)-menthyl benzoylformate with a (1S,S2)-ephedrinium derived catalyst gave (3R)-menthyl (S)-mandelate [1171]. Take note that in this case the substrate molecule is asymmetric already. Furthermore, rate constants for acetophenone reductions with catalyst **C** were determined [1172], and comments were made on structural features necessary in a PT catalyst for asymmetric reductions [1173]. Various partially protected hydroxymonosaccharides solubilize $NaBH_4$ in benzene and THF to some extent and promote optical yields up to 39% in ketone reductions [1201].

Durst [546] developed a method for reducing sulfoxides to sulfides involving methylating with methyl fluorosulfonate in methylene chloride followed by reduction with sodium cyanoborohydride. In this way one can protect other sensitive groups in the molecule against reduction. Normally, after the methylation the solvent is removed, and the residue is taken up in ethanol for reduction. In a simplified PTC version, 18-crown-6 and $NaBH_3CN$ are added to the cooled reaction mixture directly and stirring is continued for some hours.

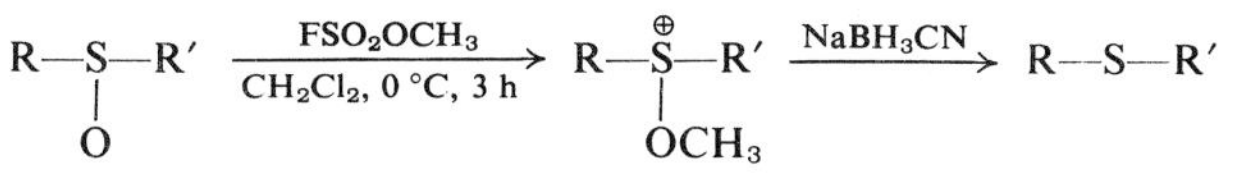

$$R{-}S(=O){-}R' \xrightarrow[CH_2Cl_2,\ 0\,°C,\ 3\ h]{FSO_2OCH_3} R{-}\overset{\oplus}{S}(OCH_3){-}R' \xrightarrow{NaBH_3CN} R{-}S{-}R'$$

Scheme 3–187

3.4.2. Other Reducing Agents

In an application of PTC to a NADPH model system, $Na_2S_2O_4$ is the reductive agent and benzyltrimethylammonium chloride the catalyst, while benzylnicotinamide chloride functions as redox model [1174].

In a first application of PTC to metal carbonyl chemistry, the reduction of nitro compounds with tri-iron dodecacarbonyl was performed in benzene/1N NaOH at room temperature with TEBA or 18-crown-6 as catalyst (Table 3–21) [547, 548].

Table 3–21. Reduction of Nitro Compounds with Triiron Dodecacarbonyl in Benzene/1 N NaOH with TEBA or 18-Crown-6 at Room Temperature.

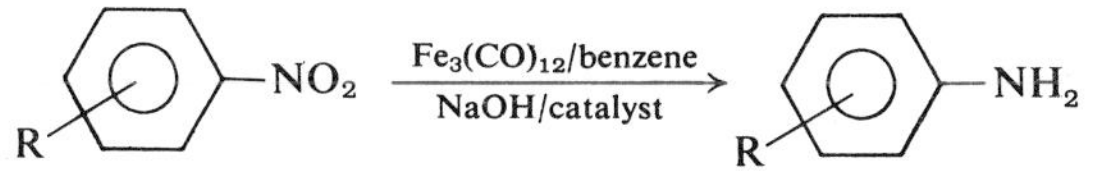

$$R{-}C_6H_4{-}NO_2 \xrightarrow[NaOH/catalyst]{Fe_3(CO)_{12}/benzene} R{-}C_6H_4{-}NH_2$$

Scheme 3–188

R	Yield with Catalyst (%)		
	Crown Ether	TEBA	None
H	60	—	—
p-CH_3	65	86	6
p-OCH_3	78	92	3
p-Cl	79	89	1
p-$COCH_3$	—	60	—

Formerly this same reduction required 10–17 hours refluxing in benzene containing some methanol. The authors assume that $OH^{\ominus}$ reacts with the iron carbonyl to form $[Fe_3(CO)_{12}]^{2\ominus}$, which in turn is protonated to give $[HFe_3(CO)_{12}]^{\ominus}$. The ion pair of this anion seems to be the actual reducing agent [547, 548]. Aniline yields are significantly lower if carbon monoxide is present as gas [1168].

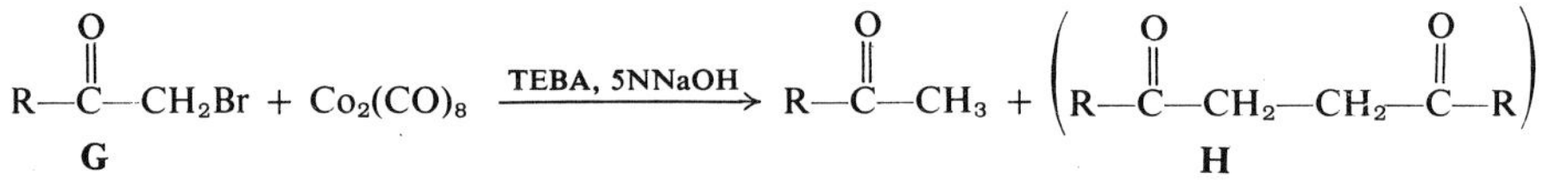

$$\underset{\mathbf{G}}{R{-}\overset{\overset{O}{\|}}{C}{-}CH_2Br} + Co_2(CO)_8 \xrightarrow{\text{TEBA, 5NNaOH}} R{-}\overset{\overset{O}{\|}}{C}{-}CH_3 + \left(\underset{\mathbf{H}}{R{-}\overset{\overset{O}{\|}}{C}{-}CH_2{-}CH_2{-}\overset{\overset{O}{\|}}{C}{-}R}\right)$$

Scheme 3–189

Treatment of an α-bromoketone **G** (R = aryl, admantyl) with an equimolar amount of dicobalt octacarbonyl, 5N sodium hydroxide, benzene, and TEBA affords the dehalogenated monoketone in almost quantitative yield after 2 hours at room temperature. Using only a 10:1 ratio of **G** to cobalt compound the PTC process gave fair yields. In some cases 1,4-diketones **H** were formed as minor by-products, rarely as major products [549]. A possible pathway for the reduction involves reaction of $OH^{\ominus}$ with the $Co_2(CO)_8$ to give $Co(CO)_4^{\ominus}$, which is extracted into the organic layer. The following series of events is then thought to take place:

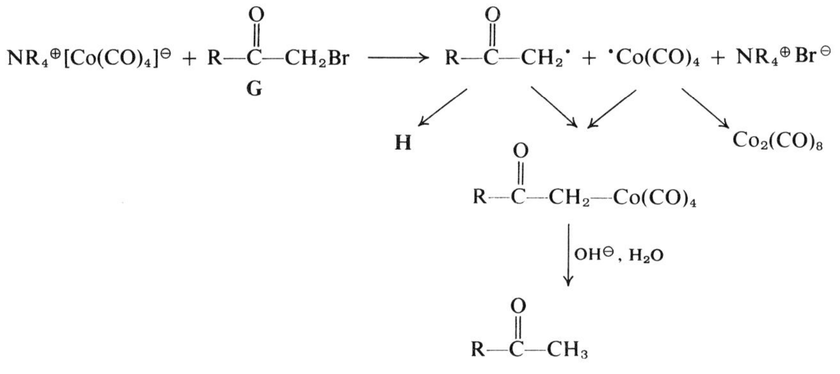

$$NR_4^{\oplus}[Co(CO)_4]^{\ominus} + \underset{\mathbf{G}}{R{-}\overset{\overset{O}{\|}}{C}{-}CH_2Br} \longrightarrow R{-}\overset{\overset{O}{\|}}{C}{-}CH_2^{\bullet} + {}^{\bullet}Co(CO)_4 + NR_4^{\oplus}Br^{\ominus}$$

$$R{-}\overset{\overset{O}{\|}}{C}{-}CH_2^{\bullet} \longrightarrow \mathbf{H} \qquad {}^{\bullet}Co(CO)_4 \longrightarrow Co_2(CO)_8$$

$$R{-}\overset{\overset{O}{\|}}{C}{-}CH_2^{\bullet} + {}^{\bullet}Co(CO)_4 \longrightarrow R{-}\overset{\overset{O}{\|}}{C}{-}CH_2{-}Co(CO)_4 \xrightarrow{OH^{\ominus},\ H_2O} R{-}\overset{\overset{O}{\|}}{C}{-}CH_3$$

Scheme 3–190

The following equation is given for the redox reaction leading to the extractable anion [511]:

$$32OH^{\ominus} + 11Co_2(CO)_8 \longrightarrow 20\ Co(CO)_4^{\ominus} + 6\ CO_3^{2\ominus} + 2\ CoCO_3 + 16H_2O$$

The hydrogenation catalyst $K_3[Co(CN)_5H]$ is easy and inexpensive to prepare. In the presence of TEBA, an anion $(H_2[Co(CN)_5H]^{\ominus}$??) can be extracted into benzene, where it can act as catalyst in atmospheric pressure hydrogenations. It seems useful mainly for reducing α,β-unsaturated ketones to saturated ketones. Thus, carvone gives 2-methyl-5-isopropenylcyclohexanone (*trans*/*cis* 6:1) in 93% isolated yield [1175].

The reduction of aromatic, including heteroaromatic, chlorides or bromides, by aqueous formates in the presence of a hydrogenating catalyst and a PT catalyst is described in a patent [553]. An example is the reduction of *o*-chloronitrobenzene which is then dehalogenated to aniline. Anionic surfactants are useful catalysts here too, implying that this is a surface reaction. A powerful reducing agent which is

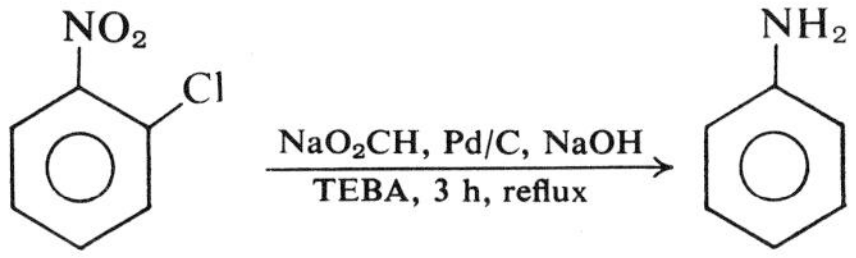

$$o\text{-}ClC_6H_4NO_2 \xrightarrow[\text{TEBA, 3 h, reflux}]{NaO_2CH,\ Pd/C,\ NaOH} C_6H_5NH_2$$

Scheme 3–191

easy to manipulate is **I**, formamidinesulfinic acid, obtained by the oxidation of thiourea with hydrogen peroxide:

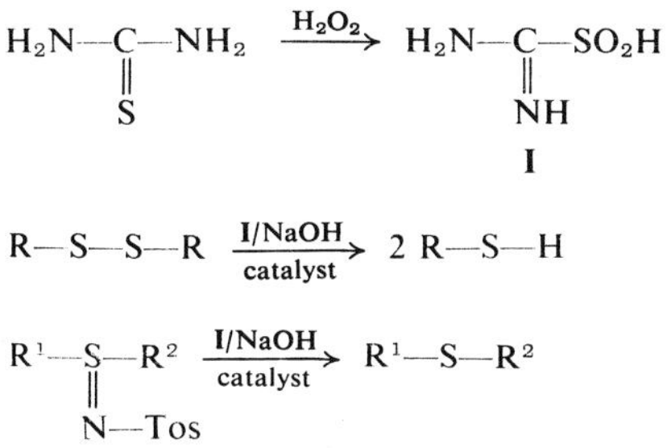

$$H_2N{-}C({=}S){-}NH_2 \xrightarrow{H_2O_2} H_2N{-}C({=}NH){-}SO_2H \quad (\mathbf{I})$$

$$R{-}S{-}S{-}R \xrightarrow[\text{catalyst}]{\mathbf{I}/NaOH} 2\ R{-}S{-}H$$

$$R^1{-}S({=}N{-}Tos){-}R^2 \xrightarrow[\text{catalyst}]{\mathbf{I}/NaOH} R^1{-}S{-}R^2$$

Scheme 3–192

Borgogno *et al.* [550] showed that the reduction of disulfides gives 60–90% yield with 1.8 equivalents **I**, 5.6 molar equivalents 3.7N NaOH and small amounts of tri-butylhexadecylphosphonium bromide without extra solvent at 80 °C. Similarly, N-tosylsulfinimines were reduced in 3–5 hours at 70 °C with diisopropyl ether as the organic phase. These PTC reductions are compared favorably to other known methods in terms of yield, simplicity, and ease of workup.

3.5. Oxidation Reactions

Many inorganic oxidants can be transferred into an organic phase by a catalyst. It must be remembered though that the stoichiometry of some oxidations requires the presence of either $H^{\oplus}$ or $OH^{\ominus}$. A few cases are known where PTC is not helpful in an oxidation, for instance with sulfides and sulfoxides [615].

3.5.1. Oxidation with Permanganate

$KMnO_4$ in aqueous solution is a well known oxidant for many substances. In most cases a rather large excess (up to more than 100%) of permanganate is consumed.

This is explained by the observation that $KMnO_4$ is partly decomposed under certain conditions to manganese dioxide with evolution of oxygen. A solution of $KMnO_4$ in 0.04N sulfuric acid decomposes nearly 20 times as fast as in neutral solution. On the other hand, both an alkaline medium and manganese dioxide also accelerate the decomposition. This is an autocatalytic process. Thus, in many preparative procedures the voluminous brown precipitate is washed with much solvent or even extracted in a Soxhlet apparatus, as the product is often firmly adsorbed to the precipitate.

The PTC version of the $KMnO_4$ oxidation is free of these difficulties. For permanganate oxidations in nonaqueous solvents, quaternary ammonium and arsonium salts as well as crown ethers have been used as catalysts.*) Solid $KMnO_4$ can be solubilized in benzene for instance by stirring it with an equimolar amount of dicyclohexano-18-crown-6 at 25 °C. The solution so prepared, dubbed "purple benzene," can be obtained in concentrations of up to 0.06 M [557]. Among the various crown ethers, the solubilizing action increases in the following way; 18-crown-6 < dibenzo-18-crown-6 < benzo-15-crown-5 < dicyclohexano-18-crown-6.†) In the beginning it was believed that only crown ethers are suitable for preparing organic $MnO_4^{\ominus}$ solutions in nonpolar media.‡) The PTC technique based on this so-called "solid-liquid phase transfer catalysis" was contrasted to "liquid-liquid PTC" with aqueous $KMnO_4$ solutions. It was thought that only the latter method was practicable both with onium salts and crown ethers. But a simple qualitative test tube experiment can demonstrate that Aliquat 336, for instance, and solid $KMnO_4$ give a colored benzene solution immediately, actually more deeply colored than with the dibenzo-18-crown-6/$KMnO_4$ pair. Thus, taking the high cost of crown ethers into account it is more reasonable to use quaternary ammonium salts as catalysts, whether or not the oxidant is applied as a solid or in aqueous solution. Alternatively, polyethyleneglycol ethers can be applied in both cases [1177].

Turning now to the extraction of $MnO_4^{\ominus}$ from aqueous solution, it is known that of the ammonium salts NMe_4Cl cannot be used, TEBA and cetyltrimethylammonium bromide are good, whereas tetrabutylammonium bromide, tetrabutylammonium chloride, and Aliquat 336 are the best [559]. Herriott and Picker [559] showed that with equal volumes of benzene and 0.02 M aqueous $KMnO_4$, 0.02 M NBu_4Br brought color to the organic phase immediately. Almost quantitative extraction of permanganate can be secured by applying an excess of this ammonium salt. The solutions obtained in this manner are rather stable; the $MnO_4^{\ominus}$ titer decreased by only 17% within 45 hours. In fact, a quantitative photometric titration of certain unsaturated hydrocarbons is performed with such solutions [910]. Oxidation reactions may be

*) Alternatively, tetrabutylammonium permanganate can be isolated. It was believed to be a stable compound used as an oxidant in pyridine solution under mild conditions [974]. Recently, however, a violent self-ignition was reported [1176].

†) The ability to bring the purple color into the organic phase can be used as a fast diagnostic tool to test complex formation and extractability. In this way new cyclic ligands of differing size and flexibility with 5 oxygen donor atoms and one extra intraanular heteroatom (*e.g.*, pyridine nitrogen, pyridine-N-oxide oxygen, methylamine) were tested in chloroform/water [558].

‡) $NaMnO_4$ can also be solubilized with 18-crown-6 [911].

carried out homogeneously at 25 °C by mixing a solution of the compound to be oxidized in benzene with purple benzene. With olefins, rapid decoloration occurs within minutes, toluene requires up to 72 hours [559]. It is more convenient, however, in a catalytic process to bypass prior formation of the reagent solution. Some of the oxidations of olefins and alcohols reviewed in Table 3–22 are highly exothermic. Overoxidation is minimized by adding some acetic acid to the organic phase [960]. Alkyl groups of aromatic compounds, however, need temperatures above 60 °C and/or longer reaction times. The following classes of compounds have been investigated extensively (see Table 3–22):

$$\mathrm{R{-}CH{=}CH{-}R'} \longrightarrow \mathrm{R{-}COOH + HOOC{-}R'}$$

$$\mathrm{R{-}CH_2OH} \longrightarrow \mathrm{R{-}COOH}$$

$$\mathrm{ArCH_3} \longrightarrow \mathrm{Ar{-}COOH}$$

A typical procedure is presented below [559].

Oxidation of an olefin: A solution of 4.8 g (0.03 mole) $KMnO_4$ in 50 ml water is stirred in a water bath while a solution of 0.01 mole olefin in 30 ml benzene containing 0.5 g NBu_4Br is added. After 3 hours of stirring at room temperature, the mixture is worked up by the addition of $NaHSO_3$, acidification, separation of the organic layer, drying, and removal of the solvent.

With solid $KMnO_4$ or $NaMnO_4$/crown [911] attention must be directed to fine grinding and high shear stirring to avoid interference by MnO_2. Some authors advocate running the reactions in a ball mill.

Some workers have used PTC permanganate oxidations with less common substrates. Dimroth [561], for example, transformed the side chain of the phosphorus heterocycle of Scheme 3–193A into the corresponding aldehyde. The heterocycle is not attacked under these conditions. Using chromic acid the conversion shown below is possible.

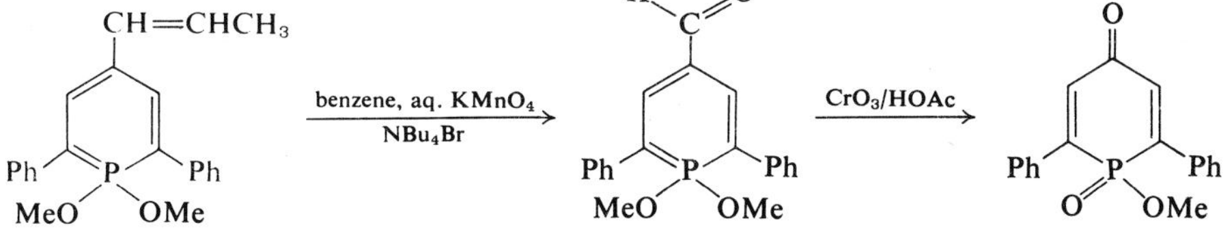

Scheme 3–193A

Nonconjugated cyclohexadienes are quantitatively aromatized in a solid/liquid PTC permanganate oxidation, whereas the conjugated analogues do not react under identical conditions [1179]. 1H-4,5-Dihydro-1,2,3-triazoles give the aromatic heterocycles in a benzene/aqueous $KMnO_4$/NBu_4Cl system [1180]. When these triazolines

Table 3–22. PTC-Oxidations with $KMnO_4$, Organic Compounds Usually Dissolved in Benzene.

Reaction	Catalyst	$KMnO_4$	Yiel (%)	Reference
olefin → acid, ketone, or aldehyde				
stilbene → benzoic acid	dicyclohexano-18-crown-6	solid	≈100	[557]
	Aliquat 336	aq. sol.	53	[38, 39]
			95	[559]
1-decene → nononoic acid	Aliquat 336	aq. sol.	91	[4]
1-octene → heptanoic acid	Aliquat 336	aq. sol.	81	[559]
	dicyclohexano-18-crown-6	solid	78	[44, 50]
cyclohexene → adipic acid	dicyclohexano-18-crown-6	solid	≈100	[557]
α-pinene → *cis*-pinonic acid	dicyclohexano-18-crown-6	solid	≈100	[557]
R—CH=CH_2 → R—COOH	Aliquat 336/benzene/HOAc	aq. sol.	≈100	[960]
R—C≡CH → R—COOH	Aliquat 336/petr. ether/HOAc	aq. sol.	≈100	[960, 1184]
9-octadecene → nonanoic acid	NBu_4Br, $C_{16}H_{11}NMe_3Br$	aq. sol.	80	[1003]
(1,4-CH_3, CH_3 epoxy-bridged bicyclic structure) → $C_6H_4(C(=O)CH_3)_2$	Aliquat 336	aq. sol.	69	[560]
(Ph, Ph, MeO, OMe-substituted P-heterocycle with CH=CH—CH_3) → (same with CHO)	NBu_4Br	aq. sol.	60	[561]
Ar—CH(C_4H_9)—O$(CH_2)_3$—CH=CH_2 → —COOH	Aliquat 336	aq. sol.		[563]

alcohol, aldehyde → acid				
Ph—CH_2OH → Ph—COOH	Aliquat 336	aq. sol.	92	[559]
	dicyclohexano-18-crown-6	solid	100	[557]
1-octanol → octanoic acid	Aliquat 336	aq. sol.	47	[559]
1-heptanol → heptanoic acid	dicyclohexano-18-crown-6	solid	70	[557]
piperonal → piperonylic acid	cetyltrimethylammonium bromide	aq. sol.	65	[475]
p-chlorobenzaldehyde → acid	NBu_4HSO_4	aq. sol.	90	[1178]
aromatic side-chains				
toluene → Ph—COOH	dicyclohexano-18-crown-6	solid	78	[557]
p-xylene → toluic acid	dicyclohexano-18-crown-6	solid	100	[557]
mesitylene → trimesic acid	NBu_4HSO_4	aq. sol.	88	[562]
other types				
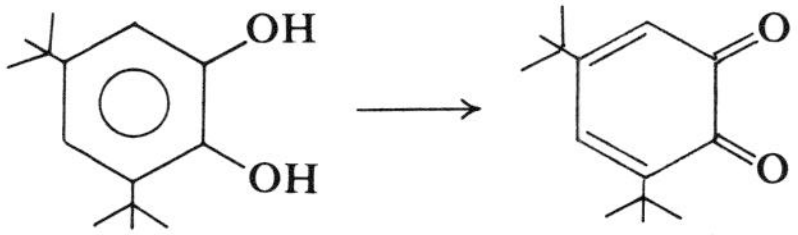	18-crown-6	solid	97	[564]
Ph—CH_2—CN → benzoic acid	Aliquat 336	aq. sol.	86	[559]
benzhydrol → benzophenone	dicyclohexano-18-crown-6	solid	≈100	[557]
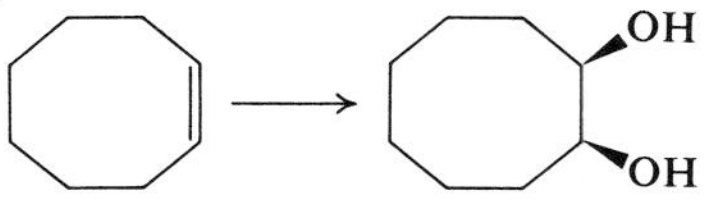	TEBA	aq. sol. + 40% aq. NaOH	50	[567]
oleyl alcohol → *erythro*-1,9,10-octadecanetriol	PBu_4Cl, NBu_4Cl	aq. sol./NaOH	40–80	[1001]
elaidyl alcohol → *threo*-1,9,10-octadecanetriol	PBu_4Cl, NBu_4Cl	aq. sol./NaOH	40–80	[1001]
9-octadecene → 9,10-dihydroxyoctadecane	NBu_4Br, $C_{16}H_{33}NMe_3Br$	aq. sol./NaOH	80	[1003]
cyclohexa-1,4-diene → benzene	dicyclohexano-18-crown-6			[1179]
triazoline → triazole	NBu_4Cl			[1180]
R—C≡C—R′ → R—CO—CO—R′	Adogen 464			[1184]
R—CH_2—OR′ → R—CO—OR′		NR_4MnO_4		[1183]
R_3C—H → R_3C—OH		NR_4MnO_4		[1182]

carry a 5-(tertiary)amino substituent, however, the side chain is oxidized in preference to the ring [1181] (Scheme 3–193B).

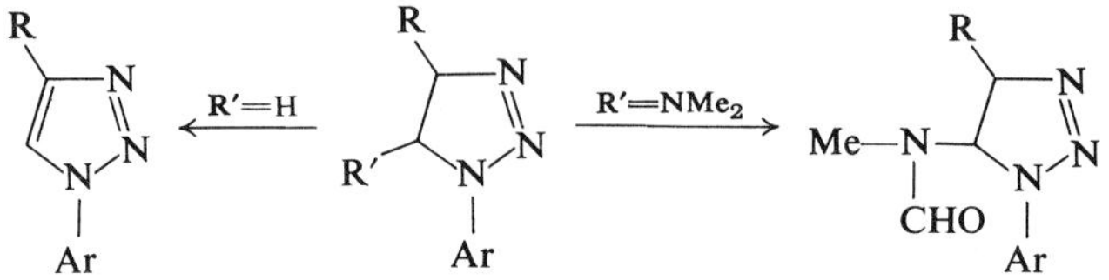

Scheme 3–193B

Using previously isolated benzyl(triethyl)ammonium permanganate (caution, *cf.*, above) in pyridine, acetic acid, or methylene chloride, the following selective oxidations become possible [1182, 1183]:

$$Ar{-}CH_2R \longrightarrow Ar{-}COR \qquad Ar{-}CHR_2 \longrightarrow Ar{-}C(OH)R_2$$

$$R_3C{-}H \longrightarrow R_3C{-}OH \qquad R{-}CH_2{-}OR' \longrightarrow R{-}CO{-}OR'$$

In the presence of a mineral acid (*e.g.*, polyphosphoric acid) tertiary amines suffice as PT catalysts for $KMnO_4$ oxidations. Thus, Starks [600] oxidized 1-tetradecene to tridecanoic acid with tridodecylamine as catalyst in the two-phase reaction. Mack and Durst [564, 565] prepared an *o*-quinone by stirring a solution of 3,5-di-*tert*-butylcatechol in methylene chloride with 18-crown-6 and pulverized $KMnO_4$ for 30 minutes at room temperature.

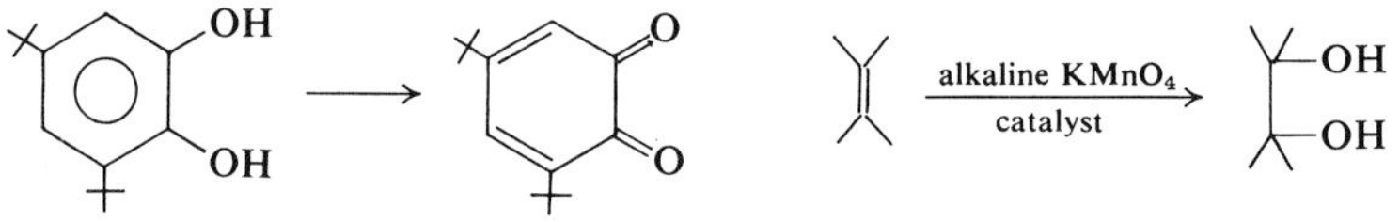

Scheme 3–194

Cis-hydroxylation of olefins with basic $KMnO_4$ is normally a reaction that gives poor yields, with a few notable exceptions such as long chain mono-unsaturated fatty acids. Before the development of PTC, therefore, standard procedures involved the use of osmium tetroxide or the Woodward technique, which involves conversion with iodine and silver acetate in moist acetic acid. Weber and Shepherd [567] performed PTC-*cis*-bishydroxylations of olefins in yields of around 50%.*) Only if the glycol formed is highly soluble in water (*e.g.*, *cis*-dihydroxycyclohexane) do lower yields occur. In addition, a small amount of acid is formed. For an application to a more complex molecule, see reference [878]. Disubstituted alkynes give α-diketones in the presence of Adogen 464 and a little acetic acid [1184].

In conclusion, it should be mentioned that methyltriphenylarsonium chloride has been used as a PT catalyst long before the concept of phase transfer catalysis came into fashion. Following up on earlier observations about the extractability of $MnO_4^{\ominus}$

*) Substrate in CH_2Cl_2, TEBA, 40% NaOH in water, solid $KMnO_4$ added at 0 °C, vigorous stirring overnight.

in $CHCl_3$ and nitrobenzene [568], Gibson and Hosking [569] shook aqueous $KMnO_4$ with organic solutions of alcohols, olefins, 1- and 2-nitropropane, and formic acid in the presence of $As(CH_3)Ph_3Cl$ and observed oxidations. *tert*-Butanol, toluene, ethyl acetate, ether, and ketones were not attacked under these conditions.

All reactions considered until now were true PTC processes. The oxidation of octene by permanganate has also been investigated with other types of catalysts [570]. Sodium stearate and laurate had a stronger rate accelerating effect than some PT catalysts under the conditions used. This is a strong indication that permanganate oxidations can occur as surface and/or micellar reactions too. Similar indications, a large rate increase, were found [475] when cetyltrimethylammonium bromide was added to piperonal/aqueous $KMnO_4$. Unlike large symmetrical quaternary ammonium ions, cetyltrimethylammonium does not even partially solubilize solid $KMnO_4$ in benzene or extract it from aqueous solution, so that the PTC mechanism does not seem to be operative here.

3.5.2. The Role of Hydrogen Peroxide in PTC Reactions

In view of the known difficulty in extracting $OH^{\ominus}$ ions from aqueous into organic media, one might expect that the $HO_2^{\ominus}$ ion is similarly hard to extract, or, in other words the anions commonly brought in with the catalysts, *i.e.*, chloride and bromide, should have larger extraction constants than the mono-anion of hydrogen peroxide. It is found, however, that the extraction of a species with oxidizing properties from 35% (≈ 10 M) H_2O_2 into methylene chloride is easy with some catalysts [57]. Iodometric titration of the organic phase after equilibration with catalyst gave the results summarized in Table 3–23.

It is immediately apparent—and not unexpected—that the hydrophilic catalysts are of little use here. After all, a catalyst partitioned predominantly to the aqueous phase cannot be of much help. Both the anion and cation of the catalyst influence

Table 3–23. Iodomitric Titration of H_2O_2 in CH_2Cl_2 after Equilibration in the Presence of a Catalyst (5 ml 35% H_2O_2, 20 ml CH_2Cl_2, and 2 mmoles catalyst).

Catalyst	Equivalents H_2O_2 per Equiv. Catalyst Present
TEBA	0.013
$(NBu_4)_2SO_4$	0.09
NBu_4HSO_4	0.10
NBu_4Cl	0.30
NBu_4Br	0.68
NBu_4 β-naphthalenesulfonate	0.92
Aliquat 336	0.88
$NHex_4Br$	0.99
$NHep_4Br$	1.00
$NOct_4Br$	1.00

its lipophilicity (*cf.*, Chapter 1), and the more lipophilic the catalyst, the greater its extraction capability. Addition of a little sulfuric acid did not bring about any change in H_2O_2 extraction properties, but the addition of 1.25 mmoles sodium hydroxide (less than the molar amount of catalyst and much less than the amount of H_2O_2) halved the extractable oxidizing equivalents. Since as hydrogen peroxide is a stronger acid (pK_a 11.6) than water (pK_a 15.7), these observations testify against the extraction of $HO_2^{\ominus}$. Instead, hydrogen peroxide must be transferred as a solvate of the catalyst. An additional proof of this was found when the organic phase of the H_2O_2 extraction with tetrahexylammonium bromide was examined for bromide: $\approx 100\%$ of the theoretical amount was found [57].

For PTC oxidations of several typical olefins with H_2O_2, a heavy metal derived catalyst plus a PT catalyst have been described [39, 599]. Tridecylmethylammonium chloride was used. It has a two-fold function in transferring hydrogen peroxide to the organic phase and preventing its decomposition to water and oxygen. The reactions were performed at room temperature to 70 °C for several hours. With cyclohexene, products **E** and **F** were formed exclusively when the heavy metal catalysts OsO_4, MoO_3, or H_2WO_4 were used. When $SeOCl_2$, V_2O_5, Cr_2O_3, TiO_2, $CeSO_4$, NiO, $MnCl_2$, $CoCl_2$, PtO_2, $FeSO_4$, $Pb(OAc)_2$, or $PdCl_2$ were applied, **G** and **H** were the

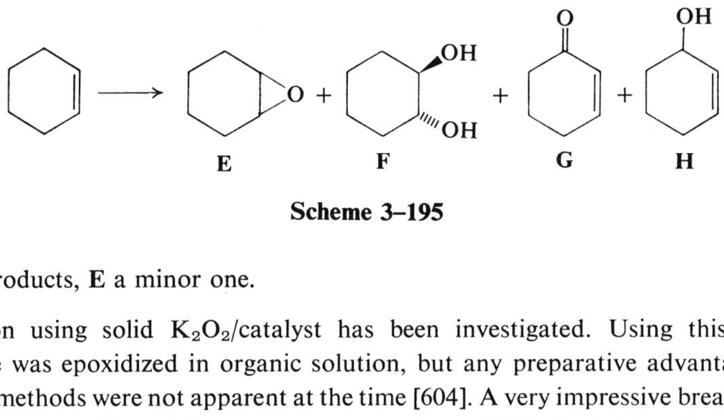

Scheme 3–195

major products, **E** a minor one.

Oxidation using solid K_2O_2/catalyst has been investigated. Using this method chalcone was epoxidized in organic solution, but any preparative advantages over existing methods were not apparent at the time [604]. A very impressive breakthrough was later reported by H. Wynberg and co-workers [603]: Optically active epoxides were obtained in excellent chemical yields and an enantiomeric excess of up to 25% by a PTC process using benzylquinium chloride (**M**) or benzylquinidinium chloride as catalyst. Epoxidation of compounds **I** to **L** was carried out by vigorous stirring of toluene solutions with 30% H_2O_2/dilute sodium hydroxide in the presence of catalytic amounts of **M** for 24 hours at room temperature. Great care seems to have been exercised to remove traces of catalysts, or their degradation products from the final epoxides. So it would appear certain that the reported effects are real. These represent the first clear-cut examples of large optical inductions in PTC processes.

No mechanism was proposed; however, since the reaction requires a base as catalyst and this is not transferred to the organic phase, it is possible that the enone picks up an $HO_2^{\ominus}$ at the phase boundary. The enolate anion formed remains anchored at the interphase until it is released by the catalyst cation exercising its chirality. Momen-

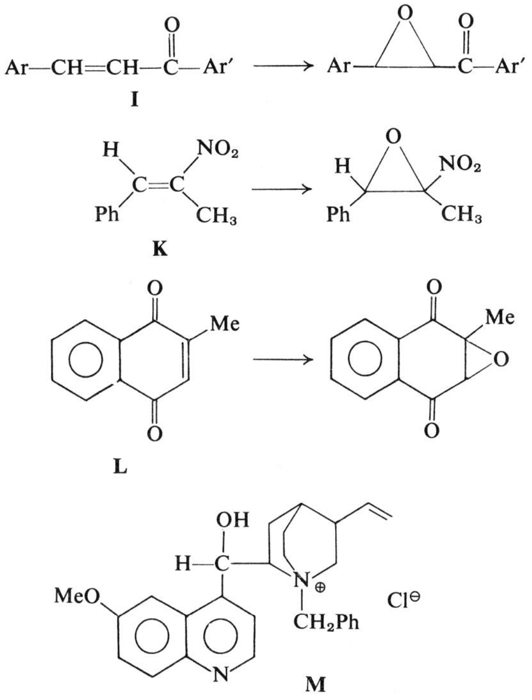

$$Ar{-}CH{=}CH{-}\overset{O}{\overset{\|}{C}}{-}Ar' \longrightarrow Ar{-}\overset{O}{\triangle}{-}\overset{O}{\overset{\|}{C}}{-}Ar'$$

I

$$\text{H(Ph)C}{=}\text{C(NO}_2\text{)(CH}_3\text{)} \longrightarrow \text{H(Ph)}\overset{O}{\triangle}\text{(NO}_2\text{)(CH}_3\text{)}$$

K

L

M

Scheme 3–196

tary loss of $OH^{\ominus}$ occurs, and the chiral epoxide is formed. This model does not require the transfer of the oxidizing power into the organic phase. The rather hydrophilic characterof **M** might be responsible for the long reaction time.

The reaction between various phenols and solid K_2O_2/Aliquat 336 in methylene chloride did not involve the oxidizing power but rather the basic character of the reagent: attack of phenolate on the solvent leads to methylene diphenol ethers [604, *cf.*, 234]:

$$ArOH \xrightarrow[\text{catalyst}]{KO_2/CH_2Cl_2} ArO{-}CH_2{-}OAr$$

Finally, a PTC halogenation of olefins is mentioned. In this method, halogen is generated and immediately consumed under ionic conditions, so that no radical side reactions disturb. This method is also advantageous where elementary chlorine is difficult to use or because of environmental reasons. To the ice-cooled alkene solution in carbon tetrachloride, stirred with concentrated hydrohalic acid and a little TEBA, 30% H_2O_2 is added dropwise [869].

$$R^1CH{=}CH{-}R^2 + 2HX + H_2O_2 \longrightarrow R^1CH(X){-}CH(X){-}R^2 + 2H_2O$$

The solid sodium salt of cumene hydroperoxide was used to oxidize and hydrate 2-arylcinnamonitriles in benzene with TEBA as catalyst [602]:

$$\text{Ph—C(CH}_3)_2\text{—OO}^{\ominus}\text{Na}^{\oplus} + \text{Ar—CH=C(Ar')—CN} \longrightarrow \text{Ar—CO—CH(Ar')—CONH}_2$$

3.5.3. Reactions with Potassium Superoxide*)

The oxidation of organic substrates with superoxide ion is of much current interest, firstly on a preparative basis and secondly because of the role $O_2^{\ominus}$ seems to play in biological systems. The superoxide ion has a rather complicated chemistry because it can act as an oxidizing or reducing agent as well as a nucleophile:

$$(1)\ R^{\ominus} + O_2^{\cdot\ominus} \longrightarrow R^{\cdot} + O_2^{2\ominus} \quad \text{oxidation}$$

$$(2)\ R^{\cdot} + O_2^{\cdot\ominus} \longrightarrow R^{\ominus} + O_2\uparrow \quad \text{reduction}$$

$$(3)\ RX + O_2^{\cdot\ominus} \longrightarrow R\text{—}O_2^{\cdot} + X^{\ominus} \quad \text{substitution.}$$

Reaction (1) requires protons to stabilize the peroxide dianion. In a protic milieu, however, decomposition occurs.

$$(4)\ 2O_2^{\cdot\ominus} \longrightarrow O_2^{2\ominus} + O_2$$

Thus, the use of this cheap reagent is limited more or less to aprotic solvents, but in the absence of crown ethers, it is partially soluble in only one organic solvent, DMSO. A 0.3 M solution of dicyclohexano-18-crown-6 in DMSO can be used to prepare a pale yellow 0.15 M solution of KO_2 [576]. In many cases benzene is better than DMSO, because it avoids potential complications by the DMSO anion [577]. The KO_2 dicyclohexano-18-crown-6 complex is soluble in benzene up to 0.05 M [577]. With 18-crown-6, KO_2 solutions in DMF, DME, and even ether have been made [578]. Stable solutions of tetraethylammonium superoxide in aprotic solvents were prepared by electrochemical generation [579, 587] and it was demonstrated recently that superoxide can be activated in the presence of Aliquat 336 as PT catalyst [1016]. The nucleophilic properties of superoxide have been utilized by a number of groups. A comparison of reactivities towards 1-bromo-octane (05. M in DMSO) by KI and KO_2 (0.5 M) in the presence of 18-crown-6 (0.05 M) showed reaction half-lives of $\approx$20 hours or $\approx$45 seconds, respectively [580]. Thus, superoxide is a "supernucleophile." Several groups have reported a diverse array of products from reactions with alkyl halides and alkyl sulfonate esters, depending on the conditions.

Johnson and Nidy [581] synthesized dialkyl peroxides from stoichiometric quantities of alkyl bromides, dicyclohexano-18-crown-6, and KO_2 in benzene (3–6 hours at room temperature). Alcohols and olefins were formed to a minor extent. If only 0.1 equivalent of crown ether was used, yields were somewhat lower and reaction times

*) For a review of KO_2 chemistry, see [913].

longer. In contrast, using excess KO_2 in DMSO, San Filippo, Chern, and Valentine [580] found alcohols as the major and carbonyl compounds as the minor products. Explosive benzo-1,2-dioxenes were prepared from *o*-bis-halomethyl aromatics and KO_2/dicyclohexano-18-crown-6 [927]. Corey *et al.* [578, *cf.*, also 861, 925] also synthesized alcohols *via* this reaction in aprotic, DMSO-containing media. It is preparatively interesting and important that the first step of all these reactions must be an SN2 substitution [578, 579, 580, 581, 861]:

$$\text{(R)- } C_6H_{13}CH(CH_3)\text{—Br} \xrightarrow[C_6H_6]{KO_2/\text{crown}} \left(\text{(S, S)- } C_6H_{13}\text{—}CH(CH_3)\text{—O}\right)_2 \quad [581]$$

$$\text{(S)- } C_6H_{13}\text{—}CH(CH_3)\text{—OTos} \xrightarrow[\text{DMSO}]{KO_2/\text{crown}} \text{(R)- } C_6H_{13}\text{—}CH(CH_3)\text{—OH} \quad [580]$$

cholesterol *p*-toluenesulfonate $\xrightarrow[\text{DMSO/DME/DMF}]{KO_2/18\text{-crown-6}}$ 3-epi-cholesterol [578]

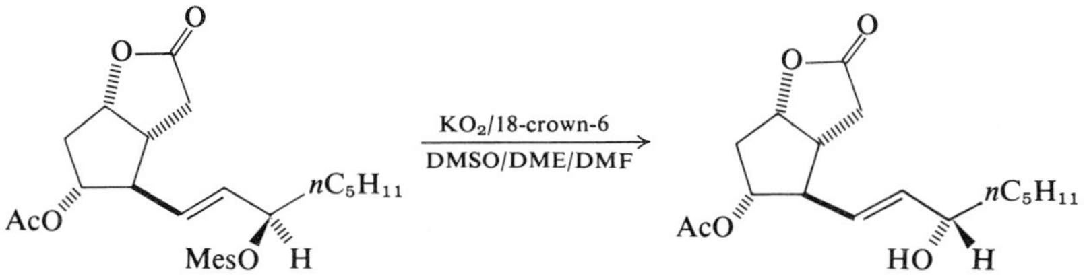

[578]

Scheme 3–197

Discussion of the interplay of the consecutive mechanistic steps and the intermediates involved is still going on [58, 582, 583, 584, 587, 1188, 1189, 1191].

Table 3–24 gives a survey of pertinent work. The yields of primary dialkyl peroxides are comparable or better than with older methods and the yields with secondary substrates (except cyclohexyl) are better. Interestingly enough, the preparation of diacyl peroxides from KO_2 and acid chlorides proceeds readily without the use of crown ether even in benzene [585]. The high yield of inverted alcohols from tosylates or mesylates in the lower part of Table 3–24 is preparatively of special significance. In some cases elimination may interfere, however [578]. *Tert*-butylbromide simultaneously gives the products of displacement (*tert*-butanol) and elimination (isobutene) [578].

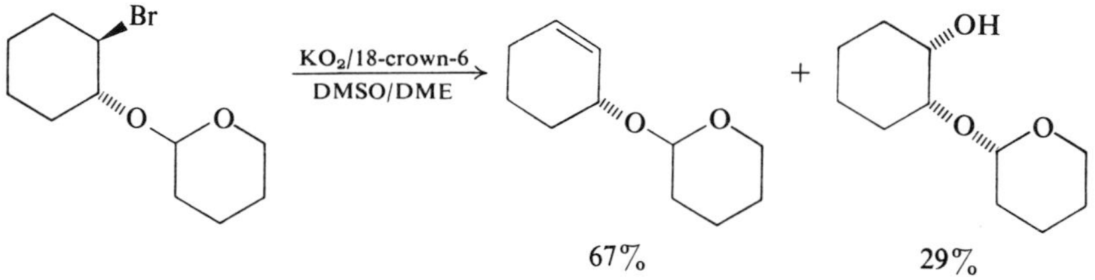

Scheme 3–198

Table 3–24. Reactions of Alkyl Halides and Sulfonates with Potassium Superoxide.

Substrate	Conditions*)	Yield Dialkyl Peroxide (%)	Alcohol	Olefin(s)	Ketone or Aldehyde	Reference
n-$C_{5-7}H_{11-15}$—Br	A	53–56			—	[581]
n-$C_{18}H_{37}$—X (X = Br, OTos, OMs)	A	46–77	21–42		—	[581]
cyclohexyl—Br	A			67	—	[581]
cyclopentyl—Br	A	42		24	—	[581]
$C_6H_{13}CH(CH_3)$—X (X = Br, OTos, OMs)	A	44–55	0–13	14–37	—	[581]
n-$C_8H_{17}X$ (X = Cl, Br, I, OTos)	B	—†	46–75†	0–3	< 1–12	[580]
n-$C_6H_{13}CH(CH_3)$—X (X = Cl, Br, I, OTos)	B	—	36–75	12–48	< 1	[580]
Ph—CH_2Cl	B	—	41	—	6	[580]
$C_3H_7C(CH_3)_2$—Br	B	—	20	30	—	[580]
OTos ⟶ OH	C	—	95	—	—	[578]
OTos ⟶ OH	C	—	96	—	—	[578]
Br, Br ⟶ OH	C	—	100	—	—	[578]

*) Conditions A: equal molar amounts of KO_2, dicyclohexano-18-crown-6, substrate in benzene, 3–6 hours, room temperature. Conditions B: 3.33 mmoles substrate, 10 mmoles KO_2, 1 mmole 18-crown-6 in 20 ml dry DMSO, 1.5–3 hours room temperature. Conditions C: 4 equivalents KO_2 and 4 equivalents 18-crown-6 in DMSO/DME at room temperature.
†) Other authors found peroxides as main products under identical conditions [1191].

Frimer and Rosenthal observed nucleophilic displacements of halogen by $O_2^{\bullet\ominus}$ in aromatic rings activated by electron-withdrawing substituents such as nitro groups [589]. An excess of both KO_2 (4-fold) and dicyclohexano-18-crown-6 (2-fold) in benzene results in fast reactions (a few hours) at room temperature:

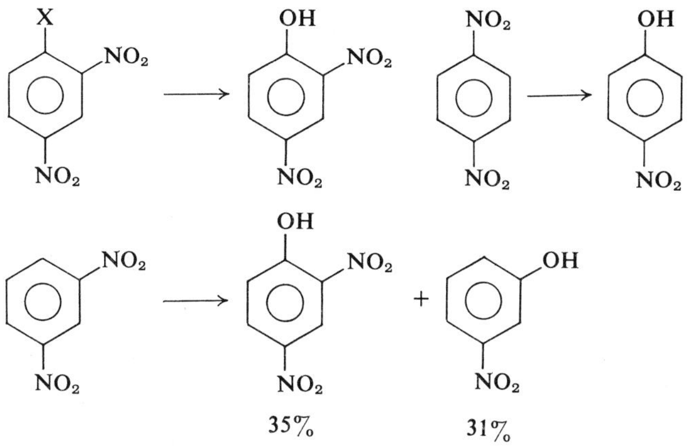

Scheme 3–199

Monohalogenodinitro compounds react 100 times faster than their monohalogeno-mononitro analogues, again faster than dinitrobenzenes. ^{18}O-tracer experiments indicate that the first step in these reactions is an electron transfer from $O_2^{\bullet\ominus}$ to the substituted benzene to give the anion radical, which is subsequently scavenged by molecular oxygen. Thus, the reaction is not a normal nucleophilic aromatic substitution.

Turning now to quite different types of reactions, San Filippo and co-workers demonstrated high yields of ester cleavages by vigorous stirring with excess KO_2 and catalytic (1/3 molar) amounts of 18-crown-6 in benzene for 8–140 (rare) hours and subsequent aqueous workup [586, 1194]. This cleavage to alcohol and acid seems to occur in many esters, including those of primary, secondary, and tertiary alcohols, as well as phenols and thiols. Phosphates are also cleaved. With DMSO as solvent the reaction time is shorter. The possibility of oxygen-alkyl-cleavage by the supernucleophile was considered and rejected, at least for secondary alcohols, as 99% retention of configuration was observed [586]. Simple amides and nitriles are unaffected.

The same group tested the oxidizing properties of solid excess KO_2/catalytic amounts 18-crown-6 in benzene with hydrazines, hydrazones, and related compounds [588]. Stirring was extended in most cases to 24 hours. Monosubstituted hydrazines, especially arylhydrazines, are converted to nitrogen-free products, often hydrocarbons, in a process apparently involving free radicals. 1,2-diarylhydrazines give the corresponding azo compounds; 1,1-disubstituted hydrazines give N-nitroso compounds. Hydrazones of aromatic ketones yield azines.

Tetracyclone (**A**) is oxidized by excess KO_2, and a molar amount of dicyclohexano-18-crown-6 in benzene to 2-hydroxy-2,4,5-triphenylfuranone-3 (**B**) [577]. A mechanism for this conversion has been formulated on the basis of ^{18}O-tracer studies in chalcone systems [590].

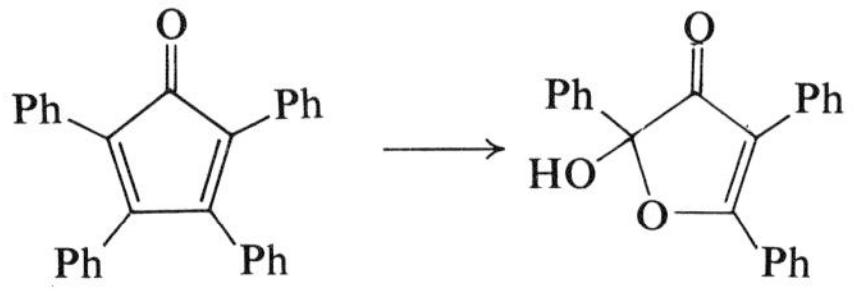

A → B

$$\mathrm{Ar{-}\overset{O}{\overset{\|}{C}}{-}CH{=}CH{-}Ar'} \xrightarrow{KO_2} \mathrm{ArCOOH + HOOC{-}CH_2{-}Ar'}$$

C D E

Scheme 3–200

Chalcones **C** are cleaved under similar conditions to give substituted benzoic acids (**D**) and aryl acetic acids (**E**) [590]. In these and the **A** → **B** reactions, the crown ethers are functioning as hydrogen donors and are eventually destroyed. Under similar conditions related compounds such as fluorenone, tetraphenylethene, and 2,5-diphenylfuran yielded only minimal amounts of polar compounds, and alkyl-substituted olefins and aromatic hydrocarbons were unreactive. Substrates having activated hydrogens, like 9,10-dihydroanthracene, 1,3- and 1,4-hexadiene, anthrone, and similar compounds, are oxidized, however [591] (KO_2: 18-crown-6: substrate 2:2:1 in dry DMSO). 2,5-di-*tert*-Butylhydroquinone is rapidly converted into the quinone [591], but 2,6-di-*tert*-butylphenoxy free radicals are reduced to the phenoxides [592].

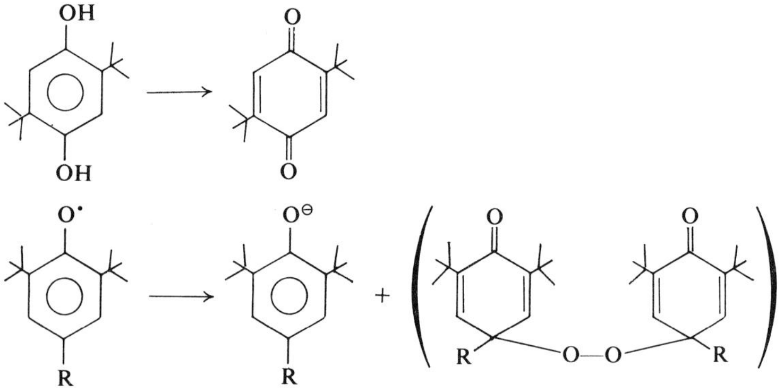

Scheme 3–201A

Another group demonstrated the formation of semiquinone radical anions both by the oxidation of 1,2- and 1,4-dihydroxyarenes and by the reduction of 1,4-quinones with KO_2/dicyclohexano-18-crown-6/THF [593]. In the case of *o*-dihydroxyarenes, further oxidation of semiquinones to dicarboxylic acids occurred.

Electron-poor olefins, α-nitro, and α,α-dicyanostyrenes are cleaved to the corresponding benzoic acids or ketones by stirring with excess KO_2/excess 18-crown-6 in dry benzene in the dark [898]. A primary formation of a radical anion is assumed in preference to attack by $O_2^{\bullet\ominus}$.

Secondary N-chloroamines were converted into imines and from there into carbonyl compounds by the action of KO_2/18-crown-6 in ether at room temperature (4–6 hours stirring) [977]. Ketones, including nonactivated ones, are cleaved into two moles of

$$R^1R^3CH{-}N(Cl)R' \longrightarrow R^1R^2C{=}N{-}R' \xrightarrow{H_3O^\oplus} R^1R^2C{=}O$$

$$R{-}CH_2{-}CO{-}R' \longrightarrow R{-}COOH + HOOC{-}R'$$

carboxylic acid by KO_2/Aliquat 336 in benzene at room temperature [1016].

An α-tocopherol model was oxidized in acetonitrile by KO_2/dicyclohexano-18-crown-6 as indicated in Scheme 3–201B [1193].

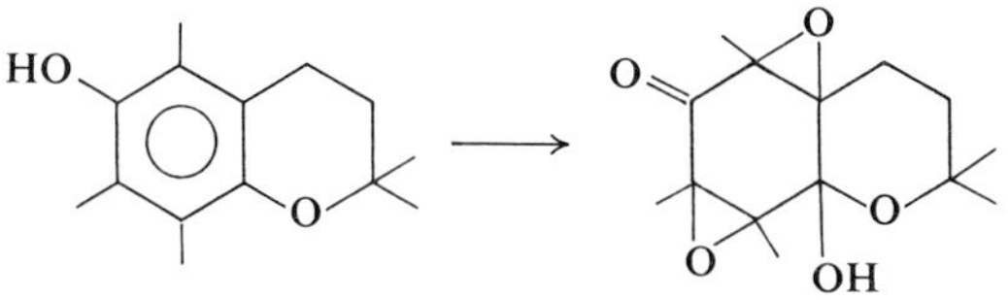

Scheme 3–201B

A new method to generate singlet oxygen consists in reacting KO_2/18-crown-6 in benzene with dibenzoyl or dilauryl peroxide [1190]:

$$2KO_2 + RCO{-}OO{-}COR \longrightarrow 2K^\oplus\ {}^\ominus O_2CR + 2O_2$$

S—S-Bonds of disulfides, thiolsulfinates, and thiosulfonates were oxidatively cleaved by KO_2/18-crown-6 in pyridine yielding sulfinic and sulfonic acids [1192]:

$$R{-}S{-}S{-}R,\ R{-}SO{-}S{-}R,\ R{-}SO_2{-}S{-}R \longrightarrow R{-}SO_2^\ominus + R{-}SO_3^\ominus$$

3.5.4. Other Oxidizing Agents

Although many other useful PTC applications to inorganic oxidants can be envisaged, until now only a few have been developed.

3.5.4.1. Chromate

Gibson *et al.* prepared organic solutions of dichromate in chloroform or nitrobenzene using methyltriphenylarsonium chloride as the counterion source [569]. Contrary to $AsMePh_3^\oplus MnO_4^\ominus$, dichromate was not an active oxidizing agent in the case of alcohols. The reason for this could be that the first step in the chromate oxidation of alcohols is the formation of chromate esters, but this is difficult to achieve in the absence of acid. Recently, Hutchins *et al.* [890] extracted $Cr_2O_7^{2-}$ (transferred ion unknown) with Adogen 464 into benzene, CCl_4, $HCCl_3$, or CH_2Cl_2. These solutions were capable of oxidizing allylic and benzylic alcohols to aldehydes or ketones in a rather slow reaction (15–18 hours at 55 °C). Again, primary alkanols were almost inert, and secondary ones reacted extremely slowly [890]. If, however, the aqueous phase contained 3–10 M H_2SO_4, a species could be extracted with tetrabutylammonium hydrogen sulfate into CH_2Cl_2 that oxidized even primary alcohols within minutes [991]. A slight excess of $K_2Cr_2O_7$ in the aqueous phase gave the best results. It is possible that $HCr_2O_7^{\ominus}$ is the active species. Primary alcohols can be converted into aldehydes with only minimal overoxidation if conditions are controlled carefully [991, 1196]. Best conditions are: −5 to 0 °C, dropwise addition of K_2CrO_4 in 30% H_2SO_4, $CHCl_3$ or CH_2Cl_2 as solvent, $NBu_4^{\oplus}HSO_4^{\ominus}$ as catalyst [1196].

Addition of tetrabutylammonium chloride to an aqueous solution of CrO_3 results in the formation of a precipitate described as $NBu_4^{\oplus}HCrO_4^{\ominus}$. It seems to be reasonably stable at room temperature, although precautions against unexpected violent decomposition may be warranted. In boiling chloroform, this reagent oxidizes secondary alcohols to ketones (3–12 hours) and allyl and benzyl alcohols to aldehydes (1–4 hours) [1198].

Formation of chromate esters is ensured if allylic or benzylic halides are reacted with chromate/dicyclohexano-18-crown-6 (or dibenzo18-crown-6) in HMPT [594]. Yields of around 80% were observed after 2 hours at 100 °C. Under these conditions, the corresponding alcohols were inert, *n*-alkyl bromides reacted only slowly (20% yield). Alternatively, the formation of aldehydes and ketones from benzyl and allyl halides was achieved by reactions with an isolated dichromate salt in chloroform [1197]. The reagent (believed to be $(NBu_4)_2Cr_2O_7$) was obtained by dichloromethane extraction from aqueous K_2CrO_4/NBu_4HSO_4. Take note that it may be difficult to differentiate this salt from the one described above as $NBu_4^{\oplus}HCrO_4^{\ominus}$ by simple C, H, N analysis or iodometric titration.

3.5.4.2. Hypochlorite

Lee and Freedman [595, 998, 1195] found that aryl carbinols are smoothly converted into aldehydes or ketones by stirring an excess of 10% aqueous NaOCl with a CH_2Cl_2 solution of substrate containing 5% NBu_4HSO_4 for 1–3 hours at room temperature. Little or no reaction is observed in the absence of catalyst. An interesting, still unexplained, solvent effect was observed for this oxidant: Benzene, carbon tetrachloride, chloroform, and methylene chloride are all suitable as solvents, but ethyl acetate in particular increases the rate of oxidation, thus permitting the oxidation of aliphatic secondary alcohols. Primary amines can also be oxidized by PTC—$OCl^{\ominus}$. Only those bearing a disubstituted α-carbon atom giving ketones are, however,

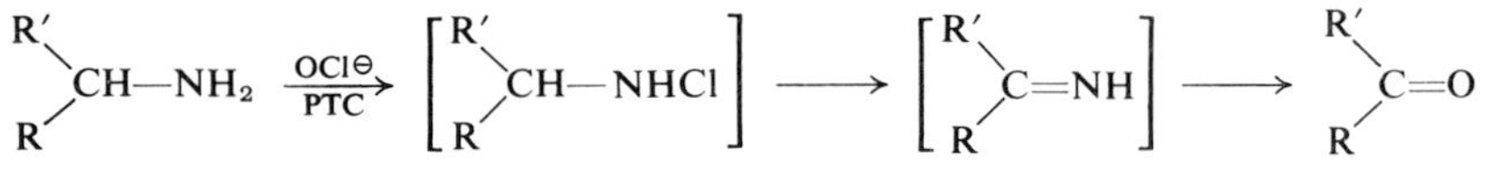

$$RR'CH{-}NH_2 \xrightarrow[PTC]{OCl^{\ominus}} [RR'CH{-}NHCl] \longrightarrow [RR'C{=}NH] \longrightarrow RR'C{=}O$$

Scheme 3–202

Table 3–25. Oxidations with $NaOCl/NBu_4HSO_4$ [595].

Reaction	Solvent	Yield (%)
alcohol → aldehyde or ketone		
$Ph—CH_2OH$	CH_2Cl_2	76
o-$CH_3O—C_6H_4CH_2OH$	CH_2Cl_2	47
	EtOAc	94
p-$CH_3O—C_6H_4CH_2OH$	CH_2Cl_2	79
	EtOAc	92
p-$CH_3—C_6H_4CH_2OH$	CH_2Cl_2	78
	EtOAc	100
p-$Cl—C_6H_4CH_2OH$	CH_2Cl_2	82
Ph_2CHOH	CH_2Cl_2	82
9-fluorenol	CH_2Cl_2	92
cycloheptanol	EtOAc	89
2-norbornanol	EtOAc	36
4-*tert*-butylcyclohexanol	EtOAc	49
amine → ketone, amine → nitrile		
cyclohexylamine → cyclohexanone	EtOAc	98
norbornylamine → norbornanone	EtOAc	84
benzhydrylamine → benzophenone	EtOAc	94
α-methylbenzylamine → acetophenone	EtOAc	98
(cyclohexylmethyl)amine → cyclohexylnitrile	EtOAc	76
1-octylamine → 1-cyanoheptane	EtOAc	60
polycyclic arenes → arene oxides, ref. [961]	$HCCl_3$	10–95

synthetically important. Primary amines monosubstituted in the α-carbon position give predominantly nitriles, with aldehydes as side-products. Amides react with $OCl^{\ominus}$/PT catalyst through successive Hofmann degradation and oxidation of the resultant amine. Generally, lower overall yields of nitriles and aldehydes or ketones respectively are observed. Table 3–25 surveys the yields. Hypochlorite oxidations have also been performed in "triphase catalysis" processes using a specially prepared polymer-bound catalyst [62]. Other oxidations where hypobromite is generated *in situ* by the electrochemical oxidation of bromide are also known, NBu_4HSO_4 is the catalyst in aqueous amyl acetate emulsions [1002].

3.5.4.3. Potassium Hexacyanoferrate(III)

$(NBu_4)_3[Fe(CN)_6]$ can be extracted into chloroform.*) It does not oxidize the phenol **A** at room temperature. Oxidation starts at once, however, when 2 equivalents of p-toluenesulfonic acid are added. **B** is formed, and the real oxidant is dihydrogen hexacyanoferrate(III) [596]. If the oxidation is performed in acidic methanol with hydrogen hexacyanoferrate, different products (**C** and **D**) arise, the proportion of which is somewhat dependent on the acidity of the medium. More of the p-methoxymethylphenol **D** is formed the less acidic the medium.

---

*) $[Fe(CN)_6]^{3\ominus}$ is more lipophilic than bromide; $[Fe(CN)_6]^{4\ominus}$ is intermediate between chloride and bromide in extractability [1185].

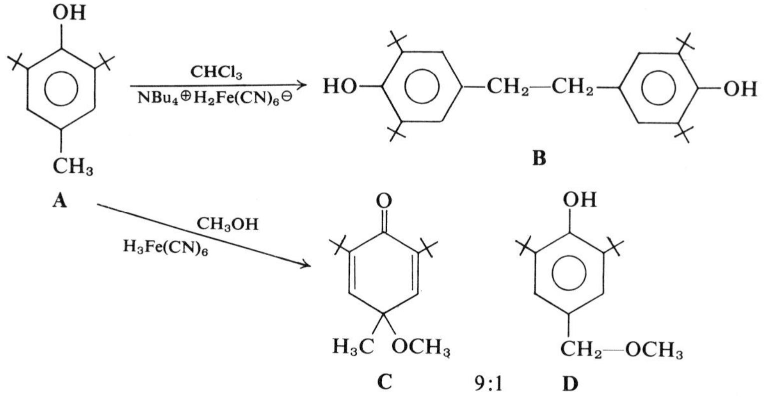

OH
CH$_3$
A
$CHCl_3$
$NBu_4^{\oplus}H_2Fe(CN)_6^{\ominus}$
HO—CH$_2$—CH$_2$—OH
B
CH_3OH
$H_3Fe(CN)_6$
O
H_3C OCH_3
C
9:1
OH
CH_2—OCH_3
D

Scheme 3–203

These results suggest that for oxidation in chloroform solution a purely free radical mechanism operates. In methanolic solution the originally formed O-radical can either be further oxidized to a phenoxonium ion (predominant process in stronger acidic solution) or may become disproportionate to quinone methide and the parent phenol:

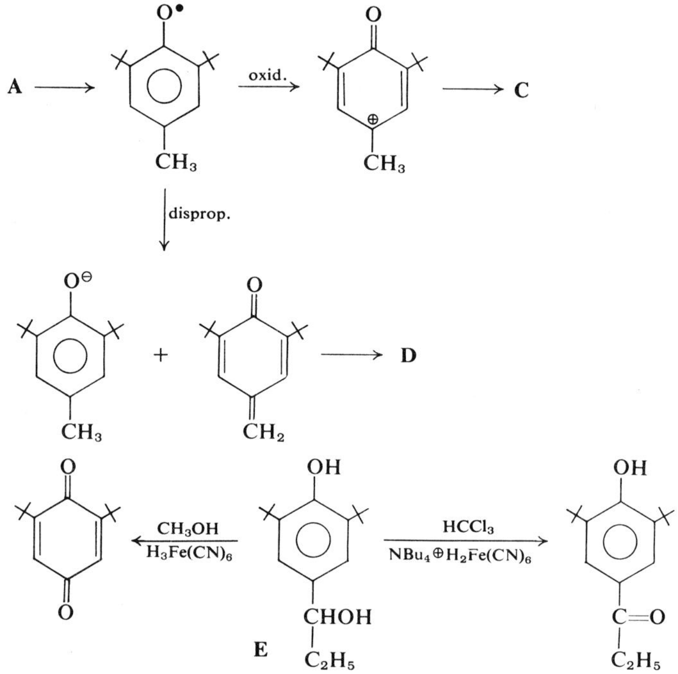

A →
O•
CH$_3$
oxid.
O
⊕
CH$_3$
→ C
disprop.
$O^{\ominus}$
CH$_3$
+
O
CH$_2$
→ D
O
O
CH_3OH
$H_3Fe(CN)_6$
OH
CHOH
E
C_2H_5
$HCCl_3$
$NBu_4^{\oplus}H_2Fe(CN)_6$
OH
C=O
C_2H_5

Scheme 3–204

A similar change in reaction path with acidity and solvent was observed in the substituted *p*-hydroxybenzylalcohol (E) [596].

3.5.4.4. Periodate

$Et_4N^{\oplus}IO_4^{\ominus}$ is a white crystalline solid. It is prepared by neutralizing $NEt_4^{\oplus}OH^{\ominus}$ with the calculated amount of paraperiodic acid. The salt is stable for weeks in a desiccator and can be handled with due caution (possible explosive! [1203]). It is very soluble in water as well as in many organic solvents. Qureshi and Sklarz [597] used this reagent to oxidize a nitrone to a nitroso compound in chloroform at room temperature. The same reaction works in water/C_2H_5OH with $NaIO_4$. Similarly, benzhydroxamic acid was oxidized to a very reactive intermediate benzoyl nitroso compound by $NProp_4^{\oplus}IO_4^{\ominus}$ [1200]. In other cases no oxidation occurred in the absence of a hydroxylic solvent [1186].

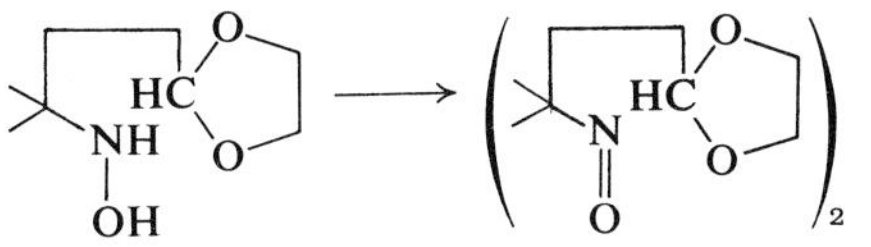

HC NH OH O O → (HC N O O O)$_2$

Scheme 3–205

The preceding reaction was carried out in homogeneous medium. Closer to our theme of PTC, and preparatively more interesting, is glycol cleavage in a two-phase medium. Lower glycols are very hydrophilic so that there is no need to work in a two-phase system. Sometimes, however, lipophilic glycols are intermediates in the well-known Lemieux-Johnson procedure [598] of oxidizing an olefin with periodate/osmium tetroxide. In this case a PTC version can be advantageous. Thus, Starks and co-workers showed that the oxidation of 1-octene in hydrocarbon solution to yield *n*-heptanal (and a little heptanoic acid) proceeds in a mildly exothermic reaction with 2 equivalents of aqueous paraperiodic acid in the presence of 0.2mole -% OsO_4 and less than 1 mole-% Aliquat 336 [39, 599].

$$R{-}CHO \xleftarrow[\text{Aliquat, 1 h}]{H_5IO_6/OsO_4} R{-}CH{=}CH_2 \xrightarrow[\text{Aliquat, 2 h}]{H_5IO_6/RuO_2} R{-}COOH$$

In a similar reaction with 4 equivalents of H_5IO_6 and ruthenium dioxide instead of osmium tetroxide, an almost quantitative yield of *n*-heptanoic acid was realized [599]. Tertiary amines (*e.g.*, tridodecylamine) can be used as catalyst in both conversions, since ion pairs of the type $NR_3H^{\oplus}$ $H_4IO_6^{\ominus}$ can be extracted into the organic medium [600]. However, it appeared that the tertiary amines were much less efficient catalysts than the quaternary ammonium salts or quaternary phosphonium salts (*e.g.*, tributylstearylphosphonium bromide) [599]. Without any catalyst these reactions were extremely slow.

3.5.4.5. Peracetic Acid

Two similar oxidative methods for the preparation of diaryldiazomethanes from diarylketone hydrazones have been reported [601], a process classically performed with mercuric oxide. In the non-PTC method (a), solvents CH_2Cl_2, EtOH, $CHCl_3$, or

$$(Ar)_2C{=}N{-}NH_2 \xrightarrow[\text{method (b) peracetic acid, trace } I_2\text{, NaOH/pH 10, PT catalyst}]{\text{method (a) peracetic acid, trace } I_2\text{, 1,1,3,3-tetramethylguanidine}} (Ar)_2C{=}N_2$$

$ClCH_2{-}CH_2Cl$ may be used. The presence of the cocatalyst iodine is essential, and its optimum concentration is 10^{-4} molar. Below that value oxidation is too slow, and much above it decomposition of the diazo compound to the azine is accelerated. The base 1,1,3,3-tetramethylguanidine proved the best, but many others were themselves oxidatively destroyed. Yields of 70–98.6% of various diaryldiazomethanes were realized. Beside peracetic acid, *m*-chloroperbenzoic acid and chloramine-T can be used as oxidants.

For the PTC method (b), various catalysts were examined for the preparation of diphenyldiazomethane.

Catalyst	Yield (%)
none	17.5
NBu_4OH	78.6
NBu_4Cl	65.5
NBu_4I	48.0
$N(C_8H_{17})_3C_3H_7Cl$	84.9

It is apparent that the highly hydrophilic peracetate needs a catalyst with a lipophilic cation and a nonlipophilic anion. Of the other oxidants tested, only chloramine-T approached peracetic acid in efficiency.

PTC-procedure: 0.1 mole benzophenone hydrazone is dissolved in 100 ml CH_2Cl_2, and 3 mmoles $(C_8H_{17})_3C_3H_7NCl$ and 4 ml 10% w/v I_2 are added. The mixture is stirred at 0 °C while 0.1 mole peracetic acid solution and aqueous NaOH are added simultaneously in such a way that the pH is always kept near 10. The reaction is finished after 15 minutes.

3.5.4.6. Oxygen as Oxidant

Oxygen is a well-known oxidant for certain carbanions. Because of the ready generation of some carbanions under PTC conditions preparative improvements of such reactions have been realized. Dietrich and Lehn [359] reported on the quantitative reaction: fluorene → fluorenone in THF with atmospheric oxygen in the presence of solid potassium hydroxide and cryptate [2.2.2] (Kryptofix [222], **5**). Durst performed the same reaction in benzene using 18-crown-6 [571]. Dicyclohexano-18-crown-6 proved unsuitable. Alneri and co-workers [572] were able to oxidize aromatics containing active methylenes and partially hydrogenated aromatics in the presence of 50% aqueous sodium hydroxide and a quaternary ammonium catalyst. The following catalyst scale of activity was found: $NEt_4^{\oplus}OH^{\ominus} \approx PhCH_2NMe_3^{\oplus}OH^{\ominus} > (C_{16}H_{33})_2Et_2N^{\oplus}Cl^{\ominus} \gg$ TEBA $\gg NEt_4^{\oplus}Cl^{\ominus} > NEt_4^{\oplus}Br^{\ominus}$. Oxidations were performed at 30–50 °C.

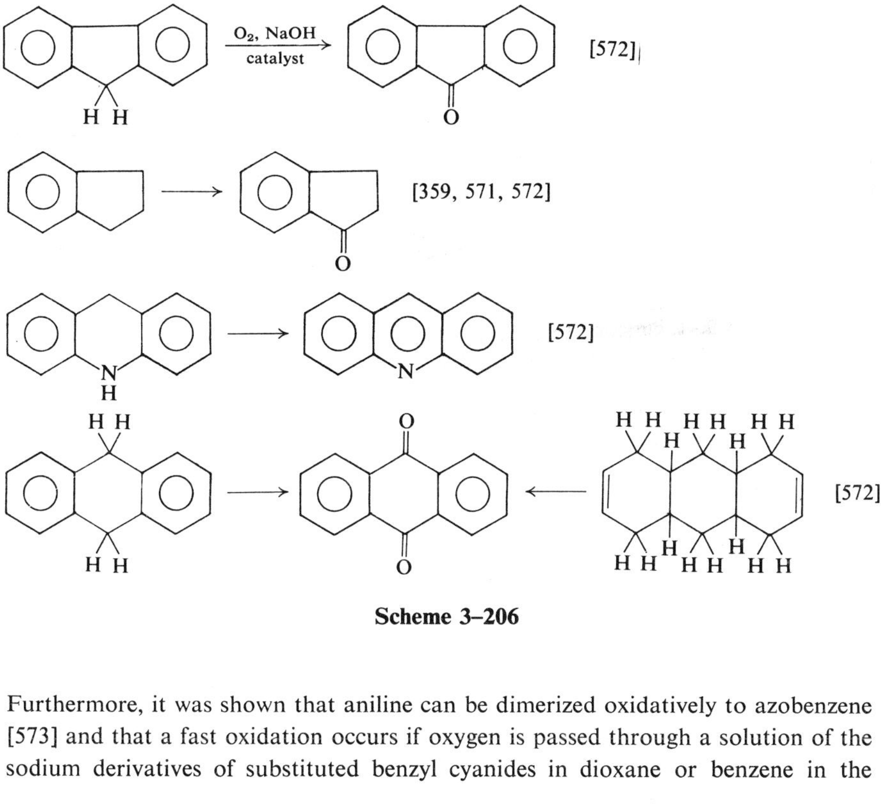

O_2, NaOH / catalyst [572]

[359, 571, 572]

[572]

[572]

Scheme 3–206

Furthermore, it was shown that aniline can be dimerized oxidatively to azobenzene [573] and that a fast oxidation occurs if oxygen is passed through a solution of the sodium derivatives of substituted benzyl cyanides in dioxane or benzene in the

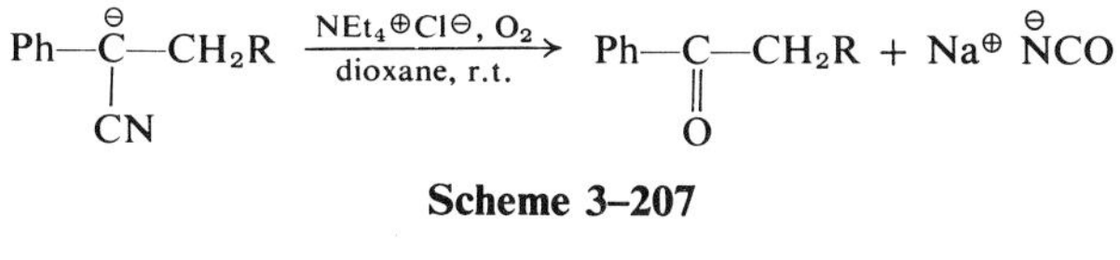

$$Ph{-}NH_2 \xrightarrow[O_2]{\text{18-crown-6, KOH}} Ph{-}N{=}N{-}Ph$$

$$Ph{-}\overset{\ominus}{C}(CN){-}CH_2R \xrightarrow[\text{dioxane, r.t.}]{NEt_4^{\oplus}Cl^{\ominus},\ O_2} Ph{-}C({=}O){-}CH_2R + Na^{\oplus}\ \overset{\ominus}{N}CO$$

Scheme 3–207

presence of PT catalysts. Aromatic ketones are formed [574, 1187]. Undoubtedly numerous other oxidations are possible under similar conditions. In a recent example, benzene solutions of Reissert-like compounds were shown to give the following products with 50% NaOH/TEBA/air [973]:

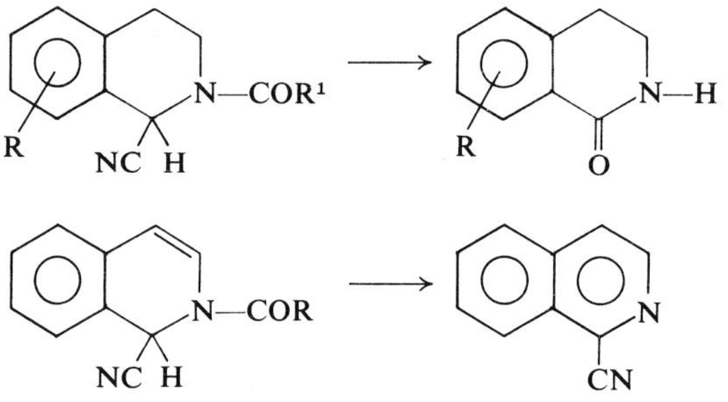

Scheme 3–208

Benzene solutions of the α,β- or β,γ-unsaturated ketones were oxidized to the single hydroperoxide with conc.NaOH/TEBA/air in 15–20% yield [981]. PTC-like conditions were used to solubilize sensitizers for the photochemical generation of singlet

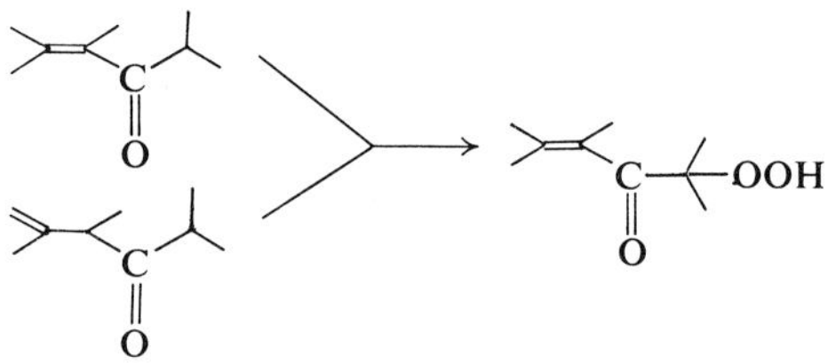

Scheme 3–209

oxygen. The authors worked in CS_2 or CH_2Cl_2 with the anionic dyes Rose Bengale and Eosin-Y applying 18-crown-6 or Aliquat 336 to make them soluble. Model reactions were the (4 + 2)cycloaddition of O_2 to anthracene and the ene-reaction with 2,3-dimethyl-2-butene [575, 1199].

References

Chapter 1

[1] J. Jarrousse, *C.R. Acad. Sci. Paris*, **232**, 1424 (1951).

[2] A. T. Babayan, N. Gambaryan, and N. P. Gambaryan, *Zh. Obshch. Khim.*, **24**, 1887 (1954); *Chem. Abstr.*, **49**, 10879 (1955); *Chem. Zentralbl. Sonderb.*, **1950/54**, 4532.

[3] G. Maerker, J. F. Carmichael, and W. Port, *J. Org. Chem.*, **26**, 2681 (1961).

[4] H. B. Copelin and G. B. Crane, U.S. Pat. 2,779,781 (1957), *Chem. Abstr.*, **51**, 10579 (1957); R. Kohler and H. Pietsch, Germ. Pat. 944,995 (1956), *Chem. Abstr.*, **53**, 1831 (1959); P. Edwards, U.S. Pat. 2,537,981 (1951), *Chem. Abstr.*, **45**, 5177 (1951); A. Conix and U. L. Laridon, Belg. Pat. 563,173 (1958) and 565,478 (1958), *Chem. Abstr.*, **55**, 25356 (1961); E. Müller, O. Bayer, and H. Morschel, Germ. Pat. 959,497 (1957), *Chem. Abstr.*, **53**, 13665 (1959); Germ. Offenl. 268 621 (1913), *Chem. Zentralbl.*, **1914**, 310.

[5] M. Mąkosza and B. Serafinowa, *Rocz. Chem.*, **39**, 1223 (1965), and subsequent papers.

[6] M. Mąkosza and W. Wawrzyniewicz, *Tetrahedron Lett.*, **1969**, 4659.

[7] A. Brändström and K. Gustavii, *Acta. Chem. Scand.*, **23**, 1215 (1969).

[8] A. Brändström, *Kem. Tidskr.*, **1970**, Nos. 5–6, 1.

[9] C. M. Starks and D. R. Napier, Ital. Pat. 832,967 (1968); Brit. Pat. 1,227,144 (1971); French Pat. 1,573,164 (1969) and further concordances; *Chem. Abstr.*, **72**, 115271 (1970).

[10] C. M. Starks, *J. Am. Chem. Soc.*, **93**, 195 (1971).

[11] A. Brändström, "Preparative Ion Pair Extraction, an Introduction to Theory and Practice," Apotekarsocieteten/Hässle Läkemedel, Stockholm, 1974.

[12] E. V. Dehmlow, *Angew. Chem.*, **86**, 187 (1974); *Angew. Chem. Int. Ed. Engl.*, **13**, 170 (1974); reprinted in: New Synthetic Methods, Vol. 1, Verlag Chemie, Weinheim, 1975, p. 1, and—in slightly shortened form—in *Chem. Technol.*, **1975**, 210.

[13] E. V. Dehmlow, *Angew. Chem.* **89**, 521 (1977), *Angew. Chem. Int. Ed. Engl.*, **16**, 493 (1977); "New Synthetic Methods," Vol. 6, Verlag Chemie, Weinheim, 1979, p. 205.

[14] M. Mąkosza, *Pure Appl. Chem.*, **43**, 439 (1975).

[15] M. Mąkosza, Naked anions—phase transfer, in "Modern Synthetic Methods 1976," R. Scheffold, Ed., Schweizerischer Chemiker-Verband, Zürich, 1976, p. 7.

[16] J. Dockx, *Synthesis*, **1973**, 441.

[17] R. A. Jones, *Aldrichimica Acta*, **9**, 35 (1976).

[18] E. Schacht, *Kontakte* (company journal of E. Merck, Germany), 3/76, p. 3.

[19] G. D. Yadav, *Chem. Ind. Dev. (Bombay)*, **9**, 16 (1975); *Chem. Abstr.*, **83**, 146585 (1975).

[20] R. Oda, *Kagaku To Kogyo (Tokyo)*, **26**, 322 (1973) [in Jap.], *Chem. Abstr.*, **81** 17066 (1974); *Hyomen*, **12**, 262 (1974) [in Jap.], *Chem. Abstr.*, **81**, 176904 (1974).

[21] L. P. Hammett, "Physical Organic Chemistry—Reaction Rates, Equilibria, and Mechanisms," McGraw-Hill, Inc., New York, 1970; "Physikalische Organische Chemie—Reaktionsgeschwindigkeiten, Gleichgewichte und Mechanismen," Verlag Chemie, Weinheim, 1973.

[22] M. Szwarc, Ed., "Ions and Ion Pairs in Organic Reactions," Vols. 1 and 2, Wiley-Interscience, New York, 1972.

[23] G. A. Olah and P. v. R. Schleyer, Eds., "Carbonium Ions, Vol. 1, General Aspects and Methods of Investigation," Interscience Publishers, New York, 1968.

[24] C. Reichardt, "Lösungsmittel-Effekte in der Organischen Chemie," Verlag Chemie, Weinheim 1969; C. Reichardt," Solvent Effects in Organic Chemistry," Monographs in Modern Chemistry, Vol. 3, Verlag Chemie, Weinheim, 1979.

[25] D. T. Copenhafer and C. A. Kraus, *J. Am. Chem. Soc.*, **73**, 4557 (1951).

[26] C. R. Witschonke and C. A. Kraus, *J. Am. Chem. Soc.*, **69**, 2472 (1947).
[27] K. H. Wong, M. Bourgoin, and J. Smid, *J. Chem. Soc. Chem. Commun.*, **1974**, 715.
[28] J. Almy, D. C. Garword, and D. J. Cram, *J. Am. Chem. Soc.*, **92**, 4321 (1970).
[29] E. J. Fendler and J. H. Fendler, Micellar catalysis in organic reactions—kinetic and mechanistic implications, in "Advances in Physical Organic Chemistry," Vol. 8, V. Gold, Ed., Academic Press, London and New York, 1970, p. 271.
[30] J. H. Fendler and E. J. Fendler, "Catalysis in Micellar and Macromolecular Systems," Academic Press, New York, 1975.
[31] J. H. Fendler, *Acc. Chem. Res.*, **9**, 153 (1976).
[32] J. Smid, *Angew. Chem.*, **84**, 127 (1972); *Angew. Chem. Int. Ed. Engl.*, **11**, 112 (1972).
[33] R. A. Sneen, *Acc. Chem. Res.*, **6**, 46 (1973).
[34] I. P. Beletskaya, *Usp. Khim.*, **44**, 2205 (1975); Engl. transl. p. 1067.
[35] D. J. McLennan, *Acc. Chem. Res.*, **9**, 281 (1976).
[36] D. J. Raber, J. M. Harris, and P. v. R. Schleyer, in ref. [22], Vol. 2, p. 247.
[37] B. S. Krumgal'z, *Zh. Obshch. Khim.*, **44**, 1617 (1974), Engl. transl. p. 1585.
[38] R. M. Fuoss and F. Accascina, "Electrolytic Conductance," Interscience Publishers, New York, 1959.
[39] M. Szwarc, "Carbanions, Living Polymers, and Electron Transfer Processes," Interscience Publishers, New York, 1968, *viz.* Chapter V.
[40] H. F. Ebel, *Fortschr. Chem. Forsch.*, **12**, 387 (1969).
[41] H. F. Ebel, "Die Acidität der CH-Säuren," G. Thieme, Verlag, Stuttgart, 1969.
[42] M. Schlosser, "Struktur und Reaktivität polarer Organometalle," Springer-Verlag, Berlin–Heidelberg, 1973.
[43] D. J. Cram, "Fundamentals of Carbanion Chemistry," Academic Press, New York, 1965.
[44] R. Modin, *Acta. Pharm. Suec.*, **9**, 157 (1972), and papers cited therein.
[45] R. Modin and G. Schill, *Acta. Pharm. Suec.*, **4**, 301 (1967), and subsequent papers of the same series.
[46] A. Brändström, ref. [112], p. 276.
[47] G. Schill, R. Modin, and B.-A. Persson, *Acta. Pharm. Suec.*, **2**, 119 (1965).
[48] B.-A. Persson and G. Schill, *Acta. Pharm. Suec.*, **3**, 281 (1966).
[49] K. Gustavii and G. Schill, *Acta. Pharm. Suec.*, **3**, 241 (1966).
[50] J. H. Blakeley and V. J. Zatka, *Anal. Chim. Acta*, **74**, 139 (1975).
[51] P.-O. Lagerström, K. O. Borg, and D. Westerlund, *Acta Pharm. Suec.*, **9**, 53 (1972).
[52] K. Gustavii, A. Brändström, and S. Allanson, *Acta Chem. Scand.*, **25**, 77 (1971).
[53] K. Gustavii, *Acta Pharm. Suec.*, **4**, 233 (1967).
[54] S. O. Jannsson, R. Modin, and G. Schill, *Talanta*, **21**, 905 (1974).
[55] R. Modin and A. Tilly, *Acta Pharm. Suec.*, **5**, 311 (1968).
[56] T. Nordgren and R. Modin, *Acta Pharm. Suec.*, **12**, 407 (1975),
[57] S. Eksborg and G. Schill, *Anal. Chem.*, **45**, 2092 (1973).
[58] K. Gustavii and G. Schill, *Acta Pharm. Suec.*, **3**, 259 (1966).
[59] B. Czapkiewicz-Tutaj and J. Czapkiewicz, *Rocz. Chem.*, **49**, 1353 (1975).
[60] R. Modin and M. Schröder-Nielsen, *Acta Pharm. Suec.*, **8**, 573 (1971).
[61] K. O. Borg and G. Schill, *Acta Pharm. Suec.*, **5**, 323 (1968).
[62] K. O. Borg and D. Westerlund, *Z. Anal. Chem.* **252**, 275 (1970).
[63] W. E. Clifford and H. Irving, *Anal. Chim. Acta*, **31**, 1 (1964).
[64] O. E. Zvyagintsev, N. M. Sinitsyn, and V. N. Pichkov, *Zh. Neorg. Khim.*, **11**, 198 (1966); Engl. transl. p. 107; *Chem. Abstr.* **64**, 8983 (1966).
[65] I. M. Ivanov, L. M. Gindin, and G. N. Chichagova, *Izv. Sib. Otd. Akad. Nauk SSSR, Ser. Khim. Nauk.*, **1967**, 100; *Chem. Abstr.*, **69**, 13377 (1968) and *Chem. Abstr.*, **78**, 8481 (1973).
[66] M. Mąkosza and E. Białecka, *Synth. Commun.*, **6**, 313 (1976).
[67] E. V. Dehmlow, M. Slopianka, and J. Heider, *Tetrahedron Lett.*, **1977**, 2361.

[68] Yu. G. Frolov, V. V. Sergievskii, A. V. Ochkin, and A. P. Zuev, *Radiokhimiya*, **14**, 643 (1972); Engl. transl. p. 664; *Chem. Abstr.*, **78**, 8479 (1973). *Solvent Extr.*, *Proc. Int. Solvent Extr. Conf.*, **2**, 1229 (1971); *Chem. Abstr.*, **83**, 153505 (1975).
[69] D. Landini, A. M. Maia, F. Montanari, and F. M. Pirisi, *J. Chem. Soc. Chem. Commun.*, **1975**, 950.
[70] Yu. G. Frolov, V. V. Sergievskii, A. P. Zuev, and L. B. Fedyanina, *Zh. Fiz. Khim.*, **47**, 1956 (1973), Engl. transl. p. 1101; *Chem. Abstr.*, **79**, 140143 (1973).
[71] Y. Marcus and A. S. Kerts, "Ion Exchange and Solvent Extraction of Metal Complexes," Wiley-Interscience, New York, 1969.
[72] "Recent Advances in Liquid-Liquid Extraction," C. Hanson, Ed., Pergamon Press, Oxford, 1971.
[73] "Solvent Extraction Chemistry," D. Dyrrssen, J. O. Liljenzin, and J. Rydberg, Ed., North-Holland Co., Amsterdam, 1967.
[74] "Solvent Extraction Research, Proc. 5th Int. Conf. on Solvent Extraction Chemistry," A. S. Kertes and Y. Marcus. Eds., Wiley-Interscience, New York, 1969.
[75] Ref. [71], p. 796; R. Bock and G. M. Beilstein, *Z. Anal. Chem.*, **192**, 44 (1963); R. Bock and J. Jainz, *Z. Anal Chem.*, **198**, 315 (1963).
[76] M. J. Gugeshashvils, M. A. Manvelyan, and L. J. Boguslavskii, *Elektrokhimiya* **10**, 819 (1974); Engl. transl. p. 782; *Chem. Abstr.*, **81**, 98514 (1974).
[77] C. J. Pedersen and H. K. Frensdorff, *Angew. Chem.* **84**, 16 (1972); *Angew. Chem. Int. Ed. Engl.*, **11**, 16 (1972).
[78] D. J. Cram and J. M. Cram, *Science*, **183**, 803 (1974).
[79] J. J. Christensen, D. J. Eatough, and R. M. Izatt, *Chem. Rev.*, **74**, 351 (1974).
[80] A. C. Knipe, *J. Chem. Educ.*, **53**, 618 (1976).
[81] G. W. Gokel and H. D. Durst, *Synthesis*, **1976**, 168.
[82] J. J. Christensen, J. O. Hill, and R. M. Izatt, *Science*, **174**, 459 (1971).
[83] B. Dietrich, J. M. Lehn, and J. P. Sauvage, *Tetrahedron Lett.*, **1969**, 2885, 2889; B. Dietrich and J. M. Lehn, *Tetrahedron Lett.*, **1973**, 1225; review: *Acc. Chem. Res.*, **11**. 49 (1978).
[84] R. C. Helgeson, K. Koga, J. M. Timko, and D. J. Cram, *J. Am. Chem. Soc.*, **95**, 3021 (1973); R. C. Helgeson, J. M. Timko, P. Moreau, S. C. Peacock, J. M. Mayer, and D. J. Cram, *ibid.*, **96**, 6762 (1974); M. Newcomb, R. C. Helgeson, and D. J. Cram, *ibid.*, **96**, 7367 (1974); L. R. Sousa, D. H. Hoffman, L. Kaplan, and D. J. Cram, *ibid.*, **96**, 7100 (1974); M. Newcomb, G. W. Gokel, and D. J. Cram, *ibid.*, **96**, 6811 (1974); G. Dotsevi, Y. Sogah, and D. J. Cram, *ibid.*, **97**, 1259 (1975), **98**, 3038 (1976); S. C. Peacock, and D. J. Cram, *J. Chem. Soc. Chem. Commun.*, **1976**, 282; G. W. Gokel, J. M. Timko, and D. J. Cram, *J. Chem. Soc. Chem. Commun.*, **1975**, 394, 444; *cf.*, W. D. Curtis, R. M. King, J. F. Stoddart, and G. H. Jones, *J. Chem. Soc. Chem. Commun.*, **1976**, 284, and further references,
[85] Recent references: F. Vögtle and B. Jansen, *Tetrahedron Lett.*, **1976**, 4895; I. J. Burden, A. C. Coxon, J. F. Stoddart, and C. M. Wheatley, *J. Chem. Soc. Perkin Trans. 1*, **1977**, 220; M. P. Mack, R. R. Hendrixson, R. A. Palmer, and R. G. Ghirardelli, *J. Am. Chem. Soc.*, **98**, 7830 (1976); R. C. Hayward, C. H. Overton, and G. H. Whitham, *J. Chem. Soc. Perkin Trans. 1*, **1976**, 2413; D. A. Laidler and J. F. Stoddart, *J. Chem. Soc. Chem. Commun.*, **1977**, 481; J. P. Behr, J. M. Lehn, and P. Vierling, *J. Chem. Soc. Chem. Commun.*, **1976**, 621; J. M. Girodeau, J. M. Lehn, and J. P. Sauvage, *Angew. Chem.*, **87**, 813 (1975), *Angew. Chem. Int. Ed. Engl.*, **14**, 764 (1975); D. G. Parsons, *J. Chem. Soc. Perkin Trans. 1*, **1975**, 245; D. A. Laidler, and J. F. Stoddart, *J. Chem. Soc. Chem. Commun.*, **1976**, 979; D. Curtis, J. F. Stoddart, and G. H. Jones, *J. Chem. Soc. Perkin Trans. I*, **1977**, 1756; N. Ando, Y. Yamamoto, J. Oda, and Y. Inouye, *Synthesis*, **1978**, 688.
[86] B. Dietrich and J. M. Lehn, *Tetrahedron Lett.*, **1973**, 1225.
[87] E. Buncel and B. Menon, *J. Am. Chem. Soc.*, **99**, 4457 (1977).
[88] E. Graf and J. M. Lehn, *J. Am. Chem. Soc.*, **98**, 6403 (1976); F. P. Schmidtchen,

Angew. Chem., **89**, 751 (1977); *Angew. Chem. Int. Engl.*, **16**, 720 (1977); B. Metz, J. M. Rosalky, and R. Weiss, *J. Chem. Soc. Chem. Commun.*, **1976**, 533.
[89] J. L. Dye, M. G. De Backer, and V. A. Nicely, *J. Am. Chem. Soc.*, **92**, 5226 (1970).
[90] F. J. Tehan, B. L. Barnett, and J. L. Dye, *J. Am. Chem. Soc.*, **96**, 7203 (1974).
[91] G. W. Gokel and D. J. Cram, *J. Chem. Soc. Chem. Commun.*, **1973**, 481.
[92] E. V. Dehmlow, *Tetrahedron Lett.*, **1976**, 91; E. V. Dehmlow and T. Remmler, *J. Chem. Res.*, **1977** (S) 72, (M) 766.
[93] A. Knöchel, J. Oehler, and G. Rudolph, *Tetrahedron Lett.*, **1975**, 3167.
[94] H. Normant, T. Cuvigny, and P. Savignac, Synthesis, **1975**, 805.
[95] G. V. Nelson and A. von Zelewsky, *J. Am. Chem. Soc.*, **97**, 6279 (1975); M. A. Kormarynski and S. I. Weissman, *ibid.*, **97**, 1589 (1975); B. Kaempf, S. Raynal, A. Collet, F. Schné, S. Boileau, and J. M. Lehn, *Angew. Chem.*, **86**, 670 (1974), *Angew. Chem. Int. Ed. Engl.*, **13**, 611 (1974).
[96] B. R. Agarwal and R. M. Diamond, *J. Phys. Chem.*, **57**, 2785 (1963).
[97] F. Vögtle and E. Weber, *Angew. Chem.*, **86**, 896 (1974), *Angew. Chem. Int. Ed. Engl.*, **13**, 814 (1974); F. Vögtle and H. Sieger, *Angew. Chem.*, **89**, 410 (1977), *Angew. Chem. Int. Ed. Engl.*, **16**, 396 (1977); B. Tümmler, G. Maass, E. Weber, W. Wehner, and F. Vögtle, *J. Am. Chem. Soc.*, **99**, 4683 (1977); F. Vögtle, W. M. Müller, W. Wehner, and E. Buhleier, *Angew. Chem.*, **89**, 564 (1977); *Angew. Chem. Int. Ed. Engl.*, **16**, 548 (1977).— For various PTC applications of open chain polyethers *cf.*, Chapter 3 and refs. [93, 94].
[98] M. Kirch and J. M. Lehn, *Angew. Chem.*, **87**, 542 (1975), *Angew. Chem. Int. Ed. Engl.*, **14**, 555 (1975); M. Newcomb, R. C. Helgeson, and D. J. Cram, *J. Am. Chem. Soc.*, **96**, 7367 (1974); J. J. Christensen, J. D. Lamb, S. R. Izatt, S. E. Starr, G. C. Weed, M. S. Astin, B. D. Stitt, and R. M. Izatt, *J. Am. Chem. Soc.*, **100**, 3219 (1978).
[99] C. L. Liotta and E. E. Grisdale, *Tetrahedron Lett.*, **1975**, 4205.
[100] J. P. Tam, W. F. Cunningham-Rundles, B. W. Erickson, and R. B. Merrifield, *Tetrahedron Lett.*, **1977**, 4001.
[101] M. J. Maskornick, *Tetrahedron Lett.*, **1972**, 1797.
[102] L. M. Thomassen, T. Ellingsen, and J. Ugelstad, *Acta Chem. Scand.*, **25**, 3024 (1971).
[103] R. A. Bartsch, E. A. Mintz, and R. M. Parlman, *J. Am. Chem. Soc.*, **96**, 4249 (1974); R. A. Bartsch, *Acc. Chem. Res.*, **8**, 239 (1975).
[104] W. Dorn, A. Knöchel, J. Oehler, and G. Rudolph, *Z. Naturforsch, Teil B*, **32**, 776 (1977).
[105] A. Knöchel and G. Rudolph, *Tetrahedron Lett.*, **1974**, 3739.
[106] *E.g.*, G. W. Gokel and B. J. Garcia, *Tetrahedron Lett.*, **1977**, 317.
[107] A. Knöchel, J. Oehler, and G. Rudolph, *Z. Naturforsch, Teil B*, **32**, 783 (1977).
[108] A. el Bosyony, J. Klimes, A. Knöchel, J. Oehler, and G. Rudolph, *Z. Naturforsch., Teil B*, **31**, 1192 (1976).
[109] T. Iwachido, M. Kimura, and K. Tôei, *Chem. Lett.*, **1976**, 1101.
[110] F. de Jong, D. N. Reinhoudt, and C. J. Smit, *Tetrahedron Lett.*, **1976**, 1371, 1375.
[111] *E.g.*, R. A. Bartsch, H. Chen, N. F. Haddock, and P. N. Juri, *J. Am. Chem. Soc.*, **98**, 6753 (1976).
[112] A. Brändström, Principles of phase transfer catalysis by quaternary ammonium salts, in "Advances in Physical Organic Chemistry," Vol. 15, V. Gold, Ed., Academic Press, London and New York, 1977, p. 267.
[113] W. P. Weber and G. W. Gokel, Phase transfer catalysis in organic synthesis, in "Reactivity and Structure," Vol. 4, K. Hafner *et al.*, Eds., Springer-Verlag, Berlin–Heidelberg, 1977.
[114] M. Mąkosza, *Uspekhi Khim.*, **46**, 2174 (1977); Engl. transl. p. 1151.
[115] J. M. McIntosh, *J. Chem. Educ.*, **55**, 235 (1978); G. W. Gokel and W. P. Weber, *ibid.*, **55**, 350, 429 (1978).
[116] C. M. Starks and C. Liotta, "Phase Transfer Catalysis; Principles and Techniques," Academic Press, London and New York, 1978.

Chapter 2

[1] A. Brändström, Principles of phase transfer catalysis by quaternary ammonium salts, in "Advances in Physical Organic Chemistry," Vol. 15, V. Gold, Ed., Academic Press, London and New York, 1977, p. 267.
[2] C. M. Starks, *J. Am. Chem. Soc.*, **93**, 195 (1971); C. M. Starks and R. M. Owens, *J. Am. Chem. Soc.*, **95**, 3613 (1973).
[3] D. H. Picker, Ph.D. dissertation, Department of Chemistry, State University of New York at Albany, 1974; *Diss. Abstr.* **35**, 3229 (1975).
[4] A. W. Herriott and D. Picker, *J. Am. Chem. Soc.*, **97**, 2345 (1975).
[5] D. Landini, A. M. Maia, F. Montanari, and F. M. Pirisi, *J. Chem. Soc. Chem. Commun.*, **1975**, 950; D. Landini, A. Maia, and F. Montanari, *J. Am. Chem. Soc.*, **100**, 2796 (1978).
[6] F. M. Menger, *J. Am. Chem. Soc.*, **92**, 5965 (1970).
[7] H. H. Freedman and R. A. Dubois, *Tetrahedron Lett.*, **1975**, 3251.
[8] A. Merz, *Angew. Chem.*, **85**, 868 (1973); *Angew. Chem. Int. Ed. Engl.* **11**, 846 (1973).
[9] E. V. Dehmlow and M. Lissel, *Tetrahedron Lett.*, **1976**, 1783.
[10] J. E. Gordon and R. E. Kutina, *J. Am. Chem. Soc.*, **99**, 3903 (1977).
[11] J. H. Fendler and E. J. Fendler, "Catalysis in Micellar and Macromolecular Systems," Academic Press, New York, 1975.
[12] A. Brändström, private communication; ref. [1], p. 307.
[13] D. Landini, A. Maia, and F. Montanari, *J. Chem. Soc. Chem. Commun.*, **1977**, 112.
[14] A. Brändström and H. Kolind-Andersen, *Acta Chem Scand.*, **29B**, 201 (1975).
[15] A. Brändström, *Acta Chem. Scand.*, **30B**, 203 (1976).
[16] M. Tomoi, T. Takubo, M. Ikeda, and H. Kakiuchi, *Chem. Lett.*, **1976**, 473.
[17] M. Mikolajczyk, S. Grzejszczak, A. Zatorski, F. Montanari, and M. Cinquini, *Tetrahedron Lett.*, **1975**, 3757.
[18] T. Tanaka and T. Mukaiyama, *Chem. Lett.*, **1976**, 1259.
[19] S. L. Regen, *J. Am. Chem. Soc.*, **98**, 6270 (1976).
[20] M. Cinquini, S. Colonna, H. Molinari, and F. Montanari, *J. Chem. Soc. Chem. Commun.*, **1976**, 394.
[21] S. L. Regen and L. Dulak, *J. Am. Chem. Soc.*, **99**, 623 (1977); S. L. Regen, J. J. Besse, and J. McLick, *J. Am. Chem. Soc.*, **101**, 116 (1979).
[22] S. L. Regen, *J. Am. Chem. Soc.*, **99**, 3838 (1977).
[23] A. W. Herriott and D. Picker, *Tetrahedron Lett.*, **1972**, 4521.
[24] H. F. Ebel, "Die Acidität der CH-Säuren," G. Thieme Verlag, Stuttgart, 1969.
[25] E. V. Dehmlow and M. Slopianka, unpublished results.
[26] M. Mąkosza, *Pure Appl. Chem.*, **43**, 439 (1975).
[27] M. Mąkosza, Naked anions—phase transfer, in "Modern Synthetic Methods," R. Scheffold, Ed., Schweizerischer Chemiker-Verband, Zürich, 1976, p. 7.
[28] E. V. Dehmlow, *Angew. Chem.* **89**, 521 (1977); *Angew. Chem. Int. Ed. Engl.*, **16**, 493 (1977).
[29] M. Lissel, Ph.D. dissertation, Technische Universität, Berlin, 1977.
[30] E. V. Dehmlow, M. Slopianka, and J. Heider, *Tetrahedron Lett.*, **1977**, 2361 (1977).
[31] M. Mąkosza and E. Białecka, *Tetrahedron Lett.*, **1977**, 183.
[32] E. V. Dehmlow and M. Lissel, unpublished results.
[33] E. V. Dehmlow and T. Remmler, *J. Chem. Res.* **1977**, (S) 72, (M) 766.
[34] S. Barahona-Naranjo, M.S. Diploma thesis, Technische Universität, Berlin, 1977.
[35] E. V. Dehmlow and S. Barahona-Naranjo, *J. Chem. Res.* **1979**, (S) 238.
[36] S. Lindenbaum and G. E. Boyd, *J. Phys. Chem.*, **68**, 911 (1964).
[37] W.-Y. Wen and S. Saito, *J. Phys. Chem.*, **68**, 2639 (1964).
[38] J. E. Gordon, J. C. Robertson, and R. L. Thorne, *J. Phys. Chem.*, **74**, 957 (1970).
[39] S. Shinkai and T. Kunitake, *J. Chem. Soc. Perkin Trans. 2*, **1976**, 980.
[40] Y. Okahata, R. Ando, and T. Kunitake, *J. Am. Chem. Soc.*, **99**, 3067 (1977).
[41] A. Tomita, N. Ebina, and Y. Tamai, *J. Am. Chem. Soc.*, **99**, 5725 (1977).

[42] A. W. Herriott and D. Picker, *Tetrahedron Lett.*, **1974**, 1511.
[43] S. Colonna and R. Fornasier, *Synthesis*, **1975**, 531.
[44] M. Mąkosza and B. Serafinowa, *Rocz. Chem.*, **39**, 1223 (1965).
[45] E. V. Dehmlow, M. Slopianka, and E. Menzel, unpublished results.
[46] E. V. Dehmlow and E. Menzel, unpublished results; E. Menzel, State Examination Thesis, Technische Universität, Berlin, 1977.
[47] J. Dockx, *Synthesis*, **1973**, 441.
[48] P. A. Verbrugge and E. W. Uurbanus, Germ. Offenl. 2,324,390 (1973); *Chem. Abstr.*, **80**, 70420 (1974).
[49] J. Kranz, State Examination Thesis, Technische Universität, Berlin, 1977.
[50] T. Hiyama, M. Tsukanaka, and H. Nozaki, *J. Am. Chem. Soc.*, **96**, 3713 (1974).
[51] T. Hiyama, H. Sawada, and M. Tsukanaka, *Tetrahedron Lett.*, **1975**, 3013.
[52] E. V. Dehmlow, M. Lissel, and J. Heider, *Tetrahedron*, **33**, 363 (1977).
[53] M. Mąkosza, private communication, July 1977.
[54] Y. Kimura, K. Isagawa, and Y. Otsuji, *Chem. Lett.*, **1977**, 951.
[55] K.-H. Wong, *J. Chem. Soc. Chem. Commun.*, **1978**, 282.
[56] S. L. Regen, *Angew. Chem.*, **91**, 464 (1979); *Angew. Chem. Int. Ed. Engl.*, **18**, 421 (1979).
[57] S. L. Regen, A. Nigam, and J. J. Besse, *Tetrahedron Lett.*, **1978**, 2757; M. Tomoi, M. Ikeda, and H. Kakiuchi, *ibid.*, **1978**, 3757.
[58] S. L. Regen and A. Nigam, *J. Am. Chem. Soc.*, **100**, 7773 (1978).
[59] S. Samaan and F. Rolla, *Phosphorus Sulfur*, **4**, 145 (1978).
[60] J. A. Hyatt, *J. Org. Chem.*, **43**, 1808 (1978).
[61] B. G. Zupančič and M. Sopčič, *Synthesis*, **1979**, 123.
[62] S. Farhat, R. Gallo, and J. Metzger, *C. R. Acad. Sci. Ser. C*, **287**, 581.
[63] S. Julia, A. Ginebreda, and J. J. Marco, *Afinidad*, **35**, 235 (1978); *Chem. Abstr.*, **90**, 22507 (1979).
[64] P. Savignac, T. Cuvigny, and H. Normant, *C. R. Acad. Sci. Ser. C*, **287**, 35 (1978).
[65] M. C. Vander Zwan and F. W. Hartner, *J. Org. Chem.*, **43**, 2655 (1978).

Chapter 3

[1] A. Brändström and B. Lamm, *Acta Chem. Scand. Ser. B*, **28**, 590 (1974).
[2] E. V. Dehmlow, unpublished results.
[3] R. Ikan, A. Markus, and Z. Goldschmidt, *Is. J. Chem.*, **11**, 591 (1973).
[4] C. M. Starks, *J. Am. Chem. Soc.*, **93**, 195 (1971).
[5] A. Brändström, "Preparative Ion Pair Extraction. An Introduction to Theory and Practice," Apotekarsocieteten/Hässle Läkemedel, Stockholm, 1974, pp. 139–140.
[6] J. Goerdeler *in* Houben-Weyl, "Methoden der Organischen Chemie," Band **11,2**, 4. Aufl., G. Thieme Verlag, Stuttgart, 1958, pp. 620–622.
[7] R. Kunin and F. X. McGarvey, *Ind. Eng. Chem.*, **41**, 1265 (1949).
[8] C. Kaiser and J. Weinstock, *Org. Synth.*, **45**, 3 (1976).
[9] Ref. [5] pp. 82, 83, 143, 144.
[10] Ref. [5] pp. 83–87, 142–144, 147–148.
[11] M. Mąkosza and E. Białecka, *Synth. Commun.*, **6**, 313 (1976).
[12] E. V. Dehmlow, M. Slopianka, and J. Heider, *Tetrahedron Lett.*, **1977**, 2361.
[13] J. Dale and P. O. Kristiansen, *Acta Chem. Scand.*, **26**, 1471 (1972).
[14] R. N. Greene, *Tetrahedron Lett.*, **1972**, 1793.
[15] G. W. Gokel, D. J. Cram, C. L. Liotta, H. P. Harris, and F. L. Cook, *J. Org. Chem.*, **39**, 2445 (1974).
[16] C. J. Pedersen, *Org. Synth.*, **52**, 66 (1972).
[17] J. Ashby, R. Hull, M. J. Cooper, and E. M. Ramage, *Synth. Commun.*, **4**, 113 (1974).
[18] I. J. Burden, A. C. Coxon, J. F. Stoddart, and C. M. Wheatley, *J. Chem. Soc. Perkin Trans.* 1, **1977**, 220.

[19] R. C. Hayward, C. H. Overton, and G. H. Whitham, *J. Chem. Soc. Perkin Trans.* 1, **1976**, 2413.
[20] G. Johns, C. J. Ransom, and C. B. Reese, *Synthesis*, **1976**, 515.
[21] F. L. Cook, T. C. Caruso, M. P. Byrne, C. W. Bowers, D. H. Speak, and C. L. Liotta, *Tetrahedron Lett.*, **1974**, 4029.
[22] J. Dale and K. Daasvatn, *J. Chem. Soc. Chem. Commun.*, **1976**, 295.
[23] B. Dietrich, J. M. Lehn, and J. P. Sauvage, *Tetrahedron Lett.*, **1969**, 2885.
[24] R. Noyori, K. Yokoyama, J. Skata, I. Kuwajima, E. Nakamura, and M. Shimizu, *J. Am. Chem. Soc.*, **99**, 1265 (1977).
[25] I. Kuwajima, T. Murofushi, and E. Nakamura, *Synthesis*, **1976**, 602.
[26] P. W. Kent and R. C. Young, *Tetrahedron*, **27**, 4057 (1971).
[27] A. W. Herriott and D. Picker, *J. Am. Chem. Soc.*, **97**, 2345 (1975).
[28] A. W. Herriott and D. Picker, *Tetrahedron, Lett.*, **1972**, 4521.
[29] A. McKillop, J.-C. Fiaud, and R. P. Hug, *Tetrahedron*, **30**, 1379 (1974).
[30] L. Dalgaard, L. Jensen, and S.-O. Lawesson, *Tetrahedron*, **30**, 93 (1974).
[31] E. V. Dehmlow and T. Remmler, *J. Chem. Res.*, **1977**, (S) 72, (M) 766.
[32] E. V. Dehmlow and M. Lissel, *Tetrahedron Lett.*, **1976**, 1783.
[33] Ref. [5], p. 141.
[34] J. v. Braun, *Justus Liebigs Ann. Chem.*, **382**, 1 (1911).
[35] D. H. Hunter and R. A. Perry, *Synthesis*, **1977**, 37.
[36] W. T. Miller, Jr., J. H. Fried, and H. Goldwhite, *J. Am. Chem. Soc.*, **82**, 3091 (1960).
[37] L. H. Andersen, Finn. Pat. 40,462 (1968); *Chem. Abstr.*, **70**, 88036 (1969).
[38] C. M. Starks and D. R. Napier, Brit. Pat. 1,227,144 (1971) = Ital. Pat. 832,967 (1968) = French Pat. 1,573,164 (1969) and further concordances; *Chem. Abstr.*, **72**, 115271 (1970).
[39] D. R. Napier and C. M. Starks, U.S. Pat. 3,992,432 (1976); *Derwent Abstr.*, 90477 (1976).
[40] D. Landini and F. Rolla, *Chem. Ind.* (*London*), **1974**, 533.
[41] D. Forster, *J. Chem. Soc. Chem. Commun.*, **1975**, 917.
[42] E. V. Dehmlow and J. Heider, unpublished results.
[43] D. J. Sam and H. E. Simmons, *J. Am. Chem. Soc.*, **96**, 2252 (1974).
[44] D. Landini, F. Montanari, and F. M. Pirisi, *J. Chem. Soc. Chem. Commun.*, **1974**, 879.
[45] D. Landini, A. M. Maia, F. Montanari, and F. M. Pirisi, *Gazz. Chim. Ital.*, **105**, 863 (1975).
[46] R. Fornasier, F. Montanari, G. Podda, and P. Tundo, *Tetrahedron Lett.*, **1976**, 1381.
[47] H. Lehmkuhl and F. Rabet, Germ. Offenl. 2,534,851 (1977).
[48] M. Cinquini, F. Montanari, and P. Tundo, (a) *J. Chem. Soc. Chem. Commun.*, **1975**, 393; (b) *Gazz. Chim. Ital.*, **107**, 11 (1977).
[49] M. S. Newman and H. M. Chung, *J. Org. Chem.*, **39**, 1036 (1974).
[50] D. Landini, A. M. Maia, F. Montanari, and F. M. Pirisi, *J. Chem. Soc. Chem. Commun.*, **1975**, 950; D. Landini, A. Maia, and F. Montanari, *J. Am. Chem. Soc.*, **100**, 2796 (1978).
[51] D. Landini, S. Quici, and F. Rolla, *Synthesis*, **1975**, 430.
[52] D. Landini, F. Montanari, and F. Rolla, *Synthesis*, **1974**, 428.
[53] C. L. Liotta and H. P. Harris, *J. Am. Chem. Soc.*, **96**, 2250 (1974).
[54] L. Fitjer, *Synthesis*, **1977**, 189.
[55] P. Ykman and H. K. Hall, Jr., *Tetrahedron Lett.*, **1975**, 2429.
[56] M. Gross and F. Peter, *Bull. Soc. Chim. Fr.*, **1975**, 871.
[57] E. V. Dehmlow and M. Slopianka, *Chem. Ber.*, **112**, 2768 (1979).
[58] D. Landini, F. Montanari, and F. Rolla, *Synthesis*, **1974**, 37.
[59] D. Bethell, K. McDonald, and K. S. Rao, *Tetrahedron Lett.*, **1977**, 1447.
[60] F. Jeanne and A. Trichet, L'Air Liquide, Centre d'Etudes Cryogéniques, F-38360 Sassenage, France, "poster" Journées de Chimie Organique d'Orsay, Société Chimique de France, Sept. 17–19, 1975.

[61] H. Lehmkuhl, F. Rabet, and K. Hauschild, *Synthesis*, **1977**, 184.
[62] S. L. Regen, *J. Org. Chem.*, **42**, 875 (1977).
[63] C. M. Starks and R. M. Owens, *J. Am. Chem. Soc.*, **95**, 3613 (1973).
[64] S. L. Regen, *J. Am. Chem. Soc.*, **98**, 6270 (1976).
[65] J. W. Zubrick, B. I. Dunbar, and H. D. Durst, *Tetrahedron Lett.*, **1975**, 71.
[66] M. E. Childs and W. P. Weber, *J. Org. Chem.*, **41**, 3486 (1976).
[67] G. Simchen and H. Kobler, *Synthesis*, **1975**, 605; H. Kobler, K.-H. Schuster, and G. Simchen, *Justus Liebigs Ann. Chem.*, **1978**, 1946.
[68] S. L. Regen, *J. Am. Chem. Soc.*, **97**, 5956 (1975).
[69] H. B. Copelin and G. B. Crane, U.S. Pat. 2,779,781 (1957); *Chem. Abstr.*, **51**, 10579 (1957).
[70] H. Leuchs and K. H. Schmidt, Brit. Pat. 1,200,970 (1970); *Chem. Abstr.*, **73**, 87662 (1970).
[71] H. Coates, R. L. Barker, R. Guest, and A. Kent, Brit. Pat. 1,336,883 (1973); *Chem. Abstr.*, **80**, 70564 (1973).
[72] W. P. Reeves and M. R. White, *Synth. Commun.*, **6**, 193 (1976).
[73] N. Sugimoto, T. Fujita, N. Shigematsu, and A. Ayada, Chem. Pharm. Bull., **10**, 427 (1962).
[74] A. S. Kende and L. S. Liebeskind, *J. Am. Chem. Soc.*, **98**, 267 (1976).
[75] M. S. Newman, T. G. Barbee, Jr., C. N. Blakesley, Z. ud Din, S. Gromelski, Jr., V. K. Khanna, L.-F. Lee, J. Radhakrishnan, R. L. Robey, V. Sankaran, S. K. Sankarappa, and J. M. Springer, *J. Org. Chem.*, **40**, 2863 (1975).
[76] K. E. Koenig and W. P. Weber, *Tetrahedron Lett.*, **1974**, 2275.
[77] F. L. Cook, C. W. Bowers, and C. L. Liotta, *J. Org. Chem.*, **39**, 3416 (1974).
[78] L. B. Engemyr, A. Martinsen, and J. Songstad, *Acta Chem. Scand. Ser. A.*, **28**, 255 (1974); J. Songstad, L. J. Stangeland, and T. Austad, *ibid.*, **24**, 355 (1970).
[79] L. Eberson and B. Helgée, *Chem. Scr.*, **5**, 47 (1974).
[80] G. W. Gokel, S. H. Korzeniowski, and L. Blum, *Tetrahedron Lett.*, **1977**, 1633.
[81] S. H. Korzeniowski and G. W. Gokel, *Tetrahedron Lett.*, **1977**, 1637.
[82] W. P. Reeves, M. R. White, R. G. Hilbrich, and L. L. Biegert, *Synth. Commun.*, **6**, 509 (1976).
[83] C. L. Liotta, E. E. Grisdale, and H. P. Hopkins, Jr., *Tetrahedron Lett.*, **1975**, 4205.
[84] A. Knöchel and G. Rudolph, *Tetrahedron Lett.*, **1974**, 3739.
[85] W. P. Reeves and M. L. Bahr, *Synthesis*, **1976**, 823.
[86] A. Brändström, B. Lamm, and I. Palmertz, *Acta Chem. Scand. Ser. B*, **28**, 699 (1974).
[87] K. J. MacNay, E. R. Rogier, and M. M. Kreevoy, Germ. Offenlegungsschrift 2, 245, 611; *Chem. Abstr.*, **81**, 63387 (1974).
[88] M. Takeishi, Y. Naito, and M. Okawara, *Angew. Makromol. Chem.*, **28**, 111 (1973).
[89] M. Takeishi, R. Kawashima, and M. Okawara, *Makromol. Chem.*, **167**, 261 (1973).
[90] B. Graham, U.S. Pat. 2,866,802 (1958); *Chem. Abstr.*, **53**, 9146 (1959).
[91] N. P. Smetankina and N. I. Miryan, *Zh. Obshch. Khim.*, **39**, 2299 (1969) (Engl. transl. p. 2239); N. P. Smetankina and N. N. Laskovenko, *Zh. Obshch. Khim.*, **42**, 617 (1972) (Engl. transl. p. 613).
[92] G. E. Vennstra and B. Zwaneburg, *Synthesis*, **1975**, 519.
[93] G. D. Hartman and S. E. Biffer, *J. Org. Chem.*, **42**, 1468 (1977).
[94] S. H. Korzeniowski, L. Blum, and G. W. Gokel, *J. Org. Chem.*, **42**, 1469 (1977).
[95] V. D. Parker and L. Eberson, *J. Chem. Soc. Chem. Commun.*, **1972**, 441.
[96] H. E. Hennis, L. R. Thompson, and J. P. Long, *Ind. Eng. Chem. Prod. Res. Dev.*, **1968,** 96; *Chem. Abstr.*, **69**, 18796 (1968).
[97] J. H. Wagenknecht, M. M. Baizer, and J. L. Chruma, *Synth. Commun.*, **2**, 215 (1972).
[98] R. H. Mills, M. W. Farrar, and O. J. Weinkauf, *Chem. Ind* (*London*), **1962**, 2144.
[99] R. L. Merker and M. J. Scott, *J. Org. Chem.*, **26**, 5180 (1961).
[100] Y. Yamashita and T. Shimamura, *Kogyo Kagaku Zasshi*, **60**, 423 (1957); *Chem. Abstr.*, **53**, 9025 (1959).

[101] R. Kay, Brit. Pat. 916,772 (1963); *Chem. Abstr.*, **59**, 2728 (1963) and Brit. Pat. 966,266 (1964); *Chem. Abstr.*, **61**, 10629 (1964).
[102] T. Toru, S. Kurozumi, T. Tanaka, S. Miura, M. Kobayashi, and S. Ishimoto, *Synthesis*, **1974**, 867.
[103] E. V. Dehmlow, unpublished results, 1974.
[104] K. Holmberg and B. Hansen, *Tetrahedron Lett.*, **1975**, 2303.
[105] Ref. 5, p. 155.
[106] Ref. 5, pp. 109–111.
[107] F. C. V. Larsson and S.-O. Lawesson, *Tetrahedron*, **28**, 5341 (1972).
[108] H. Ehrsson, *Acta Pharm. Suec.*, **8**, 113 (1971); *Chem. Abstr.*, **75**, 44,678 (1971); J. E. Greving, J. H. G. Jonkman, and R. A. de Zeeuw, *J. Chromatogr.*, **148**, 389 (1978).
[109] A. Knöchel, J. Oehler, and G. Rudolph, *Tetrahedron Lett.*, **1975**, 3167.
[110] C. L. Liotta, H. P. Harris, M. McDermott, T. Gonzalez, and K. Smith, *Tetrahedron Lett.*, **1974**, 2417.
[111] H. D. Durst, *Tetrahedron Lett.*, **1974**, 2421.
[112] S. Akabori and M. Ohtomi, *Bull. Chem. Soc. Jpn.*, **48**, 2991 (1975).
[113] E. Grushka, H. D. Durst, and E. J. Kikta, Jr., *J. Chromatogr.*, **112**, 673 (1975).
[114] H. D. Durst, M. Milano, E. J. Kikta, Jr., S. H. Connelly, and E. Grushka, *Anal. Chem.*, **47**, 1797 (1975).
[115] A. Padwa and D. Dehm, *J. Org. Chem.*, **40**, 3139 (1975).
[116] G. Gelbard and S. Colonna, *Synthesis*, **1977**, 113.
[117] H. Normant, T. Cuvigny, and P. Savignac, *Synthesis*, **1975**, 805.
[118] H. Normant and C. Laurenço, *C. R. Acad. Sci. Ser. C*, **283**, 483 (1976).
[119] P. J. Garegg, T. Iversen, and S. Oscarson, *Carbohydr. Res.*, **53**, C5 (1977).
[120] H. Schick, *Pharmazie*, **30**, 817 (1975).
[121] S. Schwarz and G. Weber, *Z. Chem.*, **15**, 270 (1975).
[122] S. Schwarz, G. Weber, M. Wahren, and M. Herrmann, *Z. Chem.*, **16**, 439 (1976).
[123] H. J. Fex, S. K. Kristensson, and A. R. Stamvik, Germ. Offenleg, 2,629,657 (1977).
[124] R. A. Bauman, *Synthesis*, **1974**, 870.
[125] R. W. Ridgway, H. S. Greenside, and H. H. Freedman, *J. Am. Chem. Soc.*, **98**, 1979 (1976).
[126] A. Zwierzak, *Synthesis*, **1975**, 507.
[127] A. Zwierzak and J. Brylikowska, *Synthesis*, **1975**, 712.
[128] A. Zwierzak and A. Sulewska, *Synthesis*, **1976**, 835.
[129] A. Zwierzak, *Synthesis*, **1976**, 305.
[130] Bayer, A. G., Germ. Offenl. 2,442,883 (1976); *Derwent Abstr.*, 22819 (1976).
[131] Bayer, A. G., Germ. Offenl. 2,512,498 (1976); *Derwent Abstr.*, 38189 (1976).
[132] Wacker-Chemie GmbH., Belg. Pat. 795,616 (1972); *Derwent Abstr.*, 50232 (1973).
[133] M. A. Johnson and K. Yang, U.S. Pat. 3,641,172 (1972); *Chem. Abstr.*, **76**, 99072 (1972).
[134] Hokko Chem. Ind., Jap. Kokai 75,14,607; *Chem. Abstr.*, **83**, 27529 (1975).
[135] W. Schoenleben, J. Datow, H. Hoffmann, and S. Winderl, Germ. Offenl. 2,149,822 (1973); *Chem. Abstr.*, **79**, 52768 (1973).
[136] R. P. Johnson, U.S. Pat. 3,941,827 (1976); *Chem. Abstr.*, **84**, 179 707 (1976).
[137] Ethyl Corp., Belg. Pat. 841,131 (1976); *Derwent Abstr.*, 83041 (1976).
[138] Dow Chemical Corp., U.S. Pat. 3,564,252 (1976); *Derwent Abstr.*, 20403 (1976).
[139] C. M. Starks, U.S. Pat. 3,725,458 (1973); *Chem. Abstr.*, **78**, 158991 (1973).
[140] C. M. Starks and R. D. Gordon, Germ. Offenl. 2,103,547 (1971); *Chem. Abstr.*, **75**, 117 993 (1971).
[141] C. Kimura, K. Kashiwaya, and K. Murai, *Chem. Abstr.*, **84**, 121027 (1976); C. Kimura, K. Murai, Y. Ishikawa, and K. Kashiwaya, *Chem. Abstr.*, **87**, 133781 (1977).
[142] M. Hiraoka, T. Nakamura, and T. Ozeki, Jap. Kokai 76,06,928; *Chem. Abstr.*, 164159 (1976).
[143] Toyo Chemical, Jap. Kokai 76,06,927; *Derwent Abstr.*, 17505 (1976).

[144] Uniroyal Inc., Germ. Offenl. 2,558,163 (1976); *Derwent Abstr.*, 54478 (1976).
[145] S. Paul and N. Rånby, *Macromolecules*, **9**, 337 (1976).
[146] A. J. Coury and D. D. Cozad, U.S. Pat. 3,873,589 (1975); *Chem. Abstr.*, **82**, 170037 (1975).
[147] K. D. MacKay, E. R. Rogier, and M. M. Kreevoy, U.S. Pat. 3,707,495 (1972); *Chem. Abstr.*, 78091 (1973).
[148] K. F. Zenner and H. Appel, Germ. Offenl. 2,126,296 (1972); *Chem. Abstr.*, **78**, 58473 (1973).
[149] Y. Nadachi and M. Kokura, Jap. Kokai 75,95,289; *Chem. Abstr.*, **84**, 164852 (1976).
[150] D. G. Brady, U.S. Pat. 3,674,750 (1972); *Chem. Abstr.*, **77**, 89339 (1972).
[151] Y. Nadachi and M. Kokura, Jap. Kokai 75,84,529; *Chem. Abstr.*, **84**, 30650 (1976).
[152] P. Edwards, U.S. Pat. 2,537,981 (1951); *Chem. Abstr.*, **45**, 5177 (1951).
[153] W. J. Heilman, U.S. Pat. 3,661,938 (1972); *Chem. Abstr.*, **77**, 49108 (1972).
[154] M. Yasuda, M. Sakai, H. Ono, and M. Tanaka, Jap. Pat. 72,41,884 (1972); *Chem. Abstr.*, **78**, 44212 (1973).
[155] Gulf Res. and Dev. Co., Jap. Pat. 72,11,365; *Derwent Abstr.*, 46931 (1972).
[156] M. Yoshino, S. Sato, H. Mochida, and M. Shibata, Jap. Pat. 71,37,326 (1971); *Chem. Abstr.*, **76**, 154695 (1972).
[157] N. Hashi and Y. Takase, Jap. Pat. 72,47,366; *Chem. Abstr.*, **78**, 125136 (1973).
[158] F. N. Bodnaryuk, M. A. Korshunov, V. M. Melekhov, *et al.*, U.S.S.R. Pat. 357,196 (1972); *Chem. Abstr.*, **78**, 160443 (1973).
[159] S. Sato, K. Kamezawa, and M. Shiba, Jap. Pat. 73,04,006; *Chem. Abstr.*, **78**, 160320 (1973).
[160] T. Horii, K. Kawamata, T. Yagi, H. Sasaki, and S. Nose, Jap. Kokai 73,39,423; *Chem. Abstr.*, **80**, 17641 (1974).
[161] Gulf Res. and Dev. Co., U.S. Pat. 3,957,831 (1976); *Derwent Abstr.*, 41873 (1976).
[162] G. Maerker, J. F. Carmichael, and W. Port, *J. Org. Chem.*, **26**, 2681 (1961).
[163] R. Köhler and H. Pietsch, Germ. Pat. 944,995 (1956); *Chem. Abstr.*, **53**, 1831 (1959).
[164] Ciba-Geigy A. G., Jap. Pat. 72,10,822; *Derwent Abstr.*, 42486 (1972).
[165] Shell Int. Res. Mij NV, Belg. Pat. 793,432 (1973); *Derwent Abstr.*, 39744 (1973); Netherl. Pat. 7,105,426 (1972); *Chem. Abstr.*, **78**, 98456 (1973).
[166] Ciba, S. A., Belg. Pat. 739,526 (1970); *Derwent Abstr.*, 21737 (1970).
[167] H. Matsuda, Y. Tange, and F. Yamauchi, Jap. Pat. 70,36,733; *Chem. Abstr.*, **74**, 126854 (1971).
[168] A. Heer and W. Schaffner, Swiss Pat. 536,835 (1973); *Chem. Abstr.*, **79**, 79726 (1973).
[169] A. Suzui and K. Matsumoto, Jap. Pat. 70,04,975 (1970); *Chem. Abstr.*, 100314 (1970).
[170] F. H. Sinnema and G. C. Vegter, Brit. Pat. 1,205,180 (1970); *Chem. Abstr.*, **73**, 130686 (1970).
[171] W. Hoffmann and K. v. Fraunberg, Germ. Offenl. 2,324,047 (1974); *Chem. Abstr.*, **82**, 73244 (1975).
[172] D. G. Brady, U.S. Pat. 3,843,719 (1974); *Chem. Abstr.*, **82**, 16570 (1975).
[173] A. Suzui and T. Horii, Jap. Pat. 70,36,737 (1970); *Chem. Abstr.*, **74**, 125 197 (1971).
[174] R. A. Gray and D. G. Brady, U.S. Pat. 3,879,445 (1975); *Chem. Abstr.*, **84**, 31962 (1976).
[175] R. Kay, French Pat. 1,391,727 (1965); *Chem. Abstr.*, **63**, 10134 (1965).
[176] H. Trautmann, East Germ. Pat. 105,620 (1974); *Derwent Abstr.*, 56825 (1974).
[177] J. G. Scruggs and D. G. Brady, U.S. Pat. 3,676,484 (1972); *Chem. Abstr.*, **77**, 126249 (1972).
[178] J. Berthoux and G. Schwachhofer, Germ. Offenl. 2,338,552 (1974); *Chem. Abstr.*, **80**, 133063 (1974).
[179] Hodoga Chem. Ind., Jap. Kokai 76,065,725; *Derwent Abstr.*, 56607 (1976).
[180] Union Camp. Corp., Netherl. Pat. Appl. 6,909,697 (1969).
[181] T. Hayashi and H. Hitoshi, Jap. Pat. 73,30,611; *Chem. Abstr.*, **80**, 26771 (1974).

[182] T. Nishikubo, S. Ukai, H. Idemitsu, and M. Kishida, Jap. Kokai 72,17,715; *Chem. Abstr.*, **77**, 139450 (1972).
[183] M. A. Korshunov, R. G. Kuzovleva, and I. V. Furaeva, U.S.S.R. Pat. 432,128 (1974); *Chem. Abstr.*, **81**, 77496 (1974).
[184] F. N. Bodnaryak, M. A. Korshunov, V. E. Lazaryants, V. M. Melekhov, A. I. Ezrielev, A. V. Lebedev, A. B. Peiznev, and L. V. Utkinn, Brit. Pat. 1,242,980 (1969); *Chem. Abstr.*, **76**, 4320 (1976); Germ. Offenl. 1,960,716 (1971); *Chem. Abstr.*, **75** 77491 (1971).
[185] Y. Inamori, Y. Yanagawa, and I. Iwasa, Jap. Pat. 74,02,098; *Chem. Abstr.*, **81**, 77493 (1974).
[186] J. E. Masters, U.S. Pat. 3,351,674 (1967); *Chem. Abstr.*, **68**, 22480 (1968).
[187] E. Müller, O. Bayer, and H. Morschel, Germ. Pat. 959,497 (1957); *Chem. Abstr.*, **53**, 13665 (1959).
[188] M. Tomikawa, H. Kaji, and K. Ueda, *Chem. Abstr.*, **68**, 40263 (1968).
[189] J. F. Yanes and O. L. Castellanos, *An. R. Soc. Esp. Fis. Quim. Ser. B*, **57**, 815 (1962); *Chem. Abstr.*, **57**, 7455 (1962).
[190] B. E. Jennings, Brit. Pat. 907,649 (1962); *Chem. Abstr.*, **58**, 5810 (1963).
[191] Z. Jedlinski and D. Sek, Polimer, **14**, 105 (1969); *Chem. Abstr.*, **71**, 39537 (1969).
[192] A. Conix and U. L. Laridon, Belg. Pat. 563,173 (1958) and 565,478 (1958); *Chem. Abstr.*, **55**, 25356 (1961).
[193] A. J. Conix and L. M. Dohmen, Germ. Pat. 1,199,500 (1962); *Chem. Abstr.*, **63**, 16502 (1965).
[194] Politechnica Warszawska, East Germ. Pat. 101,687 (1974); *Derwent Abstr.*, 13584 (1974).
[195] Gevaert Photo-Prod. N.V., Belg. Pat. 600,053 (1961); *Chem. Abstr.*, **58**, 11491 (1963); Belg. Pat. 602,793 (1961); *Chem. Abstr.*, **59**, 15405 (1963).
[196] E. P. Goldberg and F. Scardiglia, U.S. Pat. 3,271,368 (1966); *Chem. Abstr.*, **66**, 3015 (1967).
[197] Dow Chemical Co., U.S. Pat. 3,972,887 (1976); *Derwent Abstr.*, 63289 (1976).
[198] L. M. Kroposki, M. Yoshimine, and H. Freedman, U.S. Pat. 3,917,621 (1975); *Chem. Abstr.*, **84**, 121436 (1976).
[199] L. M. Kroposki and M. Yoshimine, U.S. Pat. 3,907,815 (1975); *Chem. Abstr.*, **84**, 43864 (1976).
[200] Dow Chemical Co., Belg. Pat. 832,047 (1976); *Derwent Abstr.*, 13078 (1976).
[201] A. W. Herriott and D. Picker, *J. Am. Chem. Soc.*, **97**, 2345 (1975).
[202] D. Landini and F. Rolla, *Synthesis*, **1974**, 565.
[203] A. Tozzi and P. Cassandrini, Germ. Offenl. 2,513,805 (1975); *Chem. Abstr.*, **84**, 4476 (1976).
[204] I. K. Kim and J. S. Noh, *Chem. Abstr.*, **82**, 124967 (1975).
[205] A. W. Herriott and D. Picker, *Synthesis*, **1975**, 447.
[206] S. Cabiddu, A. Maccioni, and M. Secci, *Synthesis*, **1976**, 797.
[207] N. H. Nilsson and A. Senning, *Chem. Ber.*, **107**, 2345 (1974).
[208] L. Dalgaard and S.-O. Lawesson, *Acta Chem. Scand. Ser. B*, **28**, 1077 (1974).
[209] L. Dalgaard and S.-O. Lawesson, *Tetrahedron*, **28**, 2051 (1972).
[210] T. Nakabayashi, S. Kawamura, T. Horii, and M. Hamada, *Chem. Lett.*, **1976**, 869.
[211] A. T. Babayan and M. G. Indzhikyan, *Dokl. Akad. Nauk. Arm. SSR*, **28**, 67 (1959); *Chem. Abstr.*, **64**, 1368 (1960); *Chem. Zentralbl.*, **1965**, 14–0870.
[212] H. F. Ebel, "Die Acidität der CH-Säuren," G. Thieme Verlag, Stuttgart, 1969.
[213] A. J. Gordon and R. A. Ford, "The Chemist's Companion, Handbook of Practical Data, Techniques, and References," Wiley-Interscience, New York, 1972, pp. 60–64.
[214] M. Mąkosza, *Pure Appl. Chem.*, **43**, 439 (1975).
[215] G. W. Gokel, S. A. Di Biase, and B. A. Lipisko, *Tetrahedron Lett.*, **1976**, 3495.
[216] D. J. Cram, C. A. Kingsburg, and B. Rickborn, *J. Am. Chem. Soc.*, **83**, 3688 (1961).
[217] A. Schrieschein and C. A. Rowe, Jr., *J. Am. Chem. Soc.*, **84**, 3160 (1962).

[218] D. Bethell and A. F. Cockerill, *J. Chem. Soc. B*, **1966**, 913.
[219] M. J. Maskornick, *Tetrahedron Lett.*, **1972**, 1797.
[220] R. A. Bartsch, G. M. Pruss, R. L. Buswell, and R. A. Bushaw, *Tetrahedron Lett.*, **1972**, 2621.
[221] R. A. Bartsch and K. E. Wiegers, *Tetrahedron Lett.*, **1972**, 3819.
[222] R. A. Bartsch, E. A. Mintz, and R. M. Parlman, *J. Am. Chem. Soc.*, **96**, 4249 (1974).
[223] Survey: L. F. and M. Fieser, "Reagents for Organic Syntheses," Vol. I, J. Wiley & Sons, New York, 1967, p. 1252.
[224] J. Jarrousse, *C. R. Acad. Sci. Ser. C*, **232**, 1424 (1951).
[225] A. T. Babayan, M. G. Indzhikyan, and T. A. Azizyan, *Dokl. Akad. Nauk Army. SSR*, **31**, 79 (1960); *Chem. Abstr.*, **55**, 11342 (1961); *Chem. Zentralbl.*, **1966**, 5–0867.
[226] A. T. Babayan and M. G. Indzhikyan, *Zh. Obshch. Khim.*, **27**, 1201 (1957), Engl. transl. p. 1284; *Chem. Abstr.*, **52**, 3707 (1958); *Chem. Zentralbl.*, **1961**, 2246.
[227] S. Korzeniowski, L. Blum, and G. W. Gokel, *Tetrahedron Lett.*, **1977**, 1871.
[228] W. J. Belanger, U.S. Pat. 2,940,953 (1960); *Chem. Abstr.*, **55**, 1072 (1961); U.S. Pat. 2,928,810 (1960); *Chem. Abstr.*, **54**, 13729 (1960).
[229] R. D. Gordon, U.S. Pat. 3,914,320 (1975); *Chem. Abstr.*, **84**, 30423 (1976); U.S. Pat. 3,824,295 (1974); *Chem. Abstr.*, **81**, 77464 (1974).
[230] C. M. Starks, U.S. Pat. 3,931,238 (1976); *Chem. Abstr.*, **84**, 104988 (1976).
[231] H. Baw, *J. Indian Chem. Soc.*, **3**, 101 (1926).
[232] H. H. Freedman and R. A. Dubois, *Tetrahedron Lett.*, **1975**, 3251.
[233] A. Merz, *Angew. Chem.*, **85**, 868 (1973), *Angew. Chem. Int. Ed. Engl.*, **11**, 846 (1973).
[234] E. V. Dehmlow and J. Schmidt, *Tetrahedron Lett.*, **1976**, 95.
[235] A. Merz and R. Tomahogh, *Chem. Ber.*, **110**, 96 (1977).
[236] J. Villieras, C. Baquet, and J. F. Normant, *J. Organomet. Chem.*, **97**, 355 (1975).
[237] B. Cazes and S. Julia, *Tetrahedron Lett.*, **1974**, 2077; *Bull. Soc. Chim. Fr.*, **1977**, 925.
[238] B. Cazes and S. Julia, *Synth. Commun.*, **7**, 113 (1977).
[239] P. J. Garegg, T. Ivensen, and S. Oscarson, *Carbohydr. Res.*, **50**, C12 (1976).
[240] P. di Cesare and B. Gross, *Carbohydr. Res.*, **48**, 271 (1976).
[241] G. J. H. Rall, M. E. Oberholzer, D. Ferreira and D. G. Roux, *Tetrahedron Lett.*, **1976**, 1033.
[242] J. D. Daley, J. M. Rosenfeld, and E. V. Young-Lai, *Steroids*, **27**, 481 (1976).
[243] K. Olsson, *Acta Chem. Scand. Ser. B*, **29**, 405 (1975).
[244] A. P. Bashall and J. F. Collins, *Tetrahedron Lett.*, **1975**, 3489.
[245] Takasago Perfumery, Jap. Kokai 76,023,265, *Derwent Abstr.*, 27219 (1976).
[246] D. Picker, Ph.D. Dissertation, State University of New York at Albany, 1974; *Diss. Abstr.*, **35**, 3229 (1975).
[247] Dow Chemical Co., U.S. Pat. 3,969,360 (1976); *Derwent Abstr.*, 57726 (1976).
[248] P. M. Quan and S. R. Korn, Germ. Offenl. 2,634,419 (1977).
[249] J. Ugelstad, T. Ellingsen, and A. Berge, *Acta Chem. Scand.*, **20**, 1593 (1966); L. M. Thomassen, T. Elligsen, and J. Ugelstad, *ibid.*, **25**, 3024 (1971).
[250] H. Brötell, H. Ehrsson, and O. Gyllenhaal, *J. Chromatogr.*, **78**, 293 (1973).
[251] A. T. Babayan and A. A. Grigoryan, *Izv. Akad. Nauk Arm. SSR, Fiz. Mat., Estestv. Tekh. Nauki*, **8**, 81 (1955); *Chem. Abstr.*, **50**, 10023 (1956); *Chem. Zentralbl.* **1959**, 4446.
[252] R. Kreher and K. J. Herd, *Z. Naturforsch, Teil B*, **29**, 683 (1974).
[253] K. Berg-Nielsen and E. Bernatek, *Acta. Chem. Scand.*, **26**, 4130 (1972).
[254] A. Jończyk, J. Włostowska, and M. Mąkosza, *Synthesis*, **1976**, 795.
[255] (a) Ref. [5], p. 156. (b) *ibid.*, p. 112.
[256] U. Junggren, Dissertation Göteborg, 1972, cited by Brändström, ref. [5].
[257] R. Brehme, (a) *Synthesis*, **1976**, 113; (b) East Germ. Pat. 117,446 (1976); *Derwent Abstr.*, 24554 (1976).
[258] D. Landini and F. Rolla, *Synthesis*, **1976**, 389.

[259] M.-T. Maurette, A. Lopez, R. Martino, and A. Lattes, *C. R. Acad. Sci. Ser. C*, **282**, 599 (1976).
[260] J. Paleček and J. Kuthan, *Synthesis*, **1976**, 550.
[261] A. Jończyk and M. Mąkosza, *Rocz. Chem.*, **49**, 1203 (1975).
[262] V. Bocchi, G. Gasnati, A. Dossena, and F. Villani, *Synthesis*, **1976**, 414.
[263] A. Barco, S. Benetti, G. P. Pollini, and P. G. Baraldi, *Synthesis*, **1976**, 124.
[264] N. N. Suvorov, Y. I. Smushkevich, V. S. Velezhevan, V. S. Rozhkov, and S. V. Simakov, *Khim. Geterotsikl. Soedin.*, **12**, 191 (1976); Engl. transl., p. 167.
[265] T. Greibrokk, *Acta Chem. Scand.*, **26**, 3305 (1972).
[266] H. Ehrsson and A. Tilly, *Anal. Lett.*, **6**, 197 (1973).
[267] M. Evrik and K. Gustavii, *Anal. Chem.* **46**, 39 (1974).
[268] H. Ehrsson, *Anal. Chem.*, **46**, 922 (1974).
[269] C. Hoppel, M. Garle, and M. Elander, *J. Chromatogr.*, **116**, 53 (1976).
[270] O. Gyllenhaal and H. Ehrsson, *J. Chromatogr.*, **107**, 237 (1975); O. Gyllenhaal, N. Tjärnlund, H. Ehrsson, and P. Hartvig, *ibid.*, **156**, 275 (1978).
[271] A. Arbin and P.-O. Edlund, *Acta Pharm. Suec.*, **12**, 119 (1975).
[272] A. Zwierzak and J. Brylikowska-Piotrowicz, *Angew. Chem.*, **89**, 109 (1977); *Angew. Chem. Int. Ed. Engl.*, **16**, 107 (1977).
[273] S. Cacchi, F. La Torre, and D. Misiti, *Synthesis*, **1977**, 301.
[274] H. J.-M. Dou and J. Metzger, *Bull. Soc. Chim. Fr.*, **1976**, 1861.
[275] J. Massé, *Synthesis*, **1977**, 341.
[276] A. Suzui and T. Horii, Jap. Pat. 68,29,940; *Chem. Abstr.*, **70**, 68110 (1969).
[277] R. W. Miller and J. A. Schmitt, U.S. Pat. 3,193,528 (1965); *Chem. Abstr.*, **63**, 11730 (1965).
[278] F. J. Ponder, Germ. Pat. 1,806,870 (1969); *Chem. Abstr.*, **71**, 101545 (1969).
[279] H. Wunderlich, D. Lugenheim, and A. Stark, East Germ. Pat. 51,642 (1966); *Chem. Zentralbl.*, **1969**, 14–1986.
[280] M. Mąkosza and A. Jończyk, *Org. Synth.*, **55**, 91 (1976).
[281] M. Mąkosza and B. Serafinowa, *Rocz. Chem.*, **39**, 1223 (1965).
[282] M. Mąkosza and B. Serafinowa, *Rocz. Chem.*, **39**, 1401 (1965).
[283] M. Mąkosza and B. Serafinowa, *Rocz. Chem.*, **39**, 1595 (1965).
[284] M. Mąkosza and B. Serafinowa, *Rocz. Chem.*, **39**, 1799 (1965).
[285] M. Mąkosza and B. Serafinowa, *Rocz. Chem.*, **39**, 1805 (1965).
[286] M. Mąkosza and B. Serafinowa, *Rocz. Chem.*, **40**, 1647 (1966).
[287] M. Mąkosza and B. Serafinowa, *Rocz. Chem.*, **40**, 1839 (1966).
[288] J. Lange and M. Mąkosza, *Rocz. Chem.*, **41**, 1303 (1967).
[289] M. Mąkosza, I. Kmiotek-Skarżyńska, and M. Jawdosiuk, *Synthesis*, **1977**, 56.
[290] P. Vittorelli, J. Peter-Katalinić, G. Mukherjee-Müller, H.-J. Hansen, and H. Schmid, *Helv. Chim. Acta*, **58**, 1379 (1975).
[291] M. Mąkosza, B. Serafinowa, and M. Jawdosiuk, *Rocz. Chem.*, **41**, 1037 (1967).
[292] M. Mąkosza, *Rocz. Chem.*, **43**, 79 (1965).
[293] M. Mąkosza, B. Serafinowa, and T. Bolesławska, *Rocz. Chem.*, **42**, 817 (1968).
[294] M. Mąkosza, *Rocz. Chem.*, **43**, 333 (1969).
[295] M. Mąkosza and T. Goetzen, *Rocz. Chem.*, **46**, 1239 (1972).
[296] M. Mąkosza, E. Białecka, and M. Ludwikow, *Tetrahedron Lett.*, **1972**, 2391.
[297] B. Gutkowska, *Acta Pol. Pharm.*, **30**, 109 (1973); *Chem. Abstr.*, **79**, 78552 (1973).
[298] M. Mąkosza and E. Białecka, *Tetrahedron. Lett.*, **1977**, 183.
[299] H. Komeili-Zadeh, H. J.-M. Dou, and J. Metzger, *C. R. Acad. Sci. Ser. C*, **283**, 41 (1976).
[300] M. Mąkosza and M. Fedoryński, *Synthesis*, **1974**, 274.
[301] M. Mąkosza, M. Ludwikow, and A. Urniaz, *Rocz. Chem.*, **49**, 297 (1975).
[302] M. Mąkosza and M. Ludwikow, *Angew. Chem.*, **86**, 744 (1974); *Angew. Chem. Int. Ed. Engl.*, **13**, 665 (1974).
[303] M. Mąkosza, *Tetrahedron Lett.*, **1969**, 673.

[304] M. Mąkosza, J. M. Jagusztyn-Grochowska, and M. Jawdosiuk, *Rocz. Chem.*, **45**, 851 (1971).
[305] M. Mąkosza and M. Ludwikow, *Bull. Acad. Pol. Sci. Ser. Sci. Chim.*, **19**, 231 (1971).
[306] M. Mąkosza, J. M. Jagusztyn-Grochowska, and M. Jawdosiuk, *Rocz. Chem.*, **50**, 1841 (1976).
[307] M. Mąkosza, M. Jagusztyn-Grochowska, M. Ludwikow, and M. Jawdosiuk, *Tetrahedron*, **30**, 3723 (1974).
[308] M. Mąkosza and J. M. Jagusztyn-Grochowska, *Rocz. Chem.*, **50**, 1859 (1976).
[309] M. Mąkosza and M. Jawdosiuk, *Bull. Acad. Pol. Sci. Ser. Sci. Chim.*, **16**, 597 (1968).
[310] M. Mąkosza, *Tetrahedron Lett.*, **1969**, 677.
[311] A. Jończyk, *Bull. Acad. Pol. Sci. Ser. Sci. Chim.*, **22**, 849 (1974).
[312] T. Tanaka and T. Mukaiyama, *Chem. Lett.*, **1976**, 1259.
[313] M. Mąkosza and M. Ludwikow, *Rocz. Chem.*, **51**, 829 (1977).
[314] Sumitomo Chemical KK, Netherl. Pat. Appl. 7,513,627 (1976); *Derwent Abstr.*, 45249 (1976).
[315] Sigurta Farm. Spa., Belg. Pat. 837,624 (1976); *Derwent Abstr.*, 46067 (1976).
[316] S. L. Regen and L. Dulak, *J. Am. Chem. Soc.*, **99**, 623 (1977).
[317] L. Rylski and F. Gajewski, *Acta Pol. Pharm.*, **26**, 115 (1969); *Chem. Abstr.*, **71**, 38741 (1969).
[318] G. Carenini, R. D'Ambrosio, M. Carissimi, E. Grumelli, E. Milla, and F. Ravenna, *Farmaco Ed. Sci.*, **28**, 265 (1973); *Chem. Abstr.*, **78**, 159023 (1973).
[319] M. Barreau and M. Julia, *Tetrahedron Lett.*, **1973**, 1537.
[320] M. Mąkosza and T. Goetzen, *Organ. Prep. Proced. Int.*, **5**, 203 (1973).
[321] A. Brändström and U. Junggren, *Tetrahedron Lett.*, **1972**, 473.
[322] Y. Masuyama, Y. Ueno, and Okawara, *Tetrahedron Lett.*, **1976**, 2967.
[323] E. D'Incan and J. Seyden-Penne, *Synthesis*, **1975**, 516.
[324] J. Blanchard, N. Collignon, P. Savignac, and H. Normant, *Tetrahedron*, **32**, 455 (1976).
[325] J. Blanchard, N. Collignon, P. Savignac, and H. Normant, *Synthesis*, **1975**, 655.
[326] A. Brändström and H. Junggren, *Acta Chem. Scand.*, **23**, 2203 (1969).
[327] U. Schöllkopf, D. Hoppe, and R. Jentsch, *Chem. Ber.*, **108**, 1580 (1975).
[328] A. M. van Leusen, R. J. Bouma, and O. Possel, *Tetrahedron Lett.*, **1975**, 3487.
[329] R. K. Singh and S. Danishefsky, *J. Org. Chem.*, **40**, 2969 (1975).
[330] A. Jończyk, B. Serafinowa, and J. Czyzewski, *Rocz. Chem.*, **47**, 529 (1973).
[331] A. Jończyk, B. Serafin, and M. Mąkosza, *Rocz. Chem.*, **45**, 1027 (1971).
[332] A. Jończyk, B. Serafin, and M. Mąkosza, *Rocz. Chem.*, **45**, 2097 (1971).
[333] M. Mąkosza, A. Jończyk, B. Serafinowa, and Z. Mroczek, *Rocz. Chem.*, **47**, 77 (1973).
[334] A. Jończyk, B. Serafin, and M. Mąkosza, *Tetrahedron Lett.*, **1971**, 1351.
[335] A. Jończyk and T. Pytlewski, *Roscz. Chem.*, **49**, 1425 (1975).
[336] A. Jończyk, B. Serafin, and E. Skulimowska, *Rocz. Chem.*, **45**, 1259 (1971).
[337] A. Jończyk, M. Fedoryński, and M. Mąkosza, *Rocz. Chem.*, **48**, 1713 (1974).
[338] A. Jończyk, M. Ludwikow, and M. Mąkosza, *Rocz. Chem.*, **47**, 89 (1973).
[339] M. Mąkosza and M. Fedoryński, *Rocz. Chem.*, **45**, 1861 (1971).
[340] A. T. Babayan, N. Gambaryan, and N. P. Gambaryan, *Zh. Obshch. Khim.*, **24**, 1887 (1954); *Chem. Abstr.*, **49**, 10879 (1955); *Chem. Zentralbl. Sonderb.*, **1950/54**, 4532.
[341] M. A. Boudville and H. des Abbayes, *Tetrahedron Lett.*, **1975**, 2727.
[342] B. Koutek, L. Pavličkova, and M. Souček, *Collect. Czech. Chem. Commun.*, **39**, 192 (1974).
[343] A. Jończyk, M. Ludwikow, and M. Mąkosza, *Rocz. Chem.*, **51**, 175 (1977).
[344] A. Jończyk, K. Bańco, and M. Mąkosza, *J. Org. Chem.*, **40**, 266 (1975).
[345] I. S. Aul'chenko, L. A. Kheifits, T. I. Konstantinova, and T. P. Cherkasova, *Chem. Abstr.*, **84**, 73591 (1976).
[346] J. Itakura and H. Ito, Jap. Kokai 73,00,515; *Chem. Abstr.*, **78**, 71443 (1973).

[347] Y. Tamai, T. Nishida, Y. Ohmura, T. Hosogai, Y. Ninagawa, Y. Fujita, and K. Itoi, Jap. Kokai 75,96,514; *Chem. Abstr.*, **84**, 4483 (1976).
[348] Y. Tamai, T. Nishida, F. Mori, Y. Ohmura, M. Tanomura, T. Hosogai, Y. Ninagawa, and K. Itoi, Germ. Offenl. 2,356,866 (1974); *Chem. Abstr.*, **81**, 77475 (1974).
[349] H. K. Dietl and K. C. Brannock, *Tetrahedron Lett.*, **1973**, 1273.
[350] BASF, Germ. Offenl. 2,513,996 (1976); *Derwent Abstr.*, 76050 (1976).
[351] H. des Abbayes and C. Neveau, *C. R. Acad. Sci. Ser. C*, **278**, 805 (1974).
[352] V. M. Andreev, A. I. Bibicheva, and M. I. Zhuravleva, *Zh. Org. Khim.*, **10**, 1470 (1974), Engl. transl., p. 1479.
[353] H. D. Zook and W. L. Gumby, *J. Am. Chem. Soc.*, **82**, 1386 (1960).
[354] W. Beck and M. Girnth, *Chem. Ber.*, **109**, 965 (1976).
[355] B. Samuelsson and B. Lamm, *Acta Chem. Scand.*, **25**, 1535 (1971).
[356] U. Junggren, Ph.D. Dissertation, Göteborg, 1972, as cited in ref. [5], p. 122.
[357] M. Mąkosza, *Bull. Acad. Pol. Sci. Ser. Sci. Chem.* **15**, 165 (1967).
[358] M. Mąkosza, *Tetrahedron Lett.*, **1966**, 4621.
[359] B. Dietrich and J. M. Lehn, *Tetrahedron Lett.*, **1973**, 1225.
[360] M. Mąkosza, in: "Modern Synthetic Methods 1976," R. Scheffold, Ed., Schweizerischer Chemikerverband, Zürich, 1976, p. 7.
[361] F. Guibe and G. Bram, *Bull. Soc. Chim. Fr.*, **1975**, 933.
[362] H. E. Zaugg and A. D. Schaefer, *J. Am. Chem. Soc.*, **87**, 1857 (1965).
[363] E. M. Arnett and V. M. De Palma, *J. Am. Chem. Soc.*, **98**, 2447 (1976).
[364] C. Riche, C. Pascard-Billy, C. Cambillau, and G. Bram, *J. Chem. Soc. Chem. Commun.*, **1977**, 183.
[365] E. A. Noe and M. Raban, *J. Chem. Soc. Chem. Commun.*, **1976**, 165.
[366] E. A. Noe and M. Raban, *J. Am. Chem. Soc.*, **96**, 6184 (1974).
[367] R. Gelin, S. Gelin, and A. Galliaud, *Bull. Soc. Chem. Fr.*, **1973**, 3416.
[368] Summary in: W. J. le Noble, "Highlights of Organic Chemistry," Marcel Dekker Inc., New York, 1974, p. 821 ff.
[369] A. Brändström and U. Junggren, *Acta Chem. Scand.*, **25**, 1469 (1971).
[370] H. Kolind-Andersen, R. Dyrnesli, and S.-O. Lawesson, *Bull. Soc. Chim. Belg.*, **84**, 341 (1975).
[371] B. Miller, H. Margulies, T. Drabb, and R. Wayne, *Tetrahedron Lett.*, **1970**, 3805.
[372] F. Guibe, P. Sarthou, and G. Bram, *Tetrahedron*, **30**, 3139 (1974).
[373] H. Kolind-Andersen and S. O. Lawesson, *Acta Chem. Scand. Ser. B*, **29**, 430 (1975).
[374] A. Brändström and U. Junggren, *Acta Chem. Scand.*, **23**, 2536 (1969).
[375] A. Brändström and U. Junggren, *Acta Chem. Scand.*, **23**, 3585 (1969).
[376] A. Brändström and U. Junggren, *Acta Chem. Scand.*, **23**, 2204 (1969).
[377] C. Cambillau, P. Sartou, and G. Bram, *Tetrahedron Lett.*, **1976**, 281; C. Cambillau, G. Bram, J. Corset, C. Riche, and C. Pascard-Billy, *Tetrahedron*, **34**, 2675 (1978).
[378] A. L. Kurts, S. M. Sakembaeve, I. P. Beletskaya, and O. A. Reutov, *Zh. Org. Khim.*, **10**, 1512 (1974); Engl. transl., p. 1588.
[379] A. L. Kurts, P. I. Dem'yanov, I. P. Beletskaya, and O. A. Reutov, *Zh. Org. Khim.*, **9**, 1313 (1973); Engl. transl., p. 1341.
[380] E. D'Incan and P. Viout, *Tetrahedron*, **31**, 159 (1975).
[381] J. H. Clark and J. M. Miller, *J. Chem. Soc. Chem. Commun.*, **1977**, 64; *J. Chem. Soc. Perkin Trans.*, 1, **1977**, 1743.
[382] G. Nee and B. Tchoubar, *C. R. Acad. Sci. Ser. C*, **283**, 223 (1976).
[383] J.-C. Fiaud, *Tetrahedron Lett.*, **1975**, 3495.
[384] E. V. Dehmlow, M. Lissel, and J. Heider, *Tetrahedron*, **33**, 363 (1977).
[385] G. Hata, Jap. Kokai 74,109,346; *Chem. Abstr.*, **82**, 124876 (1975).
[386] B. Cederlund and A.-B. Hörnfeldt, *Acta Chem. Scand.*, **25**, 3546 (1971).
[387] B. Cederlund, Å. Jesperson, and A.-B. Hörnfeldt, *Acta Chem. Scand*, **25**, 3656 (1971).
[388] B. Cederlund and A.-B. Hörnfeldt, *Chem. Scrip.*, **8**, 140 (1975).
[389] S. Gronowitz and T. Frejd, *Acta Chem. Scand. Ser. B* **30**, 439 (1976).

[390] B. Cederlund and A.-B. Hörnfeldt, *Acta Chem. Scand. Ser. B*, **30**, 101 (1976).
[391] L. Dalgaard, H. Kolind-Andersen, and S.-O. Lawesson, *Tetrahedron*, **29**, 2077 (1973).
[392] L. Jensen, L. Dalgaard, and S.-O. Lawesson, *Tetrahedron*, **30**, 2413 (1974).
[393] P. Hassanaly, H. J.-M. Dou, and J. Metzger, G. Assef, and J. Kister, *Synthesis*, **1977**, 253.
[394] H. J.-M. Dou, P. Hassanaly, and J. Metzger, *J. Heterocycl. Chem.*, **14**, 321 (1977).
[395] N. O. Vesterager, E. B. Pedersen, and S.-O. Lawesson, *Tetrahedron*, **29**, 321 (1973).
[396] T. Bianchi and L. A. Cate, *J. Org. Chem.*, **42**, 2031 (1977).
[397] H. D. Durst and L. Liebeskind, *J. Org. Chem.*, **39**, 3271 (1974).
[398] A. W. Burgstahler, M. E. Sanders, C. G. Shaefer, and L. O. Weigel, *Synthesis*, **1977**, 405.
[399] B. T. Golding and P. V. Ioannou, *Synthesis*, **1977**, 423.
[400] E. V. Dehmlow, unpublished results.
[401] Y. Jasor, M. Gaudry, and A. Marquet, *Tetrahedron Lett.*, **1976**, 53.
[402] K. Yamada, M. Kamai, Y. Nakano, H. Kasimura, and H. Iida, *Tetrahedron Lett.*, **1974**, 1741.
[403] W. J. Spillane, H. J.-M. Dou, and J. Metzger, *Tetrahedron Lett.*, **1976**, 2269.
[404] W. J. Spillane, P. Hassanaly, and H. J.-M.Dou, *C. R. Acad. Sci. Ser. C*, **283**, 289 (1976).
[405] T. Higgin, W. J. Spillane, H. J.-M. Dou, and J. Metzger, *C. R. Acad. Sci. Ser. C*, **284**, 929 (1977).
[406] Y. Yano, T. Okonogi, and W. Tagaki, *J. Org. Chem.*, **38**, 3912 (1973).
[407] B. A. Trofimov, S. V. Amasova, and V. V. Nosyreva, *Zh. Org. Khim.*, **12**, 1366 (1976); Engl. transl., p. 1358.
[408] M. Mąkosza, J. Czyewski, and M. Jawdosiuk, *Org. Synth.*, **55**, 99 (1976).
[409] M. Mąkosza, *Tetrahedron Lett.*, **1966**, 5489.
[410] M. Mąkosza and M. Jawdosiuk, *Bull. Acad. Pol. Sci. Ser. Sci. Chim.*, **16**, 589 (1968).
[411] L. F. and M. Fieser, "Reagents for Organic Synthesis," J. Wiley and Sons, Inc., New York, 1967, (a) p. 1252, (b) p. 933.
[412] H. A. Bruson, in: "Organic Reactions," R. Adams, Ed., Vol. **5**, J. Wiley and Sons, Inc., New York, 1949, p. 79.
[413] D. A. White and M. M. Baizer, *J. Chem. Soc. Perkin Trans.* 1, **1973**, 2230.
[414] S. Hoz, M. Albeck, and Z. Rappoport, *Synthesis*, **1975**, 162.
[415] T. Yanami, M. Kato, and A. Yoshikoshi, *J. Chem. Soc. Chem. Commun.*, **1975**, 726.
[416] I. Belsky, *J. Chem. Soc. Chem. Commun.*, **1977**, 237.
[417] C. L. Liotta, A. M. Dabdoub, and L. H. Zalkow, *Tetrahedron Lett.*, **1977**, 1117.
[418] T. Sakakibara and R. Sudoh, *J. Org. Chem.*, **40**, 2823 (1975).
[419] T. Sakakibara, M. Yamada, and R. Sudoh, *J. Org. Chem.*, **41**, 736 (1976).
[420] M. Fedoryński, I. Gorzkowska, and M. Mąkosza, *Synthesis*, **1977**, 120.
[421] A. Jończyk and M. Mąkosza, *Synthesis*, **1976**, 387.
[422] I. Artaud, J. Seyden-Penne, and P. Viout, *C. R. Acad. Sci. Ser. C*, **283**, 503 (1976).
[423] T. Wakabayashi and Y. Kato, *Tetrahedron Lett.*, **1977**, 1235.
[424] H. Wynberg and R. Helder, *Tetrahedron Lett.*, **1975**, 4057.
[425] T. Sakakibara and R. Sudoh, *Carbohydr. Res.*, **50**, 197 (1976).
[426] J. M. McIntosh and H. Khalil, *J. Org. Chem.*, **42**, 2123 (1977).
[427] S. Kurozumi, T. Toru, T. Tanaka, S. Miura, M. Kobayashi, and S. I. Ishimoto, *Bull. Chem. Soc. Jpn.*, **50**, 1357 (1977).
[428] A. Merz, *Angew. Chem.*, **89**, 484 (1977); *Angew. Chem. Int. Ed. Engl.*, **16**, 467 (1977).
[429] C. Hansson and E. Rosengreen, *Acta Chem. Scand. Ser. B*, **30**, 871 (1976).
[430] E. Buncel and B. Menon, *J. Am. Chem. Soc.*, **99**, 4457 (1977).
[431] A. M. van Leusen and J. Wildeman, *Synthesis*, **1977**, 501.
[432] W. P. Reeves and R. G. Hilbrich, *Tetrahedron*, **32**, 2235 (1976).
[433] M. Mąkosza, A. Kacprowicz, and M. Fedoryński, *Tetrahedron Lett.*, **1975**, 2119.
[434] J. N. Roitman and D. J. Cram, *J. Am. Chem. Soc.*, **93**, 2231 (1971).

[435] J. Solodar, *Tetrahedron Lett.*, **1971**, 287.
[436] J. Solodar, *Synth. Inorg. Met. Org. Chem.*, **1**, 141 (1971).
[437] S. Akabori, M. Ohtomi, and K. Arai, *Bull. Chem. Soc. Jpn.*, **49**, 746 (1976).
[438] V. Dryanska and C. Ivanov, *Tetrahedron Lett.*, **1975**, 3519.
[439] G. Cardillo, D. Savoia, and A. Umani-Ronchi, *Synthesis*, **1975**, 453.
[440] G. W. Gokel, H. M. Gerdes, and N. W. Rebert, *Tetrahedron Lett.*, **1976**, 653.
[441] I. N. Rozhkov, N. D. Kuleshova, and I. N. Knunyants, *Izv. Akad. Nauk SSSR Ser. Khim.*, **22**, 128 (1973); Engl. transl., p. 124.
[442] E. Nakamura and I. Kuwajima, *Angew. Chem.*, **88**, 539 (1976); *Angew. Chem. Int. Ed. Engl.*, **15**, 498 (1976).
[443] E. Nakamura, M. Shimizu, and I. Kuwajima, *Tetrahedron Lett.*, **1976**, 1699.
[444] D. A. Evans and L. K. Truesdale, *Tetrahedron Lett.*, **1973**, 4929.
[445] D. A. Evans, J. M. Hoffman, and L. K. Truesdale, *J. Am. Chem. Soc.*, **95**, 5822 (1973).
[446] H. des Abbayes and A. Buloup, *J. Chem. Soc. Chem. Commun.*, **1978**, 1090.
[447] E. V. Dehmlow, unpublished results.
[448] J. Jończyk, M. Fedoryński, and M. Mąkosza, *Tetrahedron Lett.*, **1972**, 2395.
[449] T. Koizumi, K. Takeda, K. Yoshida, and E. Yoshii, *Synthesis*, **1977**, 497.
[450] F. Naso and L. Ronzini, *J. Chem. Soc. Perkin Trans* 1, **1974**, 340.
[451] E. J. Corey and B. B. Snider, *J. Am. Chem. Soc.*, **94**, 2549 (1972).
[452] T. H. Chan and W. Mychajlowskij, *Tetrahedron Lett.*, **1974**, 171.
[453] J. Hayami, N. Ono, and A. Kaji, *Tetrahedron Lett.*, **1970**, 2727.
[454] R. A. Bartsch, J. R. Allaway, and J. G. Lee, *Tetrahedron Lett.*, **1977**, 779.
[455] R. A. Bartsch, *Acc. Chem. Res.*, **8**, 239 (1975).
[456] D. Landini, S. Quici, and F. Rolla, *Synthesis*, **1975**, 397.
[457] J. B. Campbell, U.S. Pat. 3,639,493 (1972); eq. to Fr. Demande 2,002,799 (1969); *Chem. Abstr.*, **72**, 78366 (1970).
[458] J. B. Campbell and R. E. Tarney, French Pat. 1,525,661 (1968); *Chem. Abstr.*, **71**, 80648 (1969).
[459] J. B. Campbell and R. E. Tarney, U.S. Pat. 3,981,937 (1976); *Derwent Abstr.*, 75550 (1976).
[460] du Pont de Nemours Co., Netherl. Pat. Appl. 6,707,711 (1967); *Bayer-Patentschnellber.*, 11-008 (1968).
[461] du Pont de Nemours Co., Netherl. Pat. Appl. 6,903,151 (1969); *Bayer-Patentschnellber.*, 51-034 (1969).
[462] R. G. Gordon, U.S. Pat. 3,664,966 (1972); *Chem. Abstr.*, **77**, 75756 (1972).
[463] K. Sennewald, K. Gehrmann, and G. Viertel, Germ. Pat. 1,271,107 (1968); *Chem. Zentralbl.*, **1969**, 18–1806.
[464] E. F. Lutz, J. T. Kelly, and D. W. Hall, Brit. Pat. 1,112,068 (1968); *Chem. Abstr.*, **69**, 10079 (1968).
[465] H. Baader, K. Sennewald, and H. Reis, Brit. Pat. 1,154,445 (1969); *Chem. Abstr.*, **71**, 30094 (1969).
[466] J. Dockx, *Synthesis*, **1973**, 441.
[467] Unpublished results by A. Verstraelen, W. Helsen, and J. Dockx, cited in ref. [466], p. 449.
[468] M. S. Newman, B. Dhawan, M. M. Hashem, V. K. Khanna, and J. M. Springer, *J. Org. Chem.*, **41**, 3925 (1976).
[469] A. Gorgues, *C. R. Acad. Sci. Ser. C.*, **278**, 287 (1974).
[470] A. Gorgues, and A. LeCoq, *Bull. Soc. Chim. Fr. II*, **1976**, 125.
[471] A. Gorgues and A. LeCoq, *Tetrahedron Lett.*, **1976**, 4723.
[472] E. T. McBee and R. O. Bolt, *Ind. Eng. Chem.*, **39**, 412 (1947).
[473] J. H. Fendler and E. J. Fendler, "Catalysis in Micellar and Macromolecular Systems," Academic Press, New York, 1975.
[474] J. H. Fendler, *Acc. Chem. Res.*, **9**, 153 (1976).
[475] F. M. Menger, J. U. Rhee, and H. K. Rhee, *J. Org. Chem.*, **40**, 3803 (1975).

[476] E. Jósefowicz and R. Soloniewicz, *Mitteilungsbl. Chem. Ges. DDR*, **Sonderheft 1959**, Katalyse, p. 215.
[477] E. V. Dehmlow and S. Barahona, unpublished results; S. Barahona, Diploma Thesis, Technical University, Berlin 1977.
[478] H. J. Bestmann, in: "Neuere Methoden des präparativen organischen Chemie," Vol. 5, W. Foerst, Ed., Verlag Chemie, Weinheim, 1967, p. 1.
[479] U. Schöllkopf, *Angew. Chem.*, **71**, 260 (1959).
[480] R. J. Anderson and C. A. Henrick, *J. Am. Chem. Soc.*, **97**, 4327 (1975).
[481] M. Schlosser, "Topics in Stereochemistry," Vol 5, E. L. Eliel and N. L. Allinger, Eds., Wiley-Interscience, New York, 1970, p. 1.
[482] W. P. Schneider, *J. Chem. Soc. Chem. Commun.*, **1969**, 785.
[483] G. Märkl and A. Merz, *Synthesis*, **1973**, 295.
[484] R. M. Boden, *Synthesis*, **1975**, 784.
[485] W. Tagaki, I. Inoue, Y. Yano, and T. Okonogi, *Tetrahedron Lett.*, **1974**, 2587.
[486] R. Broos and M. Anteunis, *Synth. Commun.*, **6**, 53 (1976).
[487] S. Hünig and I. Stemmler, *Tetrahedron Lett.*, **1974**, 3151.
[488] K. W. Lee and L. A. Singer, *J. Org. Chem.*, **39**, 3780 (1974).
[489] J. Boutagy and R. Thomas, *Chem. Rev.*, **74**, 87 (1974).
[490] T. H. Kinstle and B. Y. Mandanas, *J. Chem. Soc. Chem. Commun.*, **1968**, 1699.
[491] G. Jones and R. F. Maisey, *J. Chem. Soc. Chem. Commun.*, **1968**, 543.
[492] M. Mikolajczyk, S. Grzejszczak, W. Midura, and A. Zatorski, *Synthesis*, **1976**, 396.
[493] M. Mikolajczyk, S. Grzejszczak, W. Midura, and A. Zatorski, *Synthesis*, **1975**, 278.
[494] C. Piechucki, *Synthesis*, **1976**, 187.
[495] C. Piechucki, *Synthesis*, **1974**, 869.
[496] E. D'Incan, *Tetrahedron*, **33**, 951 (1977).
[497] Y. Yano, T. Okonogi, M. Sunaga, and W. Tagaki, *J. Chem. Soc. Chem. Commun.*, **1973**, 527.
[498] A. Merz and G. Märkl, *Angew. Chem.*, **85**, 867 (1973); *Angew. Chem. Int. Ed. Engl.*, **11**, 845 (1973).
[499] M. J. Hatch, *J. Org. Chem.*, **34**, 2133 (1969).
[500] T. Hiyama, T. Mishima, H. Sawada, and H. Nozaki, *J. Am. Chem. Soc.*, **97**, 1626 (1975).
[501] T. Hiyama, T. Mishima, H. Sawada, and H. Nozaki, *J. Am. Chem. Soc.*, **98**, 641 (1976).
[502] K. Kleveland, L. Skattebøl, and L. K. Sydnes, *Acta Chem. Scand. Ser. B*, **31**, 463 (1977).
[503] E. d'Incan and J. Seyden-Penne, *C. R. Acad. Sci. Ser. C*, **281**, 1031 (1975).
[504] C. J. Pedersen and H. K. Frensdorff, *Angew. Chem.*, **84**, 16 (1972); *Angew. Chem. Int. Ed. Engl.*, **11**, 16 (1972).
[505] C. J. Pedersen, *J. Am. Chem. Soc.*, **89**, 7017 (1967).
[506] D. H. Hunter, W. Lee, and S. K. Sim, *J. Chem. Soc. Chem. Commun.*, **1974**, 1018.
[507] J. Smid, S. Shah, L. Wong, and J. Hurley, *J. Am. Chem. Soc.*, **97**, 5932 (1975).
[508] H. Ledon, *Synthesis*, **1974**, 347.
[509] M. Nakajima and J.-P. Anselme, *Tetrahedron Lett.*, **1976**, 4421.
[510] H. Alper and D. Des Roches, *J. Organomet. Chem.*, **117**, C44 (1976).
[511] H. Alper, H. Des Abbayes, and D. Des Roches, *J. Organomet. Chem.*, **121** C31 (1976).
[512] L. Cassar, M. Foá, and A. Gordano, *J. Organomet. Chem.*, **121** C55 (1976).
[513] Dow Chemical Co., U.S. Pat. 3,974,199 (1976); *Derwent Abstr.*, 64914 (1976).
[514] S. Yanagida, Y. Noji, and M. Okahara, *Tetrahedron Lett.*, **1977**, 2337, 2893.
[515] A. C. Knipe, N. Sridhar, and A. Loughran, *J. Chem. Soc. Chem. Commun.*, **1976**, 630.
[516] H. Iida, K. Takahashi, and K. Yamada, *Nippon Kagaku Kaishi*, **1974**, 2127; *Chem. Abstr.*, **83**, 27739 (1975); *Chem. Informationsdienst*, **1975**, 11–248.
[517] K. C. Chan, S. H. Goh, S. E. Teoh, and W. H. Wong, *Aust. J. Chem.*, **27**, 421 (1974).

[518] J. R. Grunwell and S. L. Dye, *Tetrahedron Lett.*, **1975**, 1739.
[519] L. Rand and M. J. Albinak, *J. Org. Chem.*, **25**, 1837 (1960).
[520] J. Hayami, N. Ono, and A. Kaji, *Tetrahedron Lett.*, **1968**, 1385.
[521] M. Mitami, T. Tsuchida, and K. Koyama, *Chem. Lett.*, **1974**, 1209.
[522] M. Mąkosza, B. Serafinowa, and I. Gajos, *Rocz. Chem.*, **43**, 671 (1969).
[523] T. Gajda and A. Zwierzak, *Synthesis*, **1976**, 243.
[524] M. Guainazzi, G. Silvestri, and G. Serravalle, *J. Chem. Soc. Chem. Commun.*, **1975**, 200.
[525] S. J. Bakum and S. F. Ereshko, *Izv. Akad. Nauk SSSR Ser. Khim.*, **23**, 2138 (1974), Engl. transl., p. 2058.
[526] J. L. Pierre and H. Handel, *Tetrahedron Lett.*, **1974**, 2317.
[527] J. L. Pierre, H. Handel, and R. Perrand, *Tetrahedron*, **31**, 2795 (1975).
[528] H. Handel and J. L. Pierre, *Tetrahedron*, **31**, 2799 (1975).
[529] A. Loupy, J. Seiden-Penne, and B. Tschoubar, *Tetrahedron Lett.*, **1976**, 1677.
[530] E. A. Sullivan and A. A. Hinckley, *J. Org. Chem.*, **27**, 3731 (1962).
[531] A. Brändström, H. Junggren, and B. Lamm, *Tetrahedron Lett.*, **1972**, 3173.
[532] Ref. [5], p. 165.
[533] D. J. Raber and W. C. Guida, *J. Org. Chem.*, **41**, 690 (1976).
[534] D. J. Raber and W. C. Guida, *Synthesis*, **1974**, 808.
[535] N. H. Nilson and A. Senning, *Chem. Ber.*, **107**, 2345 (1974).
[536] R. O. Hutchins and D. Kandasamy, *J. Am. Chem. Soc.*, **95**, 6131 (1973).
[537] R. O. Hutchins, D. Kandasamy, C. A. Maryanoff, D. Masilamani, and B. E. Maryanoff, *J. Org. Chem.*, **42**, 82 (1977).
[538] L. O. Hutchins and D. Kandasamy, *J. Org. Chem.*, **40**, 2530 (1975).
[539] T. Matsuda and K. Koida, *Bull. Chem. Soc. Jpn.*, **46**, 2259 (1973).
[540] Ref. [5], pp. 164–166.
[541] S. Bradamante, A. Merchesi, and U. M. Pagnoni, *Ann. Chim.* (*Rome*), **65**, 131 (1975).
[542] J. Balcells, S. Colonna, and R. Fornasier, *Synthesis*, **1976**, 266.
[543] J. P. Massé and E. R. Parayre, *J. Chem. Soc. Chem. Commun.*, **1976**, 438.
[544] B. Witkop and C. M. Foltz, *J. Am. Chem. Soc.*, **79**, 197 (1957).
[545] E. V. Dehmlow, J. Heider, U. Brenner, unpublished results.
[546] H. Durst, J. W. Zubrick, and G. R. Kieczykowski, *Tetrahedron Lett.*, **1974**, 1777.
[547] H. des Abbayes and H. Alper, *J. Am. Chem. Soc.*, **99**, 98 (1977).
[548] H. Alper, D. Des Roches, and H. des Abbayes, *Angew. Chem.*, **89**, 43 (1977); *Angew. Chem. Int. Edit. Engl.*, **16**, 41 (1977).
[549] H. Alper, K. D. Logbo, and H. des Abbayes, *Tetrahedron Lett.*, **1977**, 2861.
[550] G. Borgogno, S. Colonna, and R. Fornasier, *Synthesis*, **1975**, 529.
[551] K. Y. Hui and B. L. Shaw, *J. Organomet. Chem.*, **124**, 262 (1977).
[552] R. Rebois, J. Chauveau, M. Deyris, and H. Condanne, *C. R. Acad. Sci. Ser. C*, **282**, 947 (1976).
[553] Imp. Chem. Ind. Ltd., Belg. Pat. 833,241 (1976); *Derwent Abstr.*, 22647 (1976).
[554] R. Helder, R. Arends, W. Bolt, H. Hiemstra, and H. Wynberg, *Tetrahedron Lett.*, **1977**, 2181.
[555] Kohjin, KK, Jap. Kokai 76,105,024 *Derwent Abstr.*, 82123 (1976).
[556] P. D. Woodgate, H. H. Lee, P. S. Rutledge, and R. C. Cambie, *Synthesis*, **1977**, 462.
[557] D. J. Sam and H. E. Simmons, *J. Am. Chem. Soc.*, **94**, 4024 (1972).
[558] E. Weber and F. Vögtle, *Chem. Ber.*, **109**, 1803 (1976).
[559] A. W. Herriott and D. Picker, *Tetrahedron Lett.*, **1974**, 1511.
[560] M. S. Newman, H. M. Dali, and W. M. Hung, *J. Org. Chem.*, **40**, 263 (1975).
[561] H. H. Pohl and K. Dimroth, *Angew. Chem.*, **87**, 135 (1975); *Angew. Chem. Int. Ed. Engl.*, **14**, 111 (1975).
[562] F. Vögtle and W. Offermann, *Chem. Exp. Didakt.*, **1**, 147 (1975).
[563] S. G. Boot, M. R. Boots, K. E. Guyer, and P. E. Marecki, *J. Pharm. Sci.*, **65**, 1374 (1976).

[564] G. W. Gokel and H. D. Durst, *Synthesis*, **1976**, 168.
[565] M. Mack and H. D. Durst, unpublished results, 1975, cited in ref. [564] as ref. 155.
[566] Ref. [5], p. 161.
[567] W. P. Weber and J. P. Shepherd, *Tetrahedron Lett.*, **1972**, 4907.
[568] M. L. Richardson, *Analyst* (*London*), **87**, 435 (1962); J. M. Matusek and T. T. Sugihara, *Anal. Chem.*, **33**, 35 (1961).
[569] N. A. Gibson and J. W. Hosking, *Aust. J. Chem.*, **18**, 123 (1965).
[570] F. Yamashita, A. Atsumi, and H. Inoue, *Nippon Kagaku Kaishi*, **1975**, 1102; *Chem. Abstr.*, **83**, 113378 (1975); *Chem. Informationsdienst*, **1975**, 48–162.
[571] H. D. Durst, unpublished results, cited in ref. [564] as ref. 181, 189.
[572] E. Alneri, G. Bottaccio, and V. Carletti, *Tetrahedron Lett.*, **1977**, 2117.
[573] B. J. Garcia, G. W. Gokel, and D. W. Tudor, unpublished results, cited in ref. [564] as ref. 182.
[574] J. Jarrousse and J.-C. Raulin, *C. R. Acad. Sci. Ser. C*, **284**, 503 (1977).
[575] R. M. Boden, *Synthesis*, **1975**, 783.
[576] J. S. Valentine and A. B. Curtis, *J. Am. Chem. Soc.*, **97**, 224 (1975).
[577] I. Rosenthal and A. Frimer, *Tetrahedron Lett.*, **1975**, 3731.
[578] E. J. Corey, K. C. Nicolaou, M. Shibasaki, Y. Machida, and C. S. Shiner, *Tetrahedron Lett.*, **1975**, 3183.
[579] M. V. Merritt and D. T. Sawyer, *J. Org. Chem.*, **35**, 2157 (1970).
[580] J. San Filippo, Jr., C.-I. Chern, and J. S. Valentine, *J. Org. Chem.*, **40**, 1678 (1975).
[581] R. A. Johnson and E. G. Nidy, *J. Org. Chem.*, **40**, 1680 (1975).
[582] M. J. Gibian and T. Ungermann, *J. Org. Chem.*, **41**, 2500 (1976).
[583] W. C. Danen and R. J. Warner, *Tetrahedron*, **11**, 989 (1977).
[584] M. V. Meritt and R. A. Johnson, *J. Am. Chem. Soc.*, **99**, 3713 (1977).
[585] R. A. Johnson, *Tetrahedron Lett.*, **1976**, 331.
[586] J. San Filippo, Jr., L. J. Romano, C.-I. Chern, and J. S. Valentine, *J. Org. Chem.*, **41**, 586 (1976).
[587] J. W. Peters and C. S. Foote, *J. Am. Chem. Soc.*, **98**, 873 (1976).
[588] C.-I. Chern and J. San Filippo, Jr., *J. Org. Chem.*, **42**, 178 (1977).
[589] A. Frimer and I. Rosenthal, *Tetrahedron Lett.*, **1976**, 2809.
[590] I. Rosenthal and A. Frimer, *Tetrahedron Lett.*, **1976**, 2805.
[591] Y. Moro-Oka, P. J. Chung, H. Arakawa, and T. Ikawa, *Chem. Lett.*, **1976**, 1293.
[592] A. Nishinaga, T. Shimizu, and T Matsuura, *Chem. Lett.*, **1977**, 547.
[593] E. Lee-Ruff, A. B. P. Lever, and J. Rigaudy, *Can. J. Chem.*, **54**, 1837 (1976).
[594] G. Cardillo, M. Orena, and S. Sandri, *J. Chem. Soc. Chem. Commun.*, **1976**, 190.
[595] G. A. Lee and H. H. Freedman, *Tetrahedron Lett.*, **1976**, 1641.
[596] G. Brunow and S. Sumelius, *Acta Chem. Scan. Ser. B*, **29**, 499 (1975).
[597] A. R. Qureshi and B. Sklarz, *J. Chem. Soc.*, *C*, **1966**, 412.
[598] R. Pappo, D. S. Allen, Jr., R. U. Lemieux, and W. S. Johnson, *J. Org. Chem.*, **21**, 478 (1956),
[599] C. M. Starks and D. R. Napier, S. African. Pat. 7,101,495 (1971) ≡ Brit. Pat. 1,324,763 (1973); *Chem. Abstr.*, **76**, 153191 (1972).
[600] C. M. Starks and P. H. Washecheck, U.S. Pat. 3,547,962 (1970); *Chem. Abstr.*, **74**, 140895 (1971).
[601] J. R. Adamson, R. Bywood, D. T. Eastlick, G. Gallagher, D. Walker, and E. M. Wilson, *J. Chem. Soc. Perkin Trans.* 1, **1975**, 2030; R. Bywood, G. Gallagher, G. K. Sharma, and D. Walker, *ibid.*, **1975**, 2019.
[602] K. Takahashi and H. Iida, *Synthesis*, **1979**, 301.
[603] R. Helder, J. C. Hummelen, R. W. P. M. Laane, J. S. Wiering, and H. Wynberg, *Tetrahedron Lett.*, **1976**, 1831; *cf.*, also H. Wynberg, *Chimia*, **30**, 445 (1976).
[604] E. V. Dehmlow and J. Schmidt, unpublished results (1975).
[605] W. Kirmse, "Carbene Chemistry," Second Ed., Academic Press, New York and London, 1971, Chapters 4, 8, 9, 10, and 11.

[606] W. von E. Doering and A. K. Hoffmann, *J. Am. Chem. Soc.*, **76**, 6162 (1954).
[607] G. C. Robinson, *Tetrahedron Lett.*, **1965**, 1749.
[608] M. Mąkosza and W. Wawryzniewicz, *Tetrahedron Lett.*, **1969**, 4659.
[609] P. S. Skell and M. S. Cholod, *J. Am. Chem. Soc.*, **91**, 6035, 7131 (1969).
[610] G. Köbrich, H. Büttner, and E. Wagner, *Angew. Chem.*, **82**, 177 (1970); *Angew. Chem. Int. Edit. Engl.*, **9**, 169 (1970).
[611] D. Seyferth, M. E. Gordon, J. Y.-P. Mui, and J. M. Burlitch, *J. Am. Chem. Soc.*, **89**, 959, 4953 (1967).
[612] R. A. Moss and F. G. Pilkiewicz, *J. Am. Chem. Soc.*, **96**, 5632 (1974).
[613] E. V. Dehmlow and M. Lissel, *J. Chem. Res.*, **1978**, (S)310, (M)4155.
[614] W. M. Wagner, H. Kloosterziel, S. van der Ven, and A. F. Bickel, *Rec. Trav. Chim. Pays-Bas*, **81**, 925, 933, 947 (1962).
[615] H. J.-M. Dou, H. Komeili-Zadeh, and M. Crozet, *C. R. Acad. Sci. Ser. C*, **284**, 685 (1977).
[616] P. A. Verbrügge and E. W. Uurbanus, Germ. Offenl. 2,324,390 (1973); ≡ Dutch Pat. 7,306,662 (1973) ≡ Fr. Pat. 2,184,793; *Chem. Abstr.*, **80**, 70420 (1974).
[617] K. Isagawa, Y. Kimura, and S. Kwon, *J. Org. Chem.*, **39**, 3171 (1974).
[618] R. R. Kostikov and A. P. Molchanov, *Zh. Org. Khim.*, **11**, 1767 (1975); Engl. transl., p. 1767.
[619] M. Mąkosza, private communication.
[620] Shell B.V., Brit. Pat. 1,436,854 (1976); *Derwent Abstr.*, 40579 (1976).
[621] T. Hiyama, M. Tsukanaka, and H. Nozaki, *J. Am. Chem. Soc.*, **96**, 3713 (1974).
[622] T. Hiyama, H. Sawada, M. Tsukanaka, and H. Nozaki, *Tetrahedron Lett.*, **1975**, 3013.
[623] Y. Kimura, Y. Ogaki, K. Isagawa, and Y. Otsuji, *Chem. Lett.*, **1976**, 1149.
[624] E. V. Dehmlow, *Angew. Chem.*, **89**, 521 (1977); *Angew. Chem. Int. Ed. Engl.*, **16**, 493 (1977).
[625] E. V. Dehmlow, *Angew. Chem.*, **86**, 187 (1974); *Angew. Chem. Int. Ed. Engl.*, **13**, 170 (1974).
[626] E. V. Dehmlow and J. Schönefeld, *Justus Liebigs Ann. Chem.*, **744**, 42 (1971).
[627] G. W. Gokel, J. P. Shepherd, W. P. Weber, H. G. Boettger, J. L. Holwick, and D. J. McAdoo, *J. Org. Chem.*, **38**, 1913 (1973).
[628] M. J. Aroney, K. E. Calderbank, and H. J. Stootman, *J. Chem. Soc. Perkin Trans.* 2, **1973**, 2060.
[629] E. Wada, S. Fujisaki, A. Nagashima, and S. Kajigaeshi, *Bull. Chem. Soc. Jpn.*, **48**, 739 (1975).
[630] F. Effenberger and W. Kurtz, *Chem. Ber.*, **106**, 511 (1973).
[631] K.-O. Henseling and P. Weyerstahl, *Chem. Ber.*, **108**, 2803 (1975).
[632] S. Kajigaeshi, N. Kuroda, G. Matsumoto, E. Wada, and A. Nagashima, *Tetrahedron Lett.*, **1971**, 4887.
[633] K. Kobayashi and J. B. Lambert, *J. Org. Chem.*, **42**, 1254 (1977).
[634] G. W. Gokel, J. P. Shepherd, and W. P. Weber, *J. Org. Chem.*, **38**, 1913 (1973).
[635] M. Mąkosza and I. Gajos, *Rocz. Chem.*, **48**, 1883 (1974).
[636] G. C. Joshi, N. Singh, and L. M. Pande, *Tetrahedron Lett.*, **1972**, 1461.
[637] G. C. Joshi, N. Singh, and L. M. Pande, *Synthesis*, **1972**, 317.
[638] S. Takano, K. Yuta, and K. Ogasawara, *Heterocycles*, **4**, 947 (1976).
[639] R. Ikan, A. Markus, and Z. Goldschmidt, *J. Chem. Soc. Perkin Trans.* 1, **1972**, 2423.
[640] Y. M. Sheikh, J. Leclercq, and C. Djerassi, *J. Chem. Soc. Perkin Trans.* 1, **1974**, 909.
[641] G. Hammen, T. Bässler, and M. Hanack, *Chem. Ber.*, **107**, 1676 (1974).
[642] S. S. Hixon, *J. Am. Chem. Soc.*, **97**, 1981 (1975).
[643] M. Nakazaki, K. Yamamoto, and J. Yanagi, *J. Chem. Soc. Chem. Commun.*, **1977**, 346.
[644] E. V. Dehmlow and G. C. Ezimora, *Z. Naturforsch. Teil B*, **30**, 825 (1975).

[645] K. Kawashima, T. Saraie, Y. Kawano, and I. Ishiguro, Germ. Offenl. 2,404,744 (1974); *Chem. Abstr.*, **81**, 135997 (1974); Belg. Pat. 809,816 (1973); *Derwent Abstr.*, 43495 (1974).
[646] N. I. Yakushkina, L. F. Germanova, V. D. Klebanova, L. I. Leonova, and I. G. Bolesov, *Zh. Org. Khim.*, **12**, 2141 (1976); Engl. transl., p. 2082.
[647] E. Dunkelblum and B. Singer, *Synthesis*, **1975**, 323.
[648] T. Sasaki, S. Eguchi, and M. Mizutani, *Org. Prep. Proc. Int.*, **6**, 57 (1974).
[649] T. Sasaki, S. Eguchi, and Y. Hirako, *Tetrahedron Lett.*, **1976**, 541.
[650] R. B. Miller, *Synth. Commun.*, **4**, 341 (1974).
[651] R. A. Moss and D. J. Smudin, *J. Org. Chem.*, **41**, 611 (1976).
[652] R. F. Boswell and R. G. Bass, *J. Org. Chem.*, **40**, 2419 (1975), **42**, 2342 (1977).
[653] Y. Gaoni, *Tetrahedron Lett.*, **1976**, 2167.
[654] L. Pizzala, J.-P. Aycard, and H. Bodot, *C. R. Acad. Sci. Ser. C*, **285**, 9 (1977).
[655] D. Davalian and P. J. Garratt, *J. Am. Chem. Soc.*, **97**, 6883 (1975).
[656] T. Hiyama, T. Mishima, K. Kitatani, and H. Nozaki, *Tetrahedron Lett.*, **1974**, 3297.
[657] M. G. Voronkov, S. M. Shostakovskii, V. G. Kozyrev, Y. V. Artst, L. N. Balabanova, and O. B. Bannikova, U.S.S.R. Pat. 488,814 (1976); *Chem. Abstr.*, **84**, 43839 (9176).
[658] A. A. Bredikhin and V. V. Plemenkov, *Zh. Org. Khim.*, **12**, 1001 (1976); Engl. transl., p. 1011.
[659] N. V. Kuznetsov and I. I. Krasavtsev, *Ukr. Khem. Zh.*, **42**, 968 (1976); *Chem. Informationsdienst*, **7**, 7650–293 (1976).
[660] A. K. Khusid, G. V. Kryshtal, V. F. Kucherov, and L. A. Yanoskaya, *Izv. Akad. Nauk SSSR Ser. Khim.*, **24**, 2577 (1975); Engl. transl., p. 2462.
[661] A. K. Khusid, G. V. Kryshtal, V. A. Dombrovsky, V. F. Kucherov, L. A. Yanovskaya, V. I. Kadentsev, and O. S. Chizhov, *Tetrahedron*, **33**, 77 (1977).
[662] M. F. Shostakovskii, V. I. Erofeev, and V. S. Aksenov, *Izv. Akad. Nauk SSSR Ser. Khim.*, **24**, 2577 (1975); Engl. transl., p. 2462.
[663] A. K. Khusid, G. V. Kryshtal, V. F. Kucherov, and L. A. Yanovskaya, *Synthesis*, **1977**, 428.
[664] M. Mąkosza and A. Kacprowicz, *Bull. Acad. Pol. Sci. Ser. Sci. Chem.*, **22**, 467 (1974).
[665] J. Graefe, M. Adler, and M. Mühlstädt, *Z. Chem.*, **15**, 14 (1975).
[666] S. A. G. De Graaf and U. K. Pandit, *Tetrahedron*, **29**, 4263 (1973).
[667] M. Mąkosza and A. Kacprowicz, *Rocz. Chem.*, **48**, 2129 (1974).
[668] R. R. Kostikov, A. F. Khlebnikov, and K. A. Ogloblin, U.S.S.R Pat. 482,448 (1976); *Chem. Abstr.*, **83**, 206087 (1975).
[669] J. Graefe, *Z. Chem.*, **14**, 469 (1974).
[670] D. J. Sikkema, E. Molenaar, and D. B. van Guldener, *Rec. Trav. Chim. Pays-Bas*, **95**, 154 (1976).
[671] E. V. Dehmlow, *Justus Liebigs Ann. Chem.*, **758**, 148 (1972).
[672] D. F. Hayman, Germ. Offenl. 2,201,514 (1972); *Chem. Abstr.*, **78**, 3793 (1973).
[673] B.D.H. Pharmaceuticals Ltd., Belg. Pat. 777705 (1972); *Derwent Abstr.*, 48539 (1972).
[674] R. Barlet, *C. R. Acad. Sci. Ser. C*, **278**, 621 (1974).
[675] E. V. Dehmlow, *Tetrahedron Lett.*, **1976**, 91.
[676] K. Idemori, M. Tagaki, and T. Matsuda, *Bull. Chem. Soc. Jpn.*, **50**, 1355 (1977).
[677] M. Mąkosza and I. Gajos, *Bull. Acad. Pol. Sci. Ser. Sci. Chim.*, **20**, 33 (1972).
[678] E. V. Dehmlow, S. S. Dehmlow, and F. Marschner, *Chem. Ber.*, **110**, 154 (1977).
[679] E. V. Dehmlow, *Chem. Ber.*, **100**, 3829 (1967).
[680] E. V. Dehmlow and G. Höfle, *Chem. Ber.*, **107**, 2760 (1974).
[681] G. V. Kryshtal, V. F. Kucherov, and L. A. Yanovskaya, *Izv. Akad. Nauk SSSR Ser. Khim.*, **25**, 929 (1976), Engl. transl., p. 909.
[682] G. V. Kryshtal, A. K. Khusid, V. F. Kucherov, and L. A. Yanovskaya, *Izv. Akad. Nauk SSSR Ser. Khim.*, **24**, 424 (1976); Engl. transl., p. 424.
[683] S. S. Dehmlow and E. V. Dehmlow, *Justus Liebigs Ann. Chem.*, **1973**, 1753.

[684] F. Kasper and T. Beier, *Z. Chem.*, **16**, 435 (1976).
[685] R. R. Kostikov and A. P. Molchanov, *Zh. Org. Khim.*, **11**, 1861 (1975); Engl. transl., p. 1871.
[686] W. Kraus, W. Rothenwöhrer, H. Sadlo, and G. Klein, *Angew. Chem.*, **84**, 643 (1972); *Angew. Chem. Int. Ed. Engl.*, **11**, 641 (1972).
[687] E. V. Dehmlow, *Tetrahedron*, **28**, 175 (1972).
[688] H. D. Beckhaus, J. Schoch, and C. Rüchardt, *Chem. Ber.*, **109**, 1369 (1976).
[689] W. Kuhn, H. Marschall, and P. Weyerstahl, *Chem. Ber.*, **110**, 1564 (1977).
[690] S. M. Shostakovskii, A. A. Retinskii, and A. V. Bobrov, *Izv. Akad. Nauk SSSR Ser. Khim.*, **23**, 1818 (1975); Engl. transl., p. 1736.
[691] C. B. Chapleo, C. E. Dahl, and A. S. Dreiding, *Helv. Chim. Acta*, **57**, 1876 (1975).
[692] A. De Smet, M. Anteuris, and D. Tavernis, *Bull. Soc. Chim. Belg.*, **84**, 67 (1975).
[693] T. Sasaki, K. Kanematsu, and N. Okamura, *J. Org. Chem.*, **40**, 3322 (1975).
[694] A. Busch and H. M. R. Hoffmann, *Tetrahedron Lett.*, **1976**, 2379.
[695] G. Klein and L. A. Paquette, *Tetrahedron Lett.*, **1976**, 2419.
[696] Z. Goldschmidt and U. Gutman, *Tetrahedron*, **30**, 3327 (1974).
[697] B. Cheminat and B. Mège, *C. R. Acad. Sci. Ser. C*, **280**, 1003 (1975).
[698] P. F. Ranken, B. J. Harty, L. Kapicak, and M. A. Battiste, *Synth. Commun.*, **3**, 311 (1973).
[699] T. Sasaki, S. Eguchi, and T. Kiriyama, *J. Org. Chem.*, **38**, 2230 (1973).
[700] W. Kraus, G. Klein, H. Sadlo, and W. Rothenwöhrer, *Synthesis*, **1912**, 485.
[701] C. W. Jefford, A. Sweeney, and F. Delay, *Helv. Chim. Acta*, **55**, 2214 (1972).
[702] C. W. Jefford, U. Burger, and F. Delay, *Helv. Chim. Acta*, **56**, 1083 (1973).
[703] C. W. Jefford, W. D. Graham, and U. Burger, *Tetrahedron Lett.*, **1975**, 4717.
[704] C. W. Jefford, V. de los Heros, and U. Burger, *Tetrahedron Lett.*, **1976**, 703.
[705] P. M. Kwantes and G. W. Klumpp, *Tetrahedron Lett.*, **1976**, 707.
[706] H. Hart and M. Nitta, *Tetrahedron Lett.*, **1974**, 2109.
[707] E. V. Dehmlow, *Tetrahedron Lett.*, **1975**, 203.
[708] T. Greibrokk, *Acta Chem. Scand.*, **27**, 3207 (1973).
[709] J. C. Jochims and G. Karich, *Tetrahedron Lett.*, **1974**, 4215.
[710] G. Karich and J. C. Jochims, *Chem. Ber.*, **110**, 2680 (1977).
[711] M. R. Detty and L. A. Paquette, *J. Am. Chem. Soc.*, **99**, 821 (1977).
[712] E. V. Dehmlow, H. Klabuhn, and E.-C. Hass, *Justus Liebigs Ann. Chem.*, **1973**, 1063.
[713] T. Sasaki, K. Kanematsu, and Y. Yukimoto, *J. Org. Chem.*, **39**, 455 (1974).
[714] T. Sasaki, K. Kanematsu, and Y. Yukimoto, *Heterocycles*, **1**, 1 (1973); *Chem. Abstr.*, **80**, 82615 (1974).
[715] G. Andrews and D. A. Evans, *Tetrahedron Lett.*, **1972**, 5121.
[716] T. Sasaki, S. Eguchi, T. Kiriyama, and Y. Sakito, *J. Org. Chem.*, **38**, 1648 (1973).
[717] G. Blume and P. Weyerstahl, *Tetrahedron Lett.*, **1970**, 3669.
[718] P. Weyerstahl and G. Blume, *Tetrahedron*, **28**, 5281 (1972).
[719] G. Blume, T. Neumann, and P. Weyerstahl, *Justus Liebigs Ann. Chem.*, **1975**, 201.
[720] B. Müller and P. Weyerstahl, *Tetrahedron*, **32**, 865 (1976).
[721] M. Sato, S. Ebine, and J. Tsunetsugu, *J. Chem. Soc. Chem. Commun.*, **1974**, 846.
[722] M. Sato, T. Tanaka, J. Tsunetsugu, and S. Ebine, *Bull. Chem. Soc. Jpn.*, **48**, 2395 (1975).
[723] M. Sato, J. Tsunetsugu, and S. Ebine, *Bull. Chem. Soc. Jpn.*, **49**, 2230 (1976).
[724] M. Sato, A. Uchida, J. Tsunetsugu, and S. Ebine, *Tetrahedron Lett.*, **1977**, 2151.
[725] W.-H. Gündel, *Z. Naturforsch. Teil B*, **30**, 616 (1975).
[726] W.-H. Gündel, *Z. Naturforsch. Teil B*, **31**, 807 (1976).
[727] W.-H. Gündel, *Z. Naturforsch. Teil B*, **32**, 193 (1977).
[728] F. De Angelis, A. Gambacorta, and R. Nicoletti, *Synthesis*, **1976**, 798.
[729] S. Kwon, Y. Nishimura, M. Ikeda, and Y. Tamura, *Synthesis*, **1976**, 249.
[730] E. V. Dehmlow and M. Lissel, *Liebigs Ann. Chem.*, **1979**, 181.
[731] E. V. Dehmlow and K.-H. Franke, unpublished results (1976–1977).

[732] R. A. Magarian and E. J. Benjamin, *J. Pharm. Sci.*, **64**, 1626 (1975).
[733] T. T. Coburn and W. M. Jones, *J. Am. Chem. Soc.*, **96**, 5218 (1974).
[734] J. Tsunetsugu, M. Sato, and S. Ebine, *J. Chem. Soc. Chem. Commun.*, **1973**, 363.
[735] K. Berg-Nielsen, *Acta Chem. Scand. Ser. B*, **31**, 224 (1977).
[736] I. G. Tishchenko, O. G. Kulinkovich, and Y. V. Glazkov, *Zh. Org. Khim.*, **11**, 581 (1975); Engl. transl., p. 579.
[737] I. G. Tishchenko, Y. V. Glazkov, and O. G. Kulinkovich, *Zh. Org. Khim.*, **9**, 2510 (1973); Engl. transl., p. 2530.
[738] I. G. Tishchenko, O. G. Kulinkovich, Y. V. Glazkov, and M. K. Pirshtuk, *Zh. Org. Khim.*, **11**, 576 (1975); Engl. transl., p. 574.
[739] E. V. Dehmlow, *Tetrahedron*, **27**, 4071 (1971).
[740] I. Tabushi, Z. Yoshida, and N. Takahashi, *J. Am. Chem. Soc.*, **92**, 6670 (1970).
[741] M. Mąkosza and M. Fedoryński, *Rocz. Chem.* **46**, 311 (1972).
[742] S.-H. Goh, K.-C. Chan, T.-S. Kam, and H.-C. Chong, *Austr. J. Chem.*, **28**, 381 (1975).
[743] S.-H. Goh, *J. Chem. Educ.*, **52**, 399 (1975).
[744] A. de Meijere, O. Schallner, and C. Weitemeyer, *Angew. Chem.*, **84**, 63 (1972); *Angew. Chem. Int. Ed. Engl.*, **11**, 56 (1972).
[745] A. de Meijere, C. Weitemeyer, and O. Schallner, *Chem. Ber.*, **110**, 1504 (1977).
[746] T. Greibrokk, *Tetrahedron Lett.*, **1972**, 1663.
[747] T. Tabushi, Y. Aoyama, and N. Takahashi, *Tetrahedron Lett.*, **1973**, 107.
[748] T. Saraie, T. Ishiguro, K. Kawashima, and K. Morita, *Tetrahedron Lett.*, **1973**, 2121.
[749] G. Höfle, *Z. Naturforsch. Teil B*, **28**, 831 (1973).
[750] J. Graefe, *Z. Chem.*, **15**, 301 (1975).
[751] W. P. Weber and G. W. Gokel, *Tetrahedron Lett.*, **1972**, 1637.
[752] W. P. Weber, G. W. Gokel, and I. K. Ugi, *Angew. Chem.*, **84**, 587 (1972); *Angew. Chem. Int. Ed. Engl.*, **11**, 530 (1972).
[753] T. Sasaki, S. Eguchi, and T. Katada, *J. Org. Chem.*, **39**, 1239 (1974).
[754] G. W. Gokel, R. P. Widera, and W. P. Weber, *Org. Synth.*, **55**, 96 (1976).
[755] G. Domschke, R. Beckert, and R. Mayer, *Synthesis*, **1977**, 275; East Germ. Pat., 128,531 (1977).
[756] D. T. Sepp, K. V. Scherer, and W. P. Weber, *Tetrahedron Lett.*, **1974**, 2983.
[757] J. Graefe, B. Striegler, and M. Mühlstädt, *Z. Chem.*, **16**, 356 (1976).
[758] G. V. Kryshtal, A. K. Khusid, V. F. Kucherov, and L. A. Yanovskaya, *Izv. Akad. Nauk SSSR Ser. Khim.*, **26**, 709 (1977).
[759] M. Mąkosza and A. Kacprowicz, *Rocz. Chem.* **49**, 1627 (1975).
[760] J. Graefe, I. Fröhlich, and M. Mühlstädt, *Z. Chem.*, **14**, 434 (1974).
[761] T. Hiyama, Y. Ozaki, and H. Nozaki, *Tetrahedron*, **30**, 2661 (1974).
[762] M. Mąkosza, B. Jerzak, and M. Fedoryński, *Rocz. Chem.*, **49**, 1783 (1975).
[763] I. Tabushi, Z. Yoshida, and N. Takahashi, *J. Am. Chem. Soc.*, **93**, 1820 (1971).
[764] I. Tabushi and Y. Aoyami, *J. Org. Chem.*, **38**, 3447 (1973).
[765] P. Stromquist, M. Radcliffe, and W. P. Weber, *Tetrahedron Lett.*, **1973**, 4523.
[766] I. Tabushi, Y. Kuroda, and Z. Yoshida, *Tetrahedron*, **32**, 997 (1976).
[767] M. Fedoryński and M. Mąkosza, *J. Organomet. Chem.*, **51**, 89 (1973).
[768] M. Mąkosza and M. Fedoryński, *Rocz. Chem.*, **46**, 533 (1972).
[769] M. Mąkosza and M. Fedoryński, *Rocz. Chem.*, **49**, 1779 (1975).
[770] A. Merz, *Synthesis*, **1974**, 724.
[771] P. Kuhl, M. Mühlstädt, and J. Graefe, *Synthesis*, **1976**, 825; **1977**, 502.
[772] Nisshin Flour Mill KK, Jap. Kokai 76,05,028; *Derwent Abstr.*, 82125 (1976).
[773] S. Colonna and R. Fornasier, *Synthesis*, **1975**, 531.
[774] J. Lange, *Rocz. Chem.*, **42**, 1619 (1968).
[775] M. Cinquini, S. Colonna, H. Molinari, F. Montanari, and P. Tundo, *J. Chem. Soc. Chem. Commun.*, **1976**, 394.
[776] M. Cinquini and P. Tundo, *Synthesis*, **1976**, 516.

[777] G. Cainelli, F. Manescalchi, and M. Panunzio, *Synthesis*, **1976**, 472.
[778] J. M. Saá and M. P. Cava, *J. Org. Chem.*, **42**, 347 (1977).
[779] H. Bieräugel, J. M. Akkerman, J. C. Lapierre-Armande, and U.K. Pandit, *Rec. Trav. Chim. Pays-Bas*, **95**, 266 (1976).
[780] R. A. Moss, M. A. Joyce, and J. K. Huselton, *Tetrahedron Lett.*, **1975**, 4621.
[781] M. Mąkosza and M. Fedoryński, *Bull. Acad. Pol. Sci. Ser. Sci. Chim.*, **19**, 105 (1971).
[782] J. Hine, "Divalent Carbon," The Ronald Press Co., New York, 1964.
[783] E. V. Dehmlow, J. Heider, and U. Brenner, unpublished results.
[784] L. Skattebøl, G. A. Abskharoun, and T. Greibrokk, *Tetrahedron Lett.*, **1973**, 1367.
[785] M. Mąkosza and M. Fedoryński, *Synth. Commun.*, **3**, 305 (1973).
[786] M. Mąkosza and M. Fedoryński, *Rocz. Chem.*, **50**, 2223 (1976).
[787] E. V. Dehmlow and J. Kranz, unpublished results.
[788] E. Vogel and J. Ippen, *Angew. Chem.*, **86**, 778 (1974); *Angew. Chem. Int. Ed. Engl.*, **13**, 734 (1974).
[789] J.-L. Luche, J.-C. Damiano, and P. Crabbé, *J. Chem. Res.*, **1977**, (S) 32, (M) 443.
[790] J.-C. Damiano, J.-L. Luche, and P. Crabbé, *Tetrahedron Lett.*, **1976**, 779.
[791] H. Maskill, *J. Chem. Soc. Perkin Trans.* 2, **1975**, 197.
[792] E. V. Dehmlow and M. Lissel, unpublished results.
[793] A. R. Allan and M. S. Baird, *J. Chem. Soc. Chem. Commun.*, **1975**, 172.
[794] E. Vogel, M. Königshofen, K. Müller, and J. F. M. Oth, *Angew. Chem.*, **86**, 229 (1974); *Angew. Chem. Int. Ed. Engl.*, **13**, 281 (1974).
[795] L. K. Sydnes, L. Skattebøl, C. B. Capleo, D. G. Leppard, K. L. Svanholt, and A. S. Dreiding, *Helv. Chim. Acta*, **58**, 2061 (1975).
[796] L. Sydnes and L. Skattebøl, *Tetrahedron Lett.*, **1975**, 4603.
[797] M. Braun, R. Dammann, and D. Seebach, *Chem. Ber.*, **108**, 2368 (1975).
[798] M. Braun and D. Seebach, *Chem. Ber.*, **109**, 669 (1975).
[799] I. J. Landheer, W. H. de Wolf, and F. Bickelhaupt, *Tetrahedron Lett.*, **1974**, 2813.
[800] E. V. Dehmlow and G. C. Ezimora, *Tetrahedron Lett.*, **1970**, 4047.
[801] D. Reinhard and P. Weyerstahl, *Chem. Ber.*, **110**, 138 (1977).
[802] R. Dammann, M. Braun, and D. Seebach, *Helv. Chim. Acta*, **59**, 2821 (1976).
[803] W. Kirmse, "Carbene Chemistry," Second Ed., Academic Press, New York and London, 1971.
[804] D. Seyferth, S. P. Hopper, and T. F. Jula, *J. Organomet. Chem.*, **17**, 193 (1969).
[805] M. Schlosser, B. Spahić, C. Tarchini, and L. V. Chau, *Angew. Chem.*, **87**, 346 (1975); *Angew. Chem. Int. Ed. Engl.*, **14**, 365 (1975).
[806] M. Schlosser and L. V. Chau, *Helv. Chim. Acta*, **58**, 2595 (1975).
[807] L. V. Chau and M. Schlosser, *Synthesis*, **1973**, 112.
[808] P. Weyerstahl, G. Blume, and C. Müller, *Tetrahedron Lett.*, **1971**, 3869.
[809] P. Weyerstahl, R. Mathias, and G. Blume, *Tetrahedron Lett.*, **1973**, 61.
[810] R. Mathias, Ph.D. Dissertation, Technical University, Berlin, 1977.
[811] C. Müller and P. Weyerstahl, *Tetrahedron*, **31**, 1787 (1975).
[812] C. Müller, F. Stier, and P. Weyerstahl, *Chem. Ber.*, **110**, 124 (1977).
[813] R. Mathias and P. Weyerstahl, *Angew. Chem.*, **86**, 42 (1974); *Angew. Chem. Int. Ed. Engl.*, **13**, 134 (1974).
[814] M. S. Baird, *J. Chem. Soc. Perkin Trans.* 1, **1976**, 54.
[815] B. Giese, *Angew. Chem.*, **89**, 162 (1977); *Angew. Chem. Int. Ed. Engl.*, **16**, 125 (1977).
[816] E. V. Dehmlow, unpublished results.
[817] M. Mąkosza and E. Białecka, *Tetrahedron Lett.*, **1971**, 4517.
[818] G. Boche and D. R. Schneider, *Tetrahedron Lett.*, **1975**, 4247.
[819] R. A. Moss and F. G. Pilkiewicz, *Synthesis*, **1973**, 209.
[820] M. S. Newman and T. B. Patrick, *J. Am. Chem. Soc.*, **91**, 6461 (1969) and previous work cited therein.
[821] M. S. Newman and S. J. Gromelski, *J. Org. Chem.*, **37**, 3220 (1972).
[822] M. S. Newman and Z. ud Din, *J. Org. Chem.*, **38**, 547 (1973).

[823] P. S. Stang and M. G. Mangum, *J. Am. Chem. Soc.*, **97**, 6478 (1975).
[824] M. S. Newman and M. C. Van der Zwan, *J. Org. Chem.*, **39**, 1186 (1974).
[825] M. S. Newman and M. C. Van der Zwan, *J. Org. Chem.*, **39**, 761 (1974).
[826] M. S. Newman and W. C. Liang, *J. Org. Chem.*, **38**, 2438 (1973).
[827] S. Julia, D. Michelot, and G. Linstrumelle, *C. R. Acad. Sci. Ser. C*, **278**, 1523 (1974).
[828] T. Sasaki, S. Eguchi, and T. Ogawa, *J. Org. Chem.*, **39**, 1927 (1974).
[829] T. Sasaki, S. Eguchi, M. Ohno, and F. Nakato, *J. Org. Chem.*, **41**, 2408 (1976).
[830] T. B. Patrick, *Tetrahedron Lett.*, **1974**, 1407.
[831] D. Y. Aue and M. J. Meshishnek, *J. Am. Chem. Soc.*, **99**, 223 (1977).
[832] T. Sasaki, S. Eguchi, and T. Ogawa, *Heterocycles*, **3**, 193 (1975).
[833] A. Doutheau and J. Goré, *Bull. Soc. Chim. Fr.*, **1976**, 1189.
[834] M. Schlosser and Y. Bessière, *Helv. Chim. Acta*, **60**, 590 (1977).
[835] M. Fedoryński, *Synthesis*, **1977**, 783.
[836] M. Fedoryński, M. Popławska, K. Nitschke, W. Kawalski, and M. Mąkosza, *Synth. Commun.*, **7**, 287 (1977).
[837] R. A. Jones, S. Nokkeo, and S. Singh, *Synth. Commun.*, **7**, 195 (1977).
[838] B. L. Burt, D. J. Freeman, D. G. Gray, R. K. Norris, and D. Randles, *Tetrahedron Lett.*, **1977**, 3063.
[839] S. E. Callander, Germ. Offenl. 2,656,062 (1977).
[840] S. W. Tobey and R. West, *J. Am. Chem. Soc.*, **86**, 56 (1964).
[841] E. W. Meijer and H. Wynberg, private communication.
[842] D. C. Duffey, R. C. Gueldner, B. R. Layton, and J. P. Minyard, Jr., *J. Org. Chem.*, **42**, 1082 (1977).
[843] J. Heider and E. V. Dehmlow, unpublished results.
[844] W. Kimpenhaus and J. Buddrus, *Chem. Ber.*, **110**, 1304 (1977).
[845] E. V. Dehmlow, unpublished results.
[846] R. A. Moss, M. A. Joyce, and F. G. Pilkiewicz, *Tetrahedron Lett.*, **1975**, 2425.
[847] J. Paleček and J. Kuthan, *Z. Chem.*, **17**, 260 (1977).
[848] R. O. Hutchins and F. J. Dux, *J. Org. Chem.*, **38**, 1961 (1973).
[849] T.-L. Ho, *Synth. Commun.*, **3**, 99 (1973).
[850] G. B. Fuller, R. Greenhouse, and I. Itoh, *Synth. Commun.*, **4**, 183 (1974).
[851] M. Tomoi, T. Takubo, M. Ikeda, and H. Kakiuchi, *Chem. Lett.*, **1976**, 473.
[852] E. V. Dehmlow and E. Menzel, unpublished results.
[853] E. V. Dehmlow and S. Barahona, unpublished results.
[854] E. V. Dehmlow and E. Timm, unpublished results.
[855] S. H. Korzeniowski and G. W. Gokel, *Tetrahedron Lett.*, **1977**, 3519.
[856] D. Landini and F. Rolla, *Org. Synth.*, **56** (1977), unchecked procedure number 1994.
[857] S. A. Di Biase and G. W. Gokel, *Synthesis*, **1977**, 629.
[858] H. Molinari, F. Montanari, and P. Tundo, *J. Chem. Soc. Chem. Commun.*, **1977**, 639.
[859] P. Tundo, *J. Chem. Soc. Chem. Commun.*, **1977**, 641.
[860] J. M. Brown and J. A. Jenkins, *J. Chem. Soc. Chem. Commun.*, **1976**, 458.
[861] M. Shibasaki and S. Ikeyami, *Tetrahedron Lett.*, **1977**, 4037.
[862] L. Eberson and B. Helgée, *Acta Chem. Scand. Ser. B*, **32**, 157 (1978).
[863] L. Eberson and B. Helgée, *Acta Chem. Scand. Ser. B*, **31**, 813 (1977).
[864] L. Eberson and B. Helgée, *Acta Chem. Scand. Ser. B*, **29**, 451 (1975).
[865] Y. Nakajima, R. Kinishi, J. Oda, and Y. Inouye, *Bull. Chem. Soc. Jpn.*, **50**, 2025 (1977).
[866] Y. Kimura, K. Isagawa, and Y. Otsuji, *Chem. Lett.*, **1977**, 951.
[867] J. L. Ripoll, *Tetrahedron Lett.*, **33**, 389 (1977).
[868] S. Julia and A. Ginebreda, *Synthesis*, **1977**, 682.
[869] T.-L. Ho, B. G. B. Gupta, and G. A. Olah, *Synthesis*, **1977**, 676.
[870] W. Wilczyński, M. Jawdosiuk, and M. Mąkosza, *Rocz. Chem.*, **51**, 1643 (1977).
[871] J. Renault, P. Mailliet, J. Berlot, and S. Renault, *C. R. Acad. Sci. Ser. C*, **285**, 199 (1977).

[872] CSIR, Germ. Pat. 2,653,189 (1977); *Derwent Abstr.*, 42013 (1977).
[873] P. Maggioni and C. Montevechia, Germ. Offenl. 2,703,640 (1977).
[874] A. W. Burgstahler, L. O. Weigel, M. E. Sanders, and C. G. Shaefer, *J. Org. Chem.*, **42**, 566 (1977).
[875] S. Cacchi, F. La Torre, and D. Misiti, *Chem. Ind. (London)*, **1977**, 691.
[876] Y. Masuyama, Y. Heno, and M. Okawara, *Chem. Lett.*, **1977**, 835.
[877] G. R. Kieczykowski, C. S. Pogonowski, J. E. Richman, and R. M. Schlessinger, *J. Org. Chem.*, **42**, 175 (1977).
[878] A. Chollet, C. Mahaim, C. Foetisch, M. Hardy, and P. Vogel, *Helv. Chim. Acta*, **60**, 59 (1977).
[879] M. Muhammed, J. Szabon, and E. Högfeldt, *Chem. Scr.*, **6**, 61 (1974).
[880] J. Cousseau, L. Gouin, L. V. Jones, G. Jugic, and J. A. S. Smith, *J. Chem. Soc. Faraday Trans.* 2, **1973**, 1821.
[881] J. Cousseau and L. Gouin, *J. Chem. Soc. Perkin Trans.* 1, **1977**, 1797.
[882] E. V. Dehmlow and M. Slopianka, unpublished results.
[883] R. Beugelmans, M.-T. Le Goff, J. Pusset, and G. Roussi, *Tetrahedron Lett.*, **1976**, 2305; *J. Chem. Soc. Chem. Commun.*, **1976**, 377.
[884] K. Kitatani, T. Hiyama, and H. Nozaki, *J. Am. Chem. Soc.*, **98**, 2362 (1976).
[885] A. Zwierzak and I. Padstawszyńska, *Angew. Chem.*, **89**, 737 (1977); *Angew. Chem. Int. Ed. Engl.*, **16**, 702 (1977).
[886] M. D. Rozwasowska, *Can. J. Chem.*, **55**, 164 (1977).
[887] R. A. Sheldon, P. Been, D. A. Wood, and R. F. Mason, Germ. Offenl. 2,708,590 (1977).
[888] N. Ono, R. Tamura, R. Tanikaga, and A. Kaji, *Synthesis*, **1977**, 690.
[889] G. Entenmann, *Tetrahedron Lett.*, **1975**, 4241.
[890] R. O. Hutchins, N. R. Natale, and W. J. Cook, *Tetrahedron Lett.*, **1977**, 4167.
[891] M. A. Pericás and F. Serratosa, *Tetrahedron Lett.*, **1977**, 4437.
[892] K. Hermann and H. Wynberg, *Helv. Chim. Acta*, **60**, 2208 (1977).
[893] K. Yamamura and S. Murahashi, *Tetrahedron Lett.*, **1977**, 4429.
[894] A. Brändström, P. Berntsson, S. Carlsson, A. Djurhuus, K. Gustavii, U. Junggren, B. Lamm, and B. Samuelsson, *Acta Chem. Scand.*, **23**, 2202 (1969).
[895] K. M. More and J. Wemple, *Synthesis*, **1977**, 791.
[896] L. M. Jackman and B. C. Lange, *Tetrahedron*, **33**, 2737 (1977).
[897] S. Gronowitz and T. Frejd, *Synth. Commun.*, **6**, 475 (1976).
[898] A. A. Frimer, I. Rosenthal, and S. Hoz, *Tetrahedron Lett.*, **1977**, 4631.
[899] A. Brändström and H. Kolind-Andersen, *Acta Chem. Scand. Ser. B*, **29**, 201 (1975).
[900] A. Merz and R. Tomahogh, *J. Chem. Res.*, **1977**, (S) 273 (M) 3070.
[901] P. J. Stang and D. P. Fox, *J. Org. Chem.*, **42**, 1667 (1977).
[902] E. Chiellini and R. Solaro, *J. Chem. Soc. Chem. Commun.*, **1977**, 231.
[903] P. Cazeau and B. Muckensturm, *Tetrahedron Lett.*, **1977**, 1493.
[904] R. Lantz and A.-B. Hörnfeldt, *Chem. Ser.*, **10**, 126 (1976).
[905] C. Innes and G. Lamaty, *Nouv. J. Chim.*, **1**, 503 (1977).
[906] G. Cardillo, M. Contento, M. Panunzio, and A. Umani-Ronchi, *Chem. Ind. (London)*, **1977**, 873.
[907] L. Jensen, I. Thomsen, and S.-O. Lawesson, *Bull. Soc. Chim. Belg.*, **86**, 309 (1977).
[908] A. Jończyk, J. Włostowska, and M. Mąkosza, *Bull. Soc. Chim. Belg.*, **86**, 739 (1977).
[909] J.-P. Hagenbuch and P. Vogel, *Chimia*, **31**, 136 (1977).
[910] C. Wyganowski, *Talanta*, **24**, 190 (1977).
[911] K. H. Pannell and J. McIntosh, *Chem. Ind. (London)*, **1977**, 873.
[912] G. Adembri, A. Camparini, F. Ponticelli, and P. Tedeschi, *J. Chem. Soc. Perkin Trans.* 1, **1977**, 971.
[913] E. Lee-Ruff, *Chem. Soc. Rev.*, **6**, 195 (1977).
[914] G. Märkl, H. Baier, and R. Liebl, *Synthesis*, **1977**, 842.
[915] I. Degani, R. Fochi, and M. Santi, *Synthesis*, **1977**, 873.

[916] R. Beugelmans, H. Ginsburg, A. Lecas, M.-T. Le Goff, J. Pusset, and G. Roussi, *J. Chem. Soc. Chem. Commun.*, **1977**, 885; *Tetrahedron Lett.*, **1978**, 3271.
[917] Allied Chem. Corp., Neth. Pat. Appl. 6,400,872 (1964) = U.S. Pat. 3,297,634 (1967); *Chem. Abstr.*, **62**, 6587 (1965).
[918] S. Alunni, E. Baciocchio, and P. Perucci, *J. Org. Chem.*, **41**, 2636 (1976).
[919] Z. K. Brzozowski, J. Kiełkiewicz, and Z. Gocławski, *Angew. Makromol. Chem.*, **44**, 1 (1975).
[920] L. Cassar and M. Foa, *J. Organomet. Chem.*, **134**, C 15 (1977).
[921] M. Chorev and Y. S. Klausner, *J. Chem. Soc. Chem. Commun.*, **1976**, 596; Y. S. Klausner and M. Chorev, *J. Chem. Soc. Perkin Trans.* 1, **1977**, 627.
[922] J. L. Cihonski and R. A. Levenson, *Inorg. Chem.*, **14**, 1717 (1975).
[923] S. Colonna, R. Fornasier, and U. Pfeiffer, *J. Chem. Soc. Perkin Trans.* 1, **1978**, 8.
[924] S. Colonna, H. Hiemstra, and H. Wynberg, *J. Chem. Soc. Chem. Commun.*, **1978**, 238.
[925] E. J. Corey, K. C. Nicolaou, and M. Shibasaki, *J. Chem. Soc. Chem. Commun.*, **1975**, 658.
[926] U. Curtius, V. Boellert, G. Fritz, and J. Nentwig, Belg. Pat. 616 919 (1962); *Chem. Abstr.*, **58**, 3360 (1963).
[927] L. H. Dao, A. C. Hopkinson, E. Lee-Ruff, and J. Rigaudy, *Can. J. Chem.*, **55**, 3791 (1977).
[928] D. J. Darensbourg and J. A. Froelich, *J. Am. Chem. Soc.*, **100**, 338 (1978).
[929] J. K. Rasmussen, *Chem. Lett.*, **1977**, 1295.
[930] H. des Abbayes and M.-A. Boudeville, *J. Org. Chem.*, **42**, 4104 (1977).
[931] S. A. DiBiase and G. W. Gokel, *J. Org. Chem.*, **43**, 447 (1978).
[932] H. J.-M. Dou, P. Hassanaly, J. Kister, and J. Metzger, *Phosphorus Sulfur*, **3**, 355 (1977).
[933] H. J.-M. Dou, R. Gallo, P. Hassanaly, and J. Metzger, *J. Org. Chem.*, **42**, 4275 (1977).
[934] S. O. De Silva and V. Snieckus, *Can. J. Chem.*, **56**, 1621 (1978).
[935] R. Rucman, J. Stres, and M. Jurgec, *Chem. Abstr.*, **89**, 215637 (1978).
[936] V. O. Illi, *Synthesis*, **1979**, 136.
[937] R. Boehm, *Pharmazie*, **33**, 83 (1978).
[938] N.-C. Wang, K.-E. Teo, and H. J. Anderson, *Can. J. Chem.*, **55**, 4112 (1977).
[939] M. Garle and I. Petters, *J. Chromatogr.*, **140**, 165 (1977).
[940] S. Juliá, A. Ginebreda, and J. Guixer, *J. Chem. Soc. Chem. Commun.*, **1978**, 742.
[941] J. E. Nordlander, D. B. Catalane, T. H. Eberlein, L. V. Farkas, R. S. Howe, R. M. Stevens, N. A. Tripoulas, R. E. Stansfield, J. L. Cox, M. J. Payne, and A. Viehbeck, *Tetrahedron Lett.*, **1978**, 4987.
[942] E. V. Dehmlow, unpublished results.
[943] A. Lopez, M. T. Maurette, R. Martino, and A. Lattes, *Tetrahedron Lett.*, **1978**, 2013.
[944] M. E. Fakley and A. Pidcock, *J. Chem. Soc. Dalton Trans.*, **1977**, 1444.
[945] Y. Gaoni and N. Shoef, *Bull. Soc. Chim. Fr.*, **1977**, 485.
[946] General Mills Inc., Jap. Kokai 74,66,634 (1974); *Chem. Abstr.*, **84**, 179663 (1976).
[947] R. C. Hahn and R. P. Johnson, *Tetrahedron Lett.*, **1973**, 2149.
[948] B. Hanquet, R. Guilard, and P. Fournari, *Bull. Soc. Chim. Fr.*, **1977**, 571.
[949] E. V. Dehmlow and J. Heider, unpublished results.
[950] F. R. van Heerden, J. J. van Zyl, G. J. H. Rall, E. V. Brandt, and D. G. Roux, *Tetrahedron Lett.*, **1978**, 661.
[951] J. C. Hummelen and H. Wynberg, *Tetrahedron Lett.*, **1978**, 1089.
[952] A. Jończyk, A. Kwast, and M. Mąkosza, *J. Chem. Soc. Chem. Commun.*, **1977**, 902.
[953] A. Jończyk, M. Ludwikow, and M. Mąkosza, *Angew. Chem.*, **90**, 58 (1978); *Angew. Chem. Int. Ed. Engl.*, **17**, 62 (1978).
[954] K. S. Kim and W. A. Szarek, *Synthesis*, **1978**, 48.
[955] G. Klein und W. Kraus, *Tetrahedron*, **33**, 3121 (1977).

[956] G. Klein and L. A. Paquette, *J. Org. Chem.*, **43**, 1293 (1978).
[957] M. Kluba and A. Zwierzak, *Synthesis*, **1978**, 134.
[958] H. Komeili-Zadeh, H. J.-M. Dou, and J. Metzger, *J. Org. Chem.*, **43**, 156 (1978).
[959] R. A. Kostikov, A. P. Molchanov, G. V. Golovanova, and I. G. Zenkevich, *Zh. Org. Khim.*, **13**, 1846 (1977), Engl. tranl., p. 1712.
[960] A. P. Krapcho, J. R. Larson, and J. M. Eldridge, *J. Org. Chem.*, **42**, 3749 (1977).
[961] S. Krishnan, D. G. Kuhn, and G. A. Hamilton, *J. Am. Chem. Soc.*, **99**, 8121 (1977).
[962] B. Ly, H.J.-M.Dou, P, Hassanaly, and J. Verducci, *J. Heterocycl. Chem.*, **14**, 1275 (1977).
[963] J. M. McIntosh, *Can. J. Chem.*, **55**, 4200 (1977).
[964] M. K. Meilahn, D. K. Olsen, W. J. Brittain, and R. T. Anders, *J. Org. Chem.*, **43**, 1346 (1978).
[965] A. C. Mueller, U.S. Pat. 2,772,296 (1956); *Chem. Abstr.*, **51**, 7416 (1957).
[966] D. J. Nelson and E. A. Uschak, *J. Org. Chem.*, **42**, 3308 (1977).
[967] M. S. Newman and J. O. Landers, *J. Org. Chem.*, **42**, 2556 (1977).
[968] L. A. Paquette, D. R. James, and G. Klein, *J. Org. Chem.*, **43**, 1287 (1978).
[969] K. M. Patel, H. J. Pownall, J. D. Morrisett, and J. T. Sparrow, *Tetrahedron Lett.*, **1976**, 4015.
[970] T. B. Patrick and D. J. Schmidt, *J. Org. Chem.*, **42**, 3355 (1977).
[971] V. Ratovelomanana and S. Julia, *Synth. Comm.*, **8**, 87 (1978).
[972] R. W. Roeske and P. D. Gesellchen, *Tetrahedron Lett.*, **1976**, 3369.
[973] M. D. Rozwadowska and D. Brózda, *Tetrahedron Lett.*, **1978**, 589.
[974] T. Sala and M. V. Sargent, *J. Chem. Soc. Chem. Commun.*, **1978**, 253.
[975] W. H. Saunders, Jr., S. D. Bonadies, M. Braunstein, J. K. Borchardt, and R. T. Hargreaves, *Tetrahedron*, **33**, 1577 (1977).
[976] S. Schwarz and G. Weber, East Germ Pat. 114,806 (1975); *Chem. Abstr.*, **85**, 63238 (1976).
[977] F. E. Scully, Jr., and R. C. Davies, *J. Org. Chem.*, **43**, 1467 (1978).
[978] J. E. Shaw, D. Y. Hsin, G. S. Parries, and T. K. Swayer, *J. Org. Chem.*, **43**, 1017 (1978).
[979] H. Singh and P. Singh, *Chem. Ind. (London)*, **1978**, 126.
[980] K. Steinbeck, *Tetrahedron Lett.*, **1978**, 1103.
[981] L. K. Sydnes, *Acta Chem. Scand. Ser. B*, **31**, 903 (1977).
[982] B. C. Uff, A. Al-kolla, K. E. Adamali, and V. Harutunian, *Synth. Commun.*, **8**, 163 (1978).
[983] J. Vessman, M. Johansson, P. Magnusson, and S. Strömberg, *Anal. Chem.*, **49**, 1545 (1977).
[984] E. J. Walsh, E. Derby, and J. Smegal, *Inorg. Chim. Acta*, **16**, C9 (1976).
[985] T. Wakabayashi and K. Watanabe, *Tetrahedron Lett.*, **1978**, 361.
[986] I. Willner, M. Halpern, and M. Rabinovitz, *J. Chem. Soc. Chem. Commun.*, **1978**, 155.
[987] E. V. Dehmlow and K.-H. Franke, *Liebigs Ann. Chem.*, **1979**, in press.
[988] E. V. Dehmlow and M. Lissel, unpublished results.
[989] E. V. Dehmlow and M. Slopianka, *Liebigs Ann. Chem.*, **1979**, in press.
[990] E. V. Dehmlow and A. Eulenberger, *Angew. Chem.*, **90**, 716 (1978); *Angew. Chem. Int. Ed. Engl.*, **17**, 674 (1978); *Liebigs Ann. Chem.*, **1979**, in press.
[991] D. Pletcher and S. J. D. Tait, *Tetrahedron Lett.*, **1978**, 1601; *J. Chem. Soc. Perkin Trans. 2*, **1979**, 788.
[992] D. Landini, F. Montanari, and F. Rolla, *Synthesis*, **1978**, 223.
[993] B. C. Uff and R. S. Budhram, *Heterocycles*, **6**, 1789 (1977).
[994] C. Goralski and G. A. Burk, U.S. Pat. 4,014,891 (1977); *Chem. Abstr.*, **87**, 22763 (1977).
[995] D. Martinetz and A. Hiller, *Z. Chem.*, **16**, 320 (1976).
[996] H. Alper and H. des Abbayes, *J. Organomet. Chem.*, **134**, C11 (1977).

[997] C. Kimura, K. Kashiwaya, and K. Koichi, *Chem. Abstr.*, **86**, 5112 (1977).
[998] G. A. Lee and H. H. Freedman, U.S. Pat. 3,996,259 (1976); *Chem. Abstr.*, **86**, 88453 (1977).
[999] S. Schwarz, U. Eberhardt, and H. Schick, East German Pat. 120 649 (1976); *Chem. Abstr.*, **86**, 72996 (1977).
[1000] Y. Hayashi, Jap. Kokai 77,111,486 (1976); *Chem. Abstr.*, **88**, 104653 (1978).
[1001] T. Okimoto and D. Swern, *J. Amer. Oil Chem. Soc.*, **54**, 867 A (1977); *Chem. Abstr.*, **88**, 37180 (1978).
[1002] D. Pletcher and N. Tomov, *J. Appl. Electrochem.*, **7**, 501 (1977).
[1003] T. A. Foglia, A. P. Barr, and A. J. Malley, *J. Amer. Oil Chem. Soc.*, **54**, 858 A (1977); *Chem. Abstr.*, **88**, 37210 (1978).
[1004] L. Töke and G. T. Szabo, *Acta Chim. Acad. Sci. Hung.*, **93**, 421 (1977); *Chem. Abstr.*, **88**, 38298 (1978).
[1005] A. F. Rosenthal, L. A. Vargas, and J. F. Dixon, *Chem. Phys. Lipids*, **20**, 205 (1977); *Chem. Abstr.*, **88**, 121598 (1978).
[1006] G. W. Gokel, D. J. Cram, C. L. Liotta, H. P. Harris, and F. L. Cook, *Organ. Synth.*, **57**, 30 (1977).
[1007] M. Amouyal and H. Sekiguchi, *C. R. Acad. Sci. Ser. C*, **286**, 233 (1978).
[1008] P. Tundo, *Synthesis*, **1978**, 315.
[1009] I. Degani and R. Fochi, *Synthesis*, **1978**, 365.
[1010] D. Martinetz and A. Hiller, *Z. Chem.*, **1978**, 61.
[1011] H. Wynberg and B. Greijdanus, *J. Chem. Soc. Chem. Commun.*, **1978**, 427.
[1012] E. V. Dehmlow, U. Brenner, and M. Slopianka, unpublished results.
[1013] G. W. Gokel and B. J. Garcia, *Tetrahedron Lett.*, **1978**, 1743.
[1014] S. Colonna and R. Fornasier, *J. Chem. Soc. Perkin Trans.* 1, **1978**, 371.
[1015] T. Kitazume and N. Ishikawa, *Chem. Lett.*, **1978**, 283.
[1016] M. Lissel and E. V. Dehmlow, *Tetrahedron Lett.*, **1978**, 3689.
[1017] J. J. Kaminski, K. W. Knutson, and N. Bodor, *Tetrahedron*, **34**, 2857 (1978).
[1018] H. Kobler, R. Munz, G. Al Gasser, and G. Simchen, *Justus Liebigs Ann. Chem.*, **1978**, 1937.
[1019] J. H. Clark, *J. Chem. Soc. Chem. Commun.*, **1978**, 789.
[1020] H. J.-M. Dou, P. Hassanaly, and J. Metzger, *Nouv. J. Chim.*, **2**, 445 (1978).
[1021] J. Elguero and M. Espada, *C. R. Acad. Sci. Ser. C*, **287**, 439 (1978).
[1022] W. M. McKenzie and D. C. Sherrington, *J. Chem. Soc. Chem. Commun.*, **1978**, 541.
[1023] M. Tomoi, O. Abe, M. Ikeda, K. Kihara, and H. Kakinchi, *Tetrahedron Lett.*, **1978**, 3031.
[1024] S. Akabori, S. Miyamoto, and H. Tanabe, *J. Polym. Sci. Polym. Lett. Ed.*, **16**, 533 (1978).
[1025] M. Tomoi, M. Ikeda, and H. Kakiuchi, *Tetrahedron Lett.*, **1978**, 3757; S. L. Regen, A. Nigam, and J. J. Besse, *ibid.*, **1978**, 2757.
[1026] F. Rolla, W. Roth, and L. Horner, *Naturwissenschaften*, **64**, 337 (1977).
[1027] H. J. Cristau, A. Long, and H. Christol, *Tetrahedron Lett.*, **1979**, 349.
[1028] P. Tundo, *Tetrahedron Lett.*, **1978**, 4693.
[1029] S. Samaan and F. Rolla, *Phosphorus Sulfur*, **4**, 145 (1978); *Chem. Abstr.*, **89**, 41853 (1978).
[1030] M. Ravey, L. M. Shorr, and I. Hertz, *J. Chem. Soc. Perkin Trans.* 2, **1977**, 1462.
[1031] R. L. Markezich, O. S. Zamek, P. E. Donahue, and F. J. Williams, *J. Org. Chem.*, **42**, 3435 (1977).
[1032] D. Landini, F. Montanari, and F. Rolla, *Synthesis*, **1978**, 771.
[1033] R. Lantzsch, A. Marhold, and K.-F. Lehment, *Germ. Offenl.* 2,545,644 (1977); *Chem. Abstr.*, **87**, 38874 (1977).
[1034] L. Eberson and B. Helgée, *Acta Chem. Scand. Ser. B*, **32**, 313 (1978).
[1035] L. C. Packman and D. J. H. Smith, *FEBS Lett.*, **91**, 178 (1978).
[1036] J. G. Smith and D. C. Irwin, *Synthesis*, **1978**, 894.

[1037] K. Soga, S. Hosoda, and S. Ikeda, *J. Polym. Sci. Polym. Lett. Ed.*, **15**, 611 (1977).
[1038] B. Davis, *Anal. Chem.*, **49**, 832 (1977).
[1039] K. Hermann and H. Wynberg, *J. Org. Chem.*, **44**, 2239 (1979).
[1040] C. E. Hatch III, *J. Org. Chem.*, **43**, 3953 (1978).
[1041] M. C. Vander Zwan and F. W. Hartner, *J. Org. Chem.*, **43**, 2655 (1978).
[1042] L. G. Wade Jr., J. M. Gerdes, and R. P. Wirth, *Tetrahedron Lett.*, **1978**, 731.
[1043] I. Degani, R. Fochi, and V. Regondi, *Synthesis*, **1979**, 178.
[1044] M. Kluba and A. Zwierzak, *Synthesis*, **1978**, 770.
[1045] W. P. Reeves, M. R. White, and D. Bier, *J. Chem. Educ.*, **55**, 56 (1978).
[1046] Y. Nakajima, J. Oda, and Y. Inouye, *Tetrahedron Lett.*, **1978**, 3107.
[1047] A. Jończyk, *Angew. Chem.*, **91**, 228 (1979); *Angew. Chem. Int. Ed. Engl.*, **18**, 217 (1979).
[1048] M. Fedoryński, K. Wojciechowski, Z. Matacz, and M. Mąkosza, *J. Org. Chem.*, **43**, 4682 (1978).
[1049] Y. A. Zhdanov, Y. E. Alekseev, S. S. Doroshenko, G. V. Bodganova, T. P. Sudareva, and V. G. Akekseeva, *Dokl. Akad. Nauk SSSR Ser. Khim.*, **238**, 1102 (1978); Engl. transl., p. 65.
[1050] T. D. N'Guyen, A. Deffieux, and S. Boileau, *Polymer*, **19**, 423 (1978).
[1051] V. M. Andreev, A. I. Bibicheva, and L. S. Lapikova, *Zh. Vses. Khim. Ova.*, **23**, 231 (1978); *Chem. Abstr.* **89**, 23704 (1978).
[1052] J. D. Nicholson, *Analyst*, **103**, 1, 13 (1978).
[1053] A. Arbin, *J. Chromatogr.*, **144**, 85 (1977).
[1054] P. Hartwig and C. Fagerlund, *J. Chromatogr.*, **140**, 170 (1977).
[1055] J. M. Rosenfeld and J. L. Crocco, *Anal. Chem.*, **50**, 701 (1978).
[1056] A. Berlin-Wahlén and R. Sandberg, *Acta Pharm. Suec.*, **14**, 321 (1977).
[1057] W. S. Di Menna, C. Piantadosi, and R. G. Lamb, *J. Medic. Chem.*, **21**, 1073 (1978).
[1058] H. Segawa and M. Tanyo, *Jap. Kokai*, 76,75,084 (1976); *Chem. Abstr.*, **86**, 106661 (1977).
[1059] J. P. Ferris, S. Singh, and T. A. Newton, *J. Org. Chem.*, **44**, 173 (1979).
[1060] E. V. Dehmlow and M. Lissel, unpublished results.
[1061] B. Młotkowska and A. Zwierzak, *Tetrahedron Lett.*, **1978**, 4731.
[1062] G. Vernin and J. Metzger, *Synthesis*, **1978**, 921.
[1063] D. Reuschling, H. Pietsch, and A. Linkies, *Tetrahedron Lett.*, **1978**, 615.
[1064] S. R. Fletcher and I. T. Kay, *J. Chem. Soc. Chem. Commun.*, **1978**, 903.
[1065] A. Jończyk, Z. Ochal, and M. Mąkosza, *Synthesis*, **1978**, 882.
[1066] M. Jawdosiuk, A. Jończyk, A. Kwast, M. Mąkosza, I. Kmiotek-Skarżyńska, and K. Wojciechowski, *Pol J. Chem.*, **53**, 191 (1979).
[1067] M. Jawdosiuk, M. Mąkosza, E. Malinowsky, and W. Wilczyński, *Pol. J. Chem.*, **52**, 2189 (1978).
[1068] J. W. Skiles and M. P. Cava, *Heterocycles*, **9**, 653 (1978).
[1069] E. Chiellini and R. Solaro, *J. Org. Chem.*, **43**, 2550 (1978).
[1070] M. J. O'Donnell, J. M. Boniece, and S. E. Earp, *Tetrahedron Lett.*, **1978**, 2641; M. J. O'Donnell and T. M. Eckrich, *ibid.*, **1978**, 4625.
[1071] Y. Masuyama, Y. Ueno, and M. Okawara, *Bull. Chem. Soc. Jpn.*, **50**, 3071 (1977).
[1072] J. M. J. Fréchet, M. de Smet, and M. J. Farrall, *Tetrahedron Lett.*, **1979**, 137.
[1073] T. Shioiri and Y. Hamada, *J. Org. Chem.* **43**, 3631 (1978).
[1074] M. Mąkosza, K. Wojciechowski, and M. Jawdosiuk, *Pol. J. Chem.*, **52**, 1173 (1978).
[1075] A. J. Fry and J. P. Bujanauskas, *J. Org. Chem.*, **43**, 3157 (1978).
[1076] E. Baxmann and E. Winterfeldt, *Chem. Ber.*, **111**, 3403 (1978).
[1077] A. Jończyk and T. Pytlewski, *Synthesis*, **1978**, 883.
[1078] B. Lamm and K. Ankner, *Acta Chem. Scand. Ser. B*, **32**, 193 (1978).
[1079] L. S. Hart, C. R. J. Killen, and K. D. Saunders, *J. Chem. Soc. Chem. Commun.*, **1979**, 24.

[1080] V. G. Purohit and R. Subramanian, *Chem. Ind. (London)*, **1978**, 731.
[1081] H. Kise, Y. Kaneko, T. Sato, and M. Seno, *Yukagaku*, **26**, 474 (1977); *Chem. Abstr.*, **87**, 133818 (1977).
[1082] S. Akabori and H. Tuji, *Bull. Chem. Soc. Jpn.*, **51**, 1197 (1978).
[1083] E. D'Incan and P. Viout, *Tetrahedron*, **34**, 2469 (1978).
[1084] U. Burckhardt, L. Werthemann, and R. J. Troxler, *Germ. Offenl.*, **2**, 738,588 (1978); *Chem. Abstr.*, **89**, 6,114 (1978).
[1085] I. Willner and M. Halpern, *Synthesis*, **1979**, 177.
[1086] H. J.-M. Dou, P. Hassanaly, J. Kister, G. Vernin, and J. Metzger, *Helv. Chim. Acta*, **61**, 3143 (1978).
[1087] J. Goliński and M. Mąkosza, *Synthesis*, **1978**, 823.
[1088] S. P. Markey and G. J. Shaw, *J. Org. Chem.*, **43**, 3414 (1978).
[1089] R. Durand, P. Geneste, G. Lamaty, C. Moreau, O. Pomarès, and J. P. Roque, *Recl. Trav. Chim. Pays-Bas*, **97**, 42 (1978).
[1090] V. Scherrer, M. Jackson-Mülly, Z. Zsindely, and H. Schmid, *Helv. Chim. Acta*, **61**, 716 (1978).
[1091] Y. Ali and W. A. Szarek, *Carbohydr. Res.*, **67**, C 17 (1978).
[1092] J. H. Clark, J. M. Miller, and K.-M. So, *J. Chem. Soc. Perkin Trans.* 1, **1978**, 941.
[1093] G. V. Kryshtal, V. V. Kulganek, V. F. Kucherov, and L. A. Yanovskaya, *Synthesis*, **1979**, 107.
[1094] G. A. Russel, M. Mąkosza, and J. Hershberger, *J. Org. Chem.*, **44**, 1195 (1979).
[1095] J. M. McIntosh and H. Khalil, *Can. J. Chem.*, **56**, 2134 (1978).
[1096] A. Jończyk, A. Kwast, and M. Mąkosza, *Tetrahedron Lett.*, **1979**, 541.
[1097] V. Dryanska, K. Popandova-Yambolieva, and C. Ivanov, *Tetrahedron Lett.*, **1979**, 443.
[1098] K. Takahashi, S. Kimura, Y. Ogawa, K. Yamada, and H. Iida, *Synthesis*, **1978**, 892.
[1099] Y. A. Zhdanov, Y. E. Alekseev, G. V. Zinchenko, and S. S. Doroshenko, *Dokl. Akad. Nauk SSSR Ser. Khim.*, **231**, 868 (1976); Engl. transl. p. 703.
[1100] L. Cassar, S. Panossian, and C. Giordano, *Synthesis*, **1978**, 917.
[1101] D. H. Hunter, M. Hanity, V. Patel, and R. A. Perry, *Can. J. Chem.*, **56**, 104 (1978).
[1102] A. Chollet, J.-P. Hagenbuch, and P. Vogel, *Helv. Chim. Acta*, **62**, 511 (1979).
[1103] L. T. Scott and W. R. Brunsvold, *J. Org. Chem.*, **44**, 641 (1979).
[1104] E. V. Dehmlow and M. Lissel, *Synthesis*, **1969**, 372.
[1105] E. V. Dehmlow and M. Lissel, *Liebigs Ann. Chem.*, **1979**, in press.
[1106] P. Hayden, *Germ. Offenl.* 2,633,228 (1977); *Chem. Abstr.*, **86**, 139395 (1977).
[1107] V. O. Illi, *Synthesis*, **1979**, 387.
[1108] W. Raßhofer and F. Vögtle, *Tetrahedron Lett.*, **1979**, 1217.
[1109] E. V. Dehmlow and S. Barahona-Naranjo, *J. Chem. Res.*, **1979**, (S) 238.
[1110] V. Mancini, G. Morelli, and L. Standoli, *Gazz. Chim. Ital.*, **107**, 47 (1977).
[1111] G. Wittig and U. Schoch-Grübler, *Justus Liebigs Ann. Chem.*, **1978**, 362.
[1112] C. Botteghi, G. Caccia, and S. Gladiali, *Chim. Ind. (Milan)*, **59**, 839 (1977); *Chem. Abstr.*, **88**, 152172 (1978).
[1113] E. V. Dehmlow and S. Barahona-Naranjo, unpublished work.
[1114] A. K. Khusid, G. V. Kryshtal, V. F. Kucherov, and L. A. Yanovskaya, *Izv. Akad. Nauk SSSR Ser. Khim.*, **26**, 692 (1977); Engl. transl., p. 628.
[1115] M. J. Farrall, T. Durst, and J. M. J. Fréchet, *Tetrahedron Lett.*, **1979**, 203.
[1116] B. Stanovnik, M. Tišler, M. Kunaver, D. Gabrijclčič, and M. Kočevar, *Tetrahedron Lett.*, **1978**, 3059.
[1117] P. D. Klemmensen, H. Kolind-Andersen, M. B. Madsen, and A. Svendsen, *J. Org. Chem.*, **44**, 416 (1979).
[1118] M. Seno, T. Namba, and H. Kise, *J. Org. Chem.*, **43**, 3345 (1978).
[1119] A. R. Butler and P. T. Shepherd, *J. Chem. Res.*, **1978**, (S) 339, (M) 4471.
[1120] A. Jończyk, A. Kwast, and M. Mąkosza, *J. Org. Chem.*, **44**, 1192 (1979).
[1121] S. Yanagida, K. Takahashi, and M. Okahara, *J. Org. Chem.*, **44**, 1099 (1978).
[1122] H. Alper and H.-N. Paik, *J. Am. Chem. Soc.*, **100**, 508 (1978).

[1123] A. K. Shukla and W. Preetz, *Angew. Chem.*, **91**, 160 (1979); *Angew. Chem. Int. Ed. Engl.*, **18**, 151 (1979).
[1124] P. S. Braterman, B. S. Walker, and T. H. Robertson, *J. Chem. Soc. Chem. Commun.*, **1977**, 651.
[1125] H. Alper, J. K. Currie, and H. des Abbayes, *J. Chem. Soc. Chem. Commun.*, **1978**, 311.
[1126] Y. Gaoni, *Tetrahedron Lett.*, **1978**, 3277.
[1127] T. Uyehara, A. Ichida, M. Funamizu, H. Nanbu, and Y. Kitahara, *Bull. Chem. Soc. Jpn.*, **52**, 273 (1979).
[1128] M. Seno, S. Shiraishi, Y. Suzuki, and T. Asahara, *Bull. Chem. Soc. Jpn.*, **51**, 1413 (1978).
[1129] L. Parvulescu and M. D. Gheorghiu, *Rev. Roum. Chim.*, **22**, 1089 (1977).
[1130] G. F. Weber and S. S. Hall, *J. Org. Chem.*, **44**, 447 (1979).
[1131] N. N. Labeish, E. M. Kharicheva, T. V. Mandelshtam, and R. R. Kostkov, *Zh. Org. Khim.*, **14**, 878 (1978); Engl. transl., p. 815.
[1132] P. Duchaussoy, P. DiCesare, and B. Gross, *Synthesis*, **1979**, 198.
[1133] K. Steinbeck and J. Klein, *J. Chem. Res.*, **1978**, (S) 396, (M) 4771.
[1134] K. Steinbeck, *Chem. Ber.*, in press.
[1135] K. Steinbeck, private communication.
[1136] K. Kawashima, T. Saraie, Y. Kawano, and T. Ishiguro, *Chem. Pharm. Bull.*, **26**, 942 (1978).
[1137] U. K. Pandit, *Heterocycles*, **8**, 609 (1977).
[1138] W. Ando, H. Higuchi, and T. Migata, *J. Org. Chem.*, **42**, 3365 (1977).
[1139] J. N. Shah, Y. P. Mehta, and G. M. Shah, *J. Org. Chem.*, **43**, 2078 (1978).
[1140] A. R. Shamout, Diploma Thesis, Technische Universität Berlin, 1979.
[1141] D. Landini, F. Montanari, and F. Rolla, *Synthesis*, **1979**, 26.
[1142] H. S. D. Soysa and W. P. Weber, *Tetrahedron Lett.*, **1978**, 1969.
[1143] E. V. Dehmlow and M. Lissel, *Chem. Ber.*, **111**, 3873 (1978).
[1144] R. R. Kostikov and A. P. Molchanov, *Zh. Org. Khim.*, **14**, 879 (1978); Engl. transl., p. 816.
[1145] R. R. Kostikov, A. P. Molchanov, I. A. Vasil'eva, and Y. M. Slobodin, *Zhur. Org. Khim.*, **13**, 2541 (1977); Engl. transl., p. 2361.
[1146] O. P. Vig, I. R. Trehan, G. L. Kad, and A. L. Bedi, *Indian J. Chem. Ser. B*, **16**, 455 (1978).
[1147] O. P. Vig, G. L. Cad, A. L. Bodin, and S. D. Kumar, *Indian J. Chem. Ser. B*, **16**, 452 (1978).
[1148] R. F. Heldeweg and H. Hogeveen, *J. Org. Chem.*, **43**, 1916 (1978).
[1149] W. Amman, R. A. Pfund, and C. Ganter, *Chimia*, **31**, 61 (1977).
[1150] G. I. Fray and R. G. Saxton, *Tetrahedron*, **34**, 2663 (1978).
[1151] L. K. Sydnes, *Acta Chem. Scand. Ser. B*, **31**, 823 (1977).
[1152] E. Piers and E. H. Ruediger, *J. Chem. Soc. Chem. Commun.*, **1978**, 166.
[1153] H. Sadlo and W. Kraus, *Tetrahedron*, **34**, 1965 (1978).
[1154] M. S. Baird, A. G. W. Baxter, B. R. J. Devlin, and R. J. G. Searle, *J. Chem. Soc. Chem. Commun.*, **1979**, 210.
[1155] E. Piers, I. Nagakura, and H. E. Morton, *J. Org. Chem.*, **43**, 3630 (1978).
[1156] E. Piers, I. Nagakura, and J. E. Shaw, *J. Org. Chem.*, **43**, 3431 (1978).
[1157] K. H. Holm, D. G. Lee, and L. Skattebøl, *Acta Chem. Scand. Ser. B*, **32**, 693 (1978).
[1158] J. B. Lambert, K. Kobayashi, and P. H. Mueller, *Tetrahedron Lett.*, **1978**, 4253; M. Jones, Jr., V. J. Tortorelli, P. P. Gaspar, and J. B. Lambert, *ibid.*, **1978**, 4257.
[1159] M. S. Baird, *J. Chem. Soc. Perkin Trans.* 1, **1979**, 1020.
[1160] B. Giese and J. Meister, *Angew. Chem.*, **90**, 636 (1978); *Angew. Chem. Int. Ed. Engl.*, **17**, 595 (1978).
[1161] K. P. Butin, A. N. Kashin, I. P. Beletskaya, L. S. German, and V. R. Polischchuk, *J. Organomet. Chem.*, **25**, 11 (1970).
[1162] Y. Bessière, D. N. H. Savary, and M. Schlosser, *Helv. Chim. Acta*, **60**, 1739 (1977).

[1163] M. Christl, G. Freitag, and G. Brüntrup, *Chem. Ber.*, **111**, 2307 (1978).
[1164] V. S. Aksenov and G. A. Terent'eva, *Izv. Akad. Nauk SSSR Ser. Khim.*, **26**, 623 (1977); Engl. transl., p. 560.
[1165] P. J. Stang and J. A. Bjork, *J. Chem. Soc. Chem. Commun.*, **1978**, 1057.
[1166] T. Sasaki, S. Eguchi, and F. Nakata, *Tetrahedron Lett.*, **1978**, 1999.
[1167] J. T. Martz, G. W. Gokel, and R. A. Olofson, *Tetrahedron Lett.*, **1979**, 1473.
[1168] H. Alper and H.-N. Paik, *Nouv. J. Chim.*, **2**, 245 (1978).
[1169] P. Mangeney and Y. Langlois, *Tetrahedron Lett.*, **1978**, 3015.
[1170] L. Horner and W. Brich, *Justus Liebigs Ann. Chem.*, **1978**, 710.
[1171] R. Kinishi, Y. Nakajima, J. Oda, and Y. Inouye, *Agric. Biol. Chem.*, **42**, 869 (1978); *Chem. Abstr.*, **89**, 106932 (1978).
[1172] J. Massé and E. Parayre, *Bull. Soc. Chim. Fr.*, **1978**, II 395.
[1173] J. M. McIntosh, *Tetrahedron Lett.*, **1979**, 403.
[1174] K. Nakamura, A. Ohno, S. Yasui, and S. Oka, *Tetrahedron Lett.*, **1978**, 4815.
[1175] D. L. Reger, M. M. Habib, and D. J. Fauth, *Tetrahedron Lett.*, **1979**, 115.
[1176] J. A. Morris and D. C. Mills, *Chem. Br.*, **14**, 326 (1978).
[1177] D. G. Lee and V. S. Chang, *J. Org. Chem.*, **43**, 1532 (1978).
[1178] ref. [5], p. 161.
[1179] A. Poulose and R. Croteau, *J. Chem. Soc. Chem. Commun.*, **1979**, 243.
[1180] P. K. Kadaba, *Synthesis*, 1978, 694.
[1181] L. M. Rossi and P. Trimarco, *Synthesis*, **1978**, 733.
[1182] H.-J. Schmidt and H. J. Schäfer, *Angew. Chem.*, **91**, 77 (1979); *Angew. Chem. Int. Ed. Engl.*, **18**, 77 (1979).
[1183] H.-J. Schmidt and H. J. Schäfer, *Angew. Chem.*, **91**, 78 (1979); *Angew. Chem. Int. Ed. Engl.*, **18**, 78 (1979).
[1184] D. G. Lee and V. S. Chang, *Synthesis*, **1978**, 462.
[1185] M. L. Navtanovich, L. A. Dzhanashvili, and V. L. Kheifets, *Zh. Obshch. Khim.*, **48**, 1925 (1978); Engl. transl., p. 1755.
[1186] D. R. Bender, J. Brennan, and H. Rapoport, *J. Org. Chem.*, **43**, 3354 (1978).
[1187] Y. Masuyama, Y. Ueno, and M. Okawara, *Chem. Lett.*, **1977**, 1439.
[1188] M. J. Gibian and T. Ungermann, *J. Am. Chem. Soc.*, **101**, 1291 (1979).
[1189] C.-I. Chern, R. DiCosimo, R. DeJesus, and J. S. San Filippo Jr., *J. Am. Chem. Soc.*, **100**, 7317 (1978).
[1190] W. C. Danen and R. L. Arudi, *J. Am. Chem. Soc.*, **100**, 3944 (1978).
[1191] R. A. Johnson, E. G. Nidy, and M. V. Merritt, *J. Am. Chem. Soc.*, **100**, 7960 (1978).
[1192] T. Takata, Y. H. Kim, and S. Oae, *Tetrahedron Lett.*, **1979**, 821.
[1193] M. Matsuo, S. Matsumoto, Y. Iitaka, A. Hanaki, and T. Ozawe, *J. Chem. Soc. Chem. Commun.*, **1979**, 105.
[1194] M. J. Gibian, D. T. Sawyer, T. Ungermann, R. Tangpoonpholvivat, and M. M. Morrison, *J. Am. Chem. Soc.*, **101**, 640 (1979).
[1195] G. A. Lee and H. H. Freedman, U.S. Pat. 4,079,075 (1978); *Chem. Abstr.*, **89**, 41578 (1978).
[1196] D. Landini, F. Montanari, and F. Rolla, *Synthesis*, **1979**, 134.
[1197] D. Landini and F. Rolla, *Chem. Ind.* (*London*), **1979**, 213.
[1198] S. Cacchi, F. La Torre, and D. Misiti, *Synthesis*, **1979**, 356.
[1199] G. Aksnes and B. H. Vagstad, *Acta Chem. Scand. Ser. B*, **33**, 47 (1979).
[1200] G. E. Keck and S. A. Fleming, *Tetrahedron Lett.*, **1978**, 4763.
[1201] M. Hirao, H. Mochizuki, S. Nakahama, and N. Yamazaki, *J. Org. Chem.*, **44**, 1720 (1979).
[1202] D. W. Armstrong and M. Godat, *J. Am. Chem. Soc.*, **110**, 2489 (1979).
[1203] R. C. Cookson and I. D. R. Stevens, *Chem. Br.*, **15**, 329 (1979).
[1204] Y. Yamoto, J. Oda, and Y. Inouye, *Tetrahedron Lett.*, **1979**, 2411.
[1205] M. S. Baird and M. Mitra, *J. Chem. Soc. Chem. Commun.*, **1979**, 563.

Subject Index